Signals and Systems

Signals and Systems
An Introduction

Second Edition

Leslie Balmer
School of Engineering
Coventry University

An imprint of Pearson Education

Harlow, England · London · New York · Reading, Massachusetts · San Francisco
Toronto · Don Mills, Ontario · Sydney · Tokyo · Singapore · Hong Kong · Seoul
Taipei · Cape Town · Madrid · Mexico City · Amsterdam · Munich · Paris · Milan

First published 1991

This second edition, first published 1997, by
Prentice Hall Europe
Pearson Education Limited
Edinburgh Gate
Harlow
Essex CM20 2JE
England

and Associated Companies throughout the world

Visit us on the World Wide Web at:
http://www.pearsoneduc.com

Typeset in 10/12pt Times
by Mathematical Composition Setters Ltd

Printed and bound in Great Britain by Biddles Ltd, *www.biddles.co.uk*

Library of Congress Cataloging-in-Publication Data

Available from the publisher

British Library Cataloguing-in-Publication Data

A catalogue record for this book is available from the British Library

ISBN 0-13-495672-9

9 8 7 6 5 4
03 02 01 00

To my wife, Eileen,
my sons, Richard and Brian,
and my daughter, Kim

Contents

Preface to the first edition

This book is primarily intended for first and second year undergraduate students whose courses include material on 'signals and systems'. Although primarily intended for engineering students it will also be of use to students studying mathematics and physics. It could prove useful to practising engineers and scientists whose courses did not contain some of the concepts required for modern signal processing.

The book is based on the experience gained teaching undergraduate students, on a range of different courses, over the past ten years. A major conclusion drawn from this teaching is that students find the subject of signals and systems difficult, more difficult than related subjects taught at a corresponding level. The reason for this is that the subject deals much more with concepts than with physical devices. A student studying electronics for example may not understand the operation of the integrated circuit amplifier or perhaps cannot grasp the mathematical model describing it. However the student does know that the amplifier is a physical device, knows what its uses are and appreciates why it is being studied. If one takes a topic from signals and systems, say the Fourier transform, there is immediately a shift in the level of difficulty. This is especially so, if, as is often the case, signals and systems is being taught as a basic subject for later application in such areas as control and telecommunications.

This book has been written with this difficulty very much in mind. The subject deals primarily with concepts and care has to be taken to avoid it becoming a branch of mathematics. This has been done by making clear the reasons for studying these concepts and illustrating them by reference to their applications. This does not mean that the use of mathematics has been avoided completely, very little progress could be made with such an approach. However physical reasoning can often replace mathematical rigour and such reasoning has been used in many places in the book to avoid long involved proofs. The mathematics required initially is a knowledge of the calculus and more advanced material is developed as required. An appendix on complex numbers is included for students unfamiliar with this topic.

Throughout the book the ideas are developed gradually with continual reference to the practical situations where they would be applicable. Many worked examples have been included in order to help develop the idea and to illustrate the material as it is presented. End of chapter problems are included and a solution manual giving outline solutions is available.

Throughout the book discrete-time and continuous-time signals and systems have been developed in parallel as much as possible. This has the advantage of being able to contrast and compare methods of analysis applicable to each domain. Also ideas that are easier to grasp in one domain can often help with the corresponding ideas in the other domain. This is particularly so with the use of numerical methods for discrete systems. The response of a specific discrete system is obtained numerically, this then throws light on a more general method of solution which can then be paralleled in the solution of the continuous case.

The treatment of state variable methods presents a problem in an introductory text of this nature. The state variable approach is of great importance in modern system analysis and it would be a serious omission to give it no coverage. However to cover the subject in depth would require a mathematical knowledge beyond that of the intended reader. Accordingly an introduction is given early in the book but this is not developed further in the following chapters.

A brief outline of the material in the book is as follows. Chapter 1 is an introductory chapter presenting many of the ideas of the subject in a qualitative manner. This is done by using a number of practical examples to illustrate the general theme of processing an input signal by a system to produce an output signal. Chapter 2 examines the representation and properties of signals in the time domain, this is done with some care as the foundation laid in this chapter will be used extensively in the remainder of the book. Chapter 3 examines systems from the viewpoint of signal transformation and develops differential and difference equations relating output to input signals. The idea of analogues is introduced because, although analogue computing has been replaced by more modern techniques, the analogue fits nicely into system theory and provides an insight into its digital counterpart and into digital simulation. An introduction to the state variable description is given by this chapter.

Chapter 4 investigates the solution of the system equation to obtain the response of a system to a given input. The solutions of the differential and difference equations are discussed, but not in any depth, as more convenient methods of solution (using transform methods) are introduced in a later chapter. The solution of the system equation leads to the important concepts of the convolution sum and the convolution integral. Chapter 5 considers the response of a system to a sinusoidal input signal leading to the concept of the system frequency response function, methods of portraying the frequency response function are presented. The Fourier series representation for periodic continuous signals is introduced. Chapter 6 extends the frequency response method to include Fourier transform representation of both continuous and discrete signal. The effect of representation of a continuous signal by a discrete set of samples is considered leading to the sampling theorem and the problems of aliasing and signal reconstruction.

Chapter 7 considers the Laplace transform as an extension of the Fourier transform and the corresponding z transform for the discrete case. These transforms are applied to the solution of the system equation which leads to the concept of the system transfer function. The system response is interpreted via the positions of the poles and zeros of the system transfer function in the complex place. Chapter 8 considers feedback systems and investigates how the techniques described in the previous chapters can be used to predict the performance of such systems. Chapter 9 gives an introduction to analogue and digital filtering. An Appendix on the elementary properties of complex numbers is included, also a Bibliography to enable readers to expand on the material presented if required.

I am grateful to many people for the help received in the preparation of this book. I am indebted to my colleagues at the Coventry Polytechnic for many discussions concerning the presentation of the material. I should like to thank Peter Rowe for reading the earlier chapters of the book through the eyes of what he is, a student, and asking the sort of questions students ask. I would like to thank many other students who, by their response at lectures and tutorials, have shaped the form of the book. Finally I am indebted to my wife for coping with the twin problems of my handwriting and the word processor in order to produce the typescript of the book.

Leslie Balmer

Preface to the second edition

The second edition contains significant additional material to that contained in the first edition. Two completely new chapters have been added covering the subjects of 'random signals' and 'spectral analysis.' Also, the end of chapter problems, throughout the book, have been enhanced by the inclusion of computer based exercises using MATLAB.

I have always considered the lack of some treatment of random signals as a serious omission from the first edition of this book. In the real engineering world most signals are generated by complex physical processes and must be regarded as being random. Unfortunately, the analysis of random signals is usually regarded as being on a higher mathematical plane than the corresponding analysis of deterministic signals and, therefore, not suitable for inclusion in an introductory text. However, encouraged by the reception of the first edition, I thought a lot about the presentation of such material and have included the topic using an approach that I feel is comparable to that of the earlier chapters.

Spectral analysis is one of the most important applications of digital signal processing and, therefore, the topic deserves some coverage in a book on signals. One reason why the subject was omitted from the first edition was because it required some knowledge of random processes. Another reason was the difficulty of finding suitable problems at an introductory level. Even the most basic problems in spectral estimation really require some form of computer package to aid their solution. Inclusion of the chapter on random signals and the use of MATLAB exercises have overcome these difficulties – hence the chapter on spectral analysis in this edition.

There are many reasons for the inclusion of the end of chapter MATLAB exercises in this edition.

- The student can often use the package to check the answers to the end of chapter problems. This usually involves little programming and gives a gentle introduction to MATLAB.
- More complex problems can be investigated whose solution, without computer assistance, would be impractical.

- The process of writing a program to obtain numerical results from theory often leads to a more complete understanding of the theory.
- The Student Edition of MATLAB is relatively cheap and is easily available.
- As the Student Edition of MATLAB includes many commands from the Signal Processing Toolbox and the Control Toolbox, it is ideal for application to problems in signals and systems.

Many of the exercises include notes that outline possible difficulties and give guidance towards a solution. However, the problems are often 'open ended' and the student can use the programs developed for a wider investigation of a topic. The solutions to all the exercises have been verified using Version 4 of the Student Edition of MATLAB.

In order to make room for the new chapters, some of the material in the first edition has had to be removed. The section on analogue computing was getting dated at the time of the first edition and is now mainly of historical interest. The majority of this section has been removed in this edition. The chapter on feedback and control systems has been removed completely. This chapter was always a little out of place and I feel that the new material introduced is of greater relevance to the main themes of the book.

There are many people I would like to thank for their help in the production of this second edition. I am grateful to my colleagues at Coventry University for many helpful discussions about the new material included in this edition. In particular, I would like to thank Bob Mercer for permission to include results obtained using the signal analysis package that he has developed. As in the first edition, I am grateful to the many students whose response to the presentation of the material helped shape the form of the new chapters. I would like to thank Eleanor Rimmer for typing the new material.

Leslie Balmer

1

Introduction

In this chapter some indication is given of the content and scope of the subject of signals and systems. This is done by presenting a number of examples in a qualitative manner. In these examples many of the basic concepts of the subject are introduced, although these will be re-stated in a more formal manner in later chapters. By studying these examples the student should understand the importance and scope of the subject and how its study is of fundamental importance in a range of disciplines.

EXAMPLE 1.1 REMOVAL OF NOISE FROM AN AUDIO SIGNAL

Consider the problem of re-recording the content of an old 78 rpm record into a more modern format, i.e. LP, tape, compact disc. This can be accomplished by taking the voltage from the cartridge playing the '78', amplifying this and using it to drive the recording equipment for the new format. However a feature of the '78' recordings was the characteristic 'hiss' produced owing to the large grain size of the material constituting the disc. It would be convenient if this could be removed at this stage and this can be done by using a filter, the overall scheme being as shown in Figure 1.1.

The scheme shown in Figure 1.1 is a simplified version of that which would be used in practice; it serves, however, to illustrate many basic ideas of the subject 'signals and systems'.

The voltage at the point X varies with time and a typical waveform is shown in Figure 1.1. It is this voltage and the manner in which it varies that carries information about the content of the record. The amplitude of the voltage determines the volume of the sound; its rate of change determines the pitch. Such a varying voltage is termed a *signal*. The voltage at X also represents the addition of unwelcome sound, the hiss on the record. This however, is still part of the signal at X and it is the purpose of the filter to remove it, ideally giving a signal at Y that is the required information without the hiss. In this context the filter can be regarded as a device that transforms the signal at X into that at Y and this forms a useful basis to define a *system*.

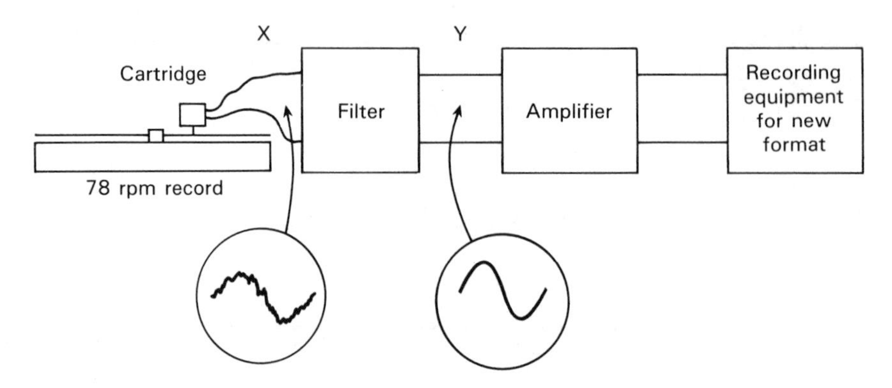

Figure 1.1 Filtering to remove 'hiss'.

Just as the variation of the voltages at X and Y carry information so does the position of the stylus in the grooves of the record. This variation in position is a signal just as the voltages at X and Y are signals; the variable involved, however, is a mechanical variable. The stylus and cartridge assembly transforms the variation in position into a variation of voltage at X – this assembly is also a system, an electromechanical system. This system could obviously be broken down further, introducing more signals, i.e. magnetic flux density or electric charge dependent upon the type of cartridge. The ability to view a system as an *interconnection of subsystems* is an important aspect of the systems viewpoint. This can be portrayed as a block diagram, the blocks representing the subsystems and their inputs and outputs representing signals. A common cartridge is of the type known as the moving-iron where the stylus is fixed to an armature within the poles of a permanent magnet. Displacement of the stylus changes the dimensions of an air gap and the magnetic flux within it. This changing flux induces a voltage in a coil, this voltage forming the cartridge output. A block diagram representing the stylus and cartridge is shown in Figure 1.2.

In this example all the signals have the property that they take a value at every instant of time; such signals are known as *continuous* signals.

It is instructive to consider heuristically a simple system that would perform the required filtering. As stated earlier, the signal from the cartridge can be considered as consisting of two components as shown in Figure 1.3.

The noise signal having been produced by the 'graininess' of the material in the groove walls, is characterized by much more rapid changes of amplitude than occur

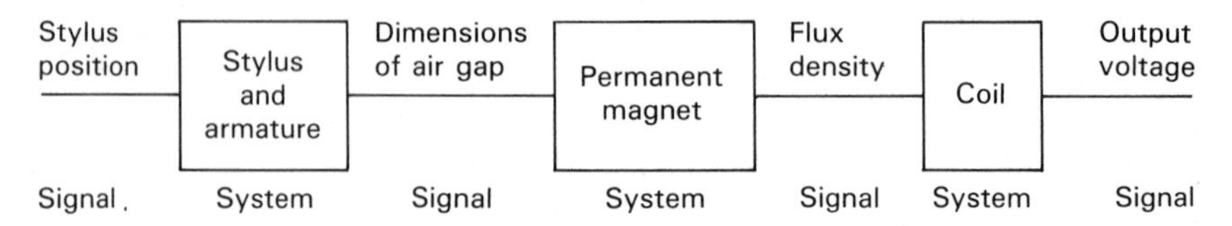

Figure 1.2 Block diagram of stylus and cartridge.

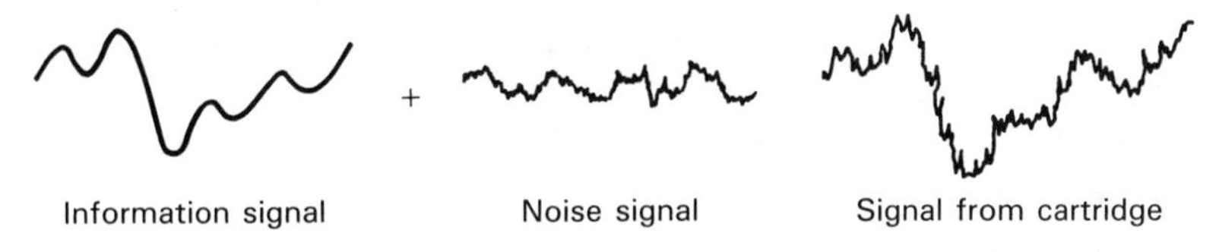

Figure 1.3 Information and noise signals.

in the information signal. Any filter used must attenuate a signal having such rapid changes more than it attenuates the information signal. By analysing the response of the system to the sum of two components the principle of *linearity* is invoked. It is assumed that the response of the system to each of the components is not affected by the presence of the other component. Systems that have this property are termed *linear systems*.

Consider the simple filter system shown in Figure 1.4.

In this circuit the difference between its output voltage v_o and its input v_i is given by the voltage drop across the resistor.

$$v_i - v_0 = iR$$

As there is no loading, the current through the capacitor is also equal to i, the current through the resistor. The basic equation describing the capacitor is one giving charge storage.

$$q = v_o C$$

However, as the current is rate of change of charge, then the relationship between current and voltage is given by

$$i = \frac{dq}{dt} = C\frac{dv_o}{dt}$$

The current is greatest when the rate of change of the signal is greatest and this in turn gives greater voltage drop across R and greater attenuation. Hence the filter will attenuate the noise signal to a greater extent than the information signal.

Considering these two signals (noise and information) they are somewhat different in character. The information signal is a much more predictable signal and it would be identical in different pressings of the disc. The noise signal is generated by the random nature of the material particles forming the disc; it is not predictable and

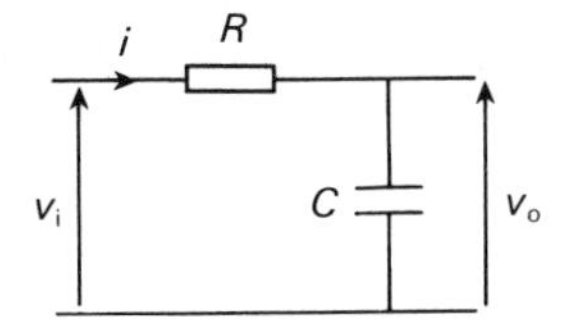

Figure 1.4 Simple filter circuit.

would differ in different pressings of the disc. In this context the information signal is called a *deterministic* signal and the noise a *random* or *stochastic* signal.

In this book, Chapters 1–8 will deal with deterministic signals; Chapters 9 and 10 will consider random signals.

EXAMPLE 1.2 PREDICTION OF SHARE PRICES

At first sight this example appears very different from the previous one, especially perhaps to students of engineering. However, as will be seen, many of the concepts introduced in the previous example can be applied here. The problem is given the price of a share at the close of the market each day, can the future prices be predicted? If it is assumed that a computer is used to perform the prediction then the problem is as shown in Figure 1.5.

This diagram is a block diagram of the form introduced in the previous example. The input and output are variables whose values vary with time and convey information; they are signals and can be portrayed graphically as in Figure 1.6.

These signals have a different character from those in the previous example; instead of being available at all instants of time, they are only available at discrete times (the close of the market each day); they are known as *discrete* signals.

In this example the computer takes the discrete values of the input signal and from them calculates the discrete values of the output signal – it transforms the input signal into the output signal. This is the definition of a system as introduced in the previous example. However in the previous example the system was constructed from 'hardware', i.e. electrical and mechanical components. In this example the computer is also constructed from components; however this aspect of the computer is of little interest here. What is of interest is the set of instructions that convert the input values into the output values, i.e. it is the software of the system that is of

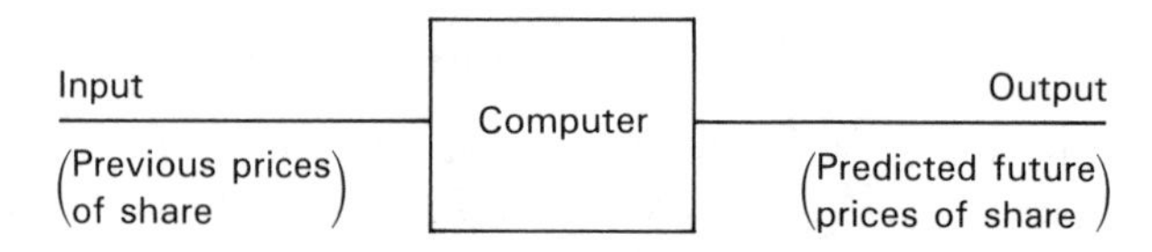

Figure 1.5 Share price prediction.

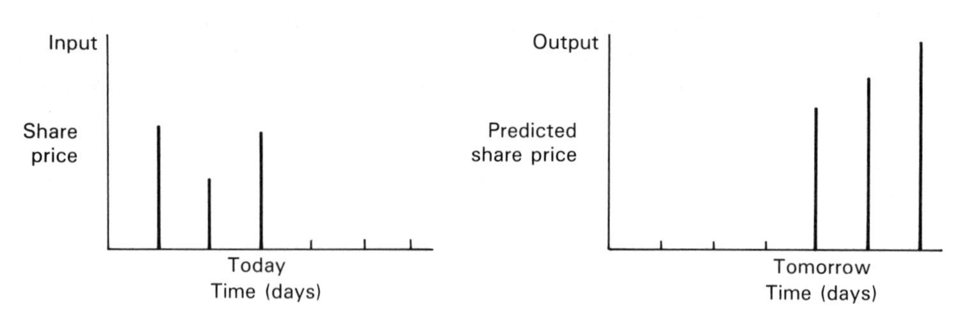

Figure 1.6 Variation of past and predicted share prices.

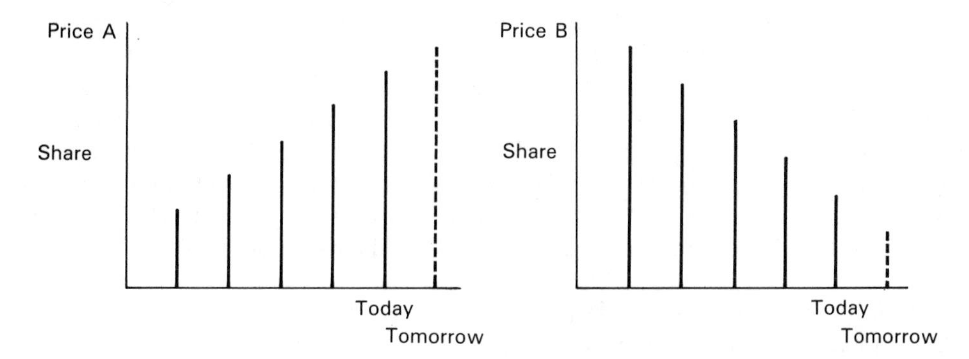

Figure 1.7 Price variation of two shares.

interest. A set of rules that transforms one discrete signal into another is known as an *algorithm.*

It is instructive to consider a simple algorithm for this problem. Consider the price of two shares A and B as shown by the signals in Figure 1.7.

As can be seen, the price of share A is rising steadily; in the absence of any other information it seems reasonable to assume that this rise will continue. Similarly, share B is falling steadily and it seems reasonable that this fall will continue. A simple algorithm would state that the change from today to tomorrow is the same as the change from yesterday to today. This would give

tomorrow's price = today's price + (today's price—yesterday's price)

Expressing this more mathematically and denoting the price yesterday, today and tomorrow by $P(n-1)$, $P(n)$, $P(n+1)$, then the required algorithm can be written

$$\begin{aligned} P(n+1) &= P(n) + [P(n) - P(n-1)] \\ &= 2P(n) - P(n-1) \end{aligned}$$

It is obvious that this is a very simple algorithm and nobody would make a fortune using it. In practice one would use other information about the market. Is the market as a whole rising or falling, how are the different sectors performing, what about other economic indicators? A more realistic scheme would take the price of many shares into consideration as shown in Figure 1.8.

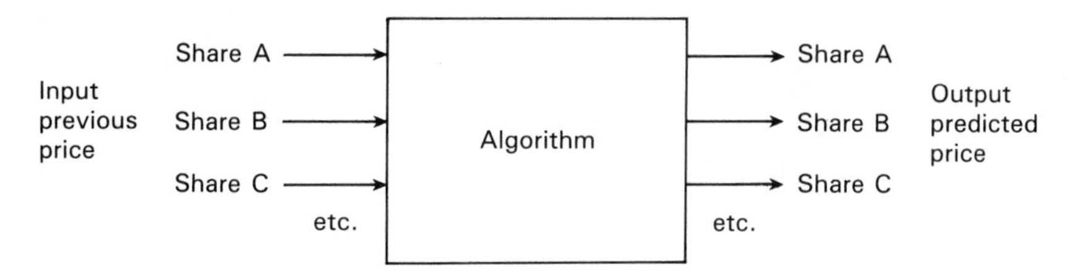

Figure 1.8 Prediction of share prices.

Note that the system could not be broken down into a number of independent systems involving only the previous and predicted prices of individual shares as the different previous prices would interact in the algorithm. A system with more than one input and output is known as a *multivariable* system; one with only one input and output is known as a *single input, single output* (SISO) system. With a multivariable system it is often convenient to regard the individual input signals as the components of one overall input vector; such a signal is termed a *vector* signal. This book is only concerned with single input, single output systems and non-vector signals.

EXAMPLE 1.3 EXAMPLE 1.1 REVISITED

In Example 1.1 it was shown how a simple filter could be constructed to transform a signal contaminated by noise into a noise free signal. This filter was constructed from electrical components and it was the relationship between the variables (current and voltage) at the terminals of these components that made it act like a filter. In Example 1.2 it was shown that a suitable algorithm implemented on a digital computer could be used to transform one discrete signal into another. It is useful to consider whether, if the information plus noise signal in Example 1.1 had been discrete, the filter could have been implemented by a suitable computer program. A filter to operate on discrete signals is a *digital filter* and the design of such a filter will be considered in a later chapter. Such a filter would appear to offer far greater flexibility, as its action is not constrained by the physical laws governing electrical components. However, its use would depend on the ability to convert continuous signals into discrete signals without loss of information. At the output of the digital filter it is then required to convert the discrete signal into a continuous signal. The overall system would be as shown in Figure 1.9.

The conversion from continuous to discrete signal can be understood by reference to Figure 1.10.

Here the continuous signal is fed into a switch that is opened and closed repetitively. However, the time when it is closed, τ, is very much less than the time

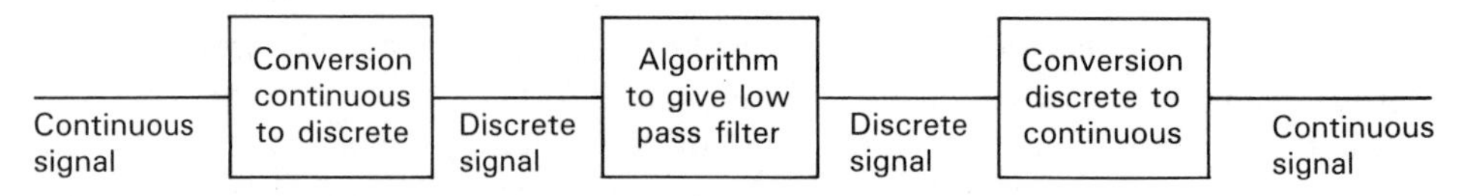

Figure 1.9 Conversion between continuous and discrete signals.

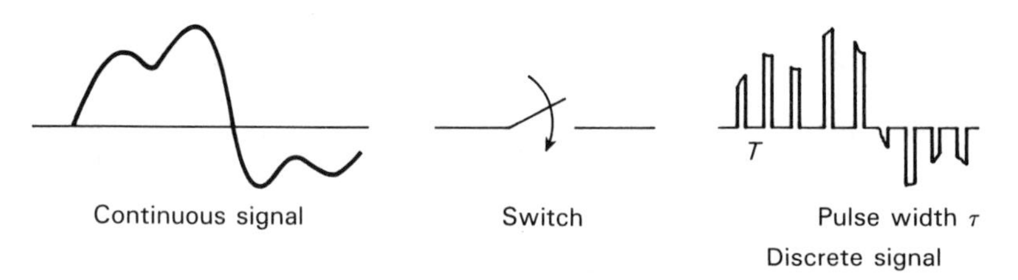

Figure 1.10 Sampling of a signal.

when it is opened, $(T - \tau)$, and the signal at the switch output will be as shown. Provided the change in the signal amplitude is small over the time τ this output signal approximates to a discrete signal. This action is known as *sampling*; T is the *sampling time*, $1/T$ the *sampling rate* and the discrete signal in this context is known as a *sampled signal*.

In the practical situation the discrete signal formed is required as the input signal to a digital computer or microprocessor. For this application it has to be represented in some form of binary code and a signal in this form is known as a *digital signal*. An alternative name for the continuous signal is an *analogue* signal and the device for converting the analogue signal to a digital signal is an *analogue to digital converter (ADC)*.

To return to the question of how well the sample signal represents the original, consider the situation shown in Figure 1.11.

Here the same signal is sampled at two different rates, the sampling time in Figure 1.11(b) being shorter. Intuitively one feels that the discrete signal in the second case is a far better representation of the continuous signal, the central peak in Figure 1.11(a) being missed completely. An indication of how well the discrete signal will represent the continuous signal is given by the maximum rate of change of the signal in comparison with the sampling time. This idea will be expressed quantitatively as the *sampling theorem* in a later chapter.

The problem of obtaining a continuous signal from a discrete signal is also a filtering problem. The discrete signal contains high rates of change (due to discontinuities) and the problem is similar to that of the removal of noise.

Returning to the noise filter, a qualitative approach can be made to the design of this filter on the assumption that the signal is varying slowly in comparison with the noise as shown in Figure 1.12. It is also assumed that the noise signal has an average value of zero.

If the sampling instant is denoted by n and the amplitude of the signal at this point by $x(n)$ then consider the average value over the last five signal points.

$$\text{average} = \frac{x(n) + x(n-1) + x(n-2) + x(n-3) + x(n-4)}{5}$$

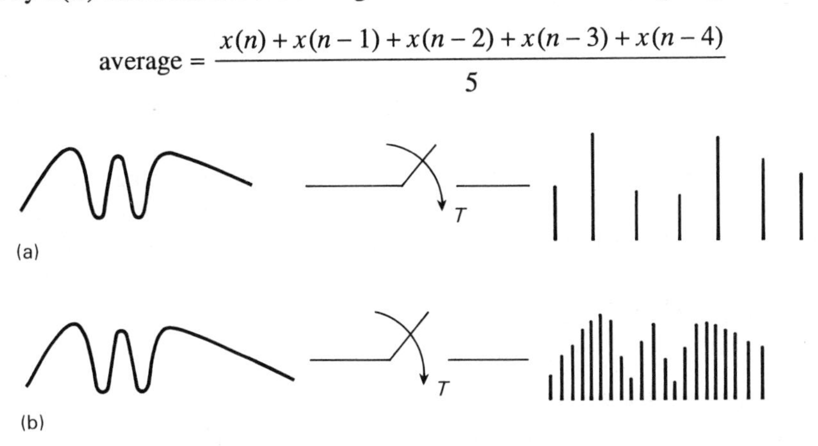

Figure 1.11 Effect of differing sampling rates.

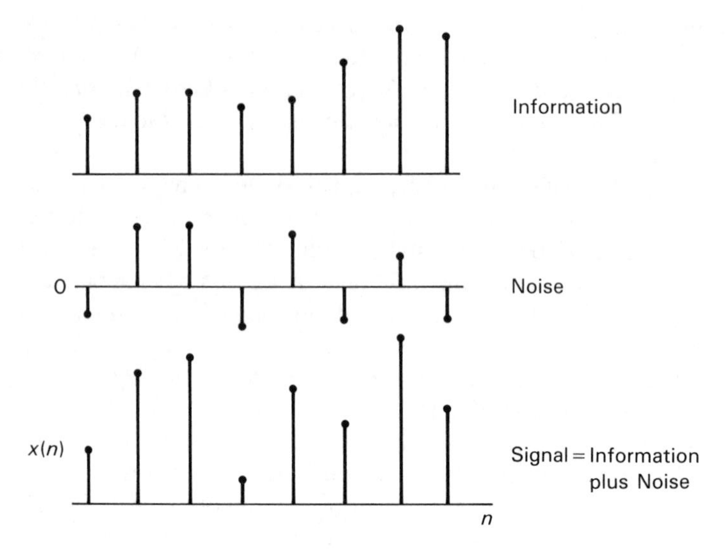

Figure 1.12 Effect of discrete noise signal.

If the information changed by a negligible amount over these points, and the average value of the noise were zero over these points, then this average would give the true value of the information. However the signal is changing and the average value of the noise is not zero over a small number of samples. The number of samples could be increased, but then the information may show an even greater change, i.e., five samples is a compromise value (in practice the number of samples chosen would depend upon how rapidly the information and noise signals are changing). Denoting the output from the filter at the nth sample by $y(n)$ and using the summation convention, the algorithm describing the filter action can be expressed.

$$y(n) = \frac{1}{5}\sum_{k=0}^{5} x(n-k)$$

The filter designs obtained by heuristic methods in this chapter and in Example 1.1 have depended upon the relative rates of change of the wanted and unwanted signals. Such design methods are termed *time domain methods*. However, an alternative and very powerful design method is based on signals that are sinusoids. A single sine wave signal is shown in Figure 1.13.

This signal can be represented by the formula.

$$x(t) = A \sin 2\pi f t$$

where A represents the maximum amplitude and f represents the frequency of the sine wave. It should be noted that $f = 1/T$, T being the periodic time, and for a given amplitude a high frequency sine wave represents a higher rate of change of

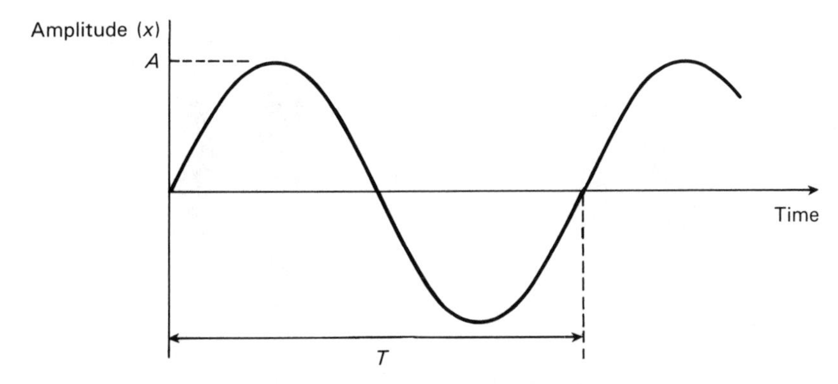

Figure 1.13 A single sine wave.

amplitude than the corresponding low frequency signal. It will be shown in a later chapter that many signals can be considered as a combination of sinusoids having differing amplitudes and frequencies. As the noise signal, in this example, is of smaller amplitude than the information signal, but has a higher rate of change of amplitude, it must be composed of components having higher frequencies than the information signal. Hence, the filter problem can be considered as the problem of designing a system that will reduce the high frequency signal components relative to the low frequency components. Such a design method is termed a *frequency domain method.*

The action of the filter derived in Example 1.1 can be explained in terms of its response to sinusoids. Referring to Figure 1.4, the circuit can be regarded as a potential divider with the top arm consisting of a resistor R and the bottom arm of a capacitor with reactance having magnitude $1/2\pi fC$. As the frequency rises the reactance falls and the attenuation of the circuit rises, hence higher frequencies (the noise) are attenuated to a greater degree than lower frequencies (the information).

The method by which signals are represented as a combination of sinusoidal signals is known as *Fourier analysis*. It is a very powerful analysis and design method that can be applied to both continuous and discrete systems. A considerable portion of later chapters will be devoted to its study.

EXAMPLE 1.4 PROCESSING OF IMAGES

Readers will be aware of the growth in the transmission of pictures in recent years. Beside the obvious example of television broadcasting, pictures of the earth are taken by satellite and transmitted to the ground for the purpose of weather forecasting and military use. Pictures of distant planets are transmitted to earth from space probes.

This example considers the manner in which the information in this type of signal may be processed by suitable systems. Consider the simple picture shown in Figure 1.14(a).

In order to transmit this picture a signal has to be constructed giving its intensity (and colour information for a non-monochrome picture) at every point on its surface.

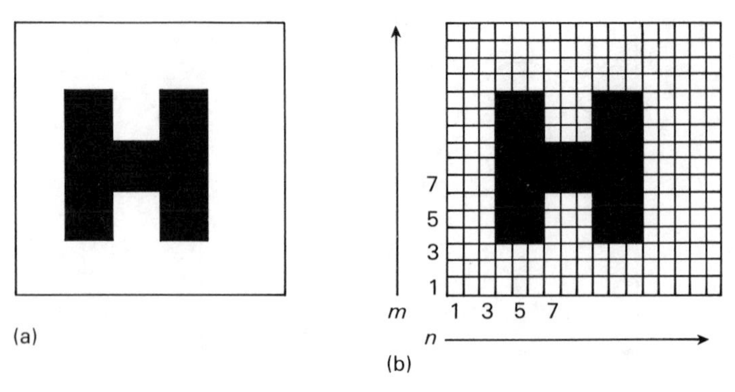

Figure 1.14 Digitization of an image.

In practice all forms of image are of a discrete nature and their representation as a discrete signal will be considered here. The picture is divided into a finite number of small areas (pixels) and the signal represents the intensity in each pixel. With reference to Figure 1.14(b) an individual pixel can be identified according to a co-ordinate system n, m as shown. Hence, the signal giving the intensity in any pixel would be represented as $I(n, m)$. This signal, unlike signals considered in previous examples, varies not with time but with distance; it is a *spatial signal*. Other examples of such signals would be the stress along the wing of an aircraft or the temperature through the core of a nuclear reactor. The signal giving intensity also differs from signals considered previously in that it is a function of two independent variables n, m. In practice the image may also be changing with time giving a third independent variable.

Systems having input signals that are functions of more than one variable are more complex to design than those where the signal is a function of time only. Consider the picture transmitted from a space probe, the effect of camera vibration is to blur the image. The blurring can be removed by passing the signal through an appropriate system. This in principle is the filtering problem similar to that investigated earlier, but of a more complex nature. Other operations that may be required on signals representing images are recognition of malignant growth on images obtained from scanners in medicine, enhancement of contrast, i.e., the marking of rivers and mountain ranges, on satellite pictures of the earth.

Problems of this nature are beyond the scope of this book. However, many of the techniques used in later chapters where only time signals are considered can be extended to solve the problem involved in image processing.

EXAMPLE 1.5 POSITION CONTROL SYSTEM

The example considered here is to control the position of a large mass by means of a low energy signal. For example, it may be required to control the position of a radio telescope weighing several hundred tonnes by means of an electrical signal having a maximum magnitude of one or two volts. The situation is shown in Figure 1.15 together with typical input and output signals.

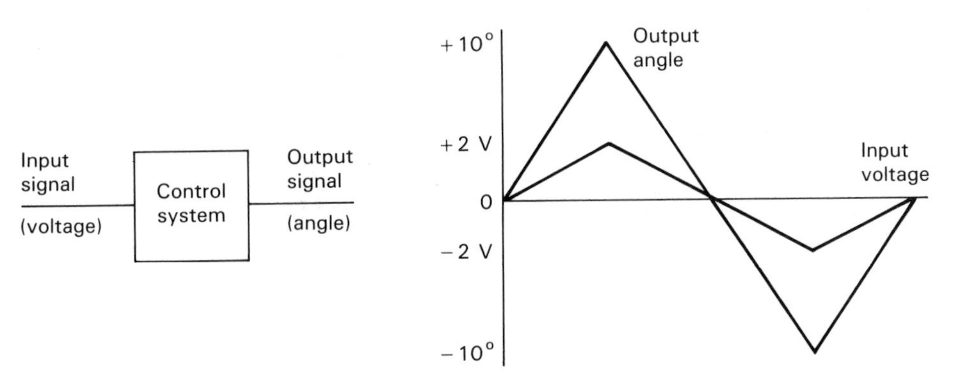

Figure 1.15 Control system with typical input and output signals.

The problem involved here is different from those in previous examples. In previous examples the systems involved have transformed an input signal into a different output signal. In this example the shapes of input and output signals are identical; however they represent different variables (electrical and mechanical), and the energy levels associated with the signals are very different (milliwatts at input, kilowatts at output). An outline of a possible scheme for the control system is shown in Figure 1.16.

Here the input signal is amplified to a level sufficient to drive a power amplifier, the output of which drives a motor. This scheme suffers from two major disadvantages:

1. The relationship between output position and input voltage depends upon all the parameters in the system. Should these change owing to ageing or temperature variation, then this will cause an error in the output position.
2. If the system is subject to an external disturbance (i.e., the force exerted by a wind gust on the aerial) this will affect the output position and cause an error.

Consider the scheme shown in Figure 1.17. The part of the system consisting of the amplifier and motor is as described previously. In this scheme, however, the output signal representing position is converted to a voltage by means of a transducer (this device presents few problems because the transformation is to a lower power level). The signal from the transducer is subtracted from the input signal to form an error signal. This system now has the following characteristics:

1. If the output and input are correct there will be no error signal and no drive to the motor – the system will remain in this condition.

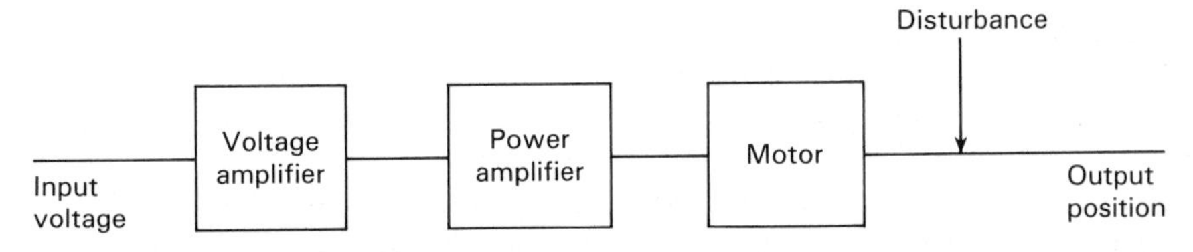

Figure 1.16 Scheme for open-loop control.

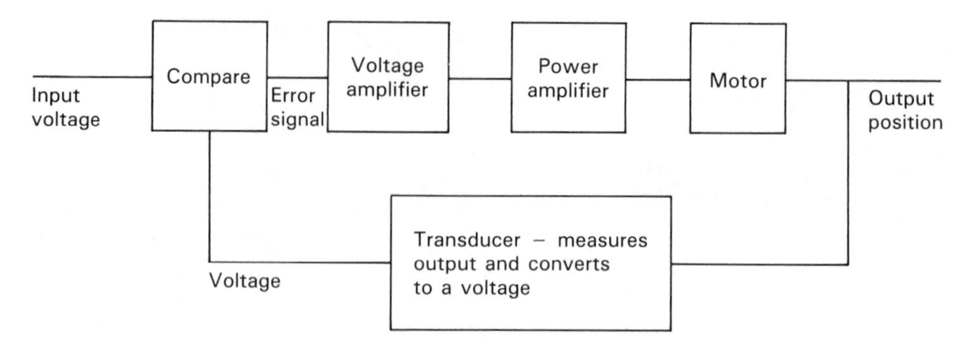

Figure 1.17 Scheme for closed-loop control.

2. If the output does not correspond to the input, for any reason, an error signal is formed. This drives the system to bring the output to the required value.

Note that the effect of external disturbances is now taken into account by characteristic 2. There are problems introduced by this type of control system, in general, however, it gives far better control than the scheme in Figure 1.16.

The method of Figure 1.16 is known as *open-loop control*, that of Figure 1.17 as *closed-loop control*. Because a signal in the latter scheme is fed back from output to input, the system is known as a *feedback system.*

To summarize, the examples in this chapter have been used in a qualitative manner to introduce many of the basic concepts in the study of signals and systems. They have also been used to give an indication of the scope of the subject. Many other fields could have been chosen to provide the examples and could have included topics such as the following:

1. The processing of signals to identify target position and velocity in radar and sonar systems.
2. The analysis of signals produced by the reflection of shock waves (from a controlled explosion) in geophysical exploration.
3. The processing of a wide range of signals in biomedical applications.

In fact any field where information is capable of being represented as a signal can be regarded as a field of application for this subject.

In order to expand the material in this chapter to form a scientific basis for study, the ideas must be expressed and developed in a more quantitive manner. This will be done in succeeding chapters with the following principal aims in view:

1. To give a mathematical description to the signals under consideration. As will be seen there are several possible descriptions and the advantages and disadvantages of alternative descriptions will be investigated.

2. To obtain a mathematical formulation for the transformation of signals by a given system and, hence, to formulate a system description or model. As more than one possible signal description can be used, more than one system model is possible.
3. To combine aims 1 and 2 such that for a given signal model and input signal the corresponding output signal can be calculated.
4. To combine the simpler systems to form more complex systems and investigate the general properties of such combinations.

2

Signal Description

2.1 Introduction

The concept of a signal has already been introduced in Chapter 1 and a signal has the fundamental property that its variation conveys information. Mathematically a signal is a function with a dependent and one or more independent variables. However the signals considered in this book will have one independent variable only and this will be restricted to time. This does not mean that signals with other independent variables are not of importance but time signals are basic to the study of signal theory.

In the introductory chapter the distinction was made between continuous and discrete signals. This chapter starts by investigating this division in more detail. The processing of these two types of signal will then form parallel themes running through the remainder of the book. Often these themes are complementary, sometimes contrasting.

Section 2.3 examines some of the basic operations that can be performed on signals, these operations lay the foundations for more complex signal processing which is covered in later chapters.

Section 2.5 considers a number of basic continuous and discrete signals. These signals have a reasonably simple mathematical description and the response of systems to such signals will form much of the material of later chapters.

2.2 Continuous and discrete signals

Some examples of time signals are as follows:

1. The variation in voltage at the collector of a transistor in an audio amplifier.
2. The variation in temperature at a point in a furnace during a smelting process.
3. The variation in pressure in the cylinder of an internal combustion engine during operation.
4. The variation in the number of shoes sold in a certain shop during the month.
5. The variation in temperature at noon on successive days at a holiday resort.
6. The concentration of a given drug in the blood stream following an injection of the drug. Samples of blood are taken and analysed every fifteen minutes.

Although all these signals have time as their independent variable, the signals in examples 1 to 3 are basically different from those in examples 4 to 6. In examples 1 to 3 the signals representing the variables are present at all instants of time; they are known as *continuous* signals. In examples 4 to 6 the signals are only available at discrete time intervals; they are known as *discrete* signals. The discreteness occurs either due to the nature of the process as in example 4 or due to the nature of the measurements as in examples 5 and 6.

It should be noted that the processes producing the variables in these latter examples are continuous processes (the resort temperature has a value at all times as

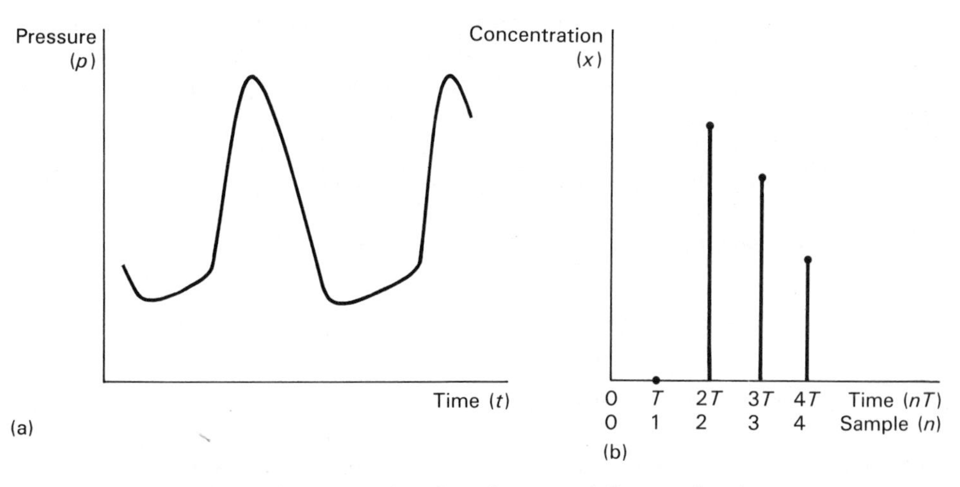

Figure 2.1 Examples of continuous and discrete signals.

does the concentration of the drug in the blood). However, because the measurements are taken at discrete intervals, they give rise to discrete signals. Most of the discrete signals encountered in this book arise due to the *sampling* of a continuous signal.

The variables in example 3 and 6 are plotted against time in Figure 2.1. Referring to Figure 2.1(a) pressure is a continuous variable, a pressure exists for every instant of time. Pressure is a function of time and this is denoted mathematically by writing it as $p(t)$. This notation can become cumbersome and when it is clear that the independent variable is time the notation will be simplified by omitting the t; for this example the signal would be written as p.

The concentration in Figure 2.1(b) is also a function of time. However time can only take discrete values and this could be shown by writing the signal $x(nT)$. Here T is the time between samples and n is the number of the sample. When a continuous process is sampled in this manner to form a discrete signal, T is known as the sampling time and $1/T$ is the sampling rate or sampling frequency. However not all discrete signals are formed by the sampling of continuous processes and in such cases the signal could be written $x(n)$. Even with time sampled signals it is often more convenient to omit the T and again to write the signal $x(n)$.

In general the notation $x(t)$, $x(n)$ will be used to denote the input signals to a system and $y(t)$, $y(n)$ will denote the corresponding output signals. In specific cases however when the signals represent variables that have accepted symbols these will be used instead, e.g., $v_i(t)$ as input voltage, $\theta(t)$ as angle.

2.3 Basic operations on signals

In the previous chapter the concept of a system was introduced. A system can be regarded as a process that transforms one signal into another signal. The next chapter will be devoted to a full discussion of systems but at this stage it is useful to consider some basic transformations. These transformations have to be performed by systems but this is not the aspect that is of importance here. As will be seen, combinations of these simple transformations enable complex signals to be constructed from simpler basic signals. The actual mechanism of the transformations will be accepted at this stage and emphasis placed on the signals produced. The operations consist of transformations on either the dependent or independent variables.

2.3.1 Amplitude scaling

This is an operation on the dependent variable. Consider a signal $x(t)$ fed into an amplifier to give an output signal $y(t)$. Assume the amplifier modifies the amplitude of the input signal by a gain factor at every time instant. This is illustrated in Figure 2.2 for an amplifier with a gain of 2. The output $y(t)$ is identical in shape to the input

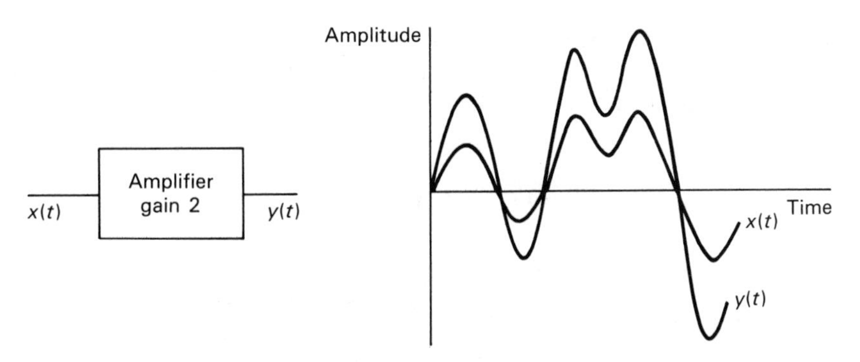

Figure 2.2 Illustration of amplitude scaling.

signal but is everywhere twice its amplitude. It is apparent that the input signal $x(t)$ could be used to represent the output signal $y(t)$ if the scale on the vertical axis were changed – hence the term amplitude scaling. In general the relationship between $x(t)$ and $y(t)$ can be written

$$y(t) = ax(t)$$

where a is a constant.

Scaling can be performed on discrete signals – a set of discrete numbers could be fed into a computer and a simple program written to multiply each by a constant.

2.3.2 Time scaling

Consider a piece of music recorded on tape, the tape is replayed but at the wrong speed. There is a very obvious difference in the sound, the signal has been compressed or expanded in time. This is shown for a much simpler signal in Figure 2.3. As with amplitude scaling the signals can be considered identical if the time axis is rescaled, hence the name time scaling.

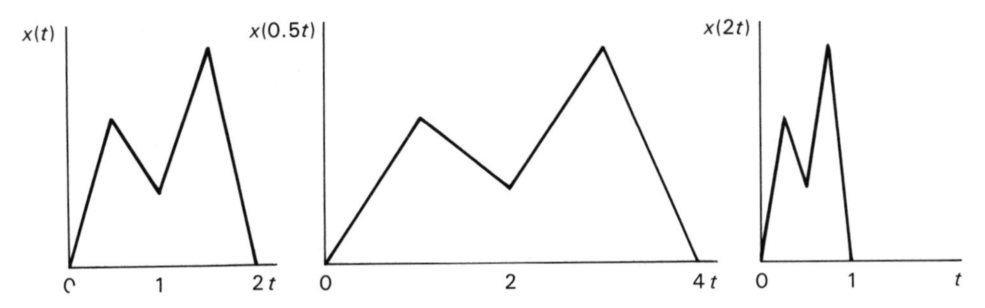

Figure 2.3 An illustration of time scaling.

Problems arise if this operation is performed on discrete signals, what meaning can be given to fractional samples, what about missing samples? For this reason the operation for discrete signals will not be discussed further.

2.3.3 Reflection and time shifting

Suppose a signal has been recorded on tape, it is then played back in reverse starting at a later time. The original and playback signals are illustrated in Figure 2.4, and the problem is to describe the playback signal in terms of the original.

The signal $y(t)$ can be obtained by performing two operations on $x(t)$:

1. The time scale of $x(t)$ is reversed.
2. The reversed signal is delayed in time.

It is convenient to consider these operations separately. The reversal of the time scale is known as reflection and is illustrated in Figure 2.5.

The reflected signal can be obtained by putting the value of the signal that was originally at t, at the point $-t$, it is hence described as the signal $x(-t)$. This is illustrated in Figure 2.5 for the specific point $t = 2$.

A time delayed signal is shown in Figure 2.6. To obtain the delayed signal the value of the signal that was originally at $(t - T)$ must now be at point t; the delayed

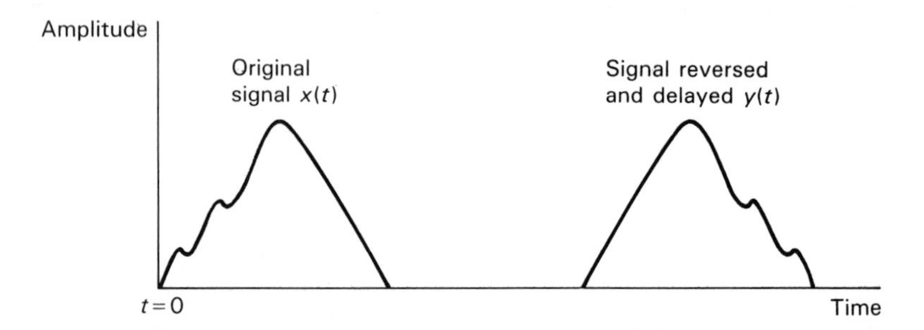

Figure 2.4 Delay and reversal of signal $x(t)$.

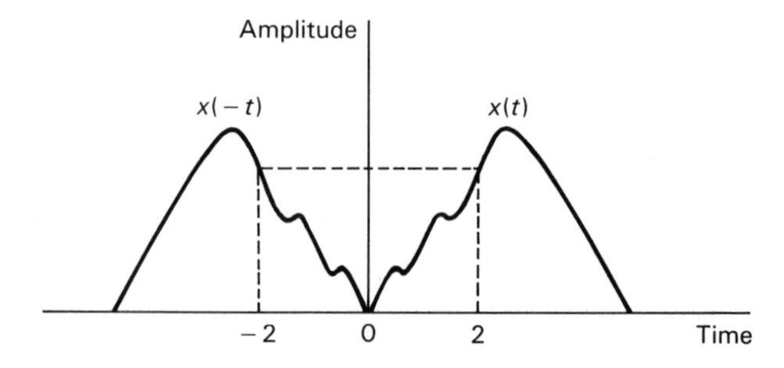

Figure 2.5 Reflection of signal $x(t)$.

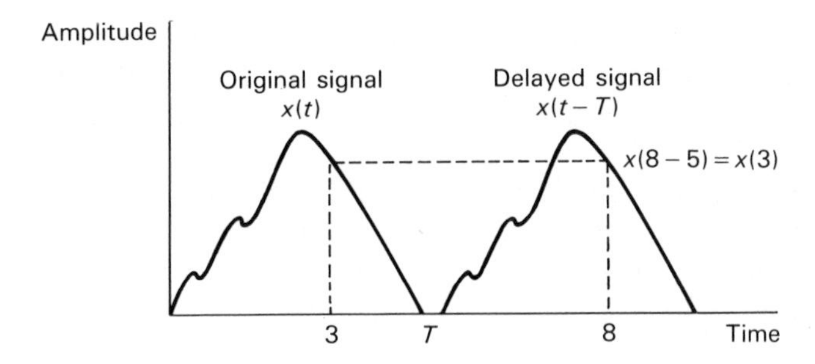

Figure 2.6 Time delay of signal $x(t)$.

signal is hence represented by $x(t-T)$. This is illustrated in Figure 2.6 for the specific point $t=8$ and for a delay $T = 5$.

An operation of time advance would be represented by the signal $x(t+T)$. The operations of time advance and reflection are not physically possible on time signals (they would involve prediction of the future). However, as will be seen, they are very useful basic operations that will be used to describe some useful properties of signals.

The operations of reflection and time shifting can also be performed on discrete signals as shown in Figure 2.7. Shifting can only be performed, in the case of a discrete signal, by an integer number of values.

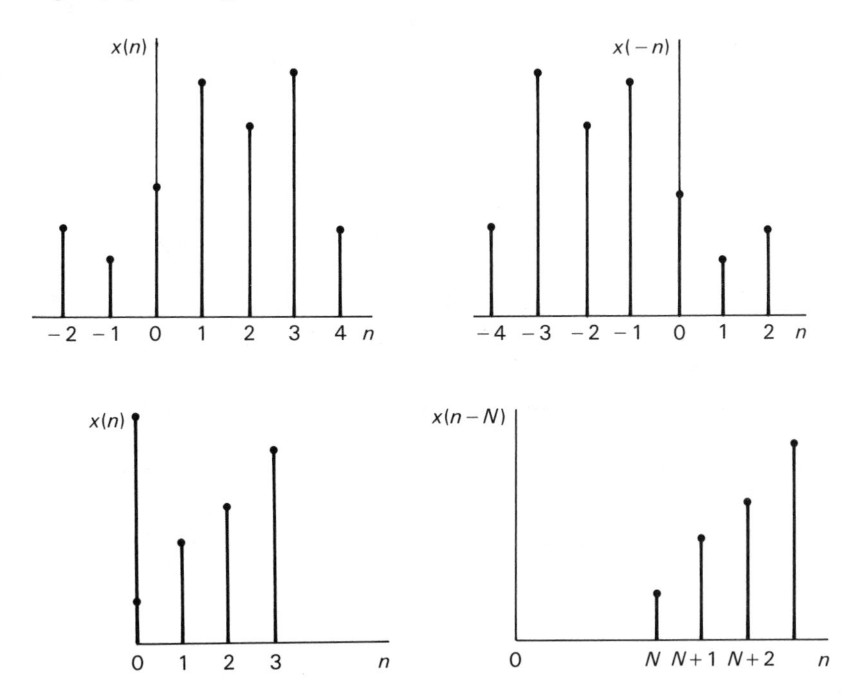

Figure 2.7 The operations of reflection and shifting for a discrete signal.

The operations of reflection and time delay can be combined to produce a delayed reversed signal as described at the start of this section. The operations can be performed in either order but some care must be taken with the details. The processes are illustrated in Figure 2.8.

In Figure 2.8(a) the signal $x(t)$ is first reflected to give $x(-t)$, this signal is then delayed by time T giving $x(-(t-T)) = x(T-t)$. In Figure 2.8(b) the signal $x(t)$ is advanced by time T giving $x(t+T)$, this signal is then reflected about the origin giving $x(T-t)$ as before.

2.3.4 *Addition of signals*

If two continuous signals are added together their values at every instant are added to form their sum at that instant. For discrete signals the definition is similar but the addition can only be performed at the integer values for which the signals are defined.

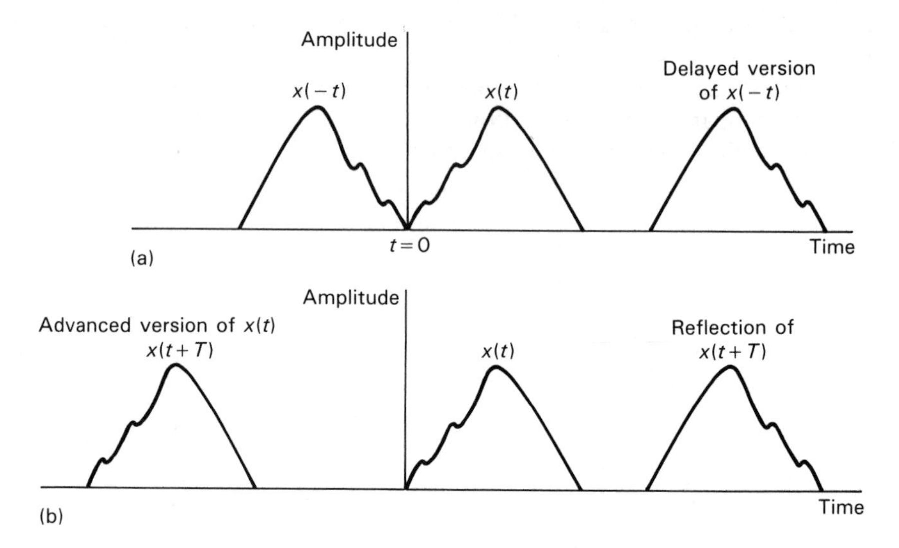

Figure 2.8 Processes of reflection and delay in: (a) reflection precedes delays; (b) an advance precedes reflection.

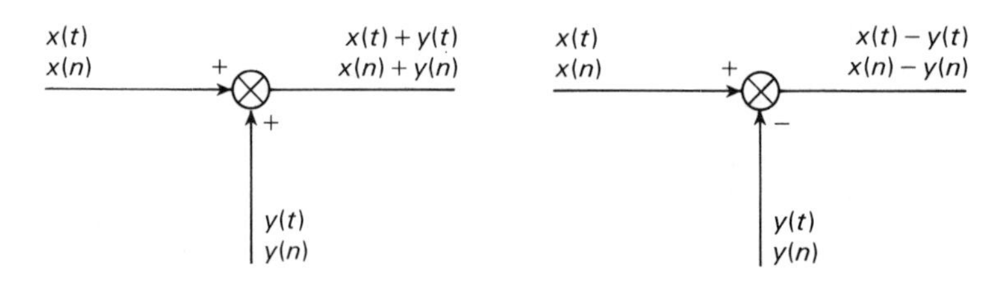

Figure 2.9 Addition and subtraction of signals.

Subtraction of signals is similarly defined. Figure 2.9 shows the symbolic representation for addition and subtraction that will be used throughout this book. The operations of amplitude scaling, time shifting and addition are basic to the theory of signals. The following example illustrates a combination of these operations.

EXAMPLE 2.3.1

(a) Figure 2.10(a) shows the signals $x(t)$, $y(t)$. Obtain the signal $z(t) = 2x(t-2) + 3y(t-1)$.
(b) Figure 2.10(b) shows the signals $x(n)$, $y(n)$. Obtain the signal $z(n) = 2x(n+2) + 0.5y(-n)$.

SOLUTION

$2x(t-2)$ represents the signal $x(t)$ delayed by 2 units and amplitude scaled by a factor of 2.

$3y(t-1)$ represents the signal $y(t)$ delayed by 1 unit and amplitude scaled by a factor of 3.

Figure 2.11 shows these signals and their addition to form the signal $z(t)$.

The student may find some difficulty in dealing with the discontinuities in these signals. This difficulty can be overcome by considering the sum of the signals at times slightly before and slightly after the discontinuity. Taking the discontinuity at $t=2$, just before this time $2x(t-2)=0$, $3y(t-1)=3$, $z(t)=3$. Just after $t=2$, $2x(t-2)=2$, $3y(t-1)=3$, $z(t)=5$.

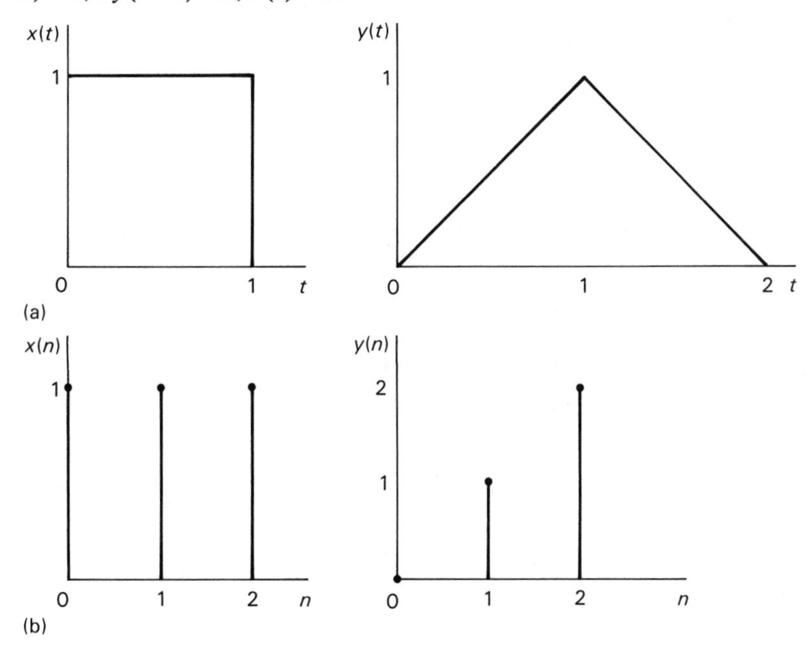

Figure 2.10 Signals for Example 2.3.1.

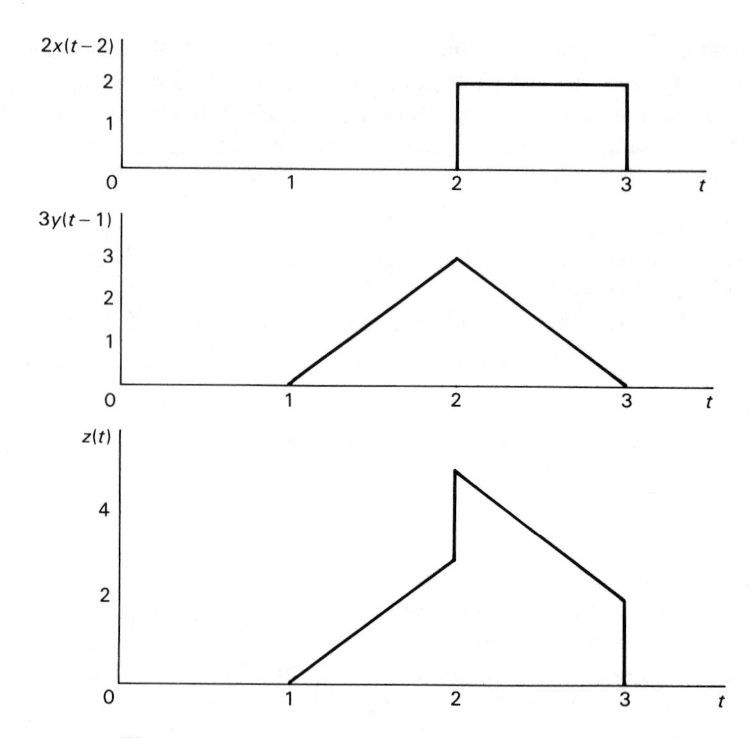

Figure 2.11 Combination of signals $x(t)$ and $y(t)$.

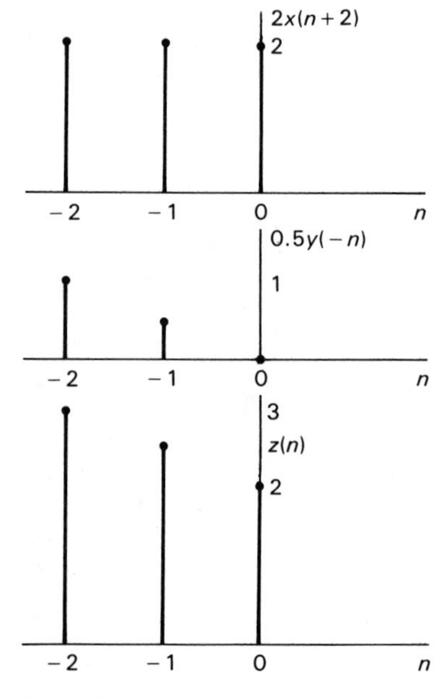

Figure 2.12 Combination of signals $x(n)$ and $y(n)$.

The second part of the example can be solved in a similar manner and the steps are shown in Figure 2.12.

2.4 Two properties of signals

In this section two properties of signals are introduced that will prove of value in later chapters.

2.4.1 Even and odd signals

These properties are based on symmetries that signals may have about $t=0$ $(n=0)$. Figure 2.13 shows both continuous and discrete signals exhibiting two forms of symmetry.

These symmetries can be conveniently expressed by using the operation of reflection. The signals in Figure 2.13(a) have the property that $x(t)=x(-t)$, $x(n)=x(-n)$, such signals are known as *even* signals. The signals in Figure 2.13(b) have the property that $x(t)=-x(-t)$, $x(n)=-x(-n)$, such signals are known as *odd* signals.

It can easily be shown that any signal can be expressed in terms of odd and even signals. An arbitrary signal $x(t)$ can be written

$$x(t)=\frac{[x(t)+x(-t)]}{2}+\frac{[x(t)-x(-t)]}{2}$$

The term $[x(t)+x(-t)]$ is even, the term $[x(t)-x(-t)]$ is odd. This can be verified by changing the sign of t in these terms, the sign of the first term is unaltered, that of the second term changes.

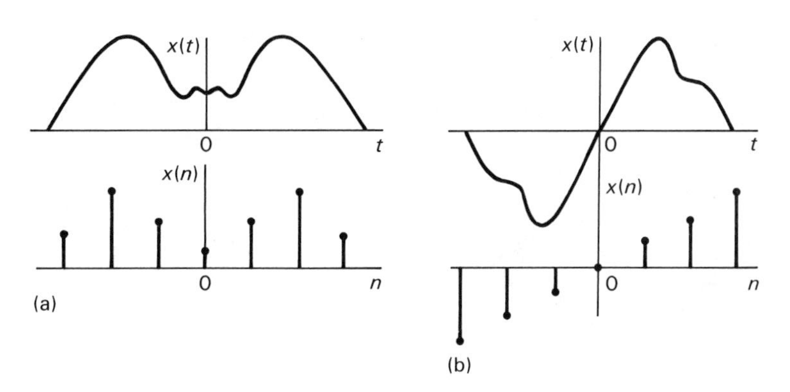

Figure 2.13 Signals showing even and odd symmetries.

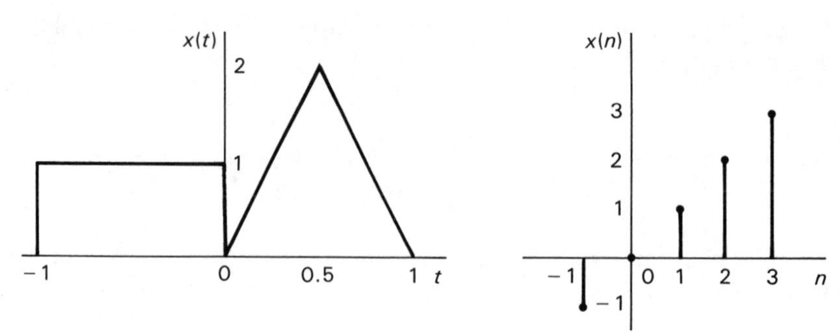

Figure 2.14 Signal for Example 2.4.1.

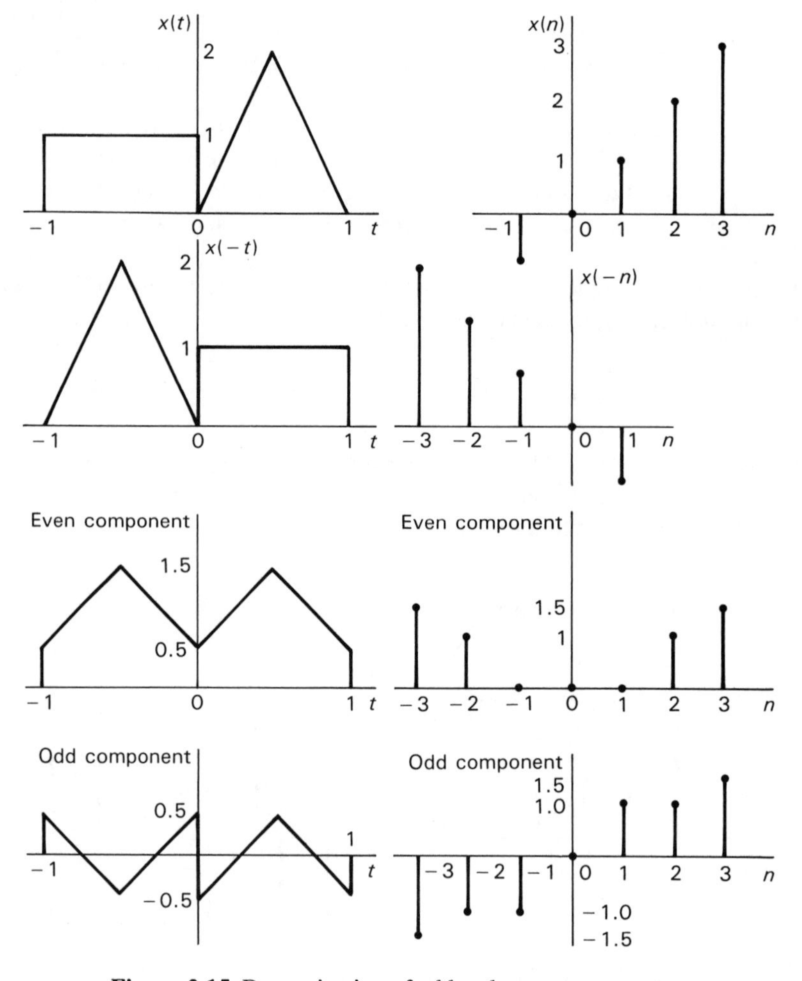

Figure 2.15 Determination of odd and even components.

The student may question why properties of a signal dependent upon its behaviour prior to $t = 0$ should be of importance. The answer to this question relates to the property of periodicity (considered in the next section) and Fourier analysis (considered in Chapters 5 and 6).

Example 2.4.1 illustrates the decomposition of a signal into odd and even components.

EXAMPLE 2.4.1

Express the signals shown in Figure 2.14 as the addition of odd and even components.

SOLUTION

The steps involved in determining the odd and even components are shown in Figure 2.15. The signals $x(-t)$, $x(-n)$ are derived; these are then added to and subtracted from, $x(t)$, $x(n)$ to give even and odd signals. The amplitude scaling factor of 0.5 must be included, otherwise adding the even and odd components will not reconstitute the original signal.

2.4.2 *Periodicity*

Figure 2.16 shows examples of periodic continuous and discrete signals. The feature that makes the signal periodic is the property that the signal repeats itself indefinitely in the future and has repeated itself indefinitely in the past. This property can be conveniently expressed by using the time shifting operation. A periodic signal $x(t)$, $x(n)$ has the property that

$$x(t+T) = x(t) \qquad \text{for all } t$$
$$x(n+N) = x(n) \qquad \text{for all } n$$

The time T (or the number N) is known as the *period* of the waveform. It should be noted that if a waveform is periodic with period T it is also periodic for any integer multiple of T (similarly for N).

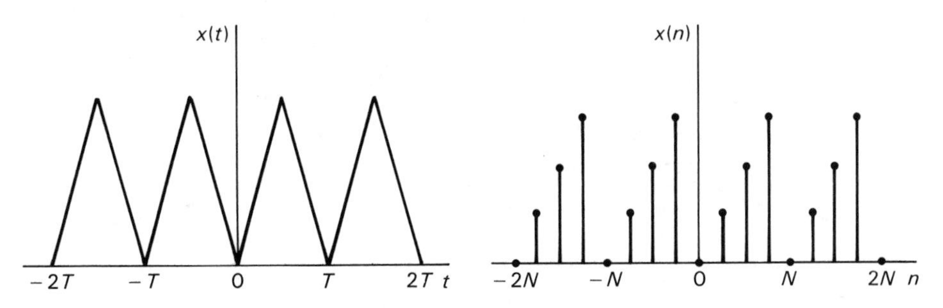

Figure 2.16 Examples of continuous and discrete periodic signals.

Periodic signals are of great importance in signal theory and will be used extensively throughout the remainder of this book. The sinusoidal signal which is the most important periodic signal will be introduced in the next section.

2.5 Some basic signals

In this section some signals are introduced that are important in the study of signal theory. This does not mean that these signals will necessarily be present as input and output signals to systems under practical working conditions. However, often these basic signals can be combined to form more complex signals that may be present in practice. Also, as will be seen later, the response of a system to some of these signals forms a very useful description of that system. Because of this, and because most of the signals are relatively easy to generate, they form very useful test signals.

2.5.1 The exponential signal

The continuous time exponential is of the form

$$x(t) = Ae^{at}$$

where A and a are constants. Figure 2.17 shows the signal for both positive and negative values of a.

The property that causes this signal to arise so frequently in systems is concerned with its rate of change or slope. For convenience, the dependence of the signal on the independent variable t will be omitted and it can be written as

$$x = Ae^{at}$$

and its slope.

$$\frac{dx}{dt} = aAe^{at} = ax$$

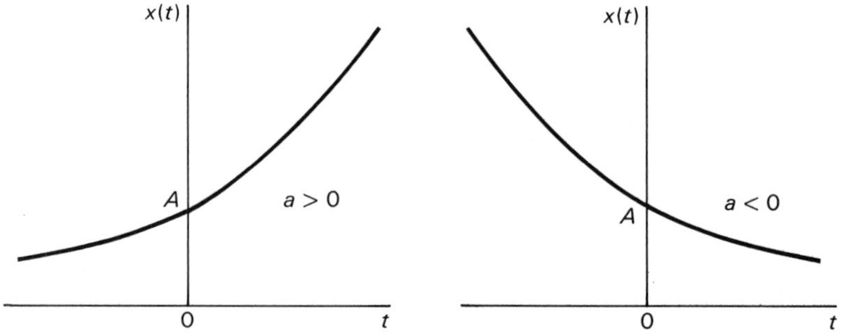

Figure 2.17 The continuous exponential signal.

The signal has the property that its slope, at any time, is proportional to its value at that time. Many systems produce signals having this property; their rate of change is proportional to their present value. Some examples are phases in bacterial growth, charge and discharge of capacitors, radioactive decay. The slope can be positive or negative depending on the sign of 'a' giving growth or decay. (Note by including the case where $a = 0$ then the constant signal can be included as a special case of the exponential).

The discrete exponential signal is of the form

$$x(n) = A\mathrm{e}^{anT}$$

where A and a are constants. Figure 2.18 shows this signal for both positive and negative values of a.

As the signal is a time signal the independent variable is the discrete time instants nT. However it is more convenient to label the independent variable as n, as stated in Section 2.2.

The signal can also be written in the form

$$\begin{aligned} x(n) = A\mathrm{e}^{anT} &= A(\mathrm{e}^{aT})^n \\ &= A(z)^n \end{aligned}$$

where

$$z = \mathrm{e}^{aT}$$

In this form it is easy to see that

$$x(n+1) = Az^{n+1} = Az^n z = x(n)z$$

At each instant the signal is formed by multiplying the signal at the previous instant by the constant z, i.e., the values form a geometric progression. If the constant z is greater than unity (a positive) this leads to growth, if it is less than unity (a negative) this leads to a decaying signal.

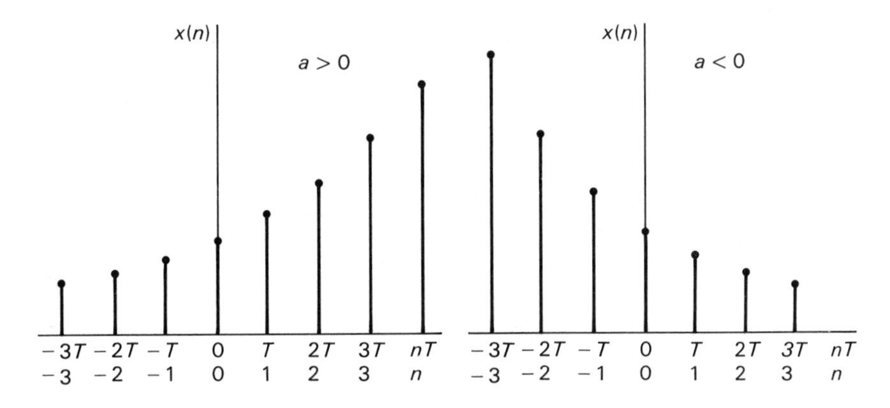

Figure 2.18 The discrete exponential signal.

The continuous exponential signal has the property that its slope is always proportional to its present value. There is no meaning to slope as a derivative when applied to a discrete signal. However, a similar property holds based upon differences between successive samples.

$$\begin{aligned} x(n+1) - x(n) &= x(n)z - x(n) \\ &= (z-1)x(n) \end{aligned}$$

As $(z-1)$ is a constant, the difference between successive samples is a constant times the nth sample.

2.5.2 The sinusoidal signal

The student will have met the sine and cosine function in mathematics and will realise that these are functions of angle; the functions are shown in Figure 2.19.

If a time signal is to be described by such a relationship it is necessary to convert the independent variable from time to angle, a relationship

$$\theta = \text{constant} \times t$$

is required. The constant can be determined by noting that the function is periodic and then relating the periodic time T to the corresponding angle 2π. This gives the constant as $2\pi/T$, and this is known as the angular frequency having units rad/s. However, $1/T$ is the number of cycles per unit time; it is the frequency f measured in hertz (Hz). Hence the relationship between θ and t can be written,

$$\theta = \omega t = 2\pi f t$$

and the sine and the cosine time signals can be represented as

$$x(t) = A \sin 2\pi f t \qquad \text{and} \qquad x(t) = A \cos 2\pi f t$$

As the maximum value of the functions $\sin\theta$ and $\cos\theta$ is unity, the A acts as a scaling factor giving maximum and minimum values $\pm A$.

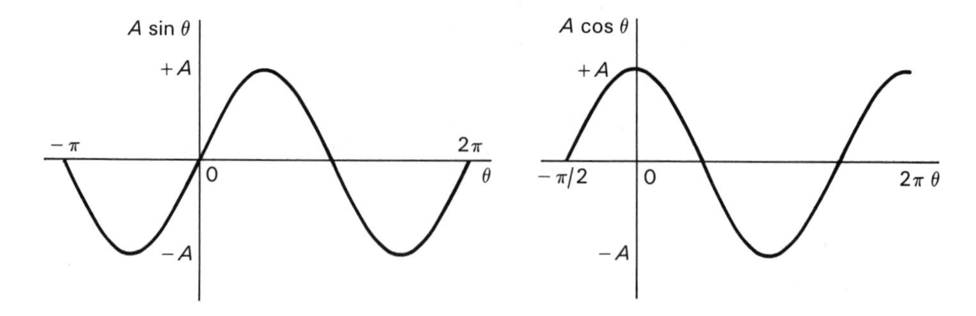

Figure 2.19 The sine and cosine functions.

Using the property of time shifting introduced in Section 2.3.3 it can be seen that a cosine wave can be obtained by advancing a sine wave by time $T/4$ (angle $\pi/2$). A more general signal can be obtained by advancing a sine wave by an arbitrary angle φ giving

$$x(t) = A\sin(\omega t + \varphi)$$

and this signal is shown as Figure 2.20. The angle φ is known as the phase angle.

By using the trigonometric identity for $\sin(A+B)$ this signal can be written,

$$\begin{aligned} x(t) &= A\sin(\omega t + \varphi) \\ &= A\sin\omega t\cos\varphi + A\cos\omega t\sin\varphi \end{aligned}$$

The signal can be regarded as the sum of a sine wave and a cosine wave with scalings $A\cos\varphi$ and $A\sin\varphi$ respectively. Throughout this book a general sine wave of this nature will be described as a *sinusoidal* signal. A property of sinusoidal signals that makes them so useful is that if two sinusoids of the same frequency (but of arbitrary amplitude and phase) are added together they will produce another sinusoid of that same frequency. The general proof of this will be left as an exercise for the student, but particular instances are given in Example 2.5.1.

Although addition of two sinusoids of the same frequency produces a further sinusoid, this is not true if the frequencies of the components are not the same. However, an important result occurs if one of the sinusoids has a frequency that is an integer multiple of the other.

Such a sinusoid is known as a harmonic, e.g., for a sinusoid of frequency 30 Hz a sinusoid of frequency 60 Hz would be its second harmonic, one of 90 Hz its third harmonic etc. In this context the 30 Hz sinusoid is termed the fundamental. If a harmonic and its fundamental are added together the resultant waveform, although not sinusoidal, will be periodic with period equal to that of the fundamental. The periodicity will remain no matter how many further harmonics are added. This result is illustrated in Figure 2.21 for a fundamental plus second harmonic.

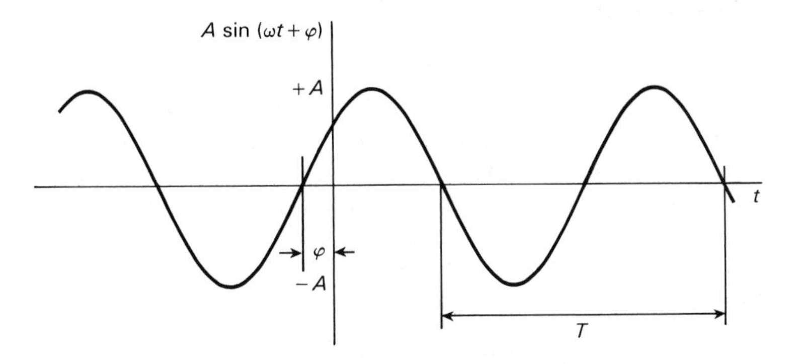

Figure 2.20 More general sine wave.

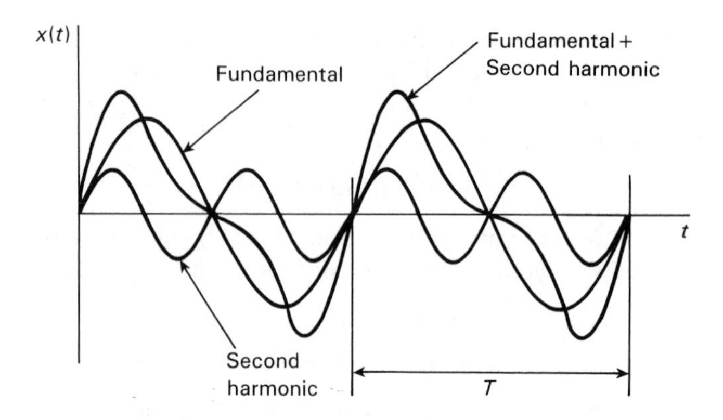

Figure 2.21 Fundamental plus second harmonic.

A discrete sinusoid can be described by the relationship

$$x(n) = A \sin(n\omega T + \varphi)$$

and a particular example is illustrated in Figure 2.22 where $\varphi = 0$, $f = 1$ $(\omega = 2\pi)$, and $T = 2/29$ (the sampling time).

Care must be taken when considering the period of a discrete sinusoid. At first glance the waveform in Figure 2.22 appears to repeat itself after time $2\pi/\omega = 1$ s. However, the definition of periodicity for a discrete waveform is that $x(n) = x(n+N)$ where N is an integer. In one second there are $29/2 = 14.5$ samples (which is obviously not an integer) and it is apparent that two seconds (two cycles) are required to obtain 29 samples.

As the period of a discrete periodic signal is an integer number N it has no units associated with it (unlike the period of a continuous signal that has the unit of time). An alternative description of a discrete sinusoid is given by

$$x(n) = A \sin(n\theta)$$

where $\theta = \omega T$ is known as the normalised frequency, it is equal to the angular frequency of the unsampled sinusoid divided by the sampling frequency.

$$\theta = \omega T = \omega / f_s$$

θ has the number of angle, radians, and can be interpreted as the number of radians/sample. As a complete cycle is 2π radians then the number of samples per cycle is given by $2\pi/\theta$. The use of normalised frequency often leads to more convenient expressions and is helpful in the design of digital filters as will be shown in a later chapter.

Great care must be taken when interpreting the results of sampling a sinusoid to produce a discrete signal. Consider the continuous signal $x(t) = A \sin(\omega t + \varphi)$ which is sampled to produce the discrete signal

$$\begin{aligned} x(n) &= A \sin(n\omega T + \varphi) \\ &= A \sin(n\theta + \varphi) \end{aligned}$$

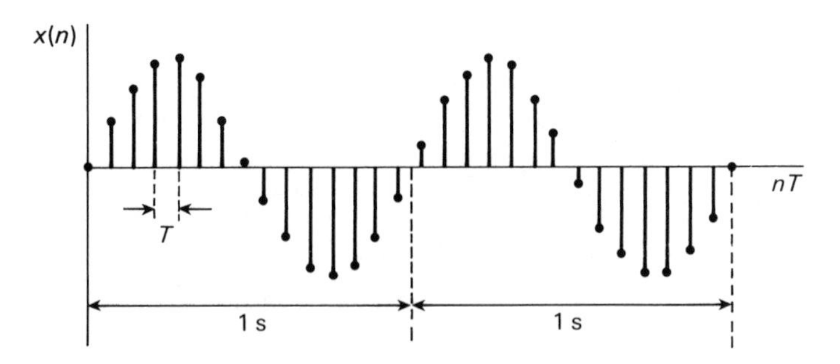

Figure 2.22 A discrete sinusoid.

Because of the periodicity of the sinusoid

$$A\ \sin(n\theta + \varphi) = A\ \sin(n[\theta + 2\pi m] + \varphi)$$

where m is an integer. Hence sampling a sinusoid of normalized frequency θ will produce exactly the same discrete signal as sampling a sinusoid of frequency $(\theta \pm 2\pi m)$. There is an ambiguity between sinusoids having angular frequencies

$$2\pi fT \text{ and } (2\pi fT \pm 2\pi m)$$

or, dividing by $2\pi T$, between frequencies

$$f \text{ and } f \pm mf_s$$

This is illustrated in Figure 2.23 for a sinusoid having a frequency 1 Hz which is sampled four times per cycle ($f_s = 4$ Hz). There is ambiguity between samples taken from this signal and samples taken from a signal of frequency 5 Hz (1 + 4). This is shown in Figure 2.23(a). There is also ambiguity between the 1 Hz signal and a signal having a frequency of −3 Hz (1 − 4). The concept of a negative frequency will be used in later chapters, at the moment, however, it is sufficient to use the relationship $\sin(-\theta) = -\sin(\theta)$ and to interpret this as a 3 Hz sine wave with a phase inversion. The ambiguity arising from this signal is shown in Figure 2.23(b).

By examination of the sampled signal it is not possible to determine which of the three sinusoids the samples represent. The identity of the original signal is lost among 'alias' signals and this effect is known as *aliasing*.

The student, although probably not realising it, has met the phenomenon of aliasing in the viewing of a film. Although the projected film appears continuous it is formed by showing in succession a series of still frames of the scene – i.e., it samples the scene. The resultant film appears continuous because of our persistence of vision, an image remains on the retina for a short period after the object has been removed. This process reconstitutes the original moving scene but there are anomalies. In the cowboy film the wagon wheels rotate as the wagon moves off, the rotational speed at first increasing as the wagon gains speed. However as the wagon speed increases further the wheels appear to slow down, then stop, and then rotate backwards. The film frame speed is sampling the rotational speed of the wagon wheel producing an incorrect alias speed.

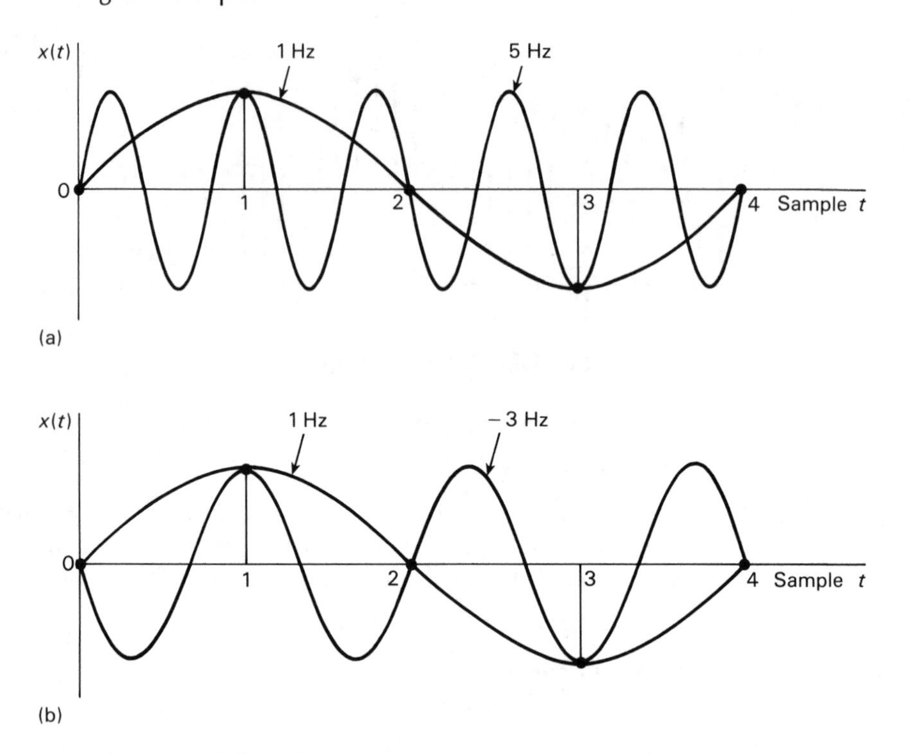

Figure 2.23 Aliasing in sampled sinusoids.

An important question concerns the choice of sampling frequency if the effect of aliasing is to be avoided when reconstituting a signal from samples. Consider a signal consisting of a combination of sinusoids but it is known that the highest frequency present is f_H Hz. Provided any alias components formed are greater than f_H then these could be removed by a filter (ideally) without affecting the signal components. This suggests that the sampling frequency f_s should be greater than the highest frequency present, i.e., greater than f_H. However no restriction has been placed on the phase of the signal and the highest frequency may be represented as $\sin(-2f_H t)$. This would produce an alias frequency of $f_H - f_s$ and for this to be greater than f_H then f_s must be greater than *twice* the sampling frequency. Sampling must occur at a frequency greater than twice the highest frequency present. This result is known as the *sampling theorem* and will be proved in a more formal manner in a later chapter.

2.5.3 The complex exponential

By use of complex numbers the exponential and sinusoidal signals can be written as special cases of a more general signal – the complex exponential. An appendix on complex numbers is given at the end of the book for students unfamiliar with this topic.

The signal

$$x(t) = Ae^{at}$$

has already been introduced in Section 2.5 for the cases where a is real. In this section the definition is widened to include the case where a is imaginary and when a is complex. Taking the case where a is imaginary, $a = \mathrm{j}\omega$, the signal becomes

$$\begin{aligned} x(t) &= Ae^{\mathrm{j}\omega t} \\ &= A \cos \omega t + \mathrm{j}A \sin \omega t \end{aligned}$$

Now $x(t)$ represents a real signal, a variable associated with a real process. However, the signal $x(t)$ contains an imaginary component and the student may wonder how it can represent a real signal. The sum of a complex number and its conjugate is real (twice the real part) hence the signal

$$\begin{aligned} x_i(t) &= Ae^{\mathrm{j}\omega t} + Ae^{-\mathrm{j}\omega t} \\ &= A \cos \omega t \end{aligned}$$

is real and can represent a signal in a physical system. Hence the signal $x(t) = Ae^{\mathrm{j}\omega t}$ can be visualised as half of a real signal (the other half being $Ae^{-\mathrm{j}\omega t}$). The reason for using the exponential form is that it is often easier to deal with mathematically than the trigonometric form and leads to more compact expressions.

The signal can be generalised to represent a sinusoid of arbitrary phase by writing

$$x(t) = Ae^{\mathrm{j}(\omega t + \varphi)} = Ae^{\mathrm{j}\omega t}e^{\mathrm{j}\varphi}$$

$$x(n) = Ae^{\mathrm{j}(n\theta + \varphi)} = Ae^{\mathrm{j}n\theta}e^{\mathrm{j}\varphi}$$

An important difference between the continuous and discrete sinusoid concerns the harmonics of a signal. The concept of a harmonic for the continuous case has been introduced in Section 2.5.2 and a similar definition applies in the discrete case. If a discrete sinusoid is given by the signal $x(n) = A \sin n\theta$ then its kth harmonic is given by $x_k(n) = A \sin nk\theta$. In the continuous case the number of harmonics of a sinusoid are infinite but this is not the case with a discrete signal, as will be shown in the following example.

Consider the harmonics of the discrete sinusoidal signal

$$x(n) = \sin(n2\pi/6 + \pi/18)$$

given by

$$x_k(n) = \sin(nk2\pi/6 + \pi/18)$$

These are plotted in Figure 2.24 for $k = 1 - 6$. As can be seen, the signals for $k = 1, 2, 3$, are periodic with increasing frequency (N, the period, decreasing 6, 3, 2). However, for $k = 4$ the frequency decreases to 3 samples/cycle and for $k = 5$ it

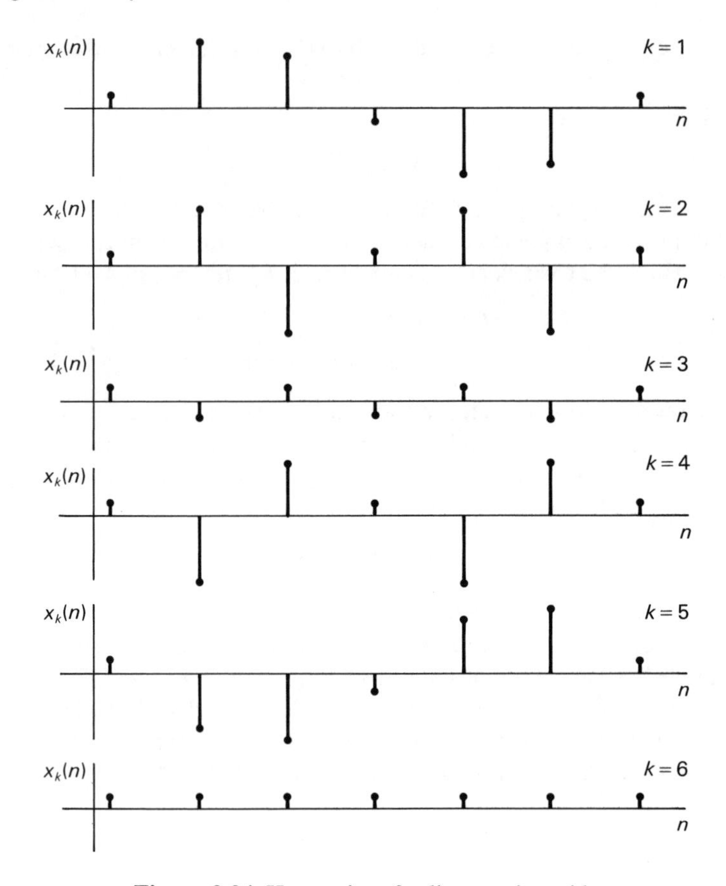

Figure 2.24 Harmonics of a discrete sinusoid.

decreases further to 2 samples/cycle. $k = 6$ gives a constant signal. It is left to the student to verify that if k is increased beyond 6 this pattern repeats.

This effect can be explained using the exponential form of the sinusoid

$$x(n) = A\mathrm{e}^{\mathrm{j}(n\theta + \varphi)}$$

and using the relationship $N = 2\pi/\theta$ this can be written

$$x(n) = A\mathrm{e}^{\mathrm{j}n2\pi/N}\mathrm{e}^{j\varphi}$$

The kth harmonic is given by

$$x_k(n) = A\mathrm{e}^{\mathrm{j}nk2\pi/N}\mathrm{e}^{\mathrm{j}\varphi}$$

Consider the $(N - k)$th harmonic, this is given by

$$\begin{aligned} x_{N-k}(n) &= A\mathrm{e}^{\mathrm{j}(N-n)k2\pi/N}\mathrm{e}^{\mathrm{j}\varphi} \\ &= A\mathrm{e}^{-\mathrm{j}nk2\pi/N}\mathrm{e}^{\mathrm{j}n2\pi}\mathrm{e}^{\mathrm{j}\varphi} \end{aligned}$$

However as $e^{jn\pi} = 1$ the frequency of the $(N-k)$th harmonic is identical to that of the kth harmonic. However, because the sign of the exponent has changed, the phase will not be identical.

If k is further increased to become $(N+k)$ then by similar reasoning the $(N+k)$th harmonic is identical to the kth harmonic. For the special case where $k = N$ then

$$x_N(n) = Ae^{j\varphi}$$

a constant value.

Hence the discrete sinusoid, unlike its continuous counterpart, does not have an infinite number of harmonics. This property will be seen to be important in the theory of the discrete Fourier series as covered in Chapter 6.

The complex exponential in general has an exponent that is complex. This gives the signals

$$x(t) = Ae^{(\sigma + j\omega)t}$$
$$x(n) = Ae^{n(\sigma + j\omega)T}$$

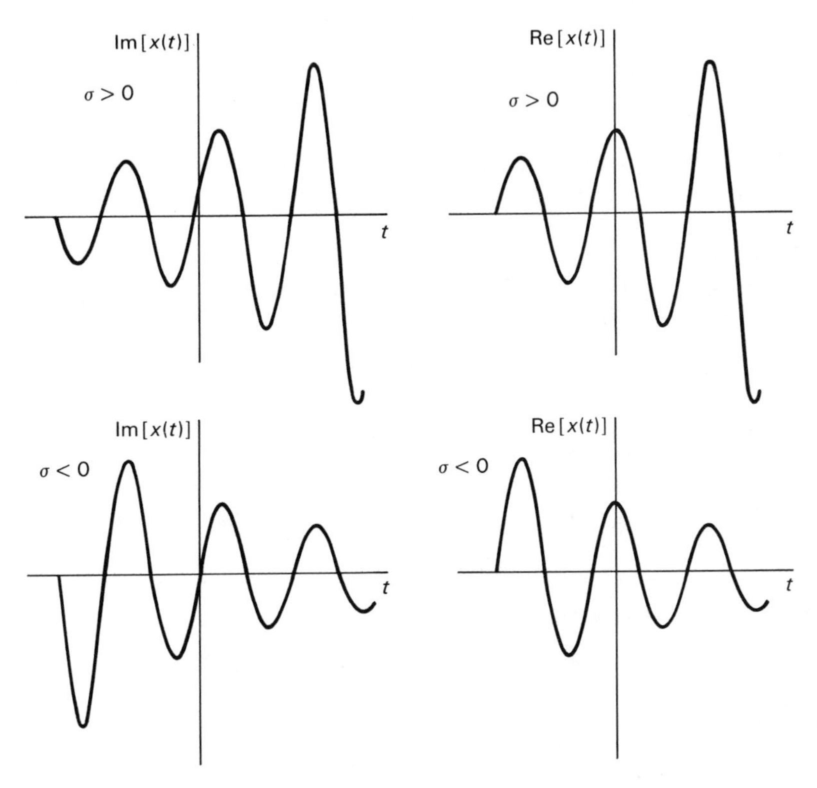

Figure 2.25 The complex exponential signal.

Considering the continuous signal, this can be written

$$x(t) = Ae^{\sigma t}e^{j\omega t}$$

The signals $e^{j\omega t}$ and $e^{\sigma t}$ have already been investigated and the resultant signal $x(t)$ can be considered as a sinusoid whose amplitude varies exponentially. Figure 2.25 shows the real and imaginary parts of this signal for the cases where $\sigma > 0$ and $\sigma < 0$. Corresponding waveforms apply to the discrete case.

This section finishes with an example which illustrates the addition of sinusoids and the respective merits of working in trigonometric and exponential form.

EXAMPLE 2.5.1

Two sinusoids $x(t)$ and $y(t)$ are represented as follows

$$x(t) = Ae^{j100t} + A^*e^{-j100t}$$

where $A = 2 + j3$ and A^* is the complex conjugate of A.

$$y(t) = 5\cos(100t + \pi/4)$$

if the signal $z(t) = x(t) + y(t)$ is represented as

$$z(t) = R\sin(100t + \varphi)$$

determine the values of R and φ.

SOLUTION

Two methods of solution will be considered; in the first both signals are expressed in trigonometric form; in the second both are expressed in exponential form.

Method 1

Considering the signal $x(t)$, this is of the form of a complex number plus its conjugate and can be written as twice its real part.

$$\begin{aligned} x(t) &= (2 + j3)e^{j100t} + \text{conjugate} \\ &= 2(\cos 100t + j\sin 100t) + j3(\cos 100t + j\sin 100t) + \text{conjugate} \\ &= 2(\cos 100t - 3\sin 100t) \\ y(t) &= 5\cos(100t + \pi/4) \\ &= 5\cos 100t\cos\pi/4 - 5\sin 100t\sin\pi/4 \\ &= \tfrac{5}{2}\cos 100t - \tfrac{5}{2}\sin 100t \end{aligned}$$

Adding $x(t)$ and $y(t)$ gives

$$z(t) = (4 + \tfrac{5}{2})\cos 100t - (6 + \tfrac{5}{2})\sin 100t$$

But

$$\begin{aligned} z(t) &= R\sin(100t + \varphi) \\ &= R\sin 100t\cos\varphi + R\cos 100t\sin\varphi \end{aligned}$$

Hence

$$R \sin \varphi = 4 + \tfrac{5}{2}$$
$$R \cos \varphi = -(6 + \tfrac{5}{2})$$

Hence

$$R = \sqrt{(4 + \tfrac{5}{2})^2 + (6 + \tfrac{5}{2})^2} = 12.15$$

$$\varphi = \tan^{-1} \frac{4 + \tfrac{5}{2}}{-(6 + \tfrac{5}{2})} = 141.7^\circ$$

Method 2

$$\begin{aligned}
x(t) &= (2 + \mathrm{j}3)\mathrm{e}^{\mathrm{j}100t} + \text{conjugate} \\
y(t) &= 5 \cos(100t + \pi/4) \\
&= \tfrac{5}{2}(\mathrm{e}^{\mathrm{j}100t}\mathrm{e}^{\mathrm{j}\pi/4} + \mathrm{e}^{-\mathrm{j}100t}\mathrm{e}^{-\mathrm{j}\pi/4}) \\
&= \tfrac{5}{2}(0.707 + \mathrm{j}0.707)\mathrm{e}^{\mathrm{j}100t} + \text{conjugate} \\
z(t) &= x(t) + y(t) = (3.767 + \mathrm{j}1.767)\mathrm{e}^{\mathrm{j}100t} + \text{conjugate} \\
&= 6.076\mathrm{e}^{\mathrm{j}51.7^\circ}\mathrm{e}^{\mathrm{j}100t} + \text{conjugate} \\
&= 12.15 \cos(100t + 51.7^\circ) \\
&= 12.15 \sin(100t + 141.7^\circ)
\end{aligned}$$

As can be seen, the second method using the exponential form produces an answer in fewer steps.

2.5.4 The unit step and unit impulse

These signals are different in form from the signals considered previously and their definition requires some care. In order to introduce the signals in the continuous case consider the circuit shown in Figure 2.26(a).

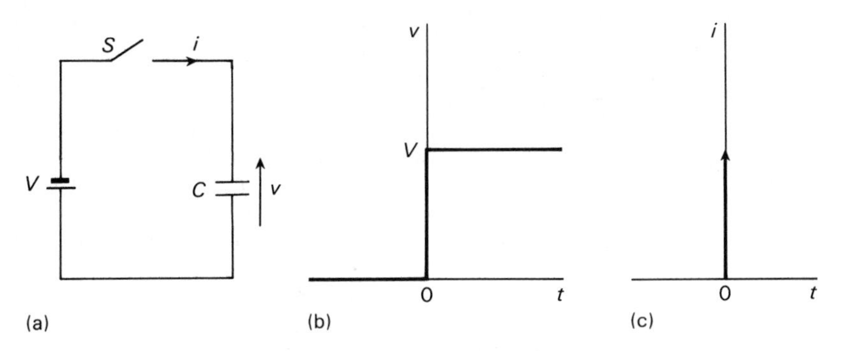

Figure 2.26 Illustration of unit step and unit impulse.

The capacitor is initially uncharged and the problem is to obtain the signals representing the voltage v across the capacitor and the current i in the circuit following the closure of the switch at $t=0$. Before $t=0$, $v=0$, after $t=0$, $v=V$, at $t=0$ the voltage is indeterminate. These conditions are shown in Figure 2.26(b).

The current can be obtained via the charge q on the capacitor using the relationship $q=Cv$; just before $t=0$ there is zero charge on the capacitor, just after $t=0$ there is charge $q=CV$. To obtain the current the relationship required is one that relates current to rate of change of charge.

$$i=\frac{\mathrm{d}q}{\mathrm{d}t}=C\frac{\mathrm{d}v}{\mathrm{d}t}$$

The current is proportional to the slope of the voltage/time curve, hence it is zero at all times before and after $t=0$ but is infinite at $t=0$. This is shown in Figure 2.26(c).

The voltage and current signals are examples of the step and impulse functions. Although the waveforms shown in Figure 2.26 serve to describe these functions in a non-rigorous manner, a more formal approach is to regard these functions as the limiting case of more conventional functions. Returning to the circuit of Figure 2.26(a), in practice, the switch would not be ideal and the waveforms of Figure 2.26(a) and (b) would not be obtained. Suppose the resistance of the switch now changes in a finite time, starting from ∞ at $t=0$ and falling to zero at $t=\varepsilon$ (the precise nature of this resistance variation is however unknown). The signals representing voltage and current will now be as shown in Figure 2.27.

Because the manner in which the switch resistance varies is not specified, the detailed shapes of these curves are unknown. However, if the voltage V and capacitor C are assumed to have unit value then the relationship

$$i=\frac{\mathrm{d}v}{\mathrm{d}t} \quad \text{and} \quad \int_{-\infty}^{+\infty} i\,\mathrm{d}t=1$$

hold. Suppose now the time ε is made smaller, the effects on i and v are shown in Figure 2.28.

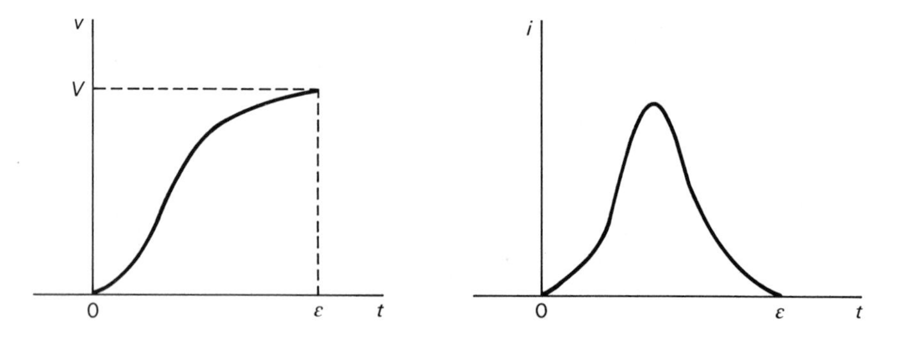

Figure 2.27 Voltage and current variation.

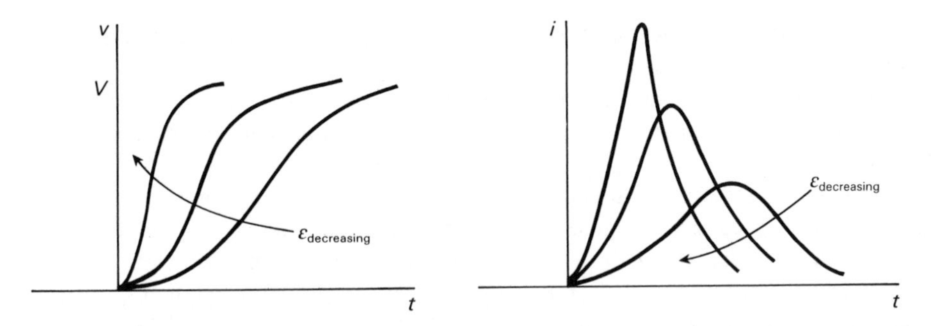

Figure 2.28 Effect on voltage and current of decrease in ε.

In the limit, as $\varepsilon \to 0$ the step and impulse functions are obtained; however an additional property of the impulse function is that its area is unity. If a simple function, a rectangular pulse, is used to describe the current variation the effect of reducing ε can be shown as in Figure 2.29.

If the amplitude of the pulse is multiplied by a constant k before the limit is taken the resulting impulse is of area k. Hence scaling can be performed on an impulse but it is the *area* of the impulse that is scaled.

The following definitions can be used to describe the step and impulse functions.

$$\text{Unit step function } u(t) \quad u(t) = 0 \qquad t < 0$$
$$u(t) = 1 \qquad t > 0$$

function is undefined for $t = 0$

$$\text{Unit impulse function } \delta(t) \quad \delta(t) = 0 \qquad t \neq 0$$
$$\delta(t) = \infty \qquad t = 0$$

area under the function is unity.

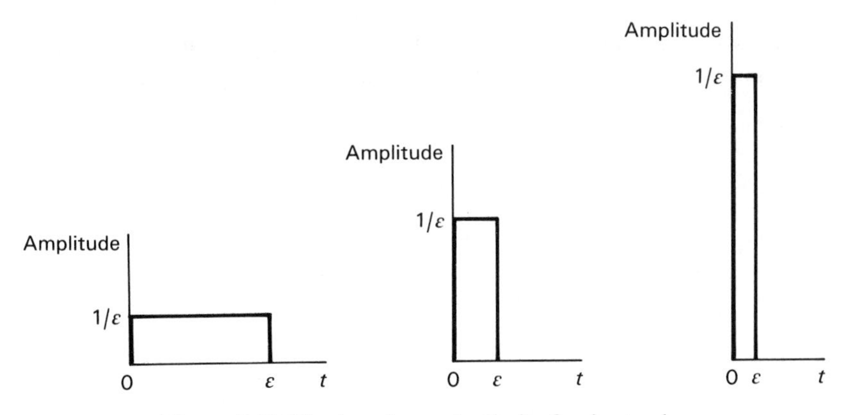

Figure 2.29 The impulse as the limit of a short pulse.

Although these definitions will serve in non-rigorous manner, it should be borne in mind that strictly these functions are obtained by taking a limiting case of more conventional functions as described earlier. This is particularly important when considering their behaviour when used in conjunction with more conventional functions.

Although the step function can be closely approximated, in practice it is apparent that an impulse cannot be generated (it would have infinite amplitude) and the student may wonder about the usefulness of such a function. However, as will be seen in later chapters, if a very short pulse of finite amplitude is applied to a system, such that the system response exhibits little change over the width of the pulse, then the response of the system approximates to the response to an input consisting of an ideal impulse. Also, as will be seen in later chapters, the impulse function is of immense theoretical value in defining system response.

Because of its infinite amplitude and undefined shape, the impulse is usually depicted as in Figure 2.30. The impulse can be delayed just as any other function and it will be met with frequently in its delayed form $\delta(t-T)$; this is also shown in Figure 2.30.

One of the most important features of the impulse is its behaviour when combined with another signal under the integral sign. Most generally this can be written

$$\int_{-\infty}^{+\infty} x(t)\delta(t-t_0)\,\mathrm{d}t$$

This expression needs care in its interpretation and the signals are as shown in Figure 2.31.

Figure 2.31(a) shows the signal $x(t)$ and an impulse $\delta(t-t_0)$. Both the operations of multiplying these signals and of integrating the result have little meaning. To interpret the integral the impulse must be regarded as a more conventional function and the limit to an impulse not taken until after the multiplication and integration. In Figure 2.31(b) the impulse has been replaced by a short pulse, width ε, height $1/\varepsilon$. Multiplication by $x(t)$ presents little difficulty, producing a pulse of width ε and varying amplitude $x(t)/\varepsilon$; integration gives the area of this pulse. As $\varepsilon \to 0$, the pulse approaches constant amplitude $x(t_0)/\varepsilon$ (approaching infinity) and

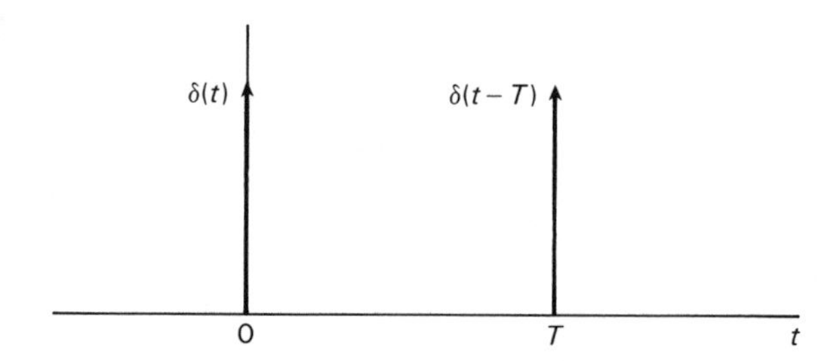

Figure 2.30 Impulse and delayed impulse functions.

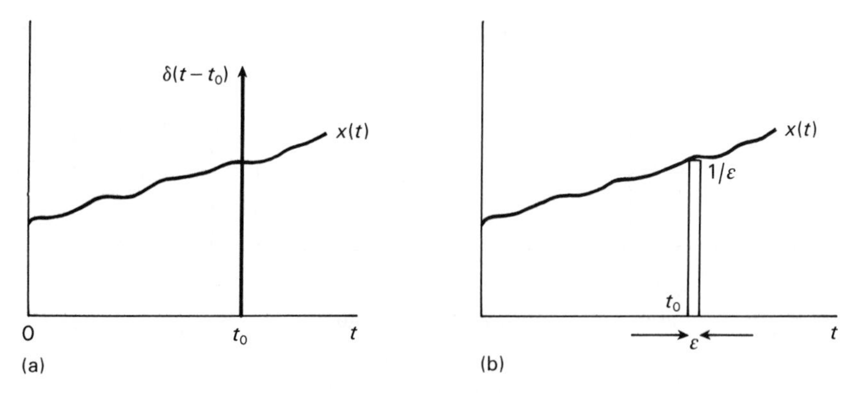

Figure 2.31 Multiplication of signal and impulse function.

the area under it approaches $x(t_0)$. Hence

$$\int_{-\infty}^{+\infty} x(t)\delta(t-t_0)\,\mathrm{d}t = x(t_0)$$

the integral takes the value of the signal $x(t)$ at the time t_0 the impulse occurs. This useful property of the impulse is known as the *sifting property*; it will be used extensively throughout the book.

The continuous time step and impulse signals have their counterparts in the discrete time unit step sequence and the unit sample sequence. There are no analytical difficulties with these signals and they are defined as follows:

Unit step sequence

$$u(n) = 0 \qquad n < 0$$
$$u(n) = 1 \qquad n \geqslant 0$$

Unit sample sequence

$$\delta(n) = 0 \qquad n \neq 0$$
$$\delta(n) = 1 \qquad n = 0$$

These functions are illustrated in Figure 2.32.

These functions have properties that are the counterpart of those of the continuous signals. In particular:

$$\delta(n) = u(n) - u(n-1)$$

$$u(n) = \sum_{k=-\infty}^{n} \delta(k)$$

$$\sum_{n=-\infty}^{+\infty} x(n)\delta(n-n_0) = x(n_0)$$

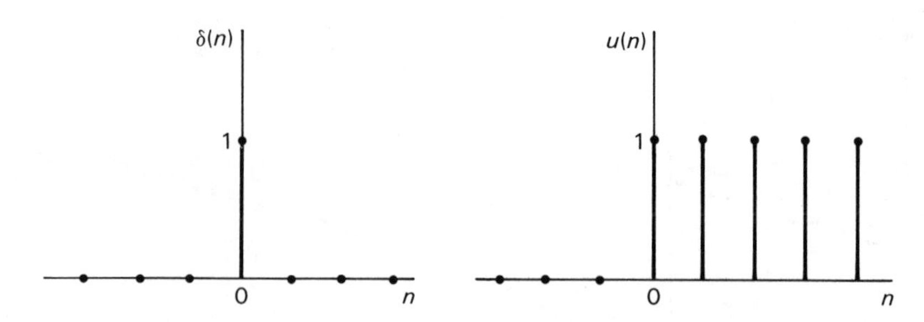

Figure 2.32 The discrete unit sample and unit step.

2.6 Summary

In this chapter many of the properties of signals have been introduced. The distinction between continuous and discrete signals has been emphasised and the effect of transformations on the dependent and independent variables demonstrated. Some properties of signals were introduced that will prove useful in later chapters. A number of basic signals were defined in both continuous and discrete forms and some of the properties of these signals examined.

Problems

2.1 Figure 2.33 shows a continuous time signal $x(t)$. Make labelled sketches of the following time signals:

(a) $2x(t)$
(b) $0.5x(-t)$

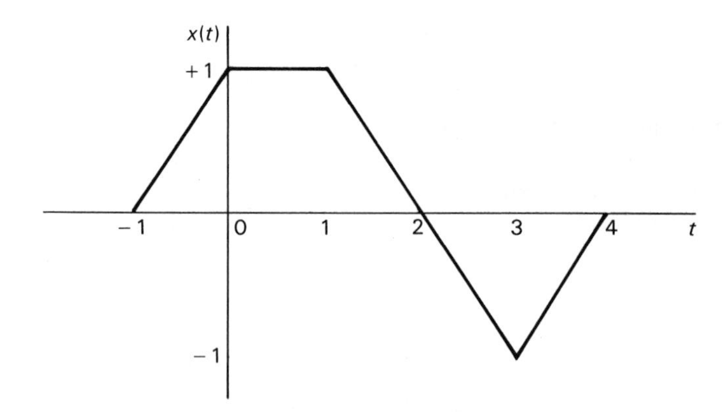

Figure 2.33

(c) $x(t-2)$
(d) $x(2t)$
(e) $x(2t+1)$
(f) $x(1-t)$

2.2 Figure 2.34 shows a discrete time signal $x(n)$. Make labelled sketches of each of the following time signals:

(a) $2x(n)$
(b) $3x(-n)$
(c) $x(n-2)$
(d) $x(2-n)$

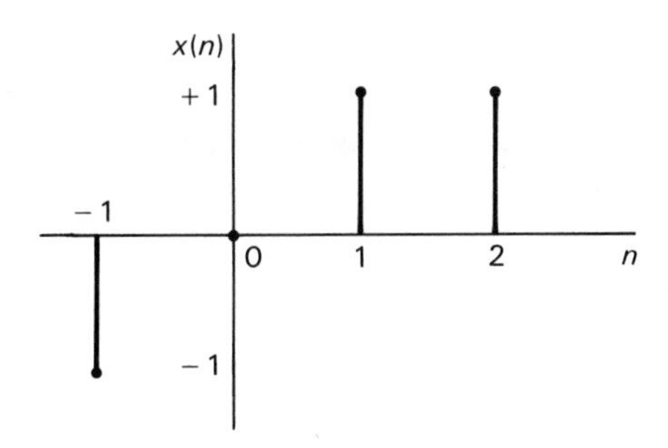

Figure 2.34

2.3 Figure 2.35 shows two continuous signals $x_1(t)$ and $x_2(t)$. Make labelled sketches of each of the following signals:

(a) $x_1(t)+x_2(t)$
(b) $x_1(t)-2x_2(t)$
(c) $0.5x_1(2t)-x_2(t)$
(d) $x_1(t-2)+x_2(4-t)$

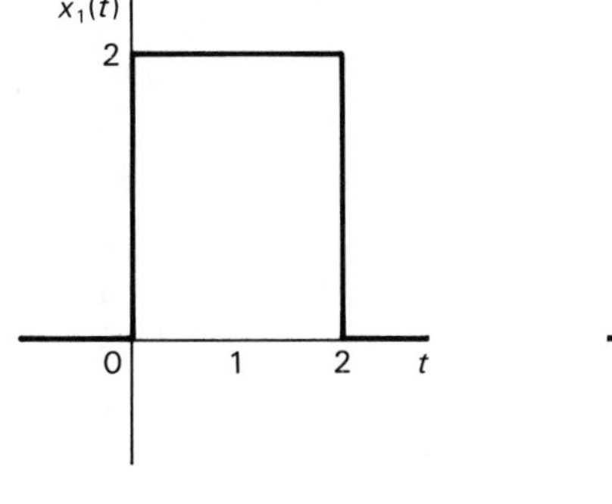

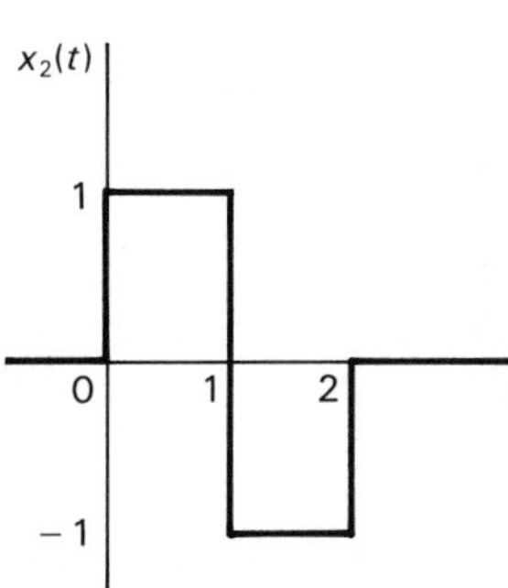

Figure 2.35

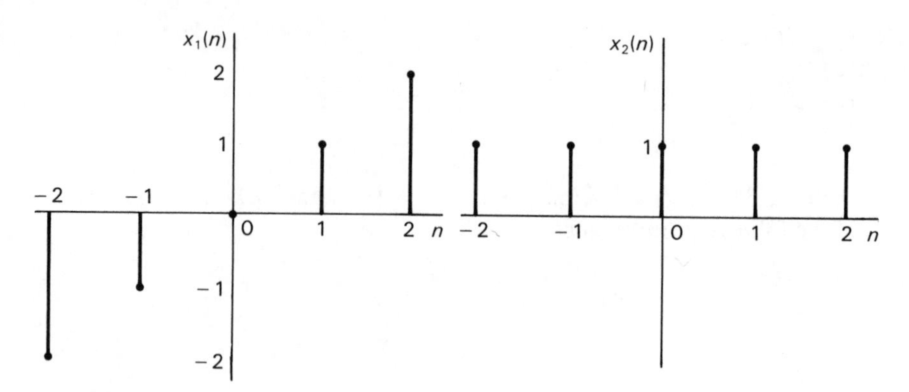

Figure 2.36

2.4 Figure 2.36 shows two discrete signals $x_1(n)$ and $x_2(n)$. Make labelled sketches of each of the following signals:

(a) $x_1(n) + x_2(n)$
(b) $x_1(-n) + x_2(n)$
(c) $x_1(n-1) + x_2(n)$
(d) $x_1(2-n) - x_2(n)$

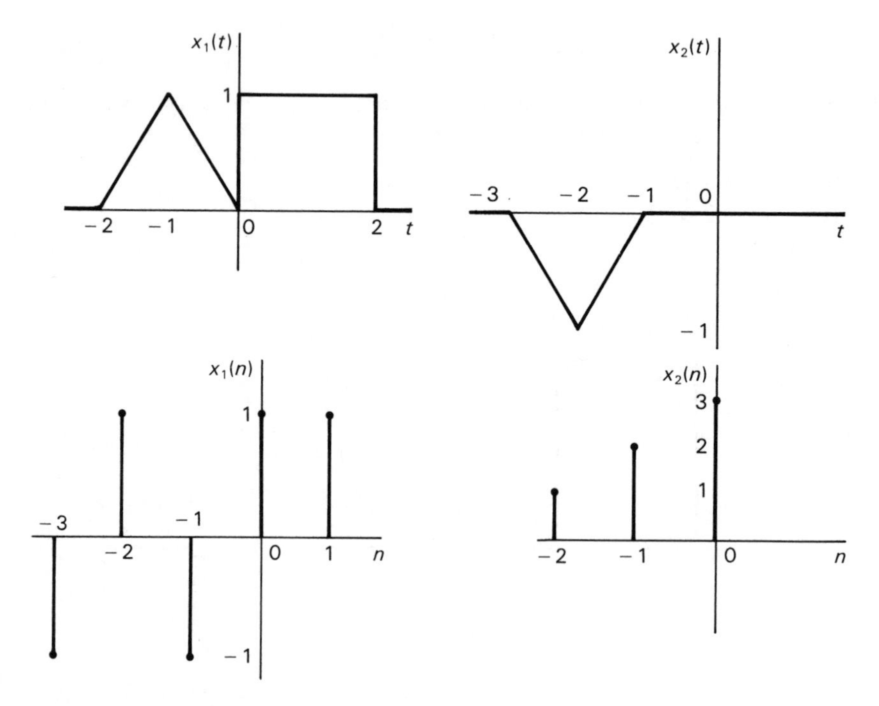

Figure 2.37

2.5 Determine and make a labelled sketch of the odd and even components of the signals shown in Figure 2.37. Verify that addition of the components produces the original signal.

2.6 If $x_1(t)$ and $x_1(n)$ are even signals and $x_2(t)$ and $x_2(n)$ are odd signals verify the following relationships:

$$\int_{-T}^{+T} x_2(t)\,\mathrm{d}t = 0 \int_{-T}^{+T} x_1(t)\,\mathrm{d}t = 2\int_{0}^{+T} x_1(t)\,\mathrm{d}t \qquad \text{arbitrary } T$$

$$\sum_{n=-N}^{+N} x_2(n) = 0 \sum_{n=-N}^{+N} = 2\sum_{n=1}^{+N} x_1(n) + x_1(0) \qquad \text{arbitrary } N$$

If $x_e(t)$ and $x_o(t)$ are the even and odd components of the signal $x(t)$ verify that:

$$\int_{-T}^{+T} x^2(t)\,\mathrm{d}t = \int_{-T}^{+T} x_e^2(t)\,\mathrm{d}t + \int_{-T}^{+T} x_o^2(t)\,\mathrm{d}t \qquad \text{arbitrary } T$$

2.7 Plot the following continuous exponential signals, use the same scales for all the signals and plot for a range of t, 0–5

$$x_1(t) = 2\mathrm{e}^{-2t},\; x_2(t) = 2\mathrm{e}^{-0.2t},\; x_3(t) = 2\mathrm{e}^{0.2t}$$

Also plot the signal $x_4(t) = 2\mathrm{e}^{2t}$ for a range of t, 0–1.

Make labelled sketches of the following discrete exponential signals:

$$x_1(n) = 2\mathrm{e}^{-2nT},\; x_2(n) = 2\mathrm{e}^{-2nT},\; x_3(n) = 2\mathrm{e}^{0.2nT}$$

Use a value of $T = 0.5$ and values of n, 0–10.

2.8 Figure 2.38 shows part of a control system where the error signal $e(t)$ is formed by the difference between the input signal $x(t)$ and the feedback signal $y(t)$.

$$e(t) = x(t) - y(t)$$

If

$$x(t) = 3\sin(\omega t + 20°)$$
$$y(t) = 5\cos(\omega t - 20°)$$

determine $e(t)$ and express it in the form $A\sin(\omega t + \varphi)$.

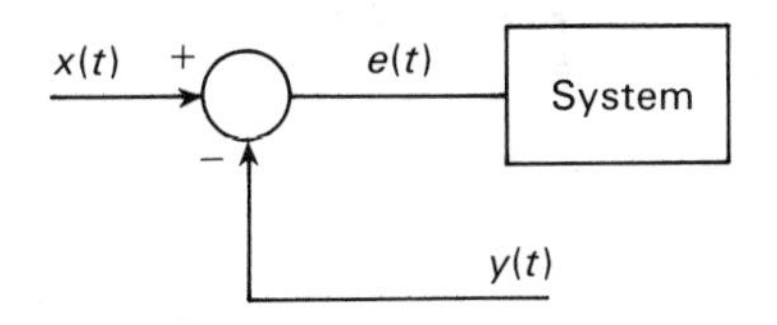

Figure 2.38

2.9 Determine the amplitude, frequency (Hz), and phase with respect to sin ωt of the following signals:

(a) $20 \sin(30t + \pi/4)$
(b) $50 \cos(100t - \pi/4)$
(c) $100 \cos 20t + 20 \sin 20t$
(d) $\mathrm{Re}\{e^{j(3t+\pi/4)}\}$
(e) $\mathrm{Im}\{e^{j(3t+\pi/4}\}$
(f) $Ae^{j2t} + A^*e^{-j2t}$, where $A = 1 + 2j$

2.10 Sketch, approximately to scale, the following two signals:

$$x_1(t) = 0.5e^{(-1+j2\pi)t} + 0.5e^{(-1-j2\pi)t}$$

$$x_2(t) = -0.5je^{(1+j2\pi)t} + 0.5je^{(1-j2\pi)t}$$

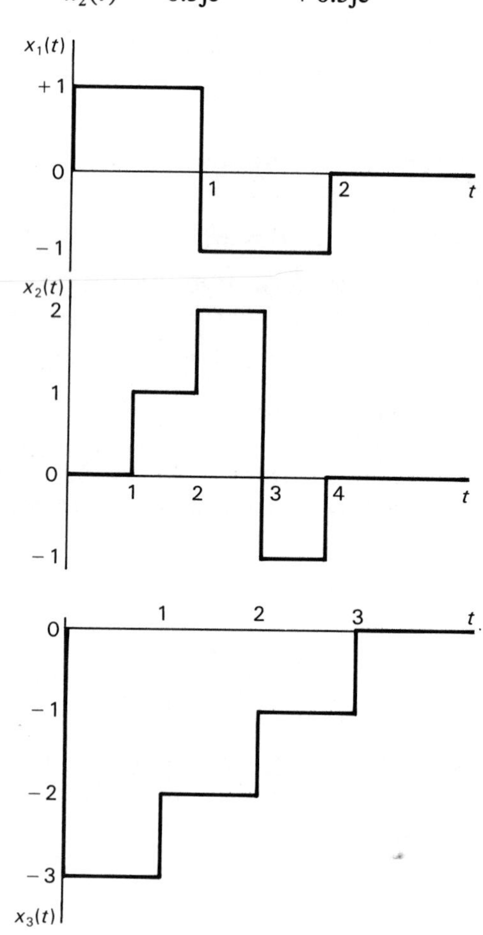

Figure 2.39

2.11 (a) Sketch one cycle of the signal $x(t) = \sin \omega_0 t$, where $\omega_0 = 2\pi/16$.
(b) On the same scale as (a) sketch the 17th harmonic of $x(t)$, $x_h(t) = \sin 17\omega_0 t$.
(c) Draw to scale the one cycle of the signal $x(n) = \sin \omega_0 nT$, where $T = 1$.
(d) On the same scale as (b) draw the 17th harmonic of $x(n)$, $x_h(n) = \sin 17\omega_0 nT$.

Compare the harmonic of the continuous signal with the corresponding harmonic of the discrete signal.

2.12 Evaluate the following integrals involving the impulse function:

(a) $\int_{-\infty}^{+\infty} 3t^2\delta(t-2)\,\mathrm{d}t$

(b) $\int_{-\infty}^{+\infty} (\delta(t)\cos t + \delta(t-1)\sin t)\,\mathrm{d}t$

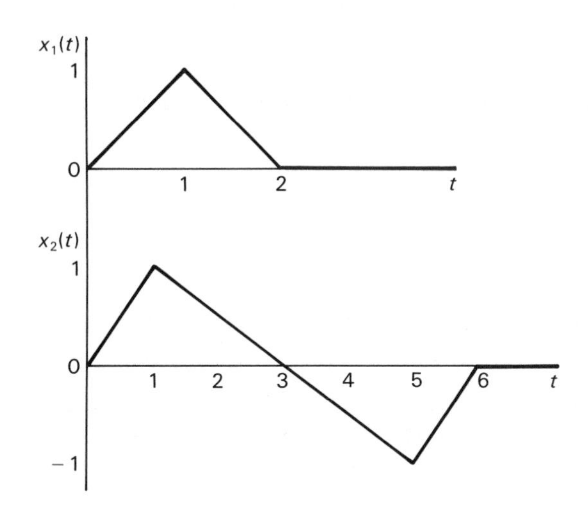

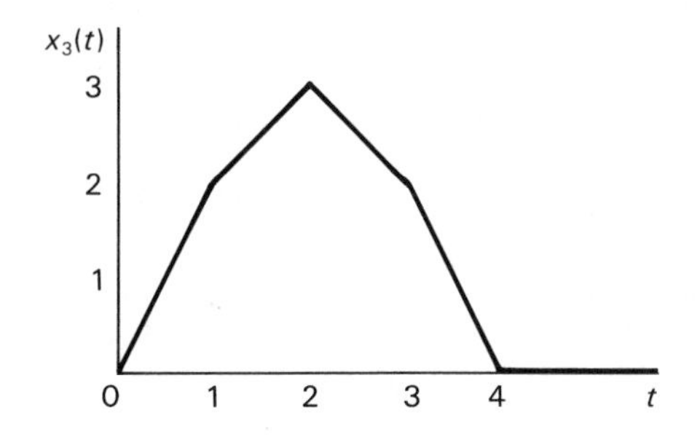

Figure 2.40

(c) $\int_{-\infty}^{+\infty} u(t)e^{-t}(\delta(t+1)+\delta(t-1))\,dt$

(d) $\int_{-\infty}^{+\infty} e^{j\omega t}\delta(t-1)\,dt$

2.13 Illustrate with labelled sketches the following signals:

$$x_1(t) = u(t) + u(t-1) - 2u(t-2)$$
$$x_2(t) = u(t) - 3u(t-2) + 2u(t-3)$$
$$x_3(t) = u(t) - 2u(t-1) + 2u(t-2) - 2u(t-3) + u(t-4)$$

Express the signals shown in Figure 2.39 in terms of scaled delayed versions of the unit step signal.

2.14 The unit ramp signal is defined as follows:

$$x(t) = t \qquad t \geqslant 0$$
$$= 0 \qquad \text{otherwise}$$

Obtain expressions, in terms of scaled unit ramps for the signals shown in Figure 2.40.

MATLAB exercises

2.15 This exercise illustrates the effect of some basic transformations on both continuous and discrete signals.

(a) Figure 2.41 (a) shows a continuous signal $x(t)$ where the time scale is -5 to $+10$ seconds. Construct a 1500 point vector to describe the amplitude of

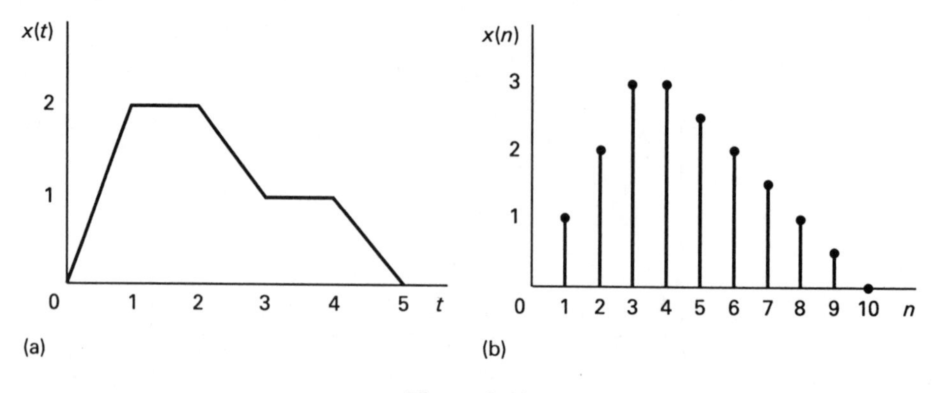

Figure 2.41

this signal over this time interval. Plot this signal showing the correct time scaling.

Obtain 1500 point vectors to represent the amplitudes of the following signals derived by transformations of $x(t)$. In each case plot the signal on the same display as $x(t)$.

(i) $y(t) = 2x(t)$ (ii) $y(t) = x(2t)$
(iii) $y(t) = x(t/2)$ (iv) $y(t) = x(t+1)$
(v) $y(t) = x(t-2.5)$ (vi) $y(t) = x(2t-3)$
(vii) $y(t) = x(-t)$ (viii) $y(t) = x(3-2t)$

(b) Figure 2.41(b) shows a discrete signal $x(n)$ where n has the range -5 to $+20$. Construct a vector to represent the amplitude of the signal over this range. Use `subplot` to display the signal as the first of two subplots and use `stem` to show it as a discrete signal.

Obtain 25 point vectors to represent the amplitudes of the following signals derived by transformations of $x(n)$. In each case, display the signal as the second subplot using n in the range -5 to $+20$.

(i) $y(n) = 2x(n)$ (ii) $y(n) = x(n+2)$
(iii) $y(n) = x(n-4)$ (iv) $y(n) = (6-n)$

Note For the continuous case, the vector representing the signal $x(t)$ is constructed most easily from a number of smaller length vectors, each representing a linear segment of the signal.

Some care is required when evaluating the transformed vector. For the continuous signal, a time scale from -5 to $+10$ seconds must be related to an index ranging from 1 to 1500. Remember that this index must be integer and positive and checks must be made on the required range. Some of the transformations produce non-integer indices and the command `round` can be used here (the error introduced is negligible for 1500 points).

The discrete transformation will always produce an integer index, but again the range must be transformed so that a positive index is always produced.

2.16 This exercise examines the properties of the exponential and sinusoidal signals in both continuous and discrete forms.

(a) The continuous signal

$$x(t) = 2e^{at}$$

(i) For values of a equal to 0.5, 1.0 and 1.5, plot the corresponding signals all on the same graph using a time scale -3 to $+3$.

(ii) Repeat (i) for values of a equal to -0.5, -1.0 and -1.5, again using the time scale -3 to $+3$.

(b) The discrete signal

$$x(n) = 2e^{anT}$$

with $T = 0.25$.

(i) For values of a equal to 0.5, 1.0 and 1.5, plot the corresponding signals using the command `stem` over a range of n from $n = -12$ to $n = +12$. Use `subplot` to show the three plots on one screen and for comparison purposes use the same vertical scales by use of `axis`.

(c) The continuous signal

$$x(t) = X \sin(\omega t + \phi)$$

(i) For $f = 50$ Hz and $\phi = 0$ plot $x(t)$ over a time scale of 0 to 100 ms for values of X equal to 1, 2 and 5 units. Show all these results on the same plot.
(ii) For $X = 1$ and $\phi = 0$ plot $x(t)$ over a time scale of 0 to 20 ms for values of f equal to 50, 100 and 200 Hz. Show all results on the same plot.
(iii) For $X = 1$ and $f = 50$ Hz plot $x(t)$ over a time scale of 0 to 40 ms for values of ϕ equal to 0, $\pi/2$ and π radians. Show all results on the same plot.

(d) The discrete signal

$$x(n) = \sin(2\pi fnT)$$

with $T = 1.25$ ms

(i) Plot two cycles of this signal for values of f equal to 50, 100 and 200 Hz. Use `stem` to show the discrete nature of the signal and display as separate subplots.
(ii) Repeat (i) for values of f equal to 200, 600 and 1000 Hz.

2.17 This exercise illustrates the effect of aliasing when a single sine wave is sampled. As signals can only be represented in MATLAB in discrete form, a large number of points (compared to the sample points) are used to approximate the continuous sine wave. The frequency of the sine wave is kept constant and the effect, on the sampled signal, of altering the sampling frequency is investigated.

Produce a signal approximating the continuous waveform given by

$$x(n) = \sin(2\pi fn/f_0)$$

where $f_0 = 5000$ Hz and $f = 5$ Hz. Choose the range of n such that five complete cycles of the waveform are generated.

Produce the sampled signal

$$x_s(k) = \sin(2\pi fk/f_s)$$

where f_s is the sampling frequency which can be set within the program and k takes the range 0 to $5f_s/f$.

Plot $x_s(k)$ and investigate the effect of varying the sampling frequency on the fundamental frequency present in the sampled signal.

Note Use `subplot` to display both $x(n)$ and $x_s(t)$ on the same screen. $x_s(k)$ should be displayed as both a discrete signal (using `stem`) and the continuous signal formed by joining the sample points. The latter display gives a good indication of the fundamental frequency present in the sampled waveform. Take particular note of the apparent frequency when f_s is such that f is below and above half the sampling frequency, $f_s/2$.

3

System Description

3.1 Introduction

In Chapter 1 the concept of a system has been introduced in a qualitative manner. In this chapter the concept is developed to give a mathematical description of systems. From this description systems are classified according to certain properties they possess.

Examples of continuous systems are given, these include electrical, mechanical, electromechanical and electronic systems. It is shown that the most general description of these systems is a differential equation relating the input and output signals. By correct choice of component values very different physical systems can be described by the same differential equation, such systems are termed analogues. An ordered method of constructing analogues based on the operational amplifier is investigated.

Discrete systems are introduced and it is shown that the most general description of such systems is the difference equation, such equations are easily realised by means of a digital computer. By approximating continuous signals by a series of sample values, continuous systems can also be approximated by a discrete realisation and this forms the basis of digital simulation languages.

The system description used for most of the chapter is based on the equations relating the input and the output quantities. An alternative description is introduced based on internal or state variables.

3.2 The system model

In Chapter 1 an example has been given of how a complex system can be represented by an interconnection of simpler subsystems. The subsystem is described by the manner in which its output signal is related to its input signal. A very simple system could relate the output $y(t)$ to the input $x(t)$ via the relationship

$$y(t) = 2x(t)$$

This would imply that the system amplifies the input and gives an output that is twice the input at all instants of time. This would be true *regardless of the shape of* the input signal. As will be seen from the work in later chapters a practical system would not follow this description when the input changes rapidly. The equation describing the system is an idealisation, it is a *mathematical model* which only approximates the true process. A more complex model could be used to describe the system which might be true for a greater range of input signals. However this may mean that the mathematical manipulation involving the system model will be so intractable that it is of little use. This problem will be re-examined in Section 3.4 and for the moment we will accept the mathematical description of the system as a *model* with the same limitations as possessed by all models. At this stage we will not inquire as to how the model is obtained, this also will be investigated in Section 3.4. This type of approach assumes the real system is hidden in a 'black box' and all that is available is a mathematical model relating output and input signals. As will be seen in later chapters, more than one description of a signal is possible and this leads to alternative system descriptions.

The mathematical model given at the start of this section describes a very simple system. A more complex example, this time of a discrete system is that of the proposed digital filter described in Example 1.3.

$$y(n) = \frac{x(n) + x(n-1) + x(n-2) + x(n-3)}{4}$$

This states that the output of the system at any discrete instant is the average of the input at that instant and the three previous instants.

The continuous counterpart of such a system might have a description

$$y(t) = \frac{1}{T}\int_{t-T}^{t} x(t)\,\mathrm{d}t \tag{3.2.1}$$

This description appears more frightening than its discrete counterpart. However it can be interpreted physically by reference to Figure 3.1. The output at time t is formed by taking the shaded area and dividing by T, it is the average value of $x(t)$ over preceding time T.

Equation (3.2.1) can appear confusing because the integral is with respect to t, yet the upper limit of the integral is also t. Often it is useful to use what is termed a 'dummy variable'. The integral concerned is a definite integral; consider the following definite integrals

$$\int_0^2 x\,\mathrm{d}x; \qquad \int_0^2 t\,\mathrm{d}t; \qquad \int_0^2 z\,\mathrm{d}t$$

It is apparent that all these integrals have the value 2 and regardless of the symbol used for the variable, provided the integral is performed with respect to that variable the result will be 2. Hence eqn (3.2.1) can be written.

$$y(t) = \frac{1}{T}\int_{t-T}^{t} x(\tau)\,\mathrm{d}\tau$$

where τ is termed a dummy variable. The use of dummy variables will prove very useful in later chapters.

In the examples used so far to illustrate this section all the mathematical models have been capable of a direct physical interpretation. Unfortunately as the systems become more complex this ceases to be the case. As will be shown by later examples the most general relationship between input and output signals for a continuous system is a differential equation. With such a relationship especially for high order systems it is difficult to see intuitively how the form of the input signal relates to the form of the output signal and in later chapters alternative descriptions are sought that will throw insight into the action of the system.

In the systems investigated so far in this chapter it has been assumed that the system input uniquely determines the system output via the mathematical description of the system. There are two reasons why this may not be the case.

Most practical systems have components that can store energy, e.g., charged capacitors in electrical systems, stretched springs in mechanical systems. The system output is not only determined by the input but also by this stored energy, even if the input were zero the stored energy would still produce an output

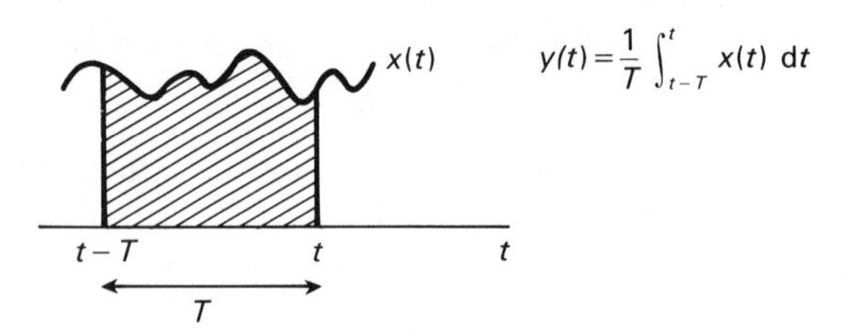

Figure 3.1 Illustration of an input/output relationship for a system.

response. The effect of initial energy storage will have to be taken into account in the system analysis.

In the examples used it has been assumed that the system only has a single input signal. Systems that have more than one input (and output) are quite common but will not be considered in this book. However in all practical systems there is always noise which consists of unpredictable or random signals that are present in the systems. These can often be modelled as additional inputs, however the signal at that input can only be described in statistical terms. Such a description will be investigated in Chapter 9. However noise has been mentioned at this point because although in many systems its effects are negligible, in others it has very important implications for the design of the system (see Example 1.1).

3.3 Classification of systems

Before considering the way in which the system model is obtained it is convenient to accept the model and use it in order to classify systems. The reason for classifying systems is the same reason for undertaking other forms of classification, e.g. plants and animals or books in a library. If one can derive properties that apply generally to a particular area of the classification then once it is established that a system belongs in this area then these properties can be used without further proof. The areas into which systems are classified are not unique (other methods of classification are possible) nor are they mutually exclusive (a system may belong to more than one area). Also the areas given are not exhaustive, but they are the most useful on which to base the material in the remainder of the book.

3.3.1 Continuous/discrete systems

The concept of continuous and discrete signals has been introduced in the previous chapter. This concept can easily be extended to cover systems. Systems where the input and output signals are continuous are continuous systems, where these signals are discrete they are discrete systems. Systems can occur where the input signal is of one form and the output signal is of the other. Such systems are hybrid systems although it is often possible to regard them as a combination of subsystems that are either continuous or discrete.

3.3.2 Linear/non-linear systems

The concept of a linear system can be illustrated by an example from everyday life. One is listening, via an audio system to some instrumental music from a tape. One instrument plays the tune, this instrument stops and another instrument plays the same tune, then both instruments play the tune together. When the instruments play together one would expect the sound produced to be a combination of the sounds of the individual instruments and if the system is linear this will be the case. However we are all familiar with the situation when the volume is turned up so loud that it almost produces distortion when one instrument plays, when the second instrument joins in distortion now occurs and the sound is not the combination of the individual instruments. Hence the basis of a linear system is that if inputs are superimposed then the responses to these individual inputs are also superimposed. To state this property more formally,

if an arbitrary input $x_1(t)$ produces output $y_1(t)$
and an arbitrary input $x_2(t)$ produces output $y_2(t)$

then if the system is linear

input $(x_1(t) + x_2(t))$ will produce output $(y_1(t) + y_2(\mathrm{t}))$

This property is one of *superposition* or additivity.

If the two input signals are identical, i.e., $x_1(t) = x_2(\mathrm{t}) = x(t)$ and each of these signals produces an output $y(t)$, then it follows that an input $2x(t)$ will produce an output $2y(t)$. It follows more generally that an input $ax(t)$ will produce an output $ay(t)$ where a is known as homogeneity.

As the property of homogeneity follows from that of superposition the student may feel that it is unnecessary to introduce this further property. However there are problems with the proof if the constant a is not rational, this condition is not likely to be met with in real engineering systems and it can be taken that either of the properties of superposition or homogeneity can be taken to define linearity. These two properties can be combined and using the previous notation:

For a linear system an input $ax_1(t) + bx_2(t)$ produces an output
$ay_1(t) + by_2(t)$ where a and b are constants.

The property of homogeneity gives an intuitive feel for the idea of linearity. Consider the amplifier introduced earlier, if the output voltage at any instant is plotted against the input voltage at that instant the result will be a straight line with slope equal to 2. Suppose this amplifier is constructed from an integrated circuit. An input of 50 mV will produce an output of 100 mV, an input of 1 V will produce an output of 2 V, an input of 6 V will probably produce an output of 12 V, however it is extremely unlikely that an input of 1000 V will produce the corresponding output of 2000 V. If the output is plotted against the input the result will be linear only for a restricted range of input as shown in Figure 3.2.

In practice in all systems the property of linearity applies only over a limited range of input and all systems cease to be linear if the input becomes large enough. Often systems are such that they can only be called linear for a small change in input signal. If real systems only approximate to being called linear the student may wonder about the usefulness of such a classification, and ask if it would not be better to try and classify according to properties of non-linear systems. However

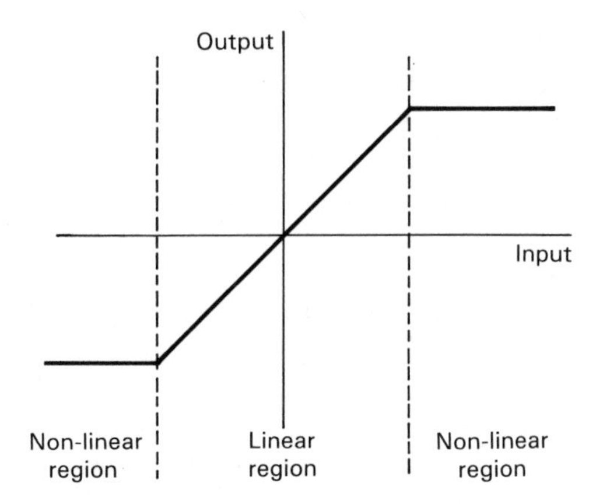

Figure 3.2 Input-output relationship for a system having a restricted linear range.

classifying such systems is difficult and the definition of a non-linear system is 'a system that is not linear'. Such a method of classification is like a biologist who divides animals into cats/non-cats. This is of great help in studying the cats but does not help very much with the study of non-cats. As will be seen in later chapters the concept of linearity is of great importance in developing a general systems theory. No such general theory is possible for non-linear systems and these systems are usually investigated by approximate methods. However a good understanding of linear system theory is required before non-linear problems can be considered.

3.3.3 Time invariant/time varying systems

Again the concept of time invariance can be illustrated with the example of the tape and audio system from Section 3.3.2. Suppose this tape is played at ten o'clock in the morning. If it were then played again at two o'clock in the afternoon one would not expect it to produce a different output from that in the morning. The property that a shift in time of an input signal causes only a corresponding time shift in the output signal is the property of time invariance. This can be expressed mathematically as follows:

If an input signal $x(t)$ causes a system output $y(t)$ then an input signal $x(t-T)$ causes a system output $y(t-T)$ for all t and arbitrary T.

Again, taking the example of the tape it may be that it does not sound the same morning and afternoon. It may be that the volume or tone controls of the audio system have been altered in between. The system is not time invariant, some of the constants (parameters) of the system have changed. Hence an alternative name for a time invariant system is a constant parameter system.

If a system is time invariant and linear it is known as a linear time invariant or LTI system. These are very important. A complex signal can often be represented as a linear combination of time shifted simpler signals. If the response of the system to the simpler signals is known the response to the more complex signal can be obtained. This will be illustrated in Example 3.3.2.

3.3.4 Instantaneous/non-instantaneous systems; causal/non-causal systems

These systems classifications are considered together in this section as they are closely related. In Section 3.3.2 an example was given of an amplifier whose output and input are related via the equation

$$y(t) = 2x(t)$$

It was pointed out in that section that this relationship implies that the output is twice the input at every time instant regardless of the form of the input signal. In this system the output at any instant depends upon the input *at that instant only*, such a system is defined as an instantaneous system.

As previously described in Section 3.2, practical systems are composed of elements that can store energy and this stored energy can give a system output even if the input is zero. Consider a LTI system that gives a response to single pulse input as shown in Figure 3.3(a).

It can be seen that the response continues after the input is zero (due to stored energy). Consider now a second pulse applied as shown in Figure 3.3(b), due to the

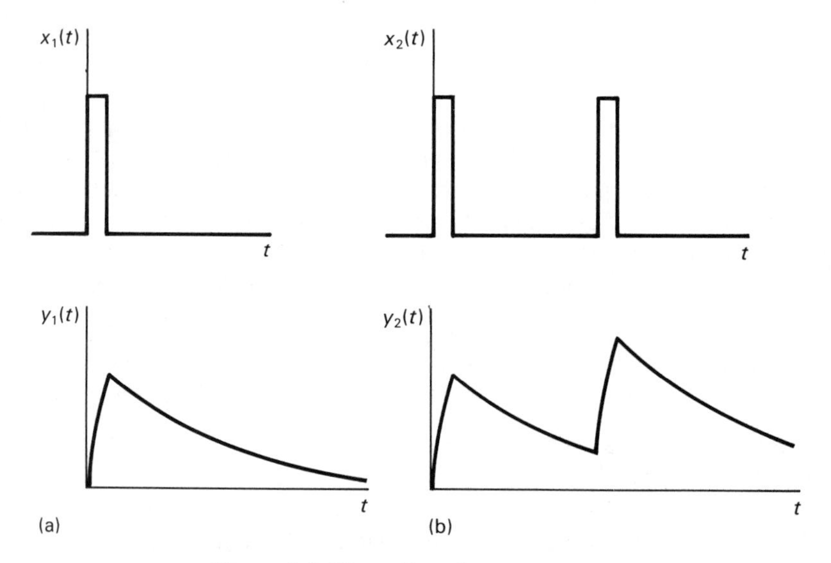

Figure 3.3 Illustration of system memory.

time invariant property of the system this second pulse gives the same response as the first but is time delayed. However the response due to the first pulse is still decaying and because of the linear property these responses add to give a combined response as shown. This system is obviously not instantaneous (otherwise the response to the second pulse would be identical to the first), it is a non-instantaneous system. Another way of viewing the non-instantaneous system is that it 'remembers' the effect of previous inputs hence non-instantaneous systems are said to have a *memory*. Instantaneous systems, from this viewpoint are termed *memoryless* systems.

Mathematically the property of memory can be expressed by saying the output $y(t)$ depends upon an input $x(t-T)$ for at least one value of T. However the definition is widened to also include the input $x(t+T)$ and at first sight this appears not to make much sense physically. How can the output of a system depend upon what is to happen in the future? Leaving aside this point for the moment, systems whose outputs do not depend upon future values of input are known as *causal* systems.

Returning now to the question of why it is necessary to consider other than causal systems, there are several reasons why non-causal systems are important.

In all the systems considered throughout this book the independent variable is time. However, as pointed out in the introductory chapter, other independent variables are possible. In image processing it may be required to process the image at a point on the image. Values of the dependent variable can be used that are from both the left and right of the point under consideration. Taking points to the right would correspond to a non-causal system.

In some applications where the independent variable is time, data are stored and then processed 'off-line' (e.g., economic analysis, analysis of seismic data). All the data record is now available and 'future' inputs can be used as the calculation of the output.

There is another reason why the concept of causality is important in system design. The description of signals and systems considered so far has been based in time. As will be seen in later chapters alternative methods of description are possible, e.g., frequency domain descriptions and often system design is carried out using this alternative description. However care must be taken to ensure that the design obtained, when translated back into the time domain, does not produce a system that is non-causal and is not physically realisable. Hence the constraint of causality in time imposes constraints in frequency.

3.3.5 Stable/unstable systems

This concept is illustrated using examples from discrete systems. Discrete systems have the advantage that it is possible to obtain numerically the output signal for a given input signal (although the process may be somewhat tedious and is best left to a computer). Consider the two discrete systems described by the following equations:

System 1 $\quad y_1(n) = x(n) + 0.5y_1(n-1)$

System 2 $\quad y_2(n) = x(n) + 2.0y_2(n-1)$

Table 3.1 Unit sample response of stable and unstable systems

n	$x(n)$	$y_1(n-1)$	$y_1(n)$	$y_2(n-1)$	$y_2(n)$
0	1	0.0000	1.0000	0	1
1	0	1.0000	0.5000	1	2
2	0	0.5000	0.2500	2	4
3	0	0.2500	0.1250	4	8
4	0	0.1250	0.0625	8	16

Consider the responses of these two systems to the same input signal consisting of a single unit sample applied at $n=0$ (the unit sample $\delta(n)$ considered in the previous chapter). Assuming the systems have no outputs initially, $y(n)=0$ for $n<0$, then the outputs for $n>0$ can be obtained by calculation from the input and previous outputs, as shown in Table 3.1.

From the results the general behaviour of the systems is apparent. In system 1 the output persists after the input is removed (the system has memory) but each output is one half of the previous output and as $n \rightarrow \infty$, $y_1(n) \rightarrow 0$. However in system 2 each output is twice the previous output and the output grows *without limit.* The first system is stable the second is unstable. The idea that an input that is bounded can lead to an output that is unbounded leads to a general definition of stability.

A system is stable if a bounded input produces a bounded output. This definition is illustrated in Figure 3.4.

It is very important to be able to predict whether a system will be unstable. Although in a practical system no signal can grow without limit (non-linearities provide limits), variables can reach magnitudes that can overload the system and may cause physical damage. The systems that are most likely to suffer instability are feedback systems because under certain conditions the feedback can change sign and reinforce the input.

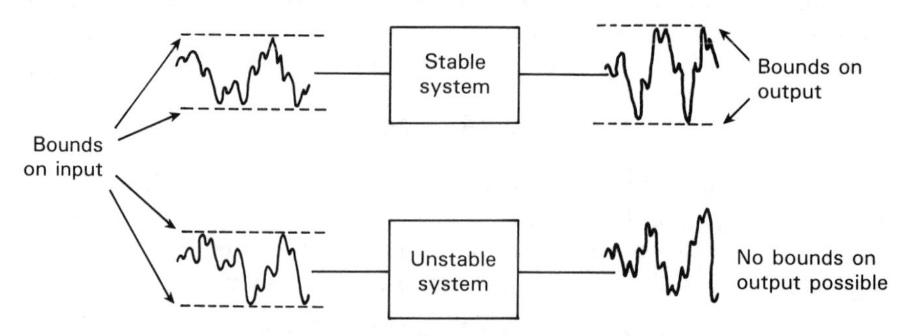

Figure 3.4 Bounded input, bounded output stability.

3.3.6 Other classifications

Two other types of system have been grouped together under this heading, these are multiple input, multiple output systems and distributed systems. Although these systems are important and occur frequently in practice their general analysis is beyond the scope of this book. In this section only a very brief outline of their properties is given.

Consider an aircraft control system where it is required to control the yaw, pitch and roll of the aircraft. The system could be represented as in Figure 3.5. It would be convenient if a given demand signal only affected the corresponding controlled signal, e.g., the demand for yaw did not affect the pitch and roll of the aircraft. If this were the case the system could be represented as three quite separate systems each with a single input and output signal. In practice however interaction occurs between the signals and this is not possible. Such a system is known as a *multiple input multiple output system.* It is obvious that the analysis of such systems is going to be far more complex that that of single input single output systems. In order to obtain a concise representation the equations describing such a system are usually presented in matrix form.

Although the analysis of simple mechanical systems is not covered until the next section most students will be familiar with a system consisting of a mass suspended from a spring. When displaced from its equilibrium position this system will describe simple harmonic motion. Consider now a vibrating string, say the string of a guitar. Each point on the string will also perform simple harmonic motion. However the system formed by the string is far more difficult to analyse because the mass and stiffness concerned are not 'lumped' into elements as they are in the case of the mass on the spring. The mass on the spring is an example of a *lumped parameter* system but the guitar string is an example of a *distributed parameter* system. In practice all systems are distributed parameter, in an electrical circuit resistance does not cease at the resistor terminals, it continues into the wiring as does the inductance and capacitance. A system can be considered lumped if its dimensions are not small compared with the shortest wavelength of interest. It will be shown in Section 3.4 that lumped parameter systems give rise to differential equations. However distributed parameter systems (not considered) can be shown to give rise to partial differential equations.

Before dealing with the manner in which the input/output relationship is derived for a given system, two examples are given which illustrate the material of this section. The first example deals with classification of systems, the second deals with properties of a linear time invariant system.

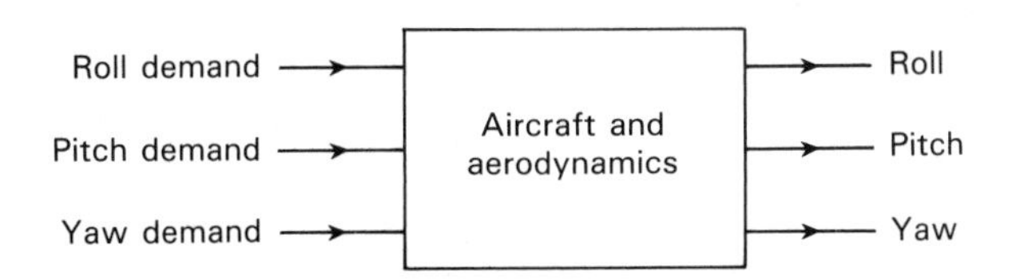

Figure 3.5 Multi-input, multi-output aircraft control system.

EXAMPLE 3.3.1

Consider the following three systems:

(a) A full wave rectifier whose output is the modulus of the input

$$y(t) = |x(t)|$$

(b) A modulator whose carrier frequency is ω_c, giving an output

$$y(t) = x(t) \cos \omega_c t$$

(c) A system whose output is the average of its input over preceding time T_1.

$$y(t) = \frac{1}{T_1} \int_{t-T_1}^{t} x(t)\, dt$$

Determine whether each of these three systems is:

(i) linear; (ii) time invariant; (iii) instantaneous.

SOLUTION

Linearity is considered by obtaining the outputs for two inputs $x_1(t)$, $x_2(t)$ applied independently. The input $ax_1(t) + bx_2(t)$ is then applied and the output should equal a linear combination of the previous outputs.

For the rectifier the outputs are $|x_1(t)|$ and $|x_2(t)|$. When the input $ax_1(t) + bx_2(t)$ is applied the output is $|ax_1(t) + bx_2(t)|$. However

$$|ax_1(t) + bx_2(t)| \neq a|x_1(t)| + b|x_2(t)|$$

and the system is not linear.

For the modulator the outputs are $y_1 = x_1(t) \cos \omega_c t$ and $y_2(t) = x_2(t) \cos x_c t$. When the input $ax_1(t) + bx_2(t)$ is applied the output is

$$\begin{aligned}(ax_1(t) + bx_2(t)) \cos \omega_c t &= ax_1(t) \cos \omega_c t + bx_2(t) \cos x_c t \\ &= ay_1(t) + by_2(t)\end{aligned}$$

and the system is linear.

For the averaging system

$$y(t) = \frac{1}{T_1} \int_{t-T_1}^{t} x_1(t)\, dt; \qquad y_2(t) = \frac{1}{T_1} \int_{t-T_1}^{t} x_2(t)\, dt$$

Input $ax_1(t) + bx_2(t)$ produces output

$$\begin{aligned}\frac{1}{T_1} \int_{t-T_1}^{t} (ax_1(t) + bx_2(t))\, dt &= \frac{a}{T_1} \int_{t-T_1}^{t} x_1(t)\, dt + \frac{b}{T_1} \int_{t-T_1}^{t} x_2(t)\, dt \\ &= ay_1(t) + by_2(t)\end{aligned}$$

The system is linear.

To consider time invariance consider an input $x(t)$ producing an output $y(t)$. To be time invariant an input $x(t-T)$ will produce an output $y(t-T)$.
For the rectifier

$$y(t) = |x(t)|$$

An input $x(t-T)$ gives output $|x(t-T)| = y(t-T)$ hence the system is time invariant. For the modulator

$$y(t) = x(t) \cos \omega_c t$$

An input $x(t-T)$ gives output $x(t-T) \cos \omega_c t$.

However

$$y(t-T) = x(t-T) \cos \omega_c (t-T)$$

and in general

$$x(t-T) \cos \omega_c t \neq x(t-T) \cos \omega_c (t-T)$$

the system is not time invariant.
For the averaging system

$$y(t) = \frac{1}{T_1} \int_{t-T_1}^{t} x(t) \, dt$$

It is easier to analyse this system if the concept of a dummy variable is used, then

$$y(t) = \frac{1}{T_1} \int_{t-T_1}^{t} x(\tau) \, d\tau$$

An input $x(t-T)$ produces output

$$y(t) = \frac{1}{T_1} \int_{t-T_1}^{t} x(\tau - T) \, d\tau$$

By introducing a new dummy variable $\lambda = \tau - T$ and making the appropriate changes to the limits of the integration then this expression can be written:

$$\frac{1}{T_1} \int_{t-T-T_1}^{t-T} x(\lambda) \, d\lambda$$

This is the expression that is obtained by delaying the output signal $y(t)$ by an amount T

$$y(t-T) = \frac{1}{T_1} \int_{t-T_1-T}^{t-T} x(\tau) \, d\tau$$

(Remember that definite integrals with respect to the different dummy variables give identical results.)

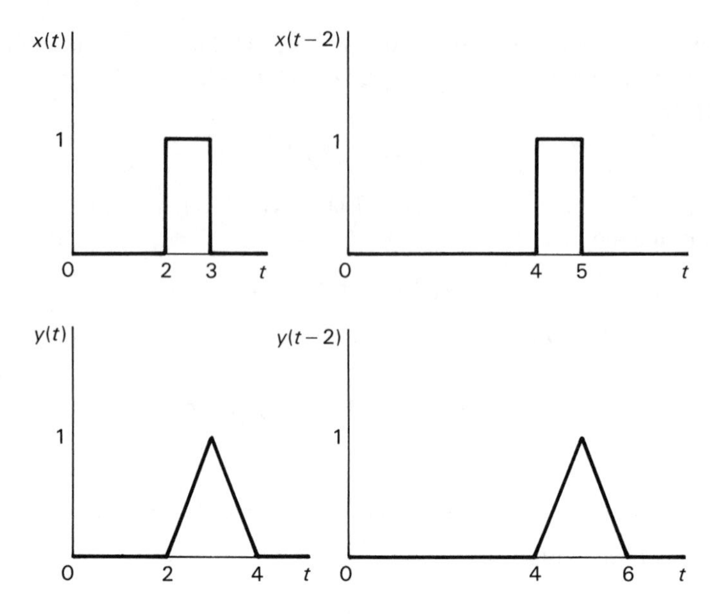

Figure 3.6 Signals illustrating a time invariant system.

If the student finds the foregoing manipulation of the integral and its limits difficult to follow it is advisable to follow the steps using a particular waveform and using specific values for T_1 and T. An example is shown in Figure 3.6 where $T_1 = 1$ and $T = 2$.

It is relatively easy to check whether a system is instantaneous. Inspection of the system equation indicates whether the output depends upon the input at any time instant other than the present.

For the rectifier only the input signal at the present instant $x(t)$ is required to form the output. The same is true of the modulator, multiplication by cos $\omega_c t$ does not alter the situation. The finite time integrator is not instantaneous, the output at any instant depends upon the input over the previous time interval T_1.

EXAMPLE 3.3.2

When the signal, $x_1(t)$, shown in Figure 3.7(a) is applied to an LTI system the response $y_1(t)$ is as shown. Determine the output if each of the signals $x_2(t)$ and $x_3(t)$ as shown in Figure 3.7(b) are applied.

SOLUTION

The response of the system is known to the input $x_1(t)$. However it is an LTI system so its response to linear combinations of time shifted versions of $x_1(t)$ can be calculated. Hence the signals $x_2(t)$ and $x_3(t)$ must be represented as suitable combinations of $x_1(t)$.

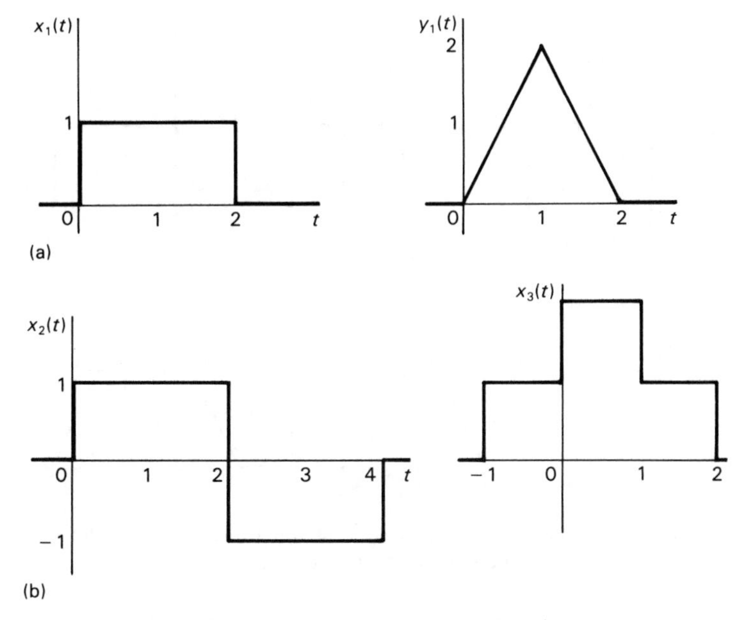

Figure 3.7 Signals for Example 3.3.2.

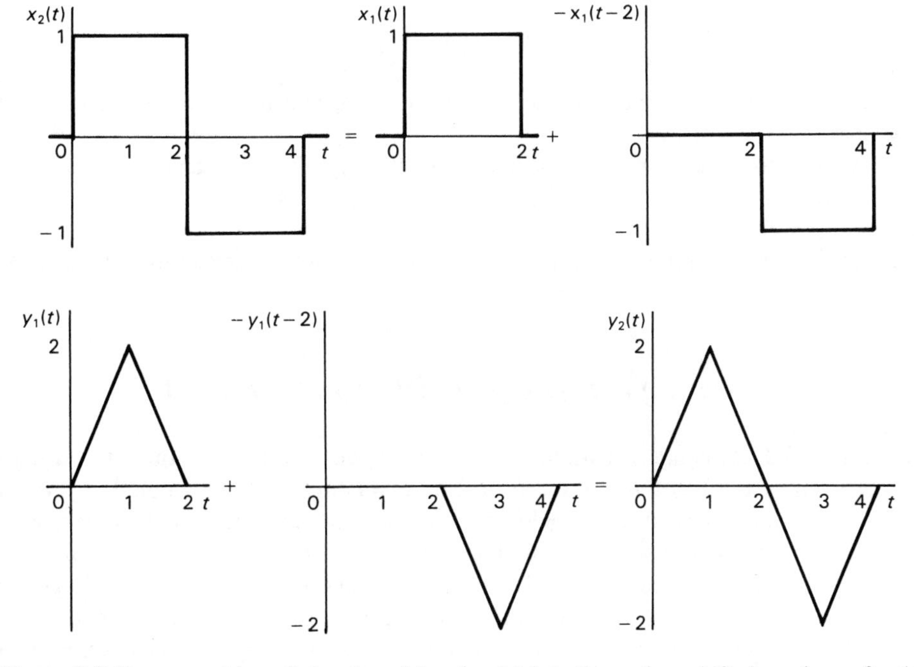

Figure 3.8 Decomposition of signals $x_2(t)$ and $y_2(t)$ into linear time shifted versions of $x_1(t)$ and $y_1(t)$.

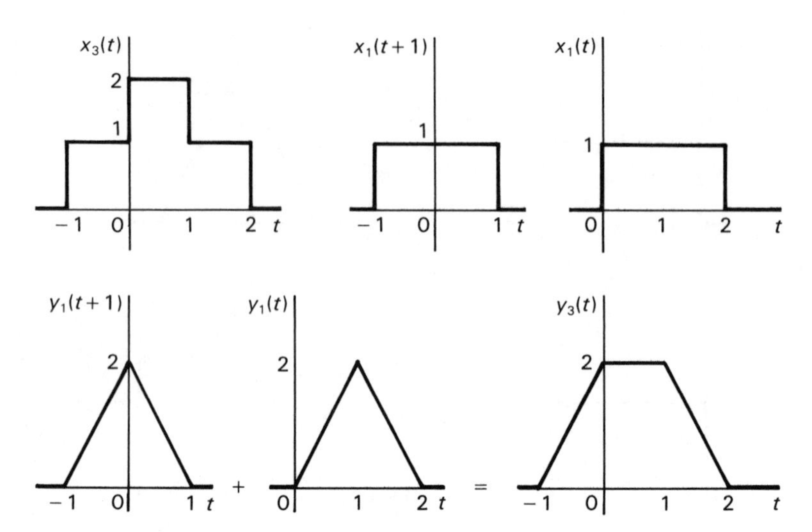

Figure 3.9 Decomposition of signals $_3(t)$ and $y_3(t)$ into linear time shifted versions of $x_1(t)$ and $y_1(t)$.

This is straightforward for $x_2(t)$, the portion of the signal up to $t = 2$ is in fact $x_1(t)$, between $t = 2$ and $t = 4$ it is a sign inverted time delayed version of $x_1(t)$. Mathematically

$$x_2(t) = x_1(t) - x_1(t-2)$$

Figure 3.8 shows the signal represented in this form and the corresponding system response $y_2(t)$.

A little more care is required with the signal $x_3(t)$ which can be expressed as

$$x_3(t) = x_1(t+1) + x_1(t)$$

Figure 3.9 shows the representation of $x_3(t)$ in this form and the resulting output $y_3(t)$.

3.4 Deriving the system model (continuous systems)

In Section 3.3 mathematical models have been used to relate the output and input signals in a system. In that section no mention was made of the methods by which such models are obtained. It is obviously of great importance to be able to derive the model for a given physical system but the derivation of such a model presents problems in a book of this nature. The range of systems to be modelled is large and varied, nuclear reactors, aircraft control systems, radar receivers etc., the reader (and the author) cannot be expected to have a detailed knowledge of these systems. Also this book is much more concerned with the overall systems approach than with the nitty-gritty of any particular system. In practice the problem is solved by leaving the derivation of the model to the specialist (the physicist, the chemical engineer, etc.) and accepting a given model when concerned with wider system aspects. However it is felt that the student should have some appreciation of the model derivation and the approach adopted is to restrict the range and complexity of systems investigated. Hence the systems considered are basic electrical, mechanical and electromechanical systems.

As stated earlier, one characteristic of the system approach, is that a complex system can be broken down into simpler interconnected subsystems. These subsystems can be further broken down into interconnected elements or components which can then be described by basic physical laws. It should be noted that these laws themselves are idealisations. For instance Ohm's Law states that voltage is proportional to current in a resistor. However this is only an approximation over a limited range even when the temperature is constant. In practice the current flow itself causes power dissipation which raises the temperature. However to write down the complete time relationship between voltage and current would lead to such a complex model that it would be too intractable to use. In practice simple models are used initially to obtain an insight into the system response. The model can then be refined becoming more complex (and leading to more complex computations) leading to a more complete description of the response.

The steps involved in the construction of the model are as follows:

1. Identify the components in the system and determine their individual describing equations relating the signals (variables) associated with them. It should be noted that no matter how the components are interconnected the equation describing the individual component does not change.
2. Write down the connecting equations for the system which relate how the individual components relate to one another. It is often possible to use more than one form of the connecting equation and it is only experience that will determine the most suitable form.
3. Eliminate all the variables except those of interest, usually these are the input and output variables. This step can be difficult because as well as containing variables that have to be eliminated the equations can also contain their differentials.

These steps will be illustrated for specific systems. The next section considers continuous systems, discrete systems will be considered in Section 3.8.

3.4.1 Electrical systems

The variables involved in electrical systems are usually current i and voltage v. Charge q is a fundamental quantity but does not usually appear in engineering systems, it is however related to the current by the equation

$$i = \frac{\mathrm{d}q}{\mathrm{d}t}$$

The basic components in electrical systems are the resistor R, the inductor L and the capacitor C. From the systems viewpoint these components are described by the relationship between the variables (voltage and current) at their terminals.

The resistor R

This stores no energy and the relationship between voltage and current at its terminals is that of an instantaneous system.

$$v(t) = i(t)R$$

The variables have been written as functions of time to emphasise that this relationship is true at every instant of time no matter what form the voltage and current waveform take. As stated in Chapter 2 such a notation is rather cumbersome and it is more convenient in general to drop the dependence on time, i.e.

$$v = iR \tag{3.4.1}$$

It should be noted that an electrical engineer would probably not be satisfied with this description of a resistor. The engineer may require the resistor's tolerance, power rating, physical size and perhaps cost. The lack of these requirements distinguishes the systems approach which accepts the equation relating voltage and current as the only description required.

The inductor L

The basic feature of inductance is that of an associated magnetic field. This field stores energy and the inductor is an energy storage element. A *change* of current in the element gives a change of flux that produces a 'back emf' across the terminals. However it is more convenient to use the voltage drop across the component rather than the back emf giving the relationship:

$$v = \frac{L\,\mathrm{d}i}{\mathrm{d}t} \tag{3.4.2}$$

where the dependence upon time in the variable has been omitted.

The capacitor C

The basic feature of the capacitor is that of an associated electric field. This field stores energy and the capacitor is an energy storage element. The charge q stored in a capacitor is given by,

$$q = vC$$

However using the relationship between current and charge gives

$$i = \frac{\mathrm{d}q}{\mathrm{d}t} = \frac{C\,\mathrm{d}v}{\mathrm{d}t}$$

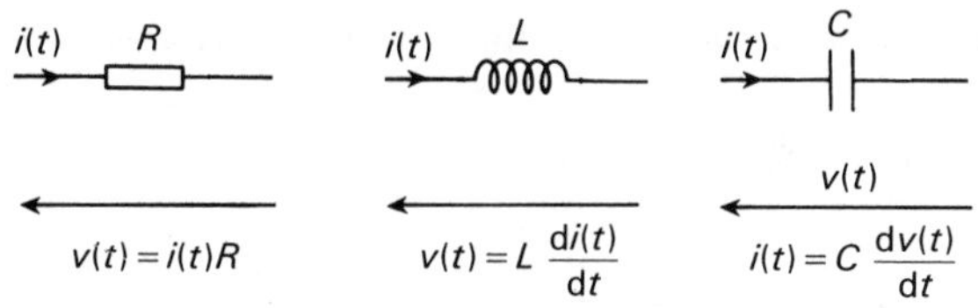

i(t) R
i(t) L
i(t) C
v(t)
v(t) = i(t)R
v(t) = L di(t)/dt
i(t) = C dv(t)/dt

Figure 3.10 Basic relationships between terminal variables for electrical components.

Hence the basic equation describing the capacitor is

$$i = \frac{C\,\mathrm{d}v}{\mathrm{d}t} \tag{3.4.3}$$

The relationships described by eqns (3.4.1), (3.4.2) and (3.4.3) are shown together with appropriate circuit symbols in Figure 3.10.

When these electrical elements are interconnected some connecting equation is required in order to analyse the circuit. The fundamental equations relating to the elements are Kirchhoff's Laws. The use of these laws for a solution of a circuit problem is best illustrated by an example.

EXAMPLE 3.4.1

(a) The circuit of Figure 3.11(a) is a simple low-pass filter (see Example 1.1). Derive the relationship between v_o and v_i for this filter.
(b) The circuit of Figure 3.11(b) is a circuit used for compensation in control systems, derive the relationship between v_o and v_i for this circuit.

SOLUTION

In order to obtain the relationship between v_o and v_i for circuit (a) it is necessary to introduce an additional variable, the current i. This variable will not be required in the answer and will have to be eliminated.

R
i
C
v_i
v_o
(a)
i_1
R_1
i_2
C_1
i_3
R_2
v_i
v_o
(b)

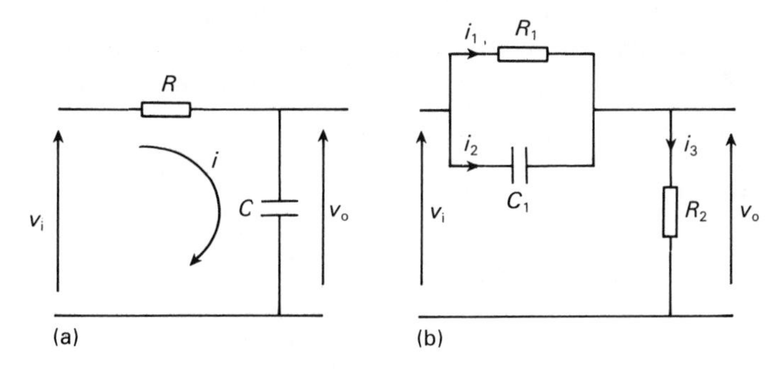

Figure 3.11 Circuits for Example 3.4.1.

The first step is to write down the relationships describing the individual elements.
The voltage across the resistor is given by

$$v = iR \tag{3.4.4}$$

The relationship between current and voltage for the capacitor is

$$i = \frac{C\,dv_o}{dt} \tag{3.4.5}$$

The second step is to write down a connecting equation for the elements. As there is only one mesh in this problem, Kirchhoff's voltage law is most convenient.

$$v_i = \text{voltage across } R + \text{voltage across } C$$

The voltage across R is given by eqn (3.4.4). The voltage across C could be written as an integral relationship involving the current. This however would be a retrograde step as v_o is a variable required in the answer, hence it is best left in the equation giving

$$v_i = iR + v_o \tag{3.4.6}$$

The third step is to eliminate the variable(s) not required, in this case the current i. This can easily be done by substituting the current from eqn (3.4.5) into eqn (3.4.6) giving

$$v_i = \frac{CR\,dv_o}{dt} + v_o$$

This is an equation containing only the variables v_i and v_o. However it also contains the rate of change of v_o, dv_o/dt. It is a *differential equation*, it is usually written with the output quantity and its derivatives on the left-hand side

$$\frac{CR\,dv_o}{dt} + v_o = v_i$$

To analyse the circuit shown in Figure 3.11(b) again additional variables are required, they are the currents i_1, i_2 and i_3 as indicated. To obtain the equations describing each element it should be noted that the voltage across the combination R_1C_1 is the difference between v_i and v_o. Hence the equation describing R_1 is

$$(v_i - v_o) = i_1R_1 \tag{3.4.7}$$

and that describing C_1 is

$$i_2 = \frac{C\,d(v_i - v_o)}{dt}$$

$$= \frac{C\,dv_i}{dt} - \frac{C\,dv_o}{dt} \tag{3.4.8}$$

The equation describing R_2 is straightforward,

$$v_o = i_3 R_2 \tag{3.4.9}$$

The circuit contains two meshes but only one node. Hence it is convenient to use the current form of Kirchhoff's laws

$$i_1 + i_2 = i_3 \tag{3.4.10}$$

The currents i_1, i_2 and i_3 can now be eliminated by substitution from eqns (3.4.7), (3.4.8) and (3.4.9) into eqn (3.4.10)

$$\frac{v_i - v_o}{R_1} + \frac{C\,\mathrm{d}v_i}{\mathrm{d}t} - \frac{C\,\mathrm{d}v_o}{\mathrm{d}t} = \frac{v_o}{R_2}$$

This equation can be written as

$$\frac{CR_1R_2}{R_1+R_2}\frac{\mathrm{d}v_o}{\mathrm{d}t} + v_o = \frac{CR_1R_2}{R_1+R_2}\frac{\mathrm{d}v_i}{\mathrm{d}t} + \frac{R_2}{R_1+R_2}v_i$$

As in the first part of the example the relationship between v_i and v_o is a differential equation. In this case as well as the differential of the output quantity $\mathrm{d}v_o/\mathrm{d}t$ appearing there is also a term depending on the differential of the input quantity.

3.5 Mechanical systems

Mechanical systems are in general concerned with the motion of a mechanical assembly in three-dimensional space. Usually the motion will combine both translational and rotational motion with respect to some reference framework. However for the purpose of this book motion will either be considered as translational along a fixed direction or rotational about a fixed co-ordinate axis. The signals (variable) involved are force, displacement, velocity and acceleration for the translational system and torque, angular displacement, angular velocity and angular acceleration for the rotational system. It is convenient to consider separately the elements involved for translational and rotational motion.

3.5.1 The spring (translational)

A spring either compresses or elongates when subject to a mechanical force. A linear spring obeys Hooke's law and relates force (f) to displacement (x) by the equation

$$f = kx$$

k is constant known as the *stiffness* of the spring and is the force required to produce unit displacement (the inverse $1/k$ is also commonly used and is known as the compliance).

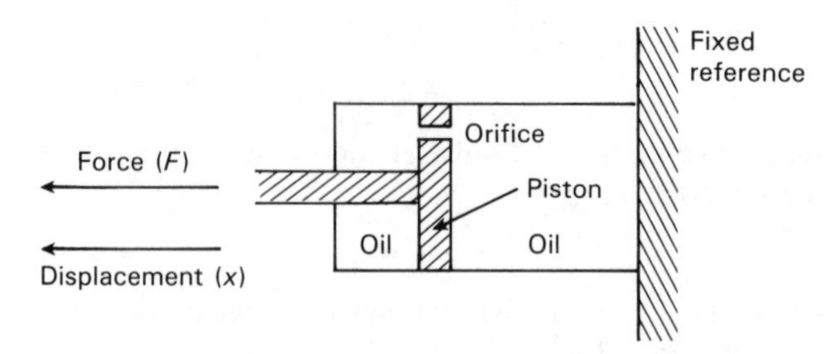

Figure 3.12 Simplified version of a practical damper.

3.5.2 The damper (translational)

A simplified version of a practical damper is shown in Figure 3.12. It consists of a piston in an oil filled cylinder. The piston has a small orifice that allows oil to flow from one side to the other. If the piston moves slowly oil can flow through the orifice and little force is required. However to move the piston quickly much more force is required. The relationship between force and velocity for an ideal damper is

$$f = Bv = \frac{B\,\mathrm{d}x}{\mathrm{d}t}$$

where B is the constant of the damper.

In practice the oil flow is very non-linear and the relationship will depart from this ideal. The student is probably aware of the use of dampers on motor vehicles where they 'damp down' the oscillation that occurs following the wheel hitting a bump.

3.5.3 The mass element

The mass of an object relates to the amount of matter in the object. Its measurement invokes the use of Newton's laws of motion and it is via these laws that mass is incorporated into overall systems. For a solid object the applied force is related to the acceleration via the equation

$$f = Ma = M\frac{\mathrm{d}v}{\mathrm{d}t} = M\frac{\mathrm{d}^2x}{\mathrm{d}t^2}$$

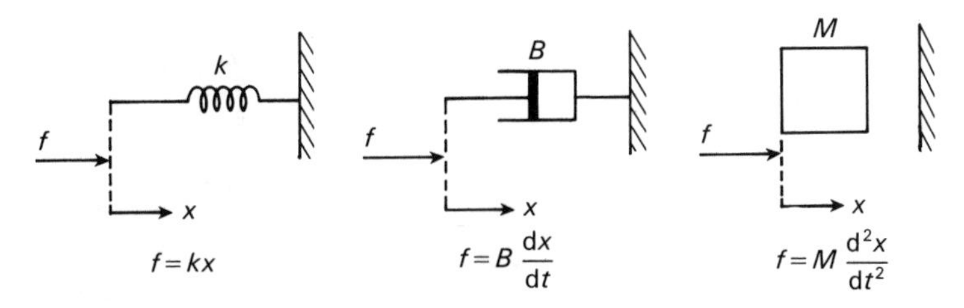

Figure 3.13 Basic relationships between variables for mechanical components (translational).

3.5.4 Interconnected mechanical systems (translational)

The three components described in the last section are represented in Figure 3.13 together with the equations relating their terminal variables. When incorporated into a more complex system each mechanical component must still obey its describing law. However the manner in which the components are interconnected produces connecting equations that define a particular system. These equations can be obtained by applying analogous methods to those used to obtain the equations for electrical systems.

Points that have the same displacement (or velocity) are regarded as nodes. A diagram can be drawn showing all the elements connected between nodes (including a reference node where the variables are zero). The describing law for each element can now be applied noting that it is the relative displacements (or velocities) between nodes that must be used in the relevant equation. At each node the net force is equated to mass times acceleration to give the equations of the system. This method will now be illustrated by means of an example.

EXAMPLE 3.5.1

For the mechanical systems shown in Figure 3.14 obtain the relationship between the displacement $x(t)$ and the applied force $f(t)$.

SOLUTION

The systems shown in Figure 3.14 have been redrawn in Figure 3.15 to show the elements connected between nodes. The system shown in Figure 3.15(a) has two nodes and the equations at these nodes can be written

$$f = B\left(\frac{dx_1}{dt} - \frac{dx}{dt}\right)$$

$$B\left(\frac{dx_1}{dt} - \frac{dx}{dt}\right) = M\frac{d^2x}{dt^2}$$

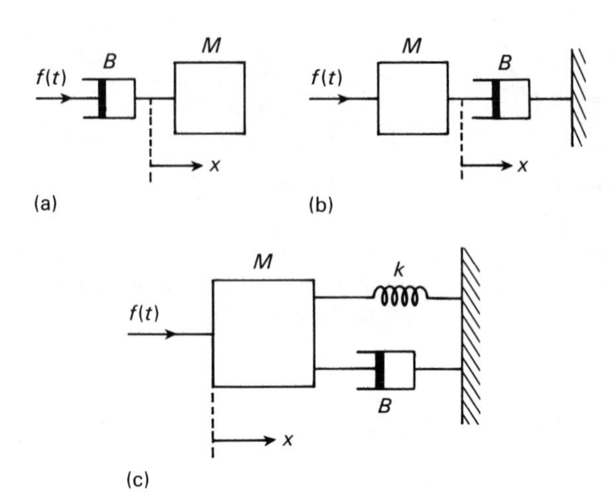

Figure 3.14 Systems for Example 3.5.1.

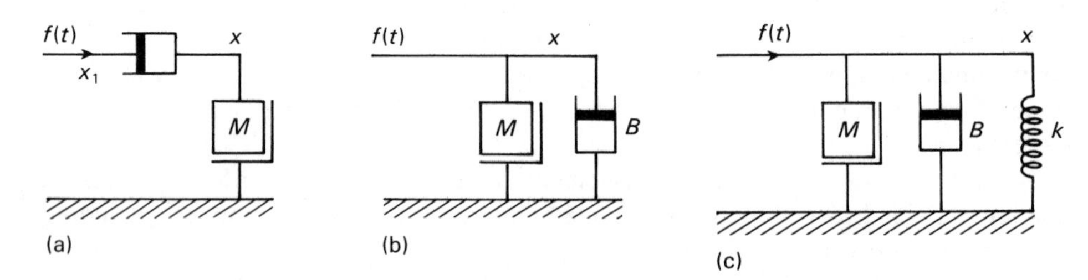

Figure 3.15 Systems re-drawn to shown interconnection of nodes.

giving the equation relating x and f as

$$\frac{d^2x}{dt^2} = \frac{f}{M}$$

The system shown in Figures 3.15(b) and 3.15(c) have only one node and the required equation can be written directly.
For Figure 3.15(b)

$$f - B\frac{dx}{dt} = M\frac{d^2x}{dt^2}$$

$$M\frac{d^2x}{dt} + B\frac{dx}{dt} = f$$

or in terms of the velocity of point, x writing $v = dx/dt$

$$M\frac{dv}{dt} + Bv = f$$

For Figure 3.15(c)

$$f - B\frac{dx}{dt} - kx = M\frac{d^2x}{dt^2}$$

$$M\frac{d^2x}{dt^2} + B\frac{dx}{dt} + kx = f$$

3.5.5 Rotational elements

The analysis of rotational systems closely parallels that of the translational systems covered in the previous sections. The variables displacement, velocity and acceleration become angular displacement θ, angular velocity ω, and angular acceleration α. Force is replaced by torque T which gives a measure of the turning moment of the force.

The elements forming a rotational system are shown in Figure 3.16 together with the relationships between the variables at their 'terminals'. The formulation of the system equations for an interconnected system follows the principles used for translational systems and these are illustrated in the following example.

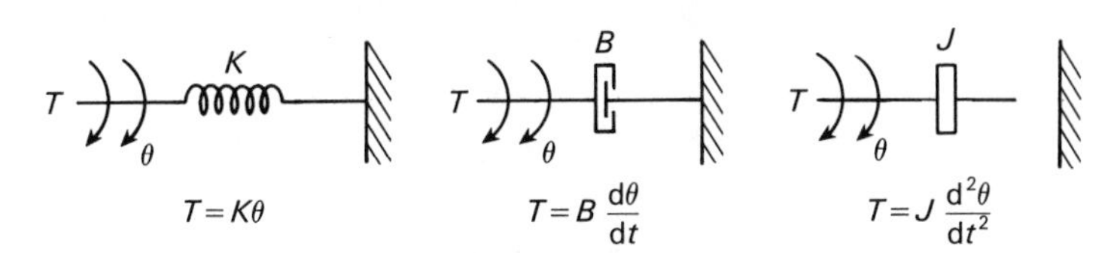

Figure 3.16 Basic relationships between variables for mechanical components (rotational).

EXAMPLE 3.5.2

Obtain the equations describing the rotational system shown in Figure 3.17.

SOLUTION

The system has three nodes and can be drawn as shown in Figure 3.18. The following equations apply at the nodes:

$$T - K_1(\theta_1 - \theta_2) = 0$$

$$K_1(\theta_1 - \theta_2) - B\left(\frac{d\theta_2}{dt} - \frac{d\theta_3}{dt}\right) = J_1\frac{d^2\theta_2}{dt^2}$$

$$B\left(\frac{d\theta_2}{dt} - \frac{d\theta_3}{dt}\right) - K_2\theta_3 = J_2\frac{d^2\theta_3}{dt^2}$$

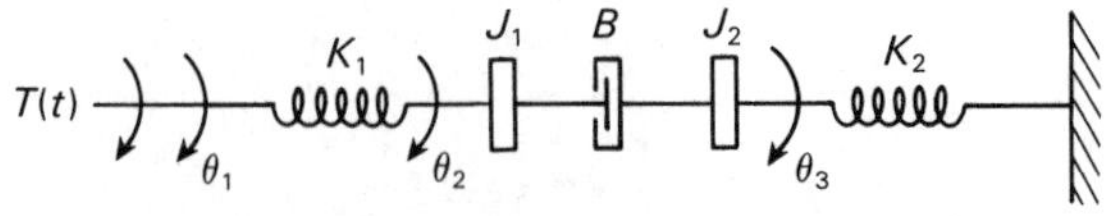

Figure 3.17 Rotational system for Example 3.5.2.

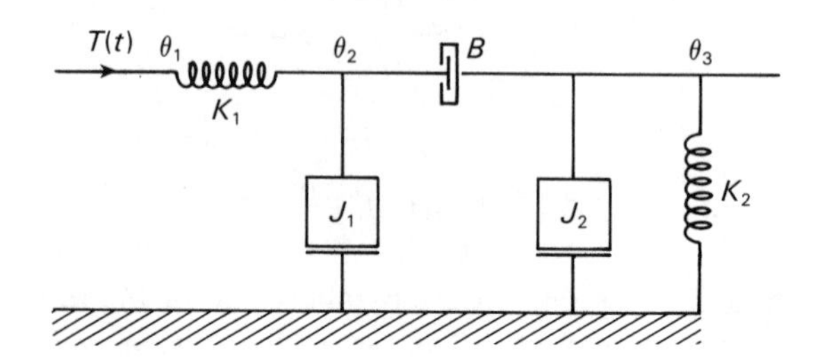

Figure 3.18 System re-drawn to show interconnection of nodes.

These equations can be re-written

$$K_1\theta_1 - K_1\theta_2 = T$$

$$-K_1\theta_1 + J_1\frac{\mathrm{d}^2\theta_2}{\mathrm{d}t^2} + B\frac{\mathrm{d}\theta_2}{\mathrm{d}t} + K_1\theta_2 - B\frac{\mathrm{d}\theta_3}{\mathrm{d}t} = 0$$

$$-B\frac{\mathrm{d}\theta_2}{\mathrm{d}t} + J_2\frac{\mathrm{d}^2\theta_3}{\mathrm{d}t^2} + B\frac{\mathrm{d}\theta_3}{\mathrm{d}t} + K_2\theta_3 = 0$$

If it is required to find a single equation relating one of the variables θ_1, θ_2, or θ_3 to the applied torque $T(t)$, these equations have to be solved as simultaneous equations. This is not easy and is best done using the Laplace transform as will be shown in Chapter 7.

3.6 Electromechanical systems: the d.c. machine

These are systems where the working signals involved are both electrical and mechanical. These systems are very common and cover such areas as loudspeakers (an acoustic system could also be included), transducers for measuring mechanical signals and producing an electrical output, and electrically operated actuators for a wide range of machinery. The electromechanical system considered here is the d.c. machine both in the form of the motor and the generator. Complete books are devoted to the study of the d.c. machine and only the basic operating features can be described here, again using the systems viewpoint where the machine description is given in terms of the signals at its input and output terminals.

3.6.1 Basic operation and description

The basic operating principles of the d.c. machine depend on the interaction between a magnetic field and a moving conductor. The machine consists of three major parts:

1. A system to produce a magnetic field – the field system.
2. A system of rotating conductors in the magnetic field – the armature.
3. A method of ensuring the torque (or e.m.f.) produced by the machine is always in the same direction – the commutator and brushes.

The machine is represented diagrammatically as in Figure 3.19.

A mathematical model for the machine is obtained by considering the basic laws relating to conductors in a magnetic field. The force on a current-carrying conductor depends on the current, the length of the conductor and the strength of the field.

The e.m.f. induced into a moving conductor in a magnetic field depends on the length of the conductor, its velocity and the strength of the field. For the machine these give rise to the equations

$$T \propto \Phi I_a$$

$$E \propto \Phi \omega$$

where

T = machine torque
Φ = field flux
E = generated e.m.f.
I_a = armature current
ω = angular velocity

When the machine is used as a generator it is driven from an external mechanical power source. If the speed is kept constant the generated e.m.f. is proportional to the field flux Φ. Except in very small machines the field is produced by an electromagnet and the flux depends upon the field current I_f. However because of the iron in the field the relationship between Φ and I_f is not linear, it follows a B/H characteristic as shown in Figure 3.20.

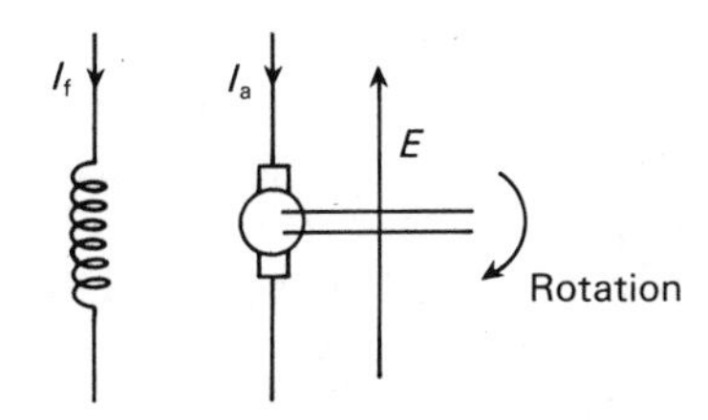

Figure 3.19 Representation of a d.c. machine.

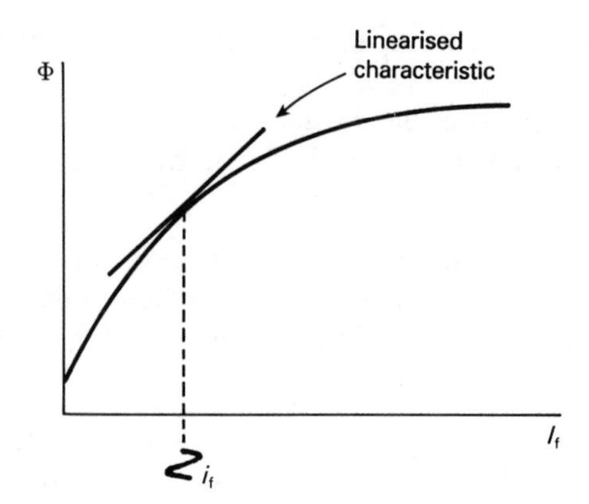

Figure 3.20 B/H characteristic of machine field.

Hence the relationship between E and I_f is non-linear. A linear model can be constructed by restricting the current I_f to a small charge i_f about a working point. A relationship that applies to small signals only can be written as

$$e = Ki_f$$

where K is a constant of the generator that depends upon the working point.

In the case of the motor the equation governing the output torque is

$$T \propto \Phi I_a$$

Two methods of controlling the output torque are possible: (a) via the field flux (assuming the field is not provided by a permanent magnet); (b) via the armature current.

Using field control the relationship between field and current is again very non-linear as in the case of the generator. If the armature current I_a is kept constant then for small signals

$$T = Ki_f$$

where T and i_f refer to small signals around a working point and K is a constant.

If the machine is controlled via its armature current then I_f is kept constant and the relationship between torque and armature current is

$$T = K_T i_a$$

where K_T is a constant of the machine known as the torque constant.

It should be noted that although the machine is being used as a motor its armature conductors are cutting a magnetic field and an e.m.f. is induced in them. As the field is constant this back-e.m.f., e_g, is proportional to speed

$$e_g = K_v \omega$$

where K_v is the e.m.f. constant of the machine (in SI units K_v and K_T are equal for a given machine). This back-e.m.f. profoundly affects the performance of the machine as will be shown in the following example.

EXAMPLE 3.6.1

Figure 3.21 shows an armature fed d.c. machine. The machine has armature resistance R_a and drives a purely inertial load J. Obtain the differential equation relating: (a) ω to v; (b) θ to v.

SOLUTION

The procedure involved in obtaining the required relationship is to write down equations governing the individual parts of the system. All the variables except those of interest are then eliminated from the equations.

Electrical equation

This can be obtained by applying Kirchhoff's mesh law around the armature circuit. It must not be forgotten that there is a back-e.m.f. e_g acting, hence

$$v - e_g = i_a R_a \qquad (3.6.1)$$

Electromechanical equations

These relate the back-e.m.f. e_g and torque T to the speed ω and armature current i_a

$$e_g = K_v \omega \qquad (3.6.2)$$

$$T = K_T i_a \qquad (3.6.3)$$

Mechanical equations

As the load driven by the machine consists only of the inertia J

$$T = J \frac{d^2\theta}{dt^2} \text{ in terms of } \theta \qquad (3.6.4)$$

$$T = J \frac{d\omega}{dt} \text{ in terms of } \omega \qquad (3.6.5)$$

The variables of interest are v and ω or v and θ depending on the relationship required, all the other variables must be eliminated between eqns (3.6.1) to (3.6.5).

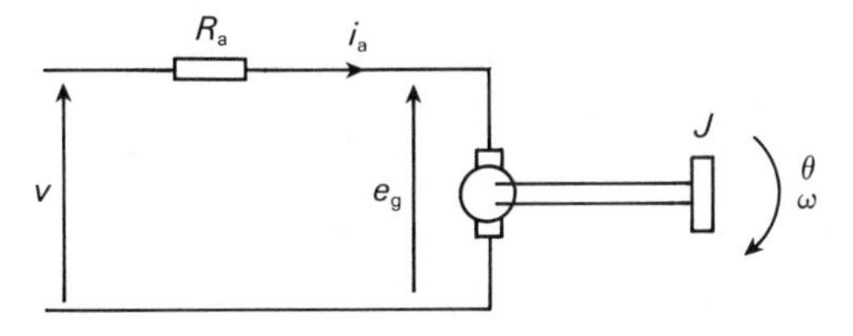

Figure 3.21 Machine systems for Example 3.6.1.

Substitution of e_g from eqn (3.6.2) and of i_a from eqn (3.6.3) into eqn (3.6.1) gives

$$v - K_v\omega = \frac{T}{K_T} R_a \tag{3.6.6}$$

Substitution of T from eqn (3.6.5) gives

$$v - K_v\omega = \frac{JR_a}{K_T}\frac{d\omega}{dt}$$

or rearranging

$$\frac{JR_a}{K_T}\frac{d\omega}{dt} + K_v\omega = v$$

This is the differential equation relating v and ω.

If the relationship between θ and v is required then using the fact that $\omega = d\theta/dt$ gives

$$\frac{JR_a}{K_T}\frac{d^2\theta}{dt^2} + K_v\frac{d\theta}{dt} = v$$

3.7 Electrical analogues and the operational amplifier

3.7.1 Electrical analogues

The systems considered in Examples 3.4.1(a) and 3.6.1 were very different physical systems. One was a very simple electrical system consisting of two components only, the other was a complex electromechanical system. The equations derived to represent these systems were:
For the electrical system:

$$CR\frac{dv_o}{dt} + v_o = v_i \tag{3.7.1}$$

For the electromechanical system:

$$\frac{JR_a}{K_T}\frac{d\omega}{dt} + K_v\omega = v$$

or

$$\frac{JR_a}{K_T K_v}\frac{d\omega}{dt} + \omega = \frac{v}{K_v} \tag{3.7.2}$$

In eqn (3.7.1) CR has the units of time, it is the time constant T of the electrical circuit. If in eqn (3.7.2) the units of the factor $JR_a/K_T K_v$ are carefully checked these are also the units of time, $JR_a/K_T K_v$ represents an electromechanical time constant. By choosing the values of CR correctly in the electrical circuit this could be made numerically equal to the electromechanical time constant

$$T = CR = \frac{JR_a}{K_T K_v}$$

The voltage applied to the machine is a function of time, however by scaling this by $1/K_v$ a voltage v' could be applied to the electrical circuit where

$$v' = \frac{v}{K_v}$$

Under these conditions the electrical circuit is represented by the equation

$$T\frac{dv}{dt} + v = v'$$

and the electromechanical system by

$$T\frac{d\omega}{dt} + \omega = v'$$

Except for the symbol used to denote the output signal these two equations are identical. For a given input the output voltage in the electrical system will be numerically equal to the speed of the machine (rad/s). Two systems that are described by the same mathematical equation are known as *analogues* (from analogous – similar in certain respects). Using an analogue can be a very useful way of investigating a system. Rather than investigate a very complex expensive system (such as a nuclear reactor) directly, one can construct a relatively simple inexpensive electrical analogue. Measurements made to the response of this analogue can be directly related to variables in the reactor.

However the problem arises as to how one constructs the analogue system, given the equations describing the real system. By interconnecting electrical components one would probably arrive at a differential equation having the same form as the required equation. However the values of the coefficients in such an equation would probably depend on several components and it would be difficult to alter the coefficients independently. What is required is an *ordered* method of arriving at the electrical system having the requisite equation and a system where alteration of one component will alter one coefficient. Such an ordered system is an *analogue computer* and at the heart of such a computer is the *computing* or *operational amplifier*.

The analogue computer formed a very useful method of analysing systems and very large analogue computer installations were built and used in the 1960s and

1970s. However the use of large analogue computers has declined since those dates. They have been superseded by digital computers using 'digital simulation languages' (see Section 3.9.1).

However, the operational amplifier forms a useful system element in its own right. Without going into analogue computing in any detail, the following section analyses operational amplifiers and shows how they can be interconnected to produce a differential equation that could be the analogue of the equation describing a physical system.

3.7.2 The operational amplifier

What makes an amplifier an 'operational amplifier' as opposed to say a 'hi-fi amplifier', a 'radio frequency amplifier', etc.? The basic requirements of an operational amplifier are as follows:

1. It should have a high voltage gain, typically, 10^6–10^{12}.
2. Its frequency response should extend down to d.c.
3. It should be phase inverting.

It is also desirable that the amplifier has a high input impedance and a low output impedance. Such amplifiers are used as operational amplifiers in a feedback configuration. Consider the amplifier used as shown in Figure 3.22.

This circuit can now be regarded as a system, it is however a system that contains an electronic element, the amplifier. If its output resistance is negligible the amplifier can be described by the equations,

$$v_o = -Ae$$

$$e = iR_i$$

where A is the amplifier voltage gain (negative due to the phase inversion) and R_i is the amplifier input resistance. The equation of the remainder of the circuit can be most easily obtained by using nodal analysis at the amplifier input terminal

$$\frac{v_i - e}{R_1} + \frac{v_o - e}{R_2} = i$$

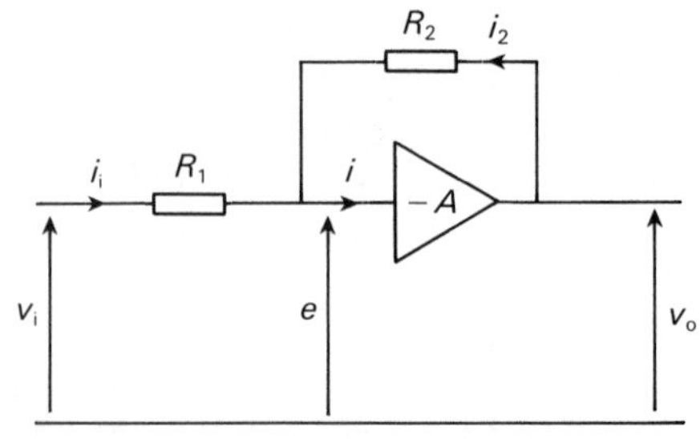

Figure 3.22 The operational amplifier.

If these equations are solved to eliminate i and e then the gain v_o/v_i is given by

$$\frac{v_o}{v_i} = \frac{-AR_2R_i}{AR_iR_1 + R_iR_2 + R_1R_i + R_1R_2}$$

This expression appears at first sight to be somewhat involved. However it can be re-written

$$\frac{v_o}{v_i} = \frac{R_2/R_1}{\left(1 + \dfrac{R_iR_2 + R_1R_i + R_1R_2}{AR_iR_1}\right)}$$

If now some typical values are substituted, say $R_1 = 1$ kΩ, $R_2 = 10$ kΩ, $R_i = 100$ kΩ, $A = 10^6$ this gives

$$\frac{v_o}{v_i} = \frac{-10}{(1 + 11.1 \times 10^{-6})}$$

As can be seen the second term in the denominator is much less than unity (due to the large value of A) and it can be ignored for most practical purposes. Under these conditions

$$\frac{v_o}{v_i} = -\frac{R_2}{R_1}$$

the overall gain of the amplifier depends only on the resistors R_1 and R_2 and is independent of the characteristics of the amplifier.

This result could have been obtained more directly by taking into account the large value of A in the initial analysis. Suppose the amplifier concerned is an integrated circuit amplifier then the output voltage swing will be limited by the supply voltage to say ±15 V. Taking the gain to be 10^6 then for linear operation (and that is all we are concerned with) the input voltage e is limited to ±15A, ±15 μV.

From Figure 3.22 the current i_1 is given by

$$i_1 = \frac{v_i - e}{R_1} \approx \frac{v_i}{R_1} \qquad \text{as } v_i \gg e$$

similarly

$$i_2 = \frac{v_o - e}{R_2} \approx \frac{v_o}{R_2} \qquad \text{as } v_o \gg e$$

The currents i_1 and i_2 sum to give the input current to the amplifier

$$i_1 + i_2 = i = e/R_i$$

As e is small, the input current to the amplifier is small (assuming a practical figure for the input resistance), i.e., $e = 15\ \mu\text{V}$, $R_i = 100\ \text{k}\Omega$, $i = 150$ pA. As this current is much smaller than i_1 and i_2 then

$$i_1 + i_2 = 0$$

and substituting $i_1 = v_i/R_1$, $i_2 = v_o/R_2$

$$v_o/v_i = -R_2/R_1$$

This method of analysis assumes the voltage e is very small compared with voltages v_i and v_o. This is equivalent to saying that the amplifier input is virtually at earth potential – it is known as the *virtual earth principle*. The input terminal cannot really be at earth potential otherwise there would be no amplifier output. The virtual earth method forms a very convenient way of analysing operational amplifier circuits especially where the input and feedback paths consist of more than one resistor.

The circuit of Figure 3.22 can be used for the operation of multiplication by a constant, the constant having value R_2/R_1. If $R_2 > R_1$ the constant will have a magnitude greater than unity, if $R_1 > R_2$ its magnitude will be less than unity. The negative sign must not be forgotten in this operation, if $R_1 = R_2$ then the circuit will give a sign inversion which is often a useful operation.

The values of coefficients required can vary widely and to cater for such a range an enormous stock of resistors R_1, R_2 would have to be kept. To avoid this the circuit of Figure 3.23 is used.

Here a potentiometer has been included at the input to the circuit and the relationship between v_o and v_i becomes

$$v_o = -\frac{kR_2 v_i}{R_1}$$

Now only a limited range of resistor values is required say R_2 to R_1 in ratios 1 : 1, 10 : 1, 100 : 1. The potentiometers used are usually ten-turn potentiometers and can be set to an accuracy of 0.1%.

The circuit of Figure 3.22 can be modified by the inclusion of additional input resistors to form the summing amplifier as shown in Figure 3.24.

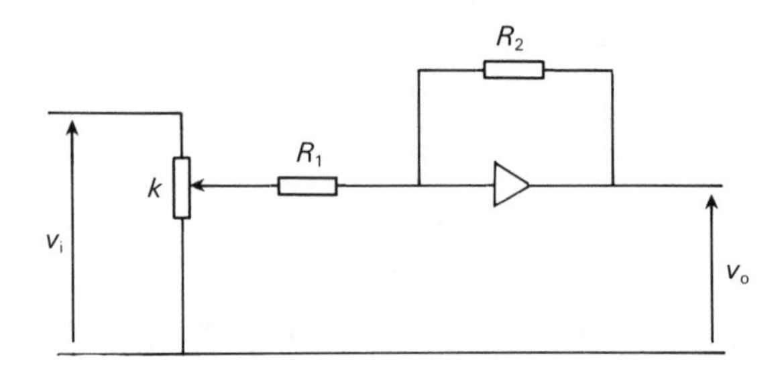

Figure 3.23 The operational amplifier with potentiometer included at input.

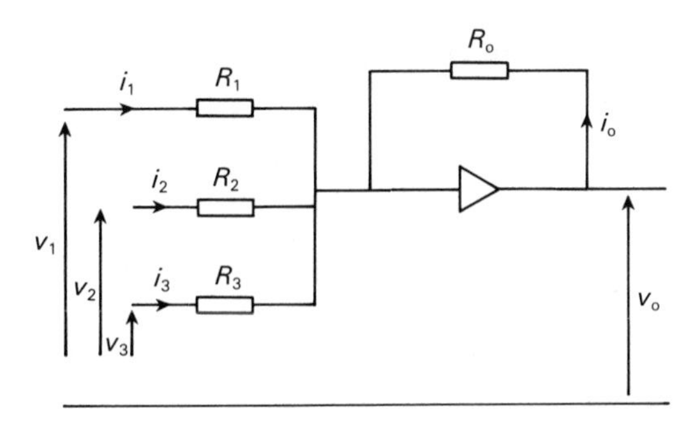

Figure 3.24 Summing amplifier.

Applying the virtual earth principle (the amplifier input terminal is assumed to be at zero input potential)

$$i_1 = \frac{v_1}{R_1}, \qquad i_2 = \frac{v_2}{R_2}, \qquad i_3 = \frac{v_3}{R_3}, \qquad i_o = \frac{v_o}{R_o}$$

Also the virtual earth principle implies that the input current to the amplifier is zero hence

$$i_1 + i_2 + i_3 + i_o = 0$$

Substituting for these currents gives the relationship

$$v_o = -R_o\left[\frac{v_1}{R_1} + \frac{v_2}{R_2} + \frac{v_3}{R_3}\right]$$

If $R_o = R_1 = R_2 = R_3$ then the output voltage is the sum of the input voltages together with a sign inversion. However by making the resistors unequal in value the operations of scaling and addition can be combined. Potentiometers can be used at each input similar to the manner used in Figure 3.23 and by using a unity gain amplifier before any of the inputs the sign of the associated signal can be inverted and subtraction performed.

Theoretically the circuit of Figure 3.25 will perform the operation of differentiation.

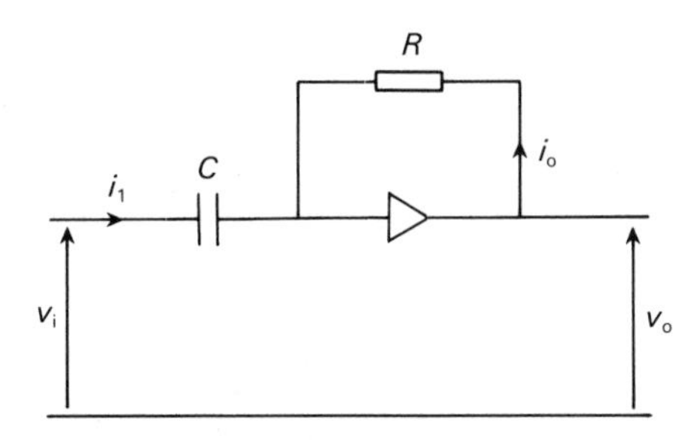

Figure 3.25 Theoretical differentiator.

Again using the virtual earth principle

$$i_1 = C\frac{dv_i}{dt}, i_2 = \frac{v_o}{R}$$

$$i_1 + i_2 = 0$$

$$C\frac{dv_i}{dt} + \frac{v_o}{R} = 0$$

$$v_o = -RC\frac{dv_i}{dt}$$

The output is a constant times the input signal differentiated with respect to time. There are two difficulties with this circuit in practice:

1. Noise on the input signal can manifest itself by rapid changes of input. Because the output is proportional to the rate of change of input the noise is accentuated at the output. It may now drive the amplifier out of its linear range and cause distortion of the input signal.
2. The circuit used to produce differentiation is a feedback circuit and a feature of such circuits is that they can go unstable by the phase shift produced in the feedback network. The feedback network used in the differentiator is particularly difficult in this respect.

The operation of differentiation can be replaced by that of integration and this can be achieved by the circuit of Figure 3.26. The positions of C and R are interchanged compared with the circuit of the differentiator. By use of the virtual earth principle it is easily shown that the relationship between v_o and v_i becomes

$$v_o = \frac{-1}{RC}\int_0^t v_i \, dt$$

The output voltage at $t = 0$ depends upon the initial charge on the capacitor. Usually additional circuitry is incorporated to set this to zero, or to any other required initial condition.

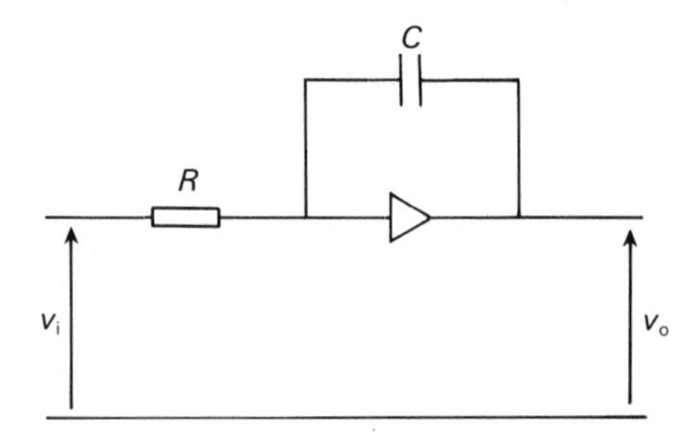

Figure 3.26 Integrator circuit.

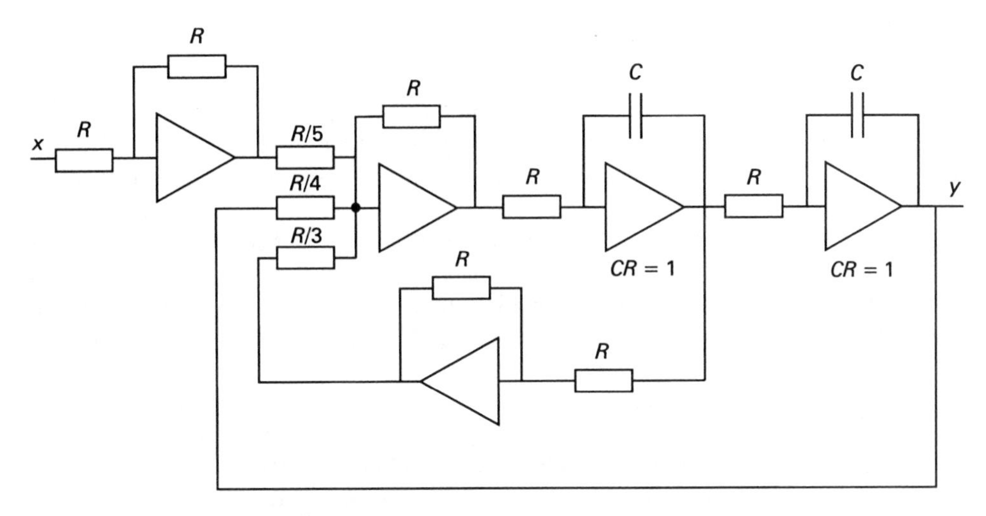

Figure 3.27 Analogue computer representation of a second order system.

Figure 3.27 shows how operational amplifier circuits performing the operations of addition, sign inversion and integration can be interconnected to produce a second order differential equation that could be the analogue of the differential equation describing a system. The student should verify that the equation relating x and y is

$$\frac{d^2y}{dt^2} + 3\frac{dy}{dt} + 4y = 5x$$

3.8 System description for discrete systems

A discrete system is described by a difference equation. The analysis of the system in order to arrive at this equation is often much easier than the corresponding analysis for a continuous system. As with the continuous system each element in the system has a describing equation and the elements are linked by some form of connecting equation. However because the system is discrete the only operations that can be performed are: (a) delay; (b) multiplication by a constant; and (c) addition and subtraction.

As described earlier discrete systems often arise from the sampling of signals in a continuous system. In this case the system is analysed as a continuous system and the time signals 'discretised'. However as the signal is only required at sample points the analysis can often be simplified to take account of this. The following example will illustrate this point and serve as an introduction to a discrete system.

3.8.1 The diode pump circuit

This is a circuit that was once used to count pulses and in a modified form as a simple frequency indicator. Although now superseded by more sophisticated digital logic circuits its relative simplicity makes it useful as an introduction to discrete systems. The circuit is shown in Figure 3.28.

The input to the circuit is the pulse waveform shown taking values 0 V and $-E$ V. The student may protest that this is not a discrete system, the input is a continuous signal. However it is not required to obtain the complete signal $v_o(t)$, all that is required is the signal as the time instant $T, 2T, \ldots$, i.e., the signal $v_o(nT)$ in terms of the input $v_i(nT)$.

The circuit of Figure 3.28 contains diodes and these are elements that have very non-linear characteristics. However for the purpose of explaining the action of this circuit their behaviour can be idealised as shown in Figure 3.29.

Figure 3.29(a) shows the actual characteristic. When forward biased the current rises rapidly with applied voltage and in order to prevent excessive current the forward voltage must be limited giving a maximum value in the order of 0.6–0.7 V. With reverse bias the current is small, measured in the pA range. This characteristic can be idealised as shown in Figure 3.29(b). In the ideal characteristic the reverse current is assumed zero and the forward current is limited only by the external circuit. Finally Figure 3.29(c) shows the diode represented by an ideal closed switch having zero forward resistance. When reversed biased the diode is represented by an ideal open switch with infinite resistance.

The other elements in the circuit are capacitors. These are elements that can be described by the equations

Charge stored $q = vC$

Current $\quad i = \dfrac{C\,dv}{dt}$

These equations enable us to formulate a simple law governing the behaviour of a capacitor when the potential at one of its plates changes rapidly (instantaneously). If

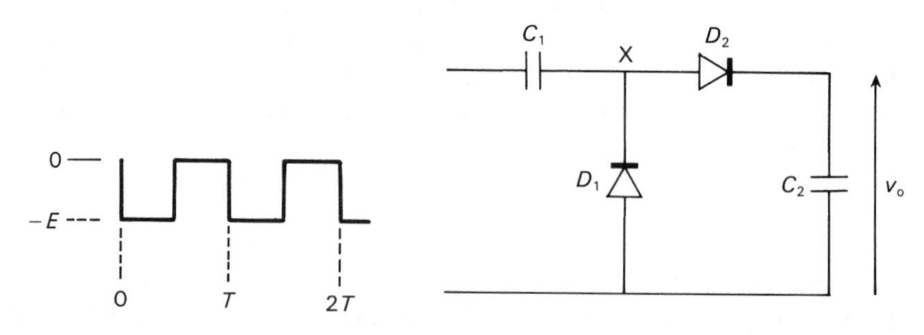

Figure 3.28 Diode pump circuit.

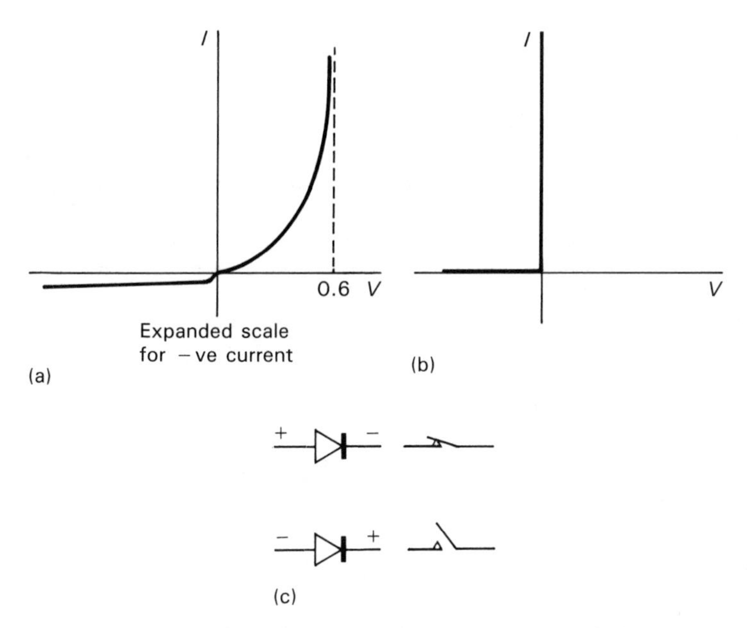

Figure 3.29 Representation of a diode by ideal switches.

the potential at one plate of a capacitor changes instantaneously then an identical instantaneous change must be produced at the other plate. If this were not the case then a change in potential would occur across the capacitor in zero time. This would infer a change of charge in zero time implying infinite current. (In practice instantaneous changes do not occur but this is an approximate model that is adequate for the purpose.)

Returning now to the circuit and assuming capacitors initially unchanged: the input voltage v_i drops instantaneously from 0 V to $-E$ V, hence the voltage at X must follow falling from 0 V to $-E$ V. D_1 is then forward biased, D_2 reverse biased and the conditions are as shown in Figure 3.30(a).

Under these conditions C_1 will charge rapidly (in practice the rate is governed by the internal resistance of the source) and the potential at X drops to zero. C_1 is now charged to voltage E.

The second portion of the input pulse now occurs, the input changing instantaneously from $-E$ to 0 volts. The same change must occur at point X hence the voltage at this point changes from *0 to +E volts*, X is at a positive potential forward biasing D_2 and reverse biasing D_1, the situation is as shown in Figure 3.30(b).

At first sight the capacitors are now connected in series. However this is not the case, the input voltage in this condition is zero and the circuit can be re-drawn as in Figure 3.30(c). At this instant the charge on C_1 is $(q = vC)$ EC_1 and the charge on C_2 is zero. This charge redistributes between the capacitors during the

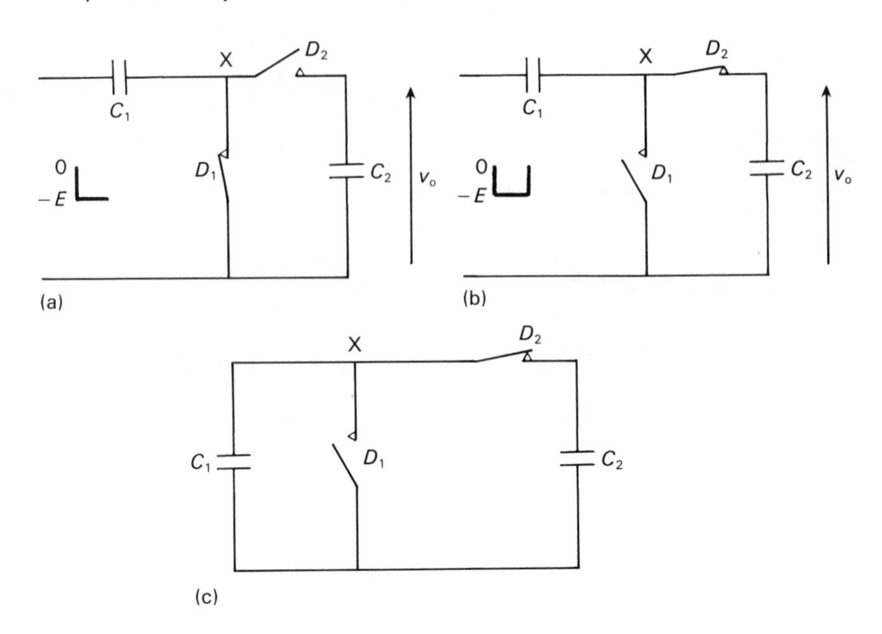

Figure 3.30 Switching actions in diode pump circuit.

second half of the input pulse. The resulting voltage v_o can be easily found as follows

$$\text{Total charge} = C_1E + 0 = C_1E$$
$$\text{Total capacitance } (\text{C}_1 \text{ and C}_2 \text{ in parallel}) = C_1 + C_2$$

hence

$$\text{Voltage } v_o(T) = \frac{q}{C} = \frac{C_1E}{C_1 + C_2}$$

This is the output voltage at time T.

During the interval from T to $2T$ the action of the circuit is very similar to that already described. During the first half of the second pulse the point X is clamped to 0 V as before. However during the second half of the pulse when the capacitors are connected in parallel the capacitor C_2 is not initially uncharged; it has charge due to voltage $v_o(T)$. Hence

$$\text{Total charge} = C_1E + \frac{C_2C_1E}{C_1 + C_2}$$

$$\text{Total capacitance} = C_1 + C_2$$

$$\text{Voltage } v_o(2T) = \frac{C_1E}{C_1 + C_2} + \frac{C_1C_2E}{(C_1 + C_2)^2}$$

This procedure can be repeated to obtain the output voltage at the end of the nth input pulse, $v_o(nT)$ or just $v_o(n)$. Such a procedure would be very tedious (although a pattern would emerge that would enable $v_o(n)$ to be written out directly). An alternative approach is to consider the circuit at the start of the nth pulse. The charge now present on C_2 is that due to the output voltage at the previous pulse, $v_o(n-1)$. Hence

$$\text{Total charge} = C_1E + C_2v_o(n-1)$$

$$\text{Total capacitance} = C_1 + C_2$$

$$\text{Voltage } v_o(n) = \frac{C_1E}{C_1+C_2} + \frac{C_2}{C_1+C_2}\,v_o(n-1)$$

This equation can be written more compactly by introducing the constants

$$k_1 = \frac{C_1}{C_1+C_2} \quad \text{and} \quad k_2 = \frac{C_2}{C_1+C_2}$$

hence

$$v_o(n) - k_2v_o(n-1) = k_1E \tag{3.8.1}$$

This is not an explicit expression for $v_o(n)$, it is an equation relating the output $v_o(n)$ to the previous output $v_o(n-1)$ and the amplitude of the input pulse E. It is a *difference equation.* In this equation the input is constant, however there is no reason

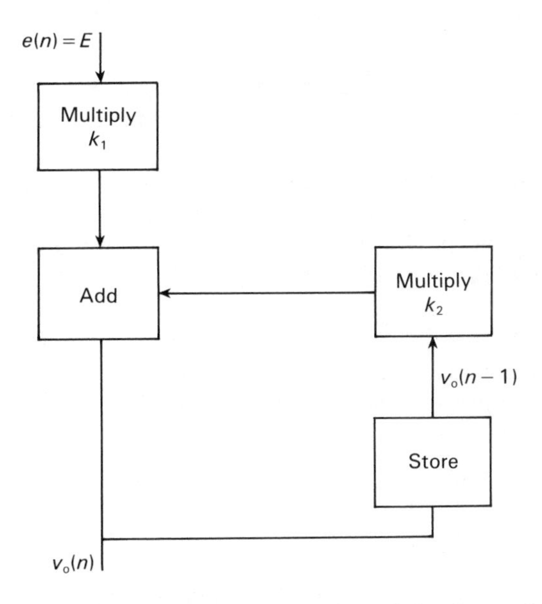

Figure 3.31 Flow chart for solution of the first order difference equation.

why the pulse of changing amplitude could not be applied, in this case the right hand side of the equation would be written $e(n)$.

The general solution of difference equations will be considered in Chapters 4 and 7. However, unlike the differential equation it is possible to obtain a numerical solution. This is done by calculating the output one point at a time starting at $n = 0$. This would be a very tedious process but it is easy to write a computer program to perform the task. The flow chart for such a program is shown in Figure 3.31.

Some initial value is required for entry in the store, this corresponds to any initial charge on the capacitor C_2 in the circuit.

3.8.2 General form of the difference equation

The difference equation derived in the last section related the present input of the system to the present output and the output at the previous time instant. It was a first order difference equation. The more general equation takes the form

$$\begin{aligned} y(n) + a_1 y(n-1) + \cdots\cdots + a_k y(n-k) \\ = b_0 x(n) + b_1 x(n-1) + \cdots\cdots + b_l x(n-l) \end{aligned} \tag{3.8.2}$$

This is a kth order difference equation. The coefficient of $y(n)$ can be taken as unity without loss of generality (division of every coefficient by a_0 will always reduce an equation to this form). Equation (3.8.1) can be rearranged as

$$\begin{aligned} y(n) = b_0 x(n) + b_1 x(n-1) + \cdots\cdots + b_l x(n-l) \\ - a_1 y(n-1) - \cdots\cdots - a_k y(n-k) \end{aligned} \tag{3.8.3}$$

In this form it indicates that the output at the present time instant is equal to the weighted sum of inputs at the present and previous time instants and of the outputs at previous time instants.

If the equation has all the a_n terms (other than a_0), zero, it is termed a *non-recursive* equation. Otherwise it is a *recursive* equation.

The following example illustrates the formulation of a difference equation for a resistive ladder network.

EXAMPLE 3.8.1

Ladder networks are circuits that are used as attenuators in communication systems. Figure 3.32 shows such a network fed by a voltage source E at its input and terminated by a short circuit at its output.

(a) Determine the difference equation relating the voltages at nodes $(n-1)$, n and $(n+1)$.
(b) Confirm that a solution to this difference equation is

$$V(n) = C_1 2^n + C_2 (0.5)^n$$

where C_1 and C_2 are arbitrary constants.

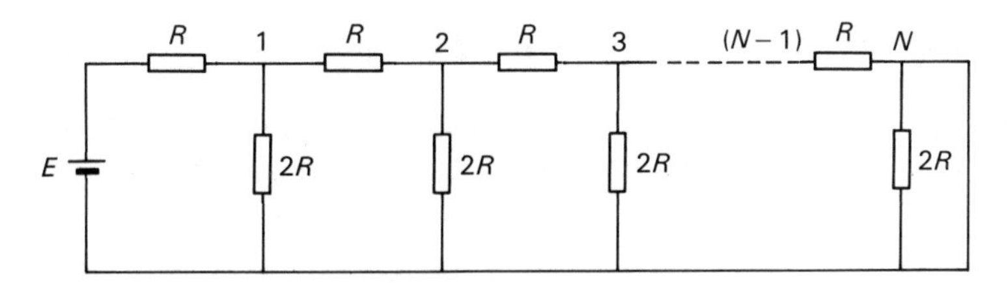

Figure 3.32 Ladder network for Example 3.8.1.

(c) Determine the values of the constants C_1 and C_2 if the number of stages $N = 5$ and the supply voltage $E = 10$ V.

SOLUTION

(a) Figure 3.33 shows how the nth node relates to the $(n-1)$th and $(n+1)$th node. Applying nodal analysis at the nth node gives

$$\frac{V(n-1)-V(n)}{R} = \frac{V(n)}{2R} + \frac{V(n)-V(n+1)}{R}$$

$$2V(n+1) - 5V(n) + 2V(n-1) = 0 \tag{3.8.4}$$

(b) This is the required difference equation. A solution is given

$$V(n) = C_1 2^n + C_2(0.5)^n \tag{3.8.5}$$

then

$$\begin{aligned} V(n+1) &= C_1 2^{n+1} + C_2(0.5)^{n+1} \\ &= 2C_1 2^n + 0.5C_2(0.5)^n \end{aligned}$$

similarly

$$V(n-1) = 2^{-1}C_1 2^n + (0.5)^{-1}C_2(0.5)^n$$

Substituting into eqn (3.8.4) gives on the left-hand side

$$(4C_1 - 5C_1 + C_1)2^n + (C_2 - 5C_2 + 4C_2)(0.5)^n$$

This is zero for all values of C_1 and C_2 hence the function $V(n)$ given by eqn (3.8.5) is a solution of eqn (3.8.4).

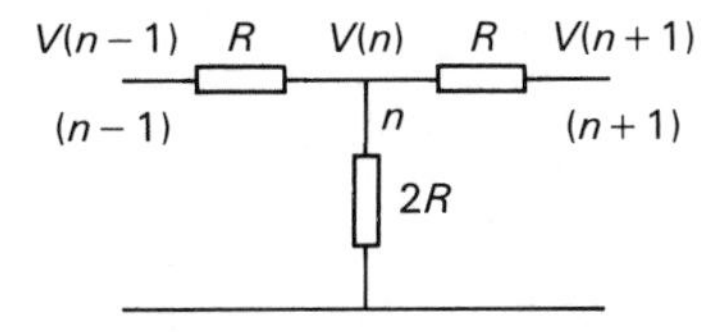

Figure 3.33 Conditions at successive nodes of the ladder network.

(c) To determine the constants C_1 and C_2 the specific nodes $n = 0$ and $n = 5$ are examined.

The node $n = 0$ is connected to the supply, hence $V(0) = 10$. The node $n = 5$ is short-circuited hence $V(5) = 0$. Substituting in turn into eqn (3.8.5)

$$10 = C_1 2^0 + C_2(0.5)^0$$

$$0 = C_1 2^5 + C_2(0.5)^5$$

giving

$$C_1 = -0.009775,\ C_2 = 10.009775$$

3.9 Discrete representation of a continuous system

In Section 3.7 it was shown how a system could be investigated by means of an analogue computer model and this required a special purpose computer. In Section 3.8 it was shown how discrete systems can be described by difference equations and these are very easy to solve numerically on a general purpose digital computer. If the differential equation describing the system could be approximated by a difference equation this could form the basis of a very convenient and flexible method of investigating continuous systems. This approach will be considered for a first order system in the next section.

3.9.1 Application to a first order system

Consider a very simple continuous system, the CR filter that was investigated in Example 3.4.1. Suppose it is required to obtain the response of this circuit to a step of input voltage, amplitude E as shown in Figure 3.34.

The differential equation describing the system has already been derived as

$$T_1 \frac{\mathrm{d}v_o}{\mathrm{d}t} + v_o = v_i \tag{3.9.1}$$

where $T_1 = CR$ and v_i is a step of amplitude E.

The method of solving this equation and deriving the response to a step input

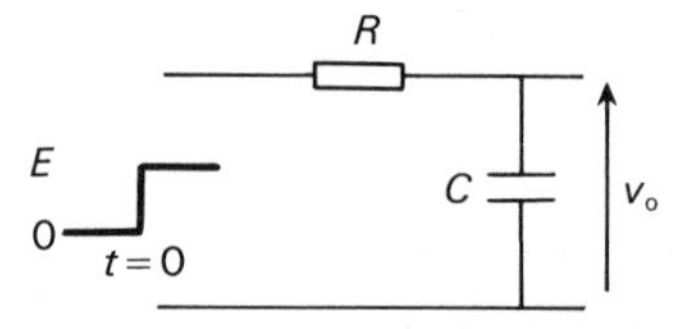

Figure 3.34 First order continuous system.

will be considered in the next chapter. However many students will be familiar with this response as the equation describing the charge of a capacitor through a resistor.

$$v_o = E(1 - e^{-t/T_1}) \tag{3.9.2}$$

(This result can be verified by substitution into eqn (3.9.1).)

Suppose now we wish to find a difference equation that approximates the differential eqn (3.9.1). Instead of the continuous variable $v_o(t)$ the variable $v_o(nT)$ or $v_o(n)$ could be substituted. Similarly the variable $v_i(t)$ could be replaced by $v_i(n)$ (for a step input $v_i(n) = E$ for $n \geqslant 0$). The problem is to find a suitable approximation for dv_o/dt, the slope of $v_o(t)$ curve.

Figure 3.35 shows how the slope can be approximated by

$$\frac{dv_o}{dt} \approx \frac{v_o(nT) - v_o(n-1)T}{T} \tag{3.9.3}$$

It can be seen that the smaller the value of T the more accurate the approximation. Making the substitutions into eqn (3.9.1) gives

$$\frac{T_1[v_o(n) - v_o(n-1)]}{T} + v_o(n) = v_i(n)$$

and rearranging gives

$$v_o(n) - \frac{T_1}{T + T_1} v_o(n-1) = \frac{T}{T + T_1} v_i(n)$$

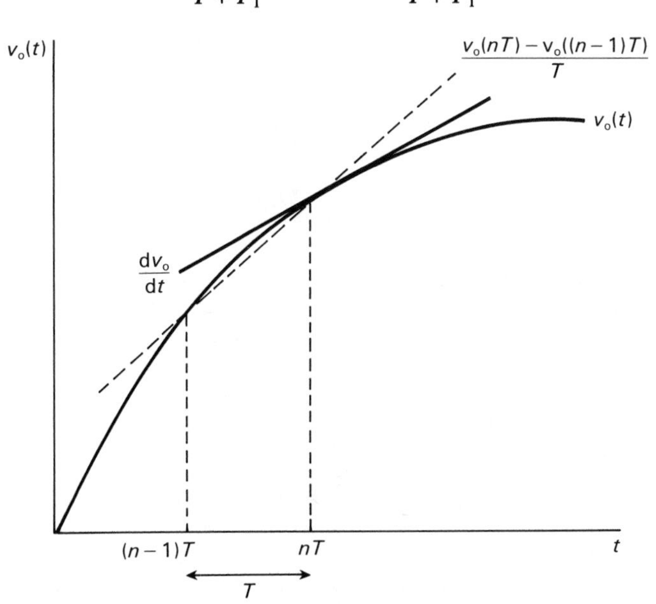

Figure 3.35 Discrete approximation to the slope of a signal.

If the substitutions are made $k_1 = T/(T+T_1)$ and $k_2 = T_1/(T+T_1)$ then this equation can be written

$$v_o(n) - k_2 v_o(n-1) = k_1 v_i(n) \tag{3.9.4}$$

It should be noted that this equation is identical to eqn (3.8.1), derived for the diode pump circuit with $v_i(n) = E$ for $n \geqslant 0$. Hence the flow chart shown in Figure 3.31 can be used to obtain a numerical solution to this problem. Table 3.2 shows the steps involved in calculating $v_o(n)$ for $n = 0$ to 5 for the case where $T_1 = 1$ and $T = 0.1$.

In this table, $v_o(n)$ has been calculated according to the relationship

$$v_o(n) = k_1 v_i(n) + k_2 v_o(n-1)$$

Each value of $v_o(n)$ calculated becomes the entry for $v_o(n-1)$ in the following row. In order to calculate the value $v_o(0)$ a value for $v_o(n-1)$ or $v_o(-1)$ has to be assumed. This represents the output voltage due to initial charge on the capacitor. This has been taken as zero.

To carry on with this calculation by hand would be extremely tedious (especially if the time between points, T, is made shorter). However it is very easy to write a program (even on a programmable pocket calculator) to perform the task. Table 3.3 was constructed in this manner and shows the response at different time

Table 3.2 Calculation of $V_o(n)$ for a first order discrete system

n	$v_i(n)$	$v_o(n-1)$	$v_o(n)$
0	10.0	0.0000	0.9091
1	10.0	0.9091	1.7355
2	10.0	1.7355	2.4869
3	10.0	2.4869	3.1699
4	10.0	3.1699	3.7908
5	10.0	3.7908	4.3552

Table 3.3 Effect of sampling time on $v_o(nT)$

t	*True response* $v_o(t)$	*Discrete approximation* $v_o(nT)$ $T = 0.1$	$T = 0.25$	$T = 0.01$
0.0	0.0000	0.9091	2.0000	0.0990
0.5	3.9347	4.3553	4.8800	3.9798
1.0	6.3212	6.4951	6.7232	6.3395
1.5	7.7687	7.8237	7.9028	7.7743
2.0	8.6465	8.6487	8.6578	8.6467
3.0	9.5021	9.4790	9.4502	9.4997
4.0	9.8168	9.7991	9.7748	9.8150
5.0	9.9326	9.9226	9.9078	9.9316

instants as well as the true response at these points. The calculations have been made for a range of sampling times T. The results for $T = 0.25$ and $T = 0.1$ are shown together with the true response in Figure 3.36 (the results for $T = 0.01$ would not be distinguishable from the true response on the scale used).

The greatest error occurs at $t = 0$. The value $v_i(n)$ has been taken as 10 V at this point. It could be argued that the value of the input step is discontinuous at $t = 0$ and taking v_i as zero would produce a better result at this point. However taking the sample $v_i(0)$ as 10 V fits into the definition of the discrete unit step and leads to solution by z transform methods as will be seen in Chapter 7.

This result appears very encouraging, however care has to be exercised. The equation being simulated is a first order differential equation. Higher differentials can be approximated by repeated use of eqn (3.9.3).

$$\frac{d^2y}{dt^2} \doteqdot \frac{\dfrac{y(n) - y(n-1)}{T} - \dfrac{y(n-1) - y(n-2)}{T}}{T}$$

$$= \frac{y(n) - 2y(n-1) + y(n-2)}{T^2}$$

However as the order of the differential equation increases the time interval T has to be decreased in order to obtain comparable accuracy. At first sight the only disadvantage is that more points are required in the simulation and the programme takes longer to run. However there is another difficulty. Because the numbers are represented in the computer by storage of finite length there is an error in the representation and each stage of the computation introduces further error. These errors are cumulative and if a large number of operations are required they can become serious. A method of decreasing this error is to use a more accurate algorithm to perform the differentiation, see Example 3.9.1.

Digital simulation

In the previous section the idea of replacing a differential equation by a difference equation has been investigated. Obtaining the difference equation requires a considerable amount of work, especially if the system is described by a high order equation.

Digital simulation languages have been developed that enable a description of the system to be entered directly. Suitable input signals can be selected and the corresponding system response displayed. The method by which the system description is entered into the computer depends upon the particular language used.

The system description can be entered as a set of differential equations. These are usually expressed as integral relationships and statements indicating integration are part of the simulation language. Suitable input signals and time scales can be selected and the corresponding system response obtained and displayed. High level language statements can be incorporated into the program to enable any required analysis of the system response.

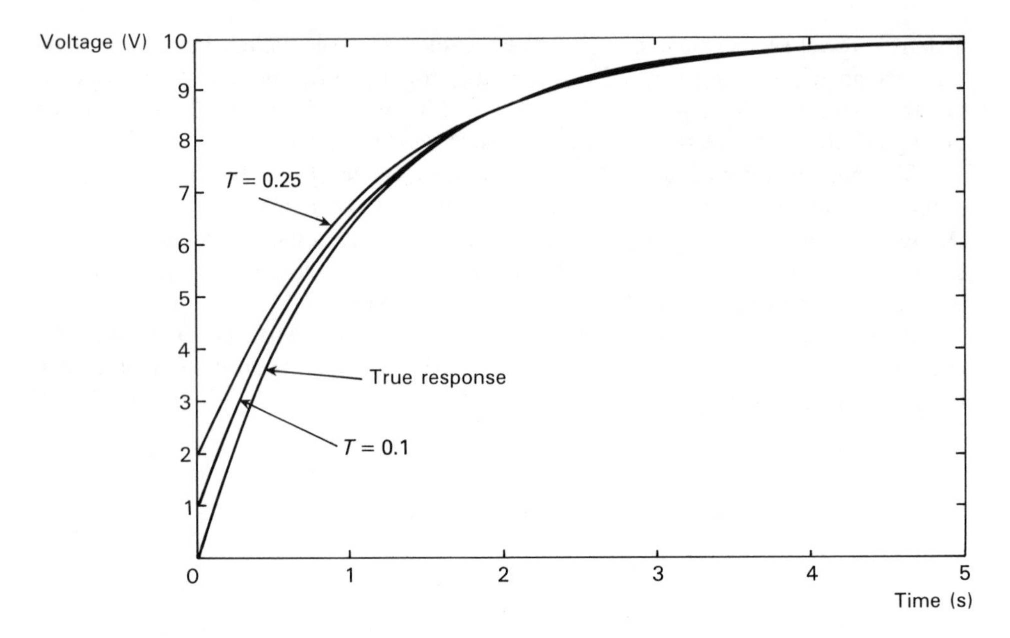

Figure 3.36 Step response of first order system, true response and discrete approximation.

Although the transfer function of a system is not covered until Chapter 7, it forms a very convenient description for simulation purposes. The differential equation of the system is replaced by an algebraic equation, relating the system output to its input, in terms of the system parameters and the Laplace operation s. The MATLAB Control Toolbox (some of which is included in the student edition of MATLAB) enables system parameters to be entered in this form. The system description can also take a state space form (see Section 3.10). As well as obtaining the system time response, this package can also obtain and plot the system's frequency response. The use of this package will also be required for the MATLAB exercises that are given at the end of most of the following chapters.

An alternative method of entering the system description is via a screen display of the system block diagram. Transfer functions can be entered into the blocks, and these can then be interconnected to form a complete system. One package using this method is SIMULINK, which is an extension of MATLAB. In this package, models can be entered and edited as screen displays. Any system responses obtained can be passed to MATLAB for analysis.

3.9.2 Implementation of the difference equation

As shown in Section 3.7 the differential equation representing a continuous system can be implemented by an interconnection of integrators and summing amplifiers. To represent a difference equation the operation of weighted summation is required together with a storage operation so that past values of the variables are available. This latter operation is usually represented by a chain of delay elements. These operations are shown symbolically in Figure 3.37.

As shown in Section 3.8.2 the general difference equation can be written in the form

$$y(n) = b_0 x(n) + b_1 x(n-1) + \cdots\cdots b_l x(n-l)$$
$$-a_1 y(n-1) - \cdots\cdots a_k y(n-k)$$

which can be re-written

$$y(n) = u(n) - a_1 y(n) - \cdots\cdots - a_k y(n-k) \tag{3.9.5}$$

where

$$u(n) = b_0 x(n) + b_1 x(n-1) + \cdots\cdots b_l x(n-l) \tag{3.9.6}$$

represents the non-recursive term. To form this term delayed (previous) values of the input $x(n)$ are required. These values can be combined by weighted addition to form the variable $u(n)$. This representation is shown in Figure 3.38.

To obtain the output $y(n)$ the signal $u(n)$ must be combined with delayed values of the output as described by eqn (3.9.5). This can be done as shown in Figure 3.39.

The student may feel uneasy about this representation, how can one delay a variable $y(n)$ when initially it does not exist? Initially, at $n=0$, what values are stored to represent the variables $y(n-1)$, $y(n-2) \ldots y(n-k)$? These values represent initial conditions and can only be obtained by consideration of the physical problem that is represented by the difference equation.

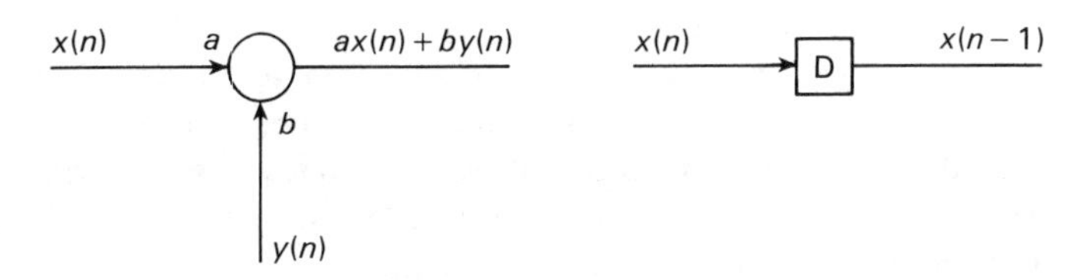

Figure 3.37 Representation of elements to perform weighted addition and signal delay.

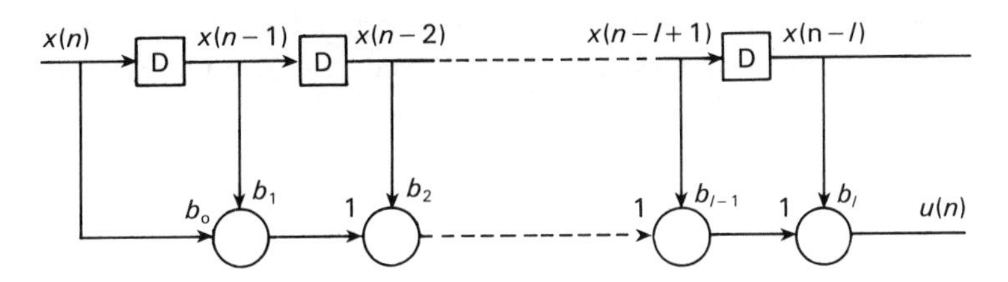

Figure 3.38 Representation of non-recursive term in a difference equation.

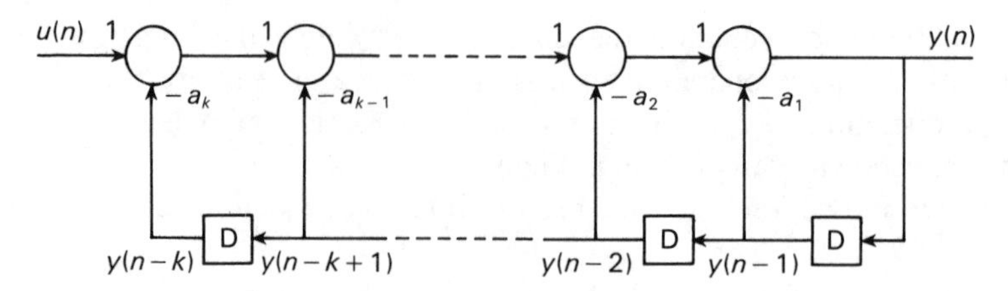

Figure 3.39 Representation of recursive term in a difference equation.

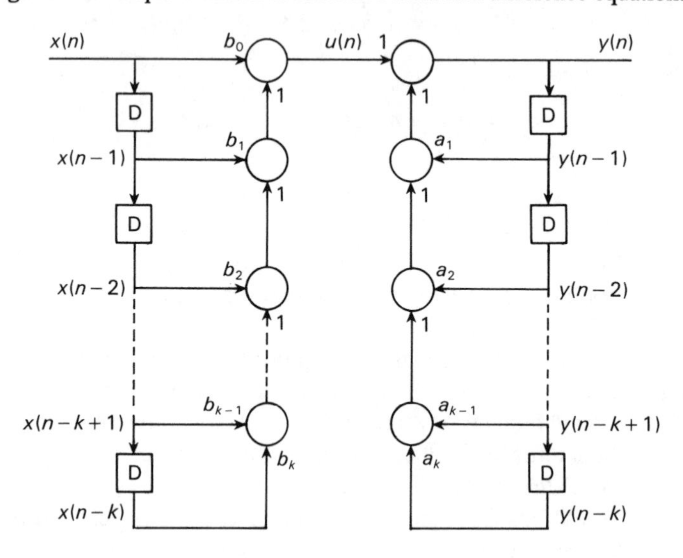

Figure 3.40 Direct realisation of the general difference equation.

The representations of Figures 3.38 and 3.39 can be combined into a single diagram, this is shown in Figure 3.40. Such a representation is known as a *realisation* and this particular method of representation is known as a *direct realisation*. As will be shown in Chapter 8, there are disadvantages with this form of realisation and alternative forms will then be discussed.

EXAMPLE 3.9.1

A more accurate algorithm to solve the first order differential equation can be obtained by re-writing the equation in terms of integration of the variables. A discrete approximation to integration can then be developed using the mid-ordinate rule (see Section 8.5.2). If

$$x = \int y \, dt$$

then this can be approximated by the relationship

$$x(n) = x(n-1) + \frac{T}{2}(y(n) + y(n-1))$$

By applying this relationship to the equation of the system considered in Section 3.9.1 obtain the step response of the system for a sampling time $T = 0.1$ s. Compare the result both with the true response and with the result obtained earlier.

Show how the resultant difference equation can be implemented by a direct realization.

SOLUTION

The equation to be solved is

$$T_1 \frac{dv_o}{dt} + v_o = v_i$$

this can be re-written as

$$v_o = \frac{1}{T_1} \int_{-\infty}^{t} (v_i - v_o)\, dt$$

Applying the discrete approximation gives

$$v_o(n) = v_o(n-1) + \frac{T}{2T_1} [v_i(n) - v_o(n) + v_i(n-1) - v_o(n-1)$$

which can be re-written as

$$v_o(n) = \left[\frac{T}{2T_1 + T}\right] v_i(n) + \left[\frac{T}{2T_1 + T}\right] v_i(n-1) + \left[\frac{2T_1 - T}{2T_1 + T}\right] v_o(n-1)$$

Inserting the values $T = 0.1$, $T_1 = 1$ gives

$$v_o(n) = 0.04762 v_i(n) + 0.04762 v_i(n-1) + 0.90476 v_o(n-1) \tag{3.9.7}$$

Table 3.4 Step response for discrete approximation of Example 3.9.1

nT	*True* $v_o(T)$	*Previous result* $v_o(nT)$	*Derived in Example 3.9.1* $v_o(nT)$
0.0	0.0000	0.9091	0.4762
0.5	3.9347	4.3553	4.2259
1.0	6.3212	6.4951	6.4993
1.5	7.7687	7.8237	7.8776
2.0	8.6465	8.6487	8.7132
3.0	9.5021	9.4790	9.5270
4.0	9.8168	9.7991	9.8261
5.0	9.9326	9.9226	9.9361

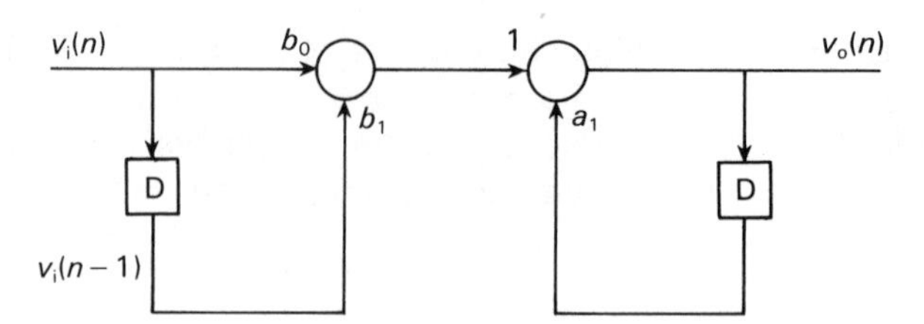

Figure 3.41 Realization of discrete approximation in Example 3.9.1.

Taking $v_i(n) = 10$ for $n \geq 0$ this equation can be easily programmed and solved recursively. The results are shown in Table 3.4 together with the time instants and the results obtained in Section 3.9.1.

Overall there is an improvement in accuracy (although this does not apply at all points) in particular at the point $n = 0$. The realization of eqn (3.9.7) by the direct method is shown in Figure 3.41.

3.10 State variable description

The system descriptions discussed so far have been based on the relationship between the input and output signals. An alternative description is based on the concept of *system state* and consists of relationships between *state variables*. A dictionary definition of state is 'a set of circumstances at any time' and the word is met in such phrases as 'state of the art', 'state of the union', etc. One would therefore expect the state variables to give a much more complete description of the system than the input/output variables.

Consider the two systems shown in Figure 3.42(a) and (b). Each is described by a differential equation relating $v_o(t)$ to $v_i(t)$. Suppose at a given time t_1 the output $v_o(t_1)$ is known – can the output $v_o(t)$ for $t > t_1$ be obtained knowing the input $v_i(t)$ for $t > t_1$? For simplicity take the input $v_i(t)$ to be zero for $t > t_1$.

To obtain the required output the differential equation relating $v_o(t)$ to $v_i(t)$ must be solved with $v_i(t) = 0$. It will be shown in the next chapter that solution of an

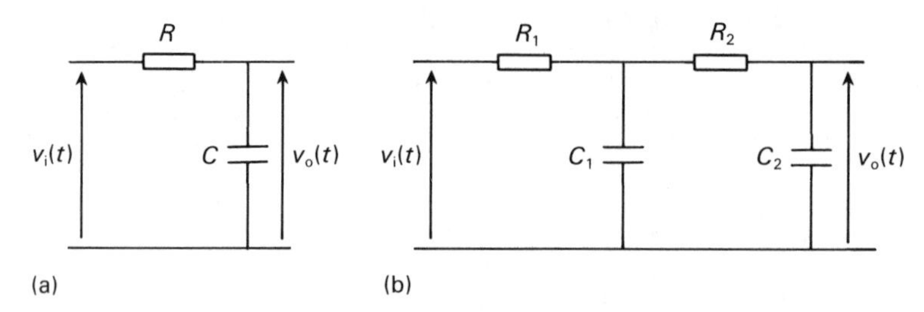

Figure 3.42 Illustration of choice of state variables.

nth order differential equation requires n initial conditions. For the first order system shown in Figure 3.42(a) one initial condition is present $v_o(t)$ and its output can be obtained. However the system in Figure 3.42(b) is a second order system, two initial conditions are required but only one is available, its output cannot be obtained. However if the voltage across C_1 were known at time t_1 this would provide the other initial condition. The voltages across both capacitors are the state variables and the state of the system at any time is the values of the state variables at that time. In general n state variables are required to describe an nth order system. The state variables are *not unique* however any set must be linearly independent, i.e., one variable must not be expressible as a linear combination of the others. (In Figure 3.42(b) one could not choose the voltage across C_2 and the charge on C_2, as state variables, they are related by the expression $q = C_2 v$.)

The following section describes how state variables are related to give a state variable description of a system.

3.10.1 *Derivation of the state equations*

Consider the circuit shown in Figure 3.43. The system has three energy storage elements therefore three state variables are required for its description. These have been chosen as x_1, the voltage across C_1, x_2, the voltage across C_2 and x_3 the current through L. (Conventionally the symbols $x_1, x_2, \ldots, x_n$ are used to denote state variables). The choice of these variables is not unique and other variables could have been chosen. The full reasons for the choice of the state variables are beyond the scope of this book. However as state variables are associated with energy storage it seems reasonable to choose voltages across capacitors (energy $= CV^2/2$) and current through inductors (energy $= LI^2/2$).

The circuit is now analysed as if an input/output relationship were required. However there are two inputs v_1, and v_2, and for the moment, the state variables are regarded as outputs.

Applying nodal analysis at node 1

$$\frac{v_1 - x_1}{R_1} = C_1 \dot{x}_1 + x_3 \tag{3.10.1}$$

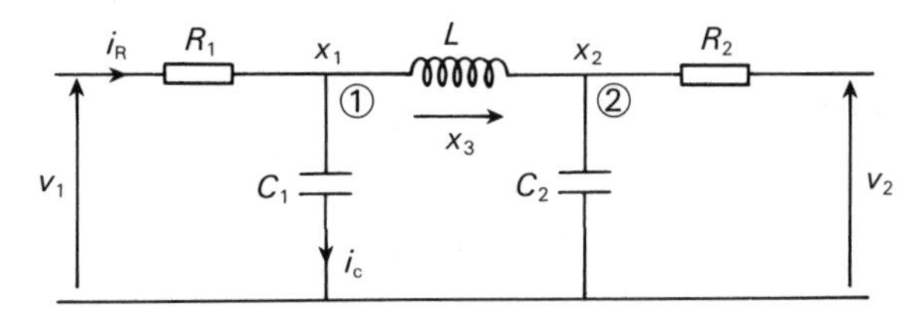

Figure 3.43 Circuit analysed by state variable method.

Applying nodal analysis at node 2

$$\frac{v_2 - x_2}{R_2} = C_2\dot{x}_2 - x_3 \tag{3.10.2}$$

For the inductor

$$x_1 - x_2 = L\dot{x}_3 \tag{3.10.3}$$

These equations are simultaneous differential equations. If an input/output relationship were required, say between x_1 and the two inputs, the variables x_2 and x_3 would have to be eliminated between the equations. However the relationship between x_1, x_2 and x_3 and the inputs v_1, v_2, eqns (3.10.1), (3.10.2) and (3.10.3), are the state equations and these equations give a description of the system. They are, however, usually written in a conventional form with the differential terms, having unity coefficients, on the left-hand side.

$$\begin{aligned}
\dot{x}_1 &= -\frac{1}{R_1C_1}x_1 \qquad\qquad -\frac{1}{C_1}x_3 + \frac{1}{R_1C_1}v_1 \\
\dot{x}_2 &= \qquad -\frac{1}{R_2C_2}x_2 + \frac{1}{C_2}x_3 \qquad + \frac{1}{R_2C_2}v_2 \\
\dot{x}_3 &= \frac{1}{L}x_1 - \frac{1}{L}x_2
\end{aligned} \tag{3.10.4}$$

In this form the equations can be written much more conveniently in terms of matrices. If the student is not that conversant with properties of matrices all that is really required is the multiplication rule.

$$\begin{bmatrix} \dot{x}_1 \\ \dot{x}_2 \\ \dot{x}_3 \end{bmatrix} = \begin{bmatrix} \frac{-1}{R_1C_1} & 0 & \frac{-1}{C_1} \\ 0 & \frac{-1}{R_2C_2} & \frac{1}{C_2} \\ \frac{1}{L} & \frac{-1}{L} & 0 \end{bmatrix} \begin{bmatrix} x_1 \\ x_2 \\ x_3 \end{bmatrix} + \begin{bmatrix} \frac{1}{R_1C_1} & 0 & 0 \\ 0 & \frac{1}{R_2C_2} & 0 \\ 0 & 0 & 0 \end{bmatrix} \begin{bmatrix} v_1 \\ v_2 \\ 0 \end{bmatrix}$$

This is now a matrix equation and can be written more compactly

$$\dot{\boldsymbol{x}} = \mathbf{A}\boldsymbol{x} + \mathbf{B}\boldsymbol{u} \tag{3.10.5}$$

A single column matrix is a vector and

$$\boldsymbol{x} = \begin{bmatrix} x_1 \\ x_2 \\ x_3 \end{bmatrix}$$

is the *state vector*.

$\dot{\boldsymbol{x}}$ is the derivative of the state vector with respect to time, it is formed by the differentiation of all the elements in the state vector.

$$\dot{\boldsymbol{x}} = \begin{bmatrix} \dot{x}_1 \\ \dot{x}_2 \\ \dot{x}_3 \end{bmatrix}$$

It should be noted that both x and $\dot{x}$ are functions of time and should strictly be written as $x(t)$ and $\dot{x}(t)$.

A is the coefficient or system matrix, **B** is the input matrix, and $\boldsymbol{u}$ is a vector of inputs.

$$\mathbf{A} = \begin{bmatrix} \frac{-1}{R_1 C_1} & 0 & \frac{-1}{C_1} \\ 0 & \frac{-1}{R_2 C_2} & \frac{1}{C_2} \\ \frac{1}{L} & \frac{-1}{L} & 0 \end{bmatrix} \quad \mathbf{B} = \begin{bmatrix} \frac{1}{R_1 C_1} & 0 & 0 \\ 0 & \frac{1}{R_2 C_2} & 0 \\ 0 & 0 & 0 \end{bmatrix}$$

$$\boldsymbol{u} = \begin{bmatrix} v_1 \\ v_2 \\ 0 \end{bmatrix}$$

The differential equations forming eqn (3.10.4) are first order differential equations. An input/output relationship for this system would have produced a third order differential equation. The state variable approach produces three first order simultaneous differential equations – why should this be preferable? Before answering this question let us consider how the state variable description could give an input/output relationship.

If the variable (or variables) of interest in the system are the state variables then the state equations give the required relationship. Suppose however we are interested in some other variables, say the currents i_R and i_C as shown in Figure 3.43.

Then

$$i_R = \frac{v_i - x_i}{R_1}$$

$$i_C = i_R - x_3$$

$$= \frac{v_i - x_1}{R_1} - x_3$$

or in matrix form

$$\begin{bmatrix} i_R \\ i_C \end{bmatrix} = \begin{bmatrix} \frac{-1}{R_1} & 0 & 0 \\ \frac{-1}{R_1} & 0 & -1 \end{bmatrix} \begin{bmatrix} x_1 \\ x_2 \\ x_3 \end{bmatrix} + \begin{bmatrix} \frac{1}{R_1} & 0 \\ \frac{1}{R_1} & 0 \end{bmatrix} \begin{bmatrix} v_1 \\ v_2 \end{bmatrix}$$

$$y = \mathbf{C}x + \mathbf{D}u \tag{3.10.6}$$

where

$$y = \begin{bmatrix} i_R \\ i_C \end{bmatrix}; \qquad \mathbf{C} = \begin{bmatrix} \frac{-1}{R_1} & 0 & 0 \\ \frac{-1}{R_1} & 0 & -1 \end{bmatrix}; \qquad \mathbf{D} = \begin{bmatrix} \frac{1}{R_1} & 0 \\ \frac{1}{R_1} & 0 \end{bmatrix}$$

In eqn (3.10.6) **C** is termed the output matrix. The elements in matrix **D** represent any direct links between the input and output signals, in many systems the matrix **D** is a null matrix showing no such paths exist. Equations (3.10.5) and (3.10.6) give a description of of the system in state variable form.

One advantage of the state variable approach is shown by the previous example. The problem had two input variables and two output variables, it was an example of a multi-input, multi-output system. In practice, complex systems are usually of this type and the state variable approach lends itself readily to the analysis of such problems. It also can be extended to cover time-varying and non-linear systems.

Another advantage of the state variable approach is that it is amenable to computer methods of solution. Any complex problem usually requires computer aid for its solution and the matrix form of the state space model is particularly suited to such an approach.

From the last two paragraphs the students will have gathered that the state space approach is most useful for complex problems when the equations can be solved by computer methods. However it is precisely these factors that make it difficult to investigate the approach further in a book of this nature. There are enough difficulties ahead in illustrating the more 'classical' methods of approach on simple systems without having to deal with complex systems. Also the knowledge of matrix theory required would be far beyond that expected from the reader of this book.

Hence, except for a few mentions of the approach, the state space method will be developed no further in the following chapters.

3.11 Summary

This chapter has been primarily concerned with the representation of a system by a mathematical model. The model used for most of the chapter is a model relating the system output to the system input. This takes the form of a differential equation for continuous systems and a difference equation for discrete systems.

By taking examples from a range of physical systems it has been shown that very different continuous systems can be described by the same differential equation. This led to the concept of the analogue computer as a method of investigating complex systems. The operations in the analogue computer were performed by operational amplifiers used in a feedback configuration in conjunction with passive components.

In contrast it was shown that the difference equations describing discrete systems could be very easily programmed on general purpose digital computers. Hence the possibility of approximating a continuous system by a difference equation was investigated. This led to the concept of digital simulation.

Finally an alternative method of system description has been introduced. This is given by the relationship between internal or state variables, leading to a set of first order differential equations, the state equations.

Problems

3.1 In the following problem $x(t)$ and $x(n)$ denote system input signals and $y(t)$ and $y(n)$ denote the corresponding output signals. Each part of the problem gives a system description relating these signals. For each system determine whether or not it is:

(a) Linear.
(b) Time invariant.
(c) Instantaneous.
(d) Causal.
(e) Stable.

(i) $y(t) = 2x(t) + x(t-1)$

(ii) $y(n) = 2x(n) + [x(n)]^2$

(iii) $y(n) = x(n) + x(n+1)$

(iv) $y(t) = \int_{-2}^{+2} x(\tau)\, d\tau$

(v) $y(t) = x(t) \cos \omega t$

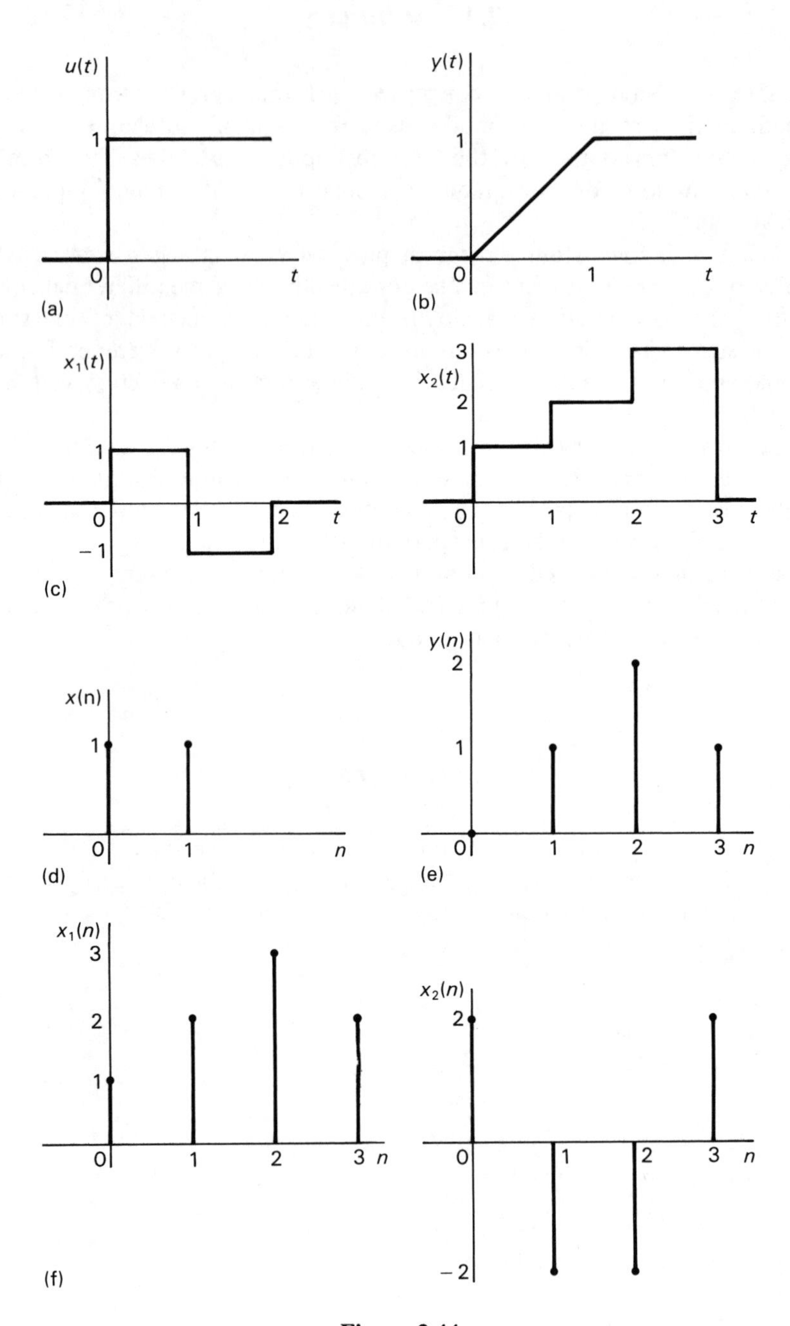
u(t)
1
0
t
(a)
y(t)
1
0
1
t
(b)
x1(t)
1
0
1
2
t
−1
(c)
3
x2(t)
2
1
0
1
2
3
t
x(n)
1
0
1
n
(d)
y(n)
2
1
0
1
2
3
n
(e)
x1(n)
3
2
1
0
1
2
3
n
(f)
x2(n)
2
0
1
2
3
n
−2

Figure 3.44

3.2 Comment on the properties of linearity and time invariance in relation to the following systems:

(a) An amplifier that limits such that its input/output relationship is described by the equation

$$\begin{aligned} v_o &= Kv_i & -V \leqslant v_i \leqslant +V \\ v_o &= KV & v_i > +V \\ v_o &= -KV & v_i < -V \end{aligned}$$

(b) A space vehicle on take-off has position x that is related to the motor thrust F by the equation

$$F = M\frac{d^2x}{dt^2}$$

M is the total mass of the vehicle, and contributing to this is the mass of the fuel which is burnt. The mass therefore decreases with time.

(c) In the space vehicle of (b) a transducer relates external air pressure to a voltage v according to the relationship

$$v = kp^2$$

The value of k is however governed by the time into flight and is switched in value at a time 20 s after launch.

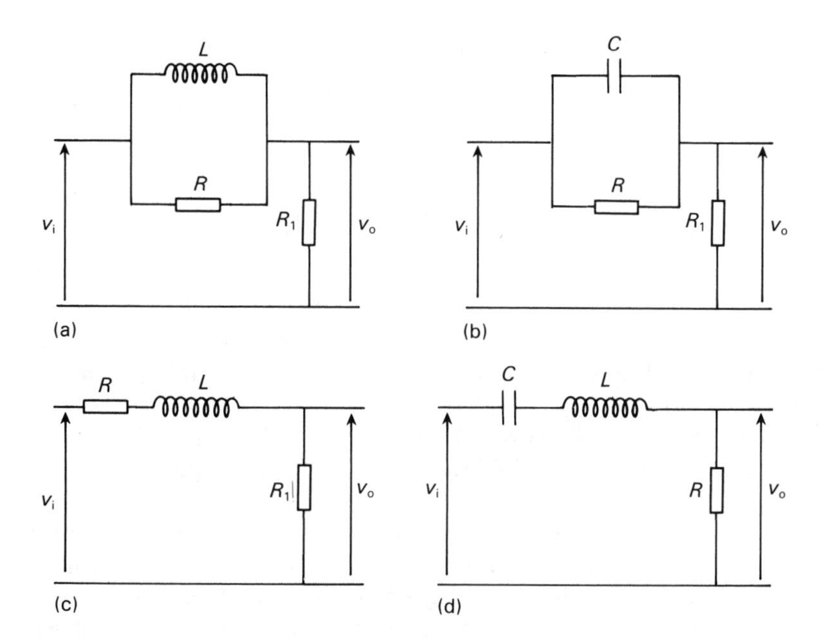

Figure 3.45

3.3 (a) The signal shown in Figure 3.44(a), a unit step, is applied as the input to a linear time invariant system. It produces an output $y(t)$ as shown in Figure 3.44(b).

Determine the system outputs if the signals $x_1(t)$ and $x_2(t)$ as shown in Figure 3.44(c) are applied in turn to the same system.

(b) The discrete signal $x(n)$ shown in Figure 3.44(d) is applied to a linear time invariant system and it produces a response as shown in Figure 3.44(e).

Determine the system output if the signals $x_1(n)$ and $x_2(n)$ as shown in Figure 3.44(f) are applied in turn to the same system.

3.4 Determine the differential equation relating v_o to v_i for the electrical systems shown in Figure 3.45.

3.5 The circuit shown in Figure 3.46 is that of an oscilloscope probe. R_i and C_i represent the input resistance and capacitance of the Y amplifier. R and C represent the resistance and capacitance of the probe (C is adjustable).

Obtain the differential equation relating v_o to v_i. If the probe capacitance C is adjusted such that $CR = C_iR_i$ show that a solution of this differential equation is $v_o = R_iv_i/(R_i + R)$. What is the practical implication of this result?

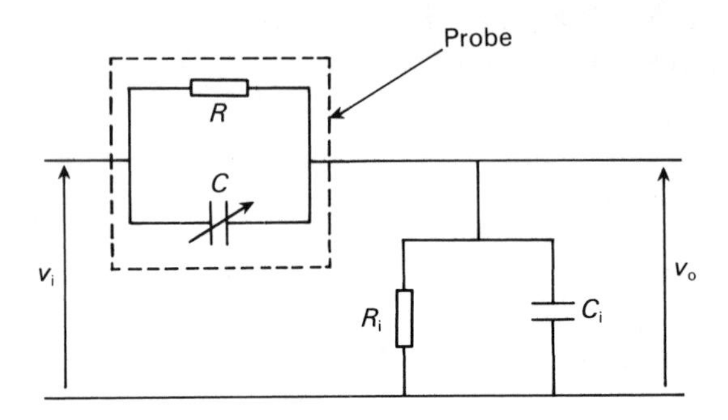

Figure 3.46

3.6 Figure 3.47 shows three mechanical systems. Derive the differential equation relating applied force $f(t)$ to displacement $x(t)$ for the system shown in Figure 3.47(a).

For the systems shown in Figures 3.47(b) and 3.47(c) derive the relationship between input displacement $x_i(t)$ and the displacement $x(t)$.

3.7 Figure 3.48 shows a simplified version of an accelerometer. A mass M is suspended by two springs (stiffness each $k/2$) within the body of the accelerometer. Damping is provided giving a damping force B per unit velocity (mass relative to body). The accelerometer output z is the difference between mass displacement and body displacement.

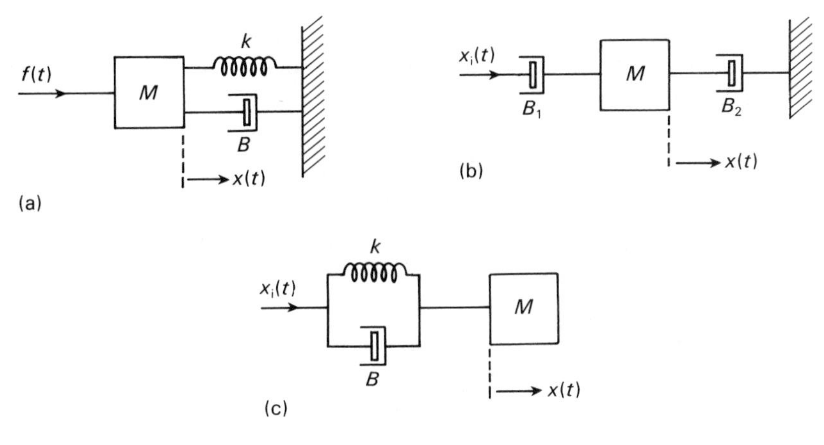

Figure 3.47

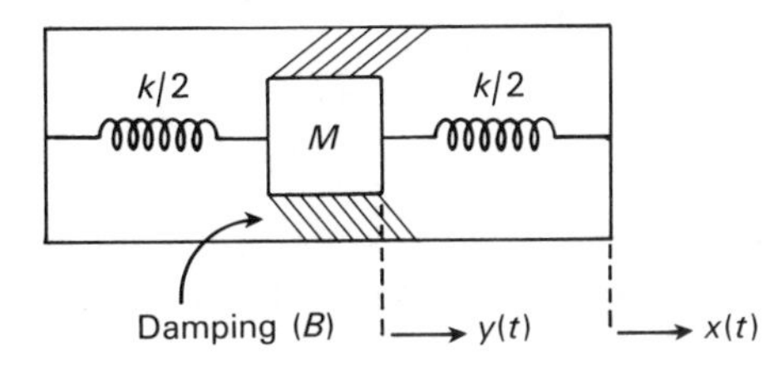

Figure 3.48

$$z = y - x$$

Derive the differential equation relating accelerometer output z to body acceleration $\mathrm{d}^2x/\mathrm{d}t^2$

3.8 Elementary thermal systems consist of elements that provide heat storage (thermal capacitance) and elements that provide a temperature drop due to heat flow (thermal resistance).

$$\text{Heat stored} = \text{thermal capacity} \times \text{temperature}$$

where the heat is measured in joules and the thermal capacity is in joules/°C.

$$\text{Heat flow through a thermal resistance} = \frac{\text{temperature difference}}{\text{thermal resistance}}$$

where heat flow has the units joules/second (watt) and the thermal resistance has the units °C/W. The following problem presents a simplified model of the heating of a voice coil in a loudspeaker.

Loudspeakers have a very low efficiency and the electrical power in the voice coil is almost entirely converted to heat. This heat is either stored or is lost to the surroundings. If this loss is considered to be by conduction through a thermal resistance R_t and the thermal capacitance is C_t, show that the differential equation relating the electrical power P in the coil, to the coil temperature θ (measured as the amount above ambient) is given by

$$P = C_t \frac{d\theta}{dt} + \frac{\theta}{R_t}$$

Verify that a solution of this equation to a step in power P_1 is given by

$$\theta = P_1 R_t (1 - e^{-t/C_t R_t})$$

For a given speaker $C_t = 2.5$ J/°C, $R_t = 3$°C/W. Calculate the time a step of 200 W in power can be applied before the speaker coil exceeds its maximum temperature rating of 300°C.

3.9 This problem is concerned with fluid flow systems and relates the storage and flow of fluids. Fluid flow through a valve can be approximated by the relationship

$$\text{Flow rate} = \frac{\text{pressure difference}}{\text{fluid resistance}}$$

The fluid systems are shown in Figure 3.49(a) and 3.49(b). In the single tank system the rate of flow out, q_o, is taken to be proportional to the liquid level h

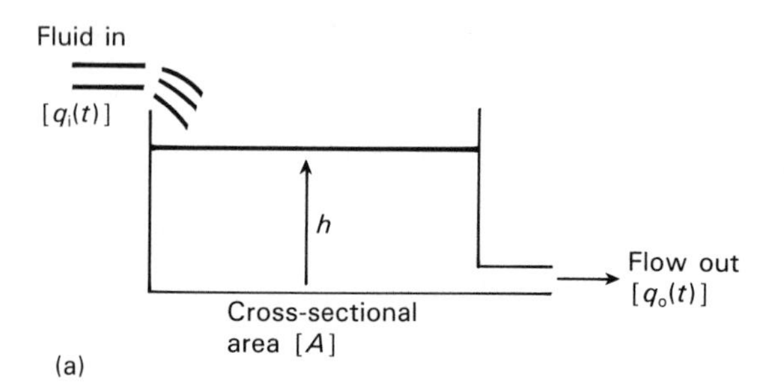

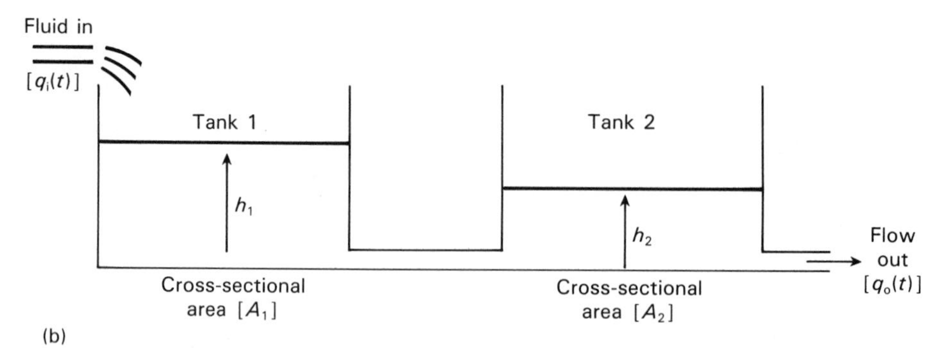

Figure 3.49

($q_o = k_1 h$). Derive a differential equation relating h to the rate of fluid flow in q_i.

In the coupled tanks of Figure 3.49(b) the rate of flow from tank 1 to tank 2 is given by

$$q = k_2(h_1 - h_2)$$

Derive the differential equation relating h_2 to q_i.

3.10 Figures 3.50(a) and 3.50(b) show two systems using operational amplifiers. Use the virtual earth principle to obtain the differential equation relating $v_o(t)$ to $v_i(t)$ in each case.

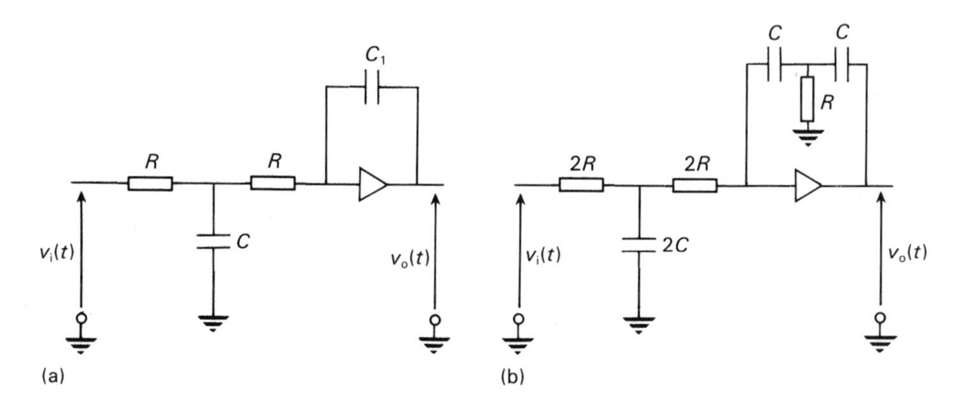

Figure 3.50

3.11 Figure 3.51 shows a large industrial vibrator. The constants of the system are as follows:

R = resistance of coil = 100 Ω
L = inductance of coil = 10 H
e = back e.m.f. in coil

$$e = K_v \frac{dx}{dt}; \qquad K_v = 15 \text{ V/ms}^{-1}$$

f = mechanical force produced

$$f = K_T i; \qquad K_T = 15 \text{ N/A}$$

M = mass of table plus load = 1000 kg
K = stiffness of table suspension = 150 000 N/m
B = damper constant = 15 000 N/ms^{-1}

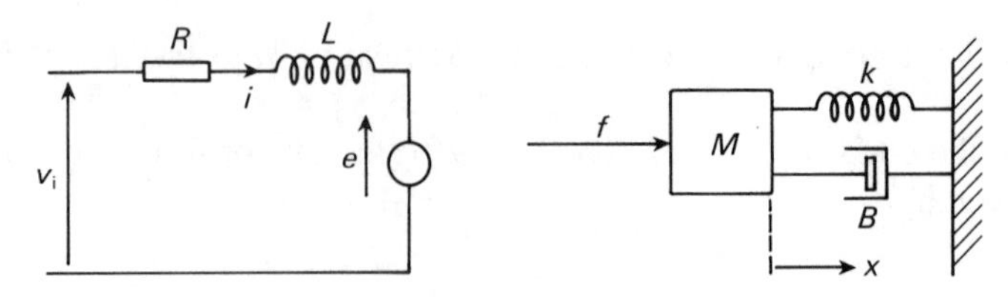

Figure 3.51

Show that the system is described by the following differential equations.

$$v_i = iR + L\frac{di}{dt} + K_v\frac{dx}{dt}$$

$$K_T i = M\frac{d^2x}{dt^2} + B\frac{dx}{dt} + kx$$

3.12 In Section 3.8.1 the difference equation describing the action of a diode pump circuit was obtained. Figure 3.52 shows the diode pump circuit modified by the addition of the resistor R across its output and this resistor causes charge to 'leak' from C_2 over the time T.

Assuming that $C_2R \gg T$ such that the fall in v_o over time T can be considered linear, derive the new difference equation describing the system. Making the assumption that $C_2 \gg C_1$ show that in the steady state the *mean value* of the output voltage is approximately proportional to the frequency of the supply.

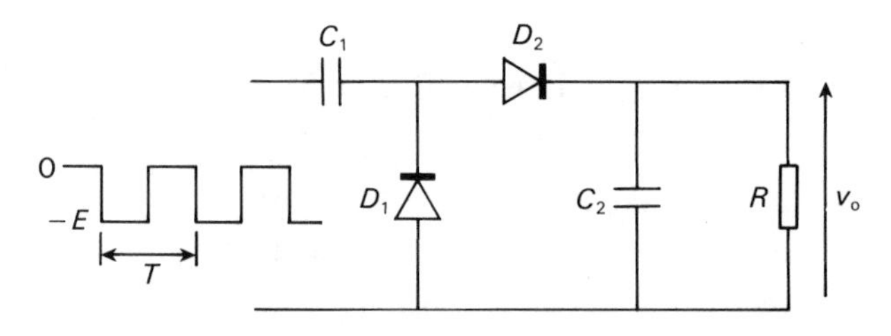

Figure 3.52

3.13 This problem involves difference equations arising from financial systems.

1. Compound interest
A person saves regularly by depositing a fixed amount annually into a savings account. Interest is calculated annually on the money in the account just prior to deposit being made. If the amount deposited annually is £A and the yearly interest rate is r% obtain a difference equation relating the total in the account (just after interest is added but before the next deposit is

made) in the nth year, $P(n)$, to that at the corresponding time in the previous year $P(n-1)$.

Verify that a solution of this equation is

$$P(n) = C(k)^n + \frac{kA}{1-k}$$

where $k = (1 + r/100)$ and C is an arbitrary constant.

Evaluate this constant if at year zero, $n = 0$, there is no money in the account.

If $A = £1000$, how many years would it take for the account to reach £15 000 with an interest rate $r = 10\%$?

2. An annuity

   To produce an annuity, a fixed sum is deposited at a financial institution. Interest is paid on the net sum held but interest and part of the initial sum is paid out at a fixed amount per year over a given number of years.

   Assuming that interest is paid just prior to the yearly pay out, obtain a difference equation relating the money in the account just after pay out in the nth year, $P(n)$, to that at the corresponding time in the previous year, $P(n-1)$. Let the initial sum be I, the interest rate r and the annual pay out £A.

   Verify that the equation is satisfied by the relationship

$$P(n) = Ck^n - \frac{A}{1-k}$$

   where C is an arbitrary constant and $k = (1 + r/100)$.

   Using the fact that at $n = 0$, $P(n) = I$, obtain the constant C. If $I = £10\ 000$ and $r = 10\%$, calculate the amount that that would be paid out yearly if equal payments were to be made over ten years.

3.14 A system is described by the differential equation

$$\frac{d^2y}{dt^2} + 2.4\frac{dy}{dt} + 4y = 4x$$

The response of this system to a unit step input is given by

$$y(t) = 1 - 1.25e^{-1.2t}\sin(1.6t + 0.9272)$$

Using the discrete approximations to the first and second differentials given in Section 3.9.1, approximate the differential equation of the system by a difference equation in terms of the sampling time T. For values of T equal to 0.5, 0.1, 0.05 obtain numerically the solution of the difference equation for a unit step input. Compare these responses with the true response given.

Note: These solutions need the use of a computer or a programmable calculator.

3.15 Give the difference equations that describe the two discrete realisations shown in Figure 3.53.

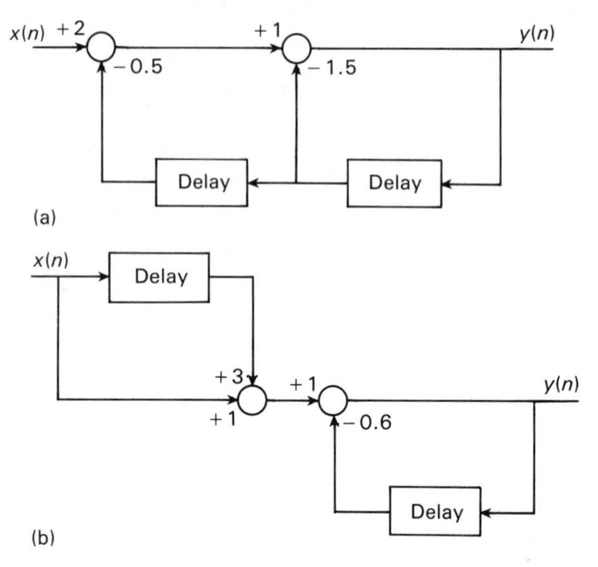

Figure 3.53

3.16 Obtain direct realisation representations of the following difference equations

$$32y(n) - 16y(n-1) - 10y(n-2) + 3y(n-3) = 8x(n)$$
$$4y(n) - y(n-2) = 2x(n) + 3x(n-1)$$

3.17 Figure 3.54 shows an armature controlled d.c. motor that is part of a position control system. The torque of the motor is related to the armature current by the constant K_T and the back e.m.f. e_g is related to the speed by the constant K_v. Using the state variables θ, ω, and i_a derive a state variable description of the system.

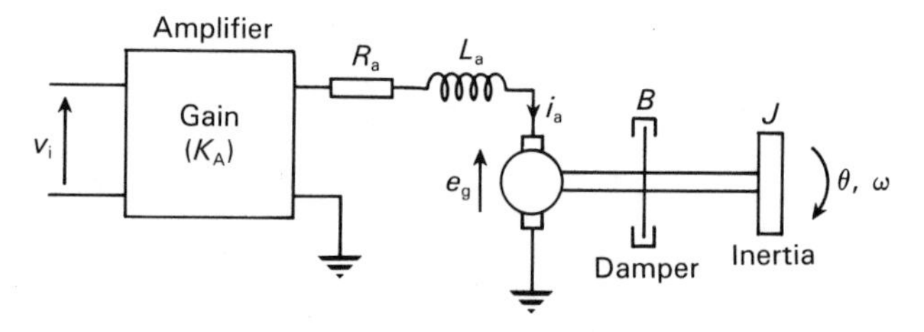

Figure 3.54

3.18 Derive the state equations for the circuit shown in Figure 3.55 and hence obtain the matrices **A** and **B** in the equation

$$\dot{x} = \mathbf{A}x + \mathbf{B}u$$

$$x = \begin{bmatrix} x_1 \\ x_2 \\ x_3 \\ x_4 \end{bmatrix} \qquad u = \begin{bmatrix} v_1 \\ 0 \\ 0 \\ v_2 \end{bmatrix}$$

The state variable x_1 is the current through the inductor as shown and the variables x_2, x_3 and x_4 are voltages measured with respect to the common capacitor point.

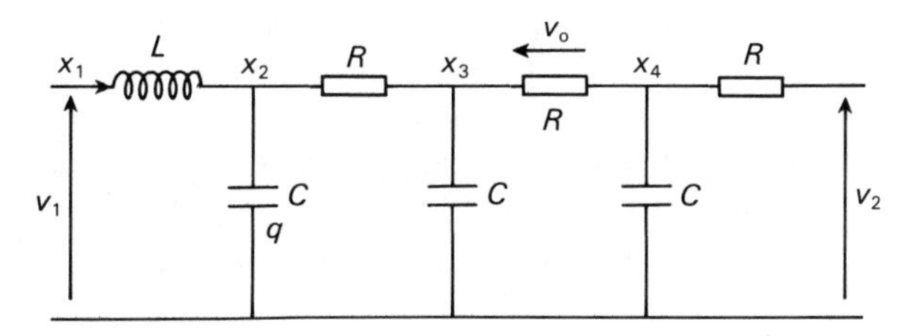

Figure 3.55

If the output quantities are the voltage v_o and the charge q as shown, obtain the output matrix **C** in the output equation

$$\begin{pmatrix} v_0 \\ q \end{pmatrix} = \mathbf{C}x$$

4

The System Response

4.1 Introduction

It has been shown in the previous chapter that a system can be described by an equation relating the input and output signals. If the input signal is known, solution of this equation will give the system response. Hence the system response can be obtained by the solution of a differential or difference equation and this can be accomplished by well established mathematical techniques. If this is the case why is there a need for a complete chapter on the system response?

Unfortunately the solution of these equations is a tedious affair even when transform methods are used (see Chapter 7). More important, from an engineering viewpoint, the problem is not usually one of obtaining the response of a given system to a specific input. Often the problem is a design problem, i.e., how can a system be modified to obtain a desired response to a given input. Unfortunately the system description consisting of a differential or difference equation does not lend itself easily to such an interpretation. There is no easy relationship between the coefficients in the equation and the form of response. To obtain a given response would require many solutions of the equation, each using different coefficients.

Hence before considering the general solution of the system equation this chapter considers an alternative method of obtaining the system response. The method relies on the properties of linearity and time invariance (LTI system). In Chapter 2 it was shown that it was possible to express a signal as a combination of scaled, time shifted basic signals. If the system response to the basic signal is known, then expressing the system input as a combination of basic signals allows the output to be expressed as a combination of basic signal responses.

It is possible to build on this concept using a variety of basic signals, however the one that is most fundamental is the unit impulse.

4.2 Use of the impulse as a basic signal

The unit impulse, and its discrete counterpart, the unit sample, have been introduced in Section 2.5.4. It will now be shown how any arbitrary signal can be represented in terms of these functions. This representation is straightforward for discrete signals but is somewhat more involved in the case of a continuous signal. Hence it is the discrete case that will be considered first.

4.2.1 Discrete signal representation

Consider the discrete signal shown in Figure 4.1. As the signal is discrete it is easy to see that it is a combination of scaled, time shifted unit samples. Figure 4.2 shows the unit sample and shows how the signal $x(n)$ can be found from it (revise Section 2.3.3 if you are unsure of the time shift operation).

The signal $x(n)$ is now obtained by adding the component signals in Figure 4.2.

$$x(n) = -2\delta(n+1) + 0\delta(n) + 2\delta(n-1) + 4\delta(n-2)$$

This is a representation of the specific signal shown in Figure 4.1. In order to obtain a more general representation that will apply to any discrete signal the expression for $x(n)$ is written

$$x(n) = x(-1)\delta(n+1) + x(0)\delta(n) + x(1)\delta(n-1) + x(2)\delta(n-2)$$

This can be written more compactly using a summation form

$$x(n) = \sum_{k=-1}^{+2} x(k)\delta(n-k)$$

and the general formula applying to any signal becomes

$$x(n) = \sum_{k=-\infty}^{+\infty} x(k)\delta(n-k) \tag{4.2.1}$$

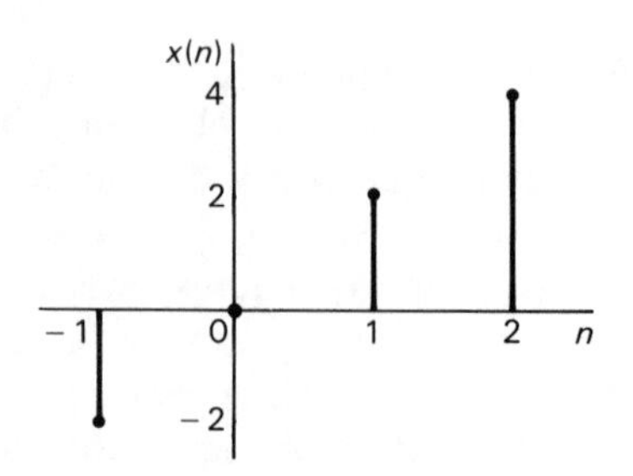

Figure 4.1 An illustrative discrete signal.

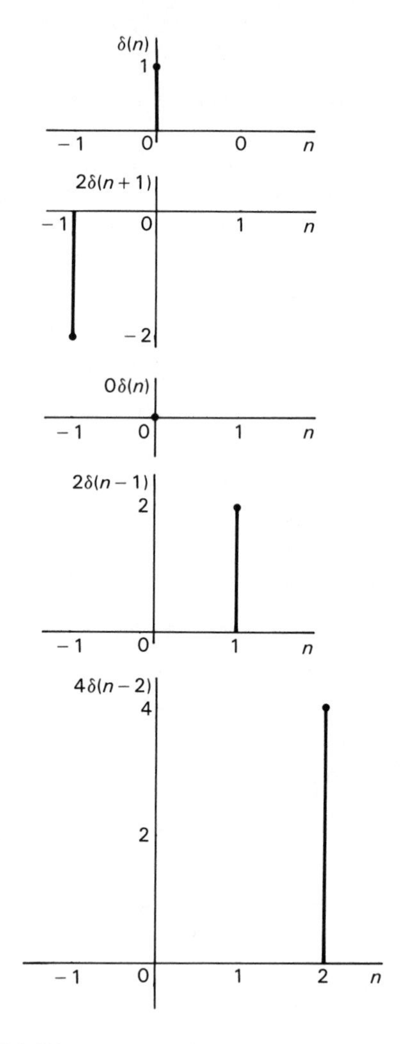

Figure 4.2 Discrete signal shown as a combination of scaled samples.

It should be noted that k in this formula could be replaced by any symbol, it is a dummy variable.

4.2.2 Representation of a continuous signal

Because this representation requires a limiting operation it is slightly more difficult than the discrete case but the argument follows very similar steps. The student at this point is advised to revise the properties of the impulse function as presented in Section 2.5.4.

Consider the signal $x(t)$ as shown in Figure 4.3(a). This signal can be approximated by the sum of a series of rectangles each of height $x(t)$ and width Δt. In turn these rectangles can be approximated by a series of impulses, each of strength $x(t)\Delta t$ as shown in Figure 4.3(b). One of these impulses, the one at time t_k can be represented

$$x(t_k)\delta(t-t_k)\Delta t_k$$

The complete signal $x(t)$ can be approximated

$$x(t) \approx \sum_{k=-\infty}^{+\infty} x(t_k)\delta(t-t_k)\Delta t_k$$

As $\Delta t \rightarrow 0$ this must become an accurate representation of the signal $x(t)$. This implies that in the limit, t_k becomes a continuous variable, this is a dummy variable and will be represented by the symbol τ. The representation of $x(t)$ becomes

$$x(t) \approx \int_{-\infty}^{+\infty} x(\tau)\delta(t-\tau)\mathrm{d}\tau \qquad (4.2.2)$$

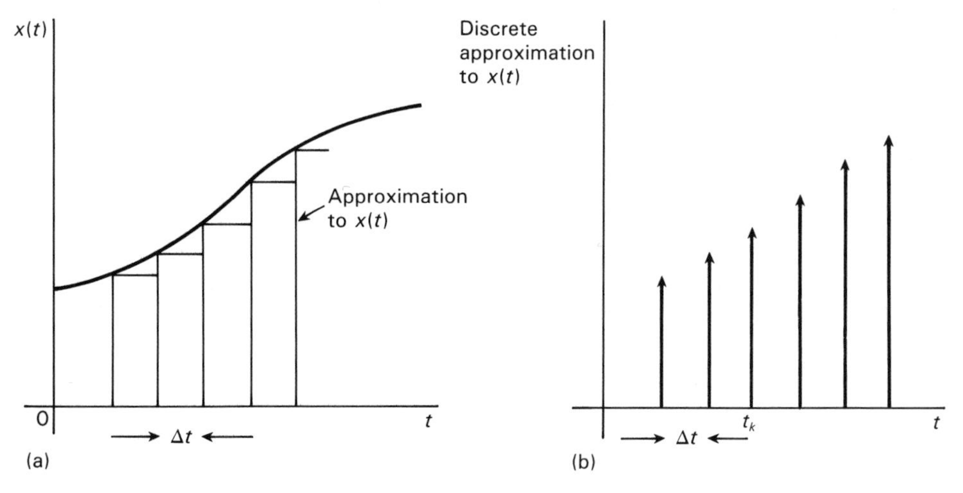

Figure 4.3 An illustrative continuous signal: (a) approximated by a combination of finite width pulses; (b) approximated by a combination of impulses.

Equations (4.2.1) and (4.2.2) express what is known as the 'sifting property' of the unit impulse (unit sample) function. The impulse acts as a sieve, selecting the value of the function at a particular instant, the integral (or the summation) reassembles the values to the original function.

The importance of eqns (4.2.1) and (4.2.2) is that an arbitrary input signal has been represented as a linear combination of the unit impulse (unit sample) function. If the response of the system to this signal, the impulse response, is known, then the response to $x(t)$ can be obtained. This process is the subject of the next section.

4.3 Convolution

4.3.1 The convolution summation formula and the convolution integral formula

Consider the signal investigated in Section 4.2.1 and suppose this is applied to a LTI system whose response, $h(n)$, to a unit sample is as given in Figure 4.4.

The response to this input will be the sum of a number of scaled, time shifted sample responses as shown in Figure 4.5.

Hence knowing the unit sample response $h(n)$ it is possible to obtain the output for an arbitrary input. This particular example can be extended to obtain an

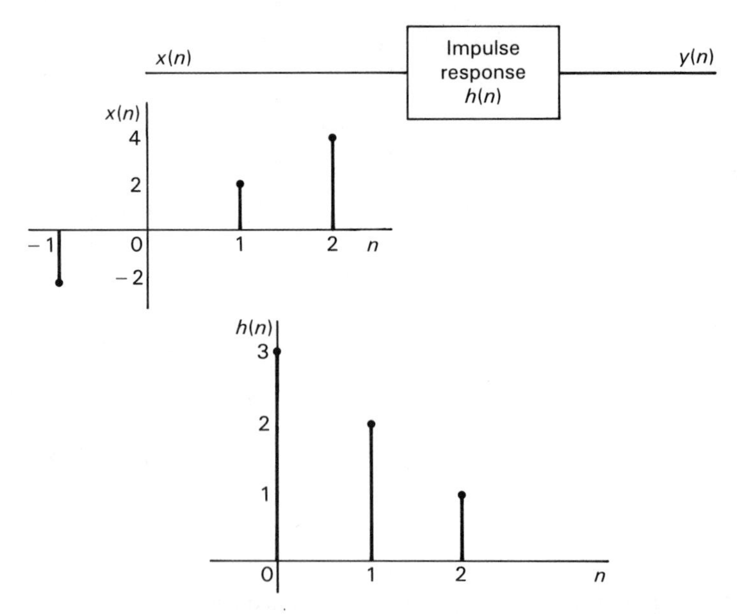

Figure 4.4 Response of a discrete system to an arbitrary input.

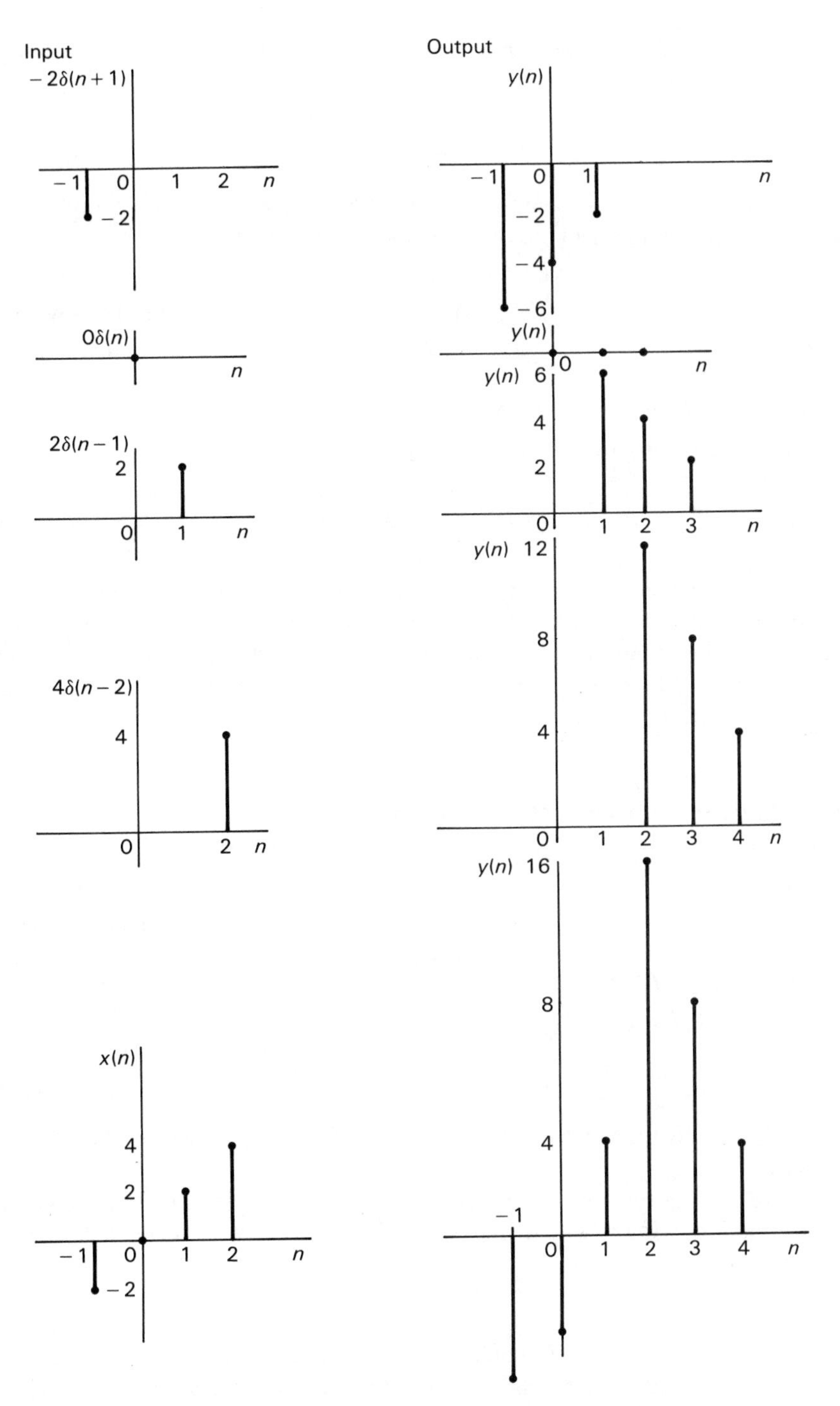

Figure 4.5 System response as a sum of impulse responses.

expression for the general case. Recalling the sifting property of the unit sample as given by eqn (4.2.1)

$$x(n) = \sum_{k=-\infty}^{+\infty} x(k)\delta(n-k)$$

and assuming a signal $x(n)$ is applied as input to a system with unit sample response $h(n)$, then listing components of the input and output signals

Input	*Output*	
$\delta(n)$	$h(n)$	Sample response
$\delta(n-k)$	$h(n-k)$	Time invariance
$x(k)\delta(n-k)$	$x(k)h(n-k)$	Linearity
$x(n) = \Sigma\, x(k)\delta(n-k)$	$y(n) = \Sigma\, x(k)h(n-k)$	Linearity

Hence the system output to the input $x(n)$ is given by

$$y(n) = \sum_{k=-\infty}^{+\infty} x(k)h(n-k) \tag{4.3.1}$$

The expression on the right-hand side of eqn (4.3.1) is known as the *convolution summation* and it is of extreme importance. Before proceeding with its interpretation the corresponding continuous case will be derived. In this case the system impulse response is $h(t)$ and the signal $x(t)$ can be expressed in terms of the impulse function by use of eqn (4.2.2)

$$x(t) = \int_{-\infty}^{+\infty} x(\tau)\delta(t-\tau)\,\mathrm{d}\tau$$

Again listing the components of input and output signals

Input	*Output*	
$\delta(t)$	$h(t)$	Impulse response
$\delta(t-\tau)$	$h(t-\tau)$	Time invariance
$x(\tau)\delta(t-\tau)$	$x(\tau)h(t-\tau)$	Linearity
$x(t) = \int x(\tau)\delta(t-\tau)\,\mathrm{d}\tau$	$y(t) = \int x(\tau)h(t-\tau)\,\mathrm{d}\tau$	Linearity

Hence the system output to input $x(t)$ is

$$x(t) = \int_{-\infty}^{+\infty} x(\tau)h(t-\tau)\mathrm{d}\tau \tag{4.3.2}$$

The expression on the right-hand side of eqn (4.3.2) is known as the convolution integral.

Before considering a very useful graphical interpretation of the convolution formulae the following example shows how they can be used to calculate the system output for a given input.

EXAMPLE 4.3.1

Figure 4.6(a) shows the input signal $x(n)$ to a discrete system having a unit sample response $h(n)$. Figure 4.6(b) shows the corresponding signals $x(t)$ and $h(t)$ for a continuous system. For both systems obtain expressions for the resulting output signals.

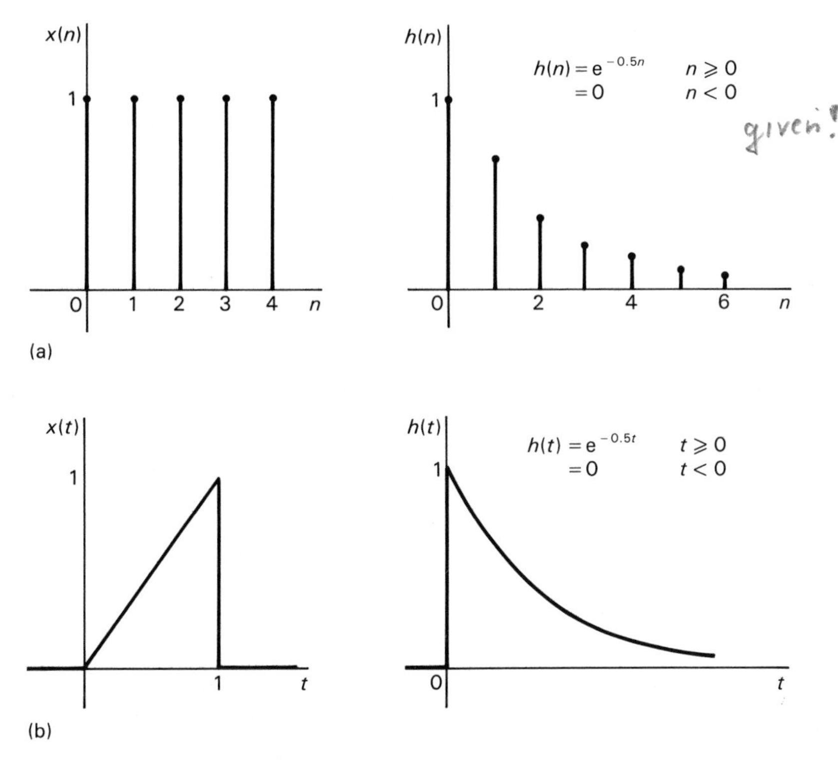

Figure 4.6 Signals applying to Example 4.3.1.

SOLUTION

(a) Discrete system. To use the convolution summation formula the input $x(n)$ has to be expressed mathematically.

$$\begin{aligned} x(n) &= 0 \qquad n<0 \\ x(n) &= 1 \qquad 0 \leqslant n \leqslant 4 \\ x(n) &= 0 \qquad n>4 \end{aligned}$$

The convolution summation formula (eqn 4.3.1) gives the output as

$$y(n) = \sum_{k=-\infty}^{+\infty} x(k)h(n-k)$$

As $x(k)$ is zero for $k<0$ and $k>4$ the limits on the summation can be modified accordingly and within this range $x(k) = 1$.

Care must be taken in interpreting the delayed sample response $h(n-k)$.

$$\begin{aligned} h(n-k) &= e^{-0.5(n-k)} \qquad k \leqslant n \\ &= 0 \qquad\qquad\quad k>n \end{aligned}$$

hence

$$\begin{aligned} y(n) &= \sum_{k=0}^{4} h(n-k) \\ &= \sum_{k=0}^{n} e^{-0.5(n-k)} \qquad n \leqslant 4 \\ &= \sum_{k=0}^{4} e^{-0.5(n-k)} \qquad n > 4 \end{aligned}$$

The student should convince himself/herself that these expressions are correct by writing out the summations in full for different values of n.

As the summation is with respect to k it can be written

$$\begin{aligned} \sum_{k} e^{-0.5(n-k)} &= \sum_{k} e^{-0.5n} e^{0.5k} \\ &= e^{-0.5n} \sum_{k} e^{0.5k} \end{aligned}$$

The expression under the summation sign represents a geometrical progression and a closed form can be obtained for its summation (be careful about n and the number of terms). This gives the following expressions for $y(n)$.

$$\begin{aligned} y(n) &= 0 && n<0 \\ y(n) &= 2.542(1-e^{-0.5(n+1)}) && 0 \leqslant n \leqslant 4 \\ y(n) &= 17.23e^{-0.5n} && n>4 \end{aligned}$$

(b) This part uses the convolution integral giving

$$y(t) = \int_{-\infty}^{+\infty} x(\tau)h(t-\tau)\,d\tau$$

The input $x(\tau)$ can be expressed mathematically

$$\begin{aligned} x(\tau) &= 0 && \tau<0 \\ x(\tau) &= \tau && 0 \leqslant \tau \leqslant 1 \\ x(\tau) &= 0 && \tau>1 \end{aligned}$$

and the limits on the integral can be restricted to the range 0 to 1.

Again care must be taken in interpreting the delayed impulse response.

$$\begin{aligned} h(t-\tau) &= e^{-0.5(t-\tau)} && \text{for } \tau \leqslant t \\ &= 0 && \text{for } \tau > t \end{aligned}$$

hence

$$y(t) = \int_0^1 \tau h(t-\tau)\,\mathrm{d}\tau$$

$$= \int_0^t \tau \mathrm{e}^{-0.5(t-\tau)}\,\mathrm{d}\tau \qquad t \leqslant 1$$

$$= \int_0^1 \tau \mathrm{e}^{-0.5(t-\tau)}\,\mathrm{d}\tau \qquad t > 1$$

This integral can be written

$$\int \tau \mathrm{e}^{-0.5(t-\tau)}\,\mathrm{d}\tau = \int \tau \mathrm{e}^{-0.5t}\mathrm{e}^{0.5\tau}\,\mathrm{d}\tau$$

$$= \mathrm{e}^{-0.5t}\int \tau \mathrm{e}^{0.5\tau}\,\mathrm{d}\tau$$

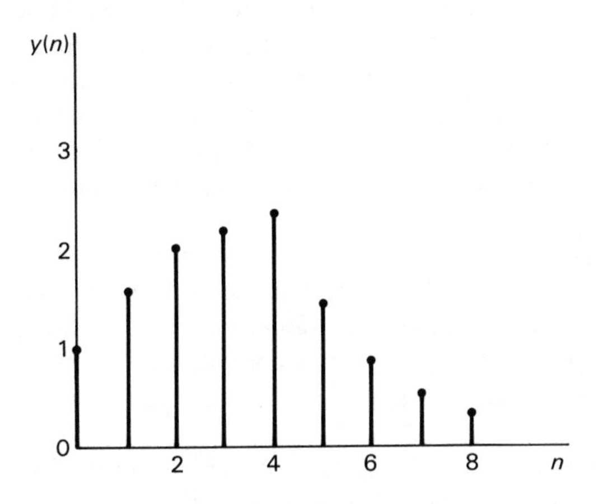

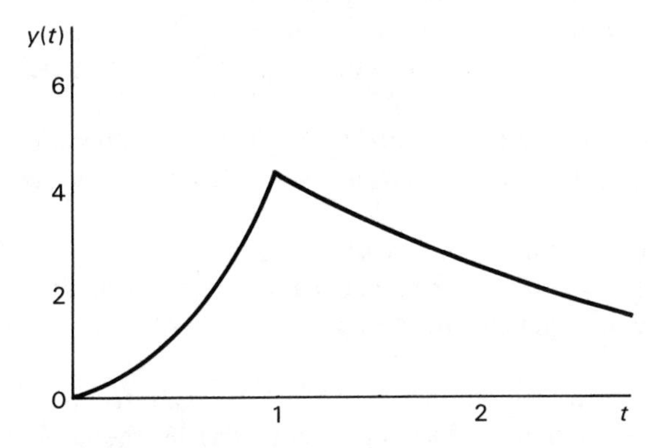

Figure 4.7 System responses for Example 4.3.1.

Integrating by parts gives

$$\int \tau e^{0.5\tau}\, d\tau = e^{0.5\tau}(2\tau - 4)$$

Inserting the limits gives the result

$$\begin{aligned} y(t) &= 0 && t<0 \\ y(t) &= 2t + 4(1 - e^{-0.5t}) && 0 \leqslant t \leqslant 1 \\ y(t) &= 0.703e^{-0.5t} && t>1 \end{aligned}$$

These results are shown graphically in Figure 4.7.

4.3.2 Graphical interpretation of convolution

The student working through Example 4.1 will probably realise that pitfalls lurk for the unwary when using the convolution formula. In particular care is needed to interpret the delayed function correctly over the ranges of integration (summation). The example used the convolution formula without any physical interpretation of the operations involved. By making a graphical interpretation of the steps the chances of error can be very much reduced. Also if only a rough sketch of the output waveform is required this can often be obtained by graphical means without needing to perform the integrations.

The system response has already been interpreted graphically for a specific discrete system in Section 4.3.1. However this method of representation is not very convenient and it cannot be used for continuous systems. An alternative method is to consider the graphical interpretation of the convolution formula, not with graphs plotted against the independent variable, t or n, but against the dummy variable, τ or k.

Consider the convolution of the signals in the first part of Example 4.3.1, the graphical steps involved are shown in Figure 4.8(a)–(d).

Figure 4.8(a) shows the variable n replaced by the dummy variable k. As the functions are now plotted against k this makes no difference to their shapes (compare with Figure 4.6(a)).

Figure 4.8(b) shows the function $h(-k)$, this is the operation of reflection and produces a folding of the function around $k = 0$. It is this folding that produces the name convolution.

Figure 4.8(c) shows the function $h(n-k)$, this produces a shift of the function such that the origin is at point n. Remember n is a variable and the function can only be shown for a specific value of n, in this case $n = 2$. Also shown in Figure 4.8(c) is the function $x(k)$.

Figure 4.8(d) shows the product $x(n)h(n-k)$, again for the specific value $n = 2$. It is the summation of this product that gives $y(n)$ by the convolution summation formula for $n = 2$.

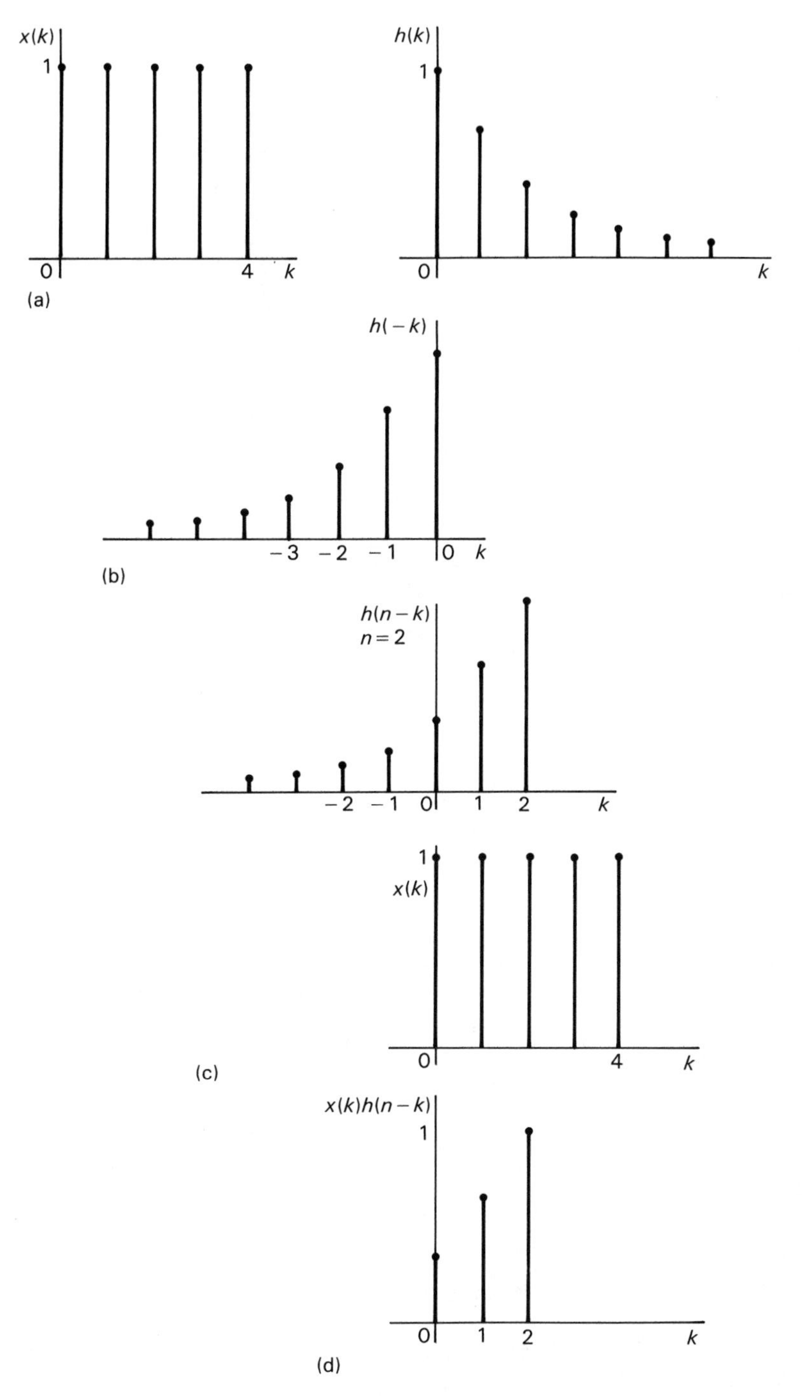

Figure 4.8 Graphical interpretation of discrete convolution.

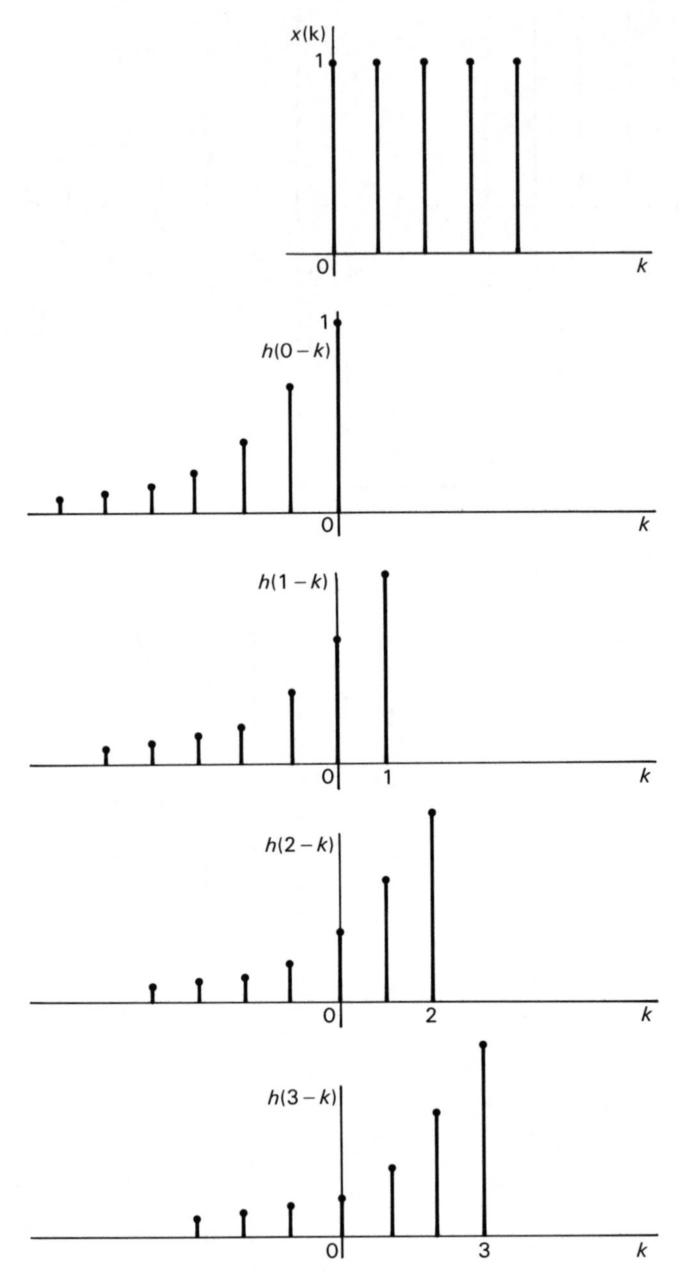

Figure 4.9 Graphical convolution showing the effect of varying n.

In order to obtain the function $y(n)$ for varying n one has to imagine the function $h(n-k)$ moving from left to right as n increases. This is shown in Figure 4.9 for n varying from 0 to 3. To obtain the corresponding $y(n)$ one must perform a multiplication and summation.

The interpretation for the continuous variable is developed along similar lines. Based on Example 4.3.1 (a) this leads to the interpretation of $x(\tau)$ and $h(t-\tau)$ as shown in Figure 4.10.

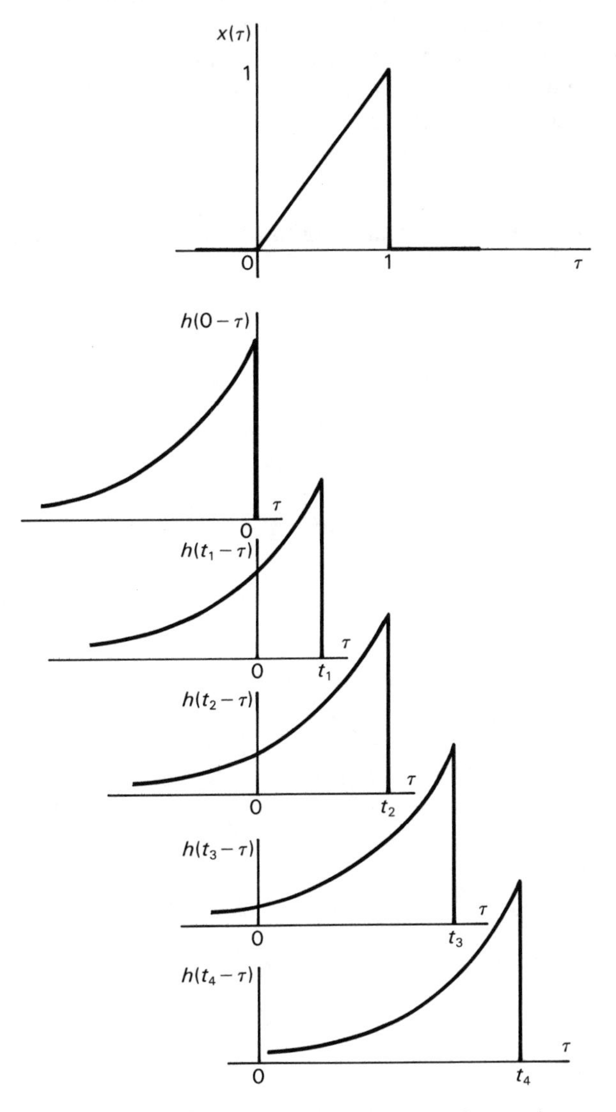

Figure 4.10 Graphical interpretation of continuous convolution.

$h(t-\tau)$ is shown for a range of values of t. Multiplication of this function by $x(\tau)$ and integration of the result gives the corresponding value of $y(t)$, this is illustrated in Figure 4.11 for a specific time t.

There are many advantages in using this graphical interpretation even if only in the form of a rough sketch:

1. It enables the limits for the integration to be established easily.
2. If the functions are simple then multiplication and addition will often give the answer in the discrete case. In the continuous case often the function to be integrated has a shape whose area is easily calculated.
3. Often a complete analytical solution to the problem is not required, just a general idea of how the output behaves. In Figure 4.11 it is apparent without any calculation that the output is zero for $t \leq 0$, reaches a maximum around $t = 1$ and then falls away.

If the student finds difficulty in visualising the function $h(t-\tau)$ it might be helpful

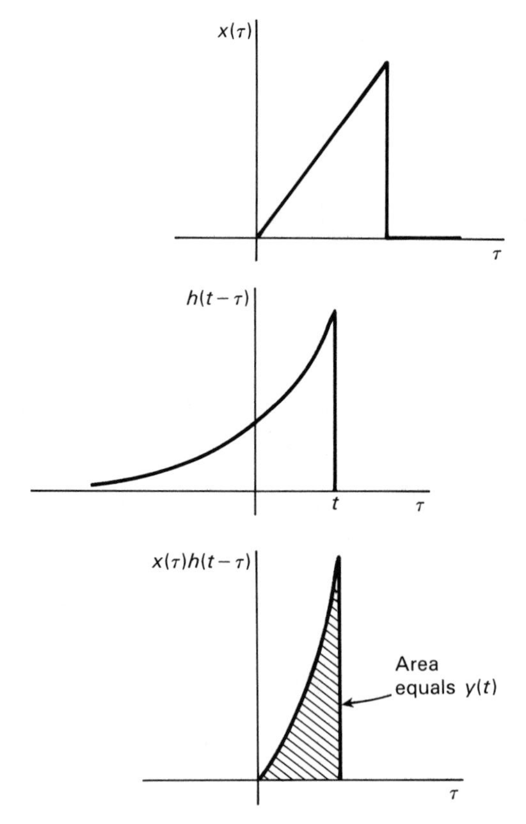

Figure 4.11 Formation of $y(t)$ by multiplication and integration.

to draw the function $h(-\tau)$ on a separate piece of paper to that of $x(\tau)$. This paper can then be slid along with respect to $x(\tau)$ giving a function $h(t-\tau)$. The following example illustrates some of the advantages of a graphical interpretation of convolution.

EXAMPLE 4.3.2

A system has an impulse response $h(t)$ as shown in Figure 4.12 where the input signal $x(t)$ shown is applied. Obtain an expression for the resulting output and sketch its waveform.

SOLUTION

Figure 4.13 shows the waveforms $x(\tau)$ and $h(t-\tau)$. Figure 4.14 shows the product $x(\tau)h(t-\tau)$ for different ranges of t. The area of the product, which represents the system output can always be calculated by adding the areas of rectangles and

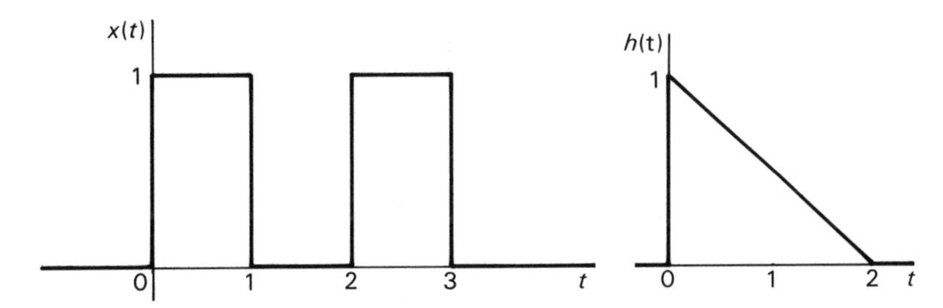

Figure 4.12 Signals applying to Example 4.3.2.

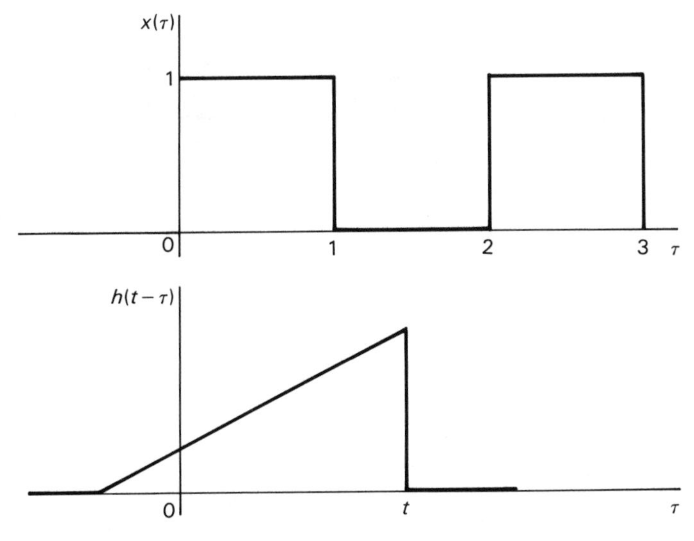

Figure 4.13 Waveform $x(\tau)$ and $h(t-\tau)$.

triangles. Because the slope of the line forming $h(t)$ is $1/2$ the triangles all have a height equal to one half of their base width.

Hence the following expressions for the output $y(t)$ can be obtained.

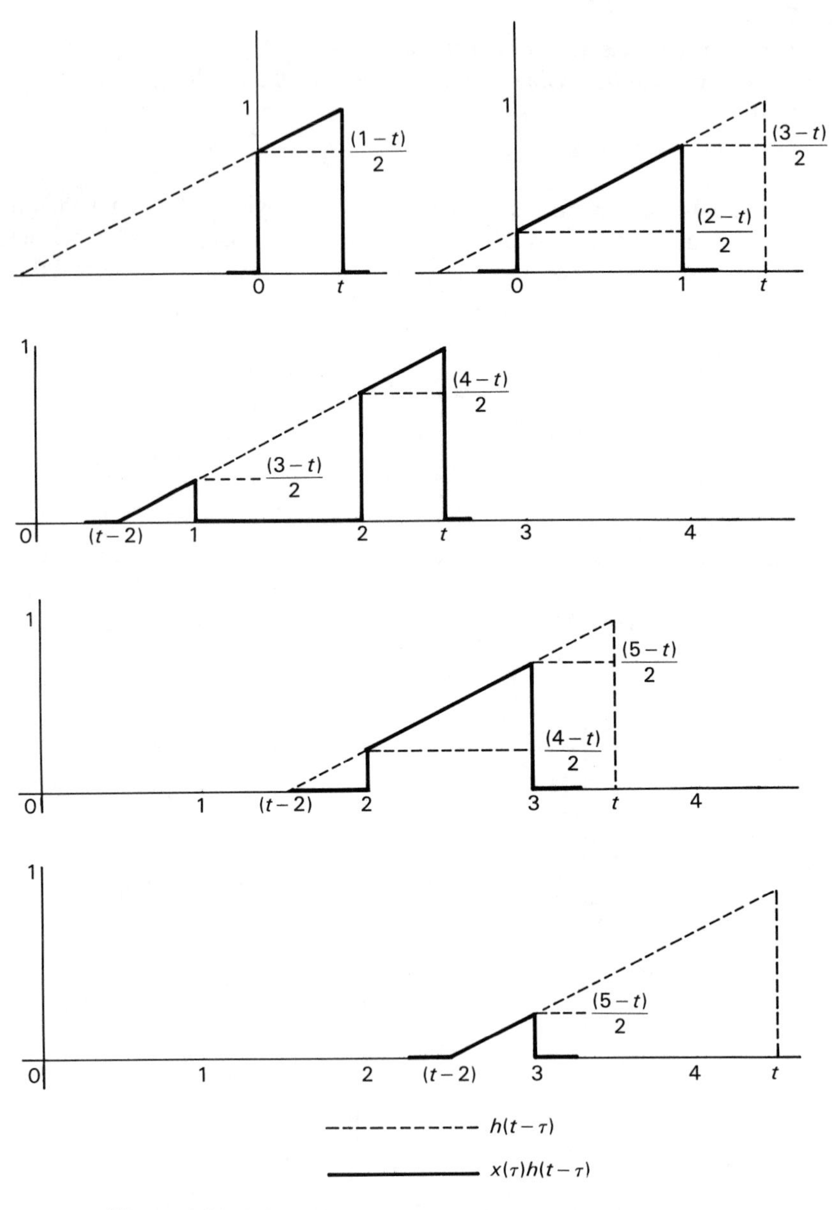

Figure 4.14 Calculation of the area of the product $x(\tau)h(t-\tau)$.

$$y(t) = 0 \qquad t \leqslant 0$$

$$y(t) = t\left[1 - \frac{t}{2}\right] + \frac{1}{2}\cdot t\cdot\frac{t}{2} \qquad 0 \leqslant t \leqslant 1$$

$$= t - \frac{t^2}{4}$$

$$y(t) = \frac{(2-t)}{2} + \frac{1}{2}\cdot\frac{1}{2} \qquad 1 \leqslant t \leqslant 2$$

$$= \frac{(5-2t)}{4}$$

$$y(t) = \frac{(4-t)(t-2)}{2} + \frac{(t-2)^2}{4} + \frac{(3-t)^2}{4} \qquad 2 \leqslant t \leqslant 3$$

$$= \frac{(2t-3)}{4}$$

$$y(t) = \frac{(4-t)}{2} + \frac{1}{4} \qquad 3 \leqslant t \leqslant 4$$

$$= \frac{(9-2t)}{4}$$

$$y(t) = \frac{(5-t)^2}{4} \qquad 4 \leqslant t \leqslant 5$$

$$y(t) = 0 \qquad 5 \leqslant t$$

These results are plotted in Figure 4.15.

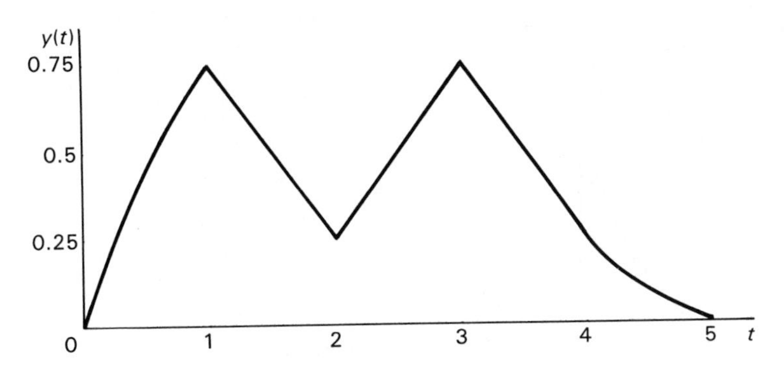

Figure 4.15 Output response for Example 4.3.2.

4.3.3 More convolution

Consider the convolution of a system input $x(t)$ and a system impulse response $h(t)$, these have been shown as $x(\tau)$ and $h(\tau)$ in Figure 4.16.

Figure 4.16(a) shows the functions $x(\tau)$, $h(t-\tau)$ and their product. Figure 4.16(b) shows the functions $h(\tau)$, $x(t-\tau)$ and their product. With a little thought it is apparent that the areas under these two products are equal for all values of t. This means

$$\int_{-\infty}^{+\infty} x(\tau)h(t-\tau)\,\mathrm{d}\tau = \int_{-\infty}^{+\infty} h(\tau)x(t-\tau)\,\mathrm{d}\tau$$

the operation of convolution is said to be commutative. This can be proved somewhat more formally by substituting a new dummy variable λ as $t-\tau$ into one of the above integrals. To avoid having to write the convolution integral out in full the symbol $*$ is used to denote convolution. $x(t) * h(t)$ is the convolution of $x(t)$ and $h(t)$. Hence the commutative property can be written $x(t) * h(t) = h(t) * x(t)$.

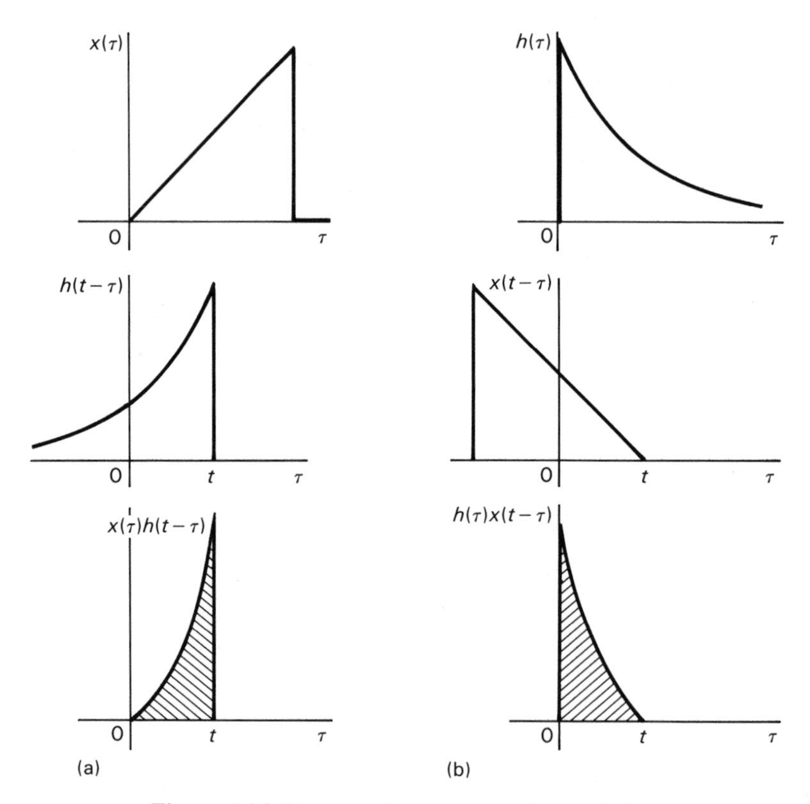

Figure 4.16 Commutative property of convolution.

The limits that have been placed on the convolution integral are in general $\pm\infty$. However these limits can be restricted further:

1. When $h(t) = 0,\ t < 0$. This will always be the case when the system is causal (only causal systems are considered in this book).
2. When $x(t) = 0,\ t < 0$. This will often be the case as input prior to $t = 0$ is usually incorporated via initial conditions (see Section 4.5).

The student should check by using a graphical approach to convolution that the following restrictions can be placed on the limits:

1. $h(t) = 0,\ t < 0$

$$\int_{-\infty}^{+\infty} x(\tau)h(t-\tau)\,d\tau = \int_{-\infty}^{t} x(\tau)h(t-\tau)\,d\tau$$

$$\int_{-\infty}^{+\infty} h(\tau)x(t-\tau)\,d\tau = \int_{0}^{+\infty} h(\tau)x(t-\tau)\,d\tau$$

2. $h(t) = 0,\ t < 0$ and $x(t) = 0,\ t < 0$

$$\int_{-\infty}^{+\infty} x(\tau)h(t-\tau)\,d\tau = \int_{0}^{t} x(\tau)h(t-\tau)\,d\tau$$

$$\int_{-\infty}^{+\infty} h(\tau)x(t-\tau)\,d\tau = \int_{0}^{t} h(\tau)x(t-\tau)\,d\tau$$

A physical interpretation of the impulse response can be obtained via the concept of convolution. Figure 4.17 shows a system with input $x(t)$ and output $y(t)$ and their relationship via the convolution integral is also shown.

The concept of system memory has been introduced in Chapter 3. Convolution gives a quantitive measure to this concept. Referring to Figure 4.17 it is the translated reflected impulse response $h(t-\tau)$, that gives gives a measure of how the input

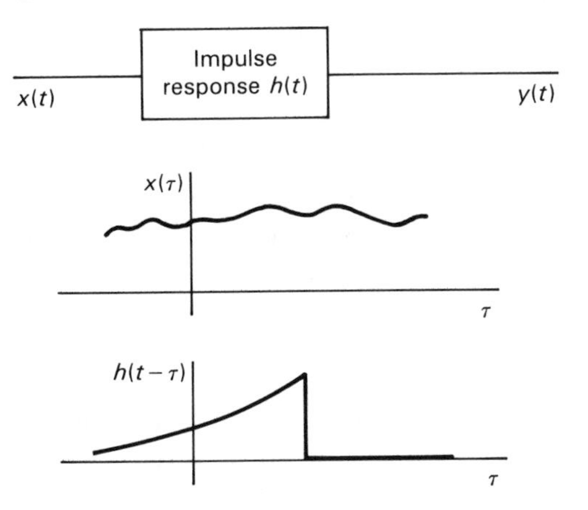

Figure 4.17 Impulse response interpreted as system memory.

previous to time t affects the output at time t. The impulse response is a form of memory; it weights previous inputs to form the present output. In older books the impulse response is often referred to as the *weighting function.*

Another system property that can be obtained from the impulse response is that of system stability. The concept of system stability was introduced in Section 3.3.5 as the bounded-input, bounded-output criterion. Suppose the input $x(t)$ is bounded such that $|x(t)| < B$ for all t. The output $y(t)$ of a system with impulse response $h(t)$ is given by

$$y(t) = \int_{-\infty}^{+\infty} h(\tau)x(t-\tau)\,\mathrm{d}\tau$$

$$|y(t)| = \left|\int_{-\infty}^{+\infty} h(\tau)x(t-\tau)\,\mathrm{d}\tau\right|$$

Now the magnitude of an integral can never be greater than the integral of the magnitude. Also the modulus of a product is the product of the moduli.

$$|y(t)| \leqslant \int_{-\infty}^{+\infty} |h(\tau)||x(t-\tau)|\,\mathrm{d}\tau$$

$$\leqslant B\int_{-\infty}^{+\infty} h(\tau)\,\mathrm{d}\tau$$

This is bounded provided

$$\int_{-\infty}^{+\infty} |h(t)|\,\mathrm{d}t < \infty$$

If the impulse response meets this condition it is said to be absolutely integrable. The corresponding condition for the discrete system is

$$\sum_{n=-\infty}^{\infty} |h(n)| < \infty$$

The student who has worked through this section should now be familiar with the concept of convolution. It is a concept that will prove very useful in the remainder of this book. The actual mechanism of performing convolution in a specific case is often very tedious and the result can often be obtained more easily by other methods. The following example illustrates the use of convolution in a communications application.

EXAMPLE 4.3.3

The 'matched filter' is an important concept in communication theory. Such a filter provides maximum signal at its output for a given transmitted symbol waveform. Suppose that two symbols represented by the signals $x_1(t)$ and $x_2(t)$, as shown in Figure 4.18(a), are to be transmitted and they are to be detected by matched filters as shown in Figure 4.18(b).

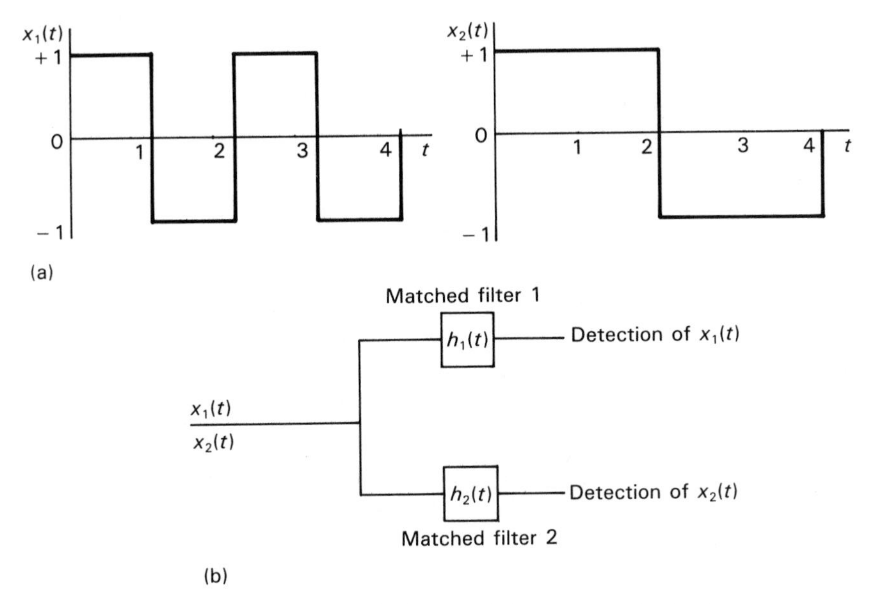

Figure 4.18 Matched filters applying to Example 4.3.3.

Determine the impulse responses $h_1(t)$ and $h_2(t)$ of the filters if their outputs to the input signals $x_1(t)$ and $x_2(t)$ respectively are to be maximized at $t=4$. The impulse responses are subject to the following constraints:

(a) $h(t)=0 \quad t<0$
(b) $|h(t)| \leqslant 1$ for all t

Determine the filter outputs (at $t=4$) if the 'other' signal is applied, i.e., $x_2(t)$ to filter 1 and $x_1(t)$ to filter 2.

SOLUTION

Considering filter 1, Figure 4.19 shows an arbitrary impulse response together with the input signal $x_1(t)$. The output at $t=4$ is given by the convolution of $h_1(t)$ and $x_1(t)$ evaluated at $t=4$.

$$y(t) = \int_0^4 x(\tau)h(t-\tau)\,\mathrm{d}\tau$$

The signal $h_1(4-\tau)$ is also shown in Figure 4.19 and the output at $t=4$ is the area under the product of this signal and the signal $x_1(\tau)$. The required impulse response is that which maximises this area. As the impulse response is subject to the constraint that $|h_1(t)| \leqslant 1$ then a little thought will show that the required function, $h_1(4-\tau)$, is as shown in Figure 4.19. Figure 4.20 shows the required functions $h_1(t)$ and $h_2(t)$.

The situation where the signal $x_1(t)$ is applied to filter 2 and vice versa is shown in Figure 4.21. Here the signals $x_1(\tau)$, $h_2(4-\tau)$ and $x_2(\tau)$, $h_1(4-\tau)$ are shown. The

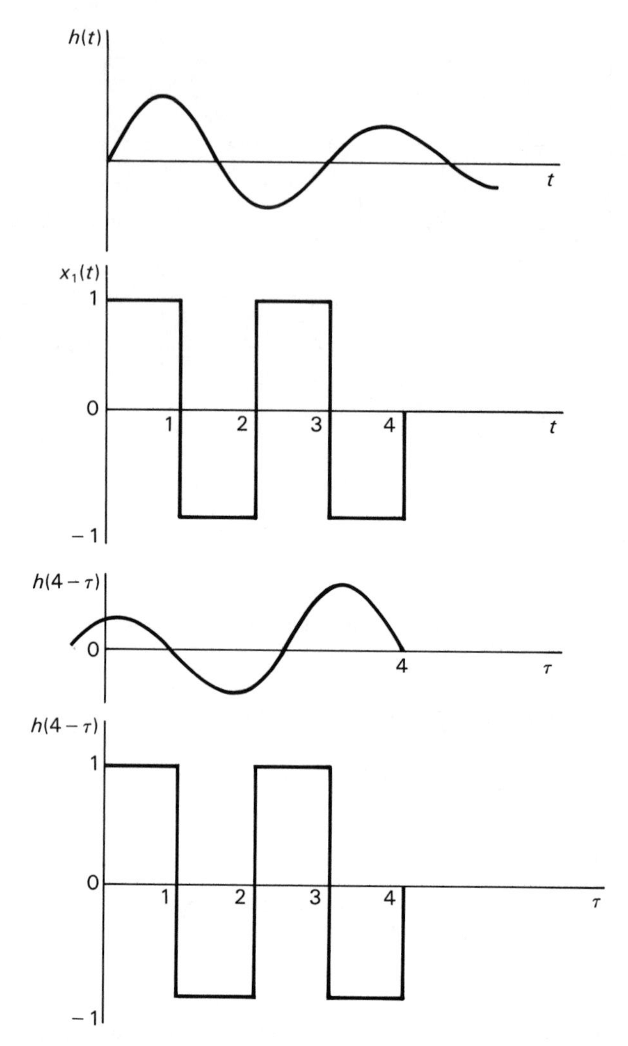

Figure 4.19 Optimum impulse response to detect signal $x_1(t)$.

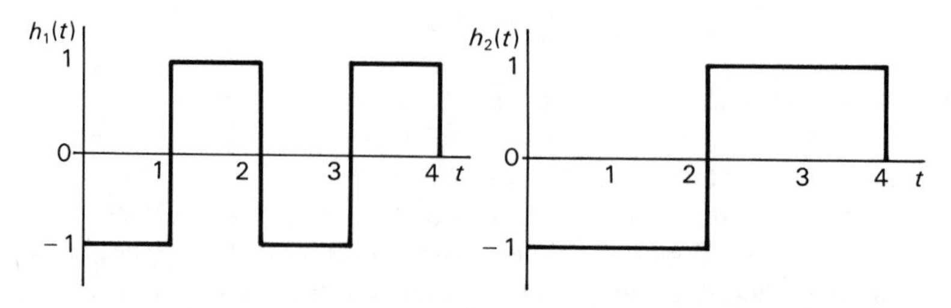

Figure 4.20 Optimum impulse responses for Example 4.3.3.

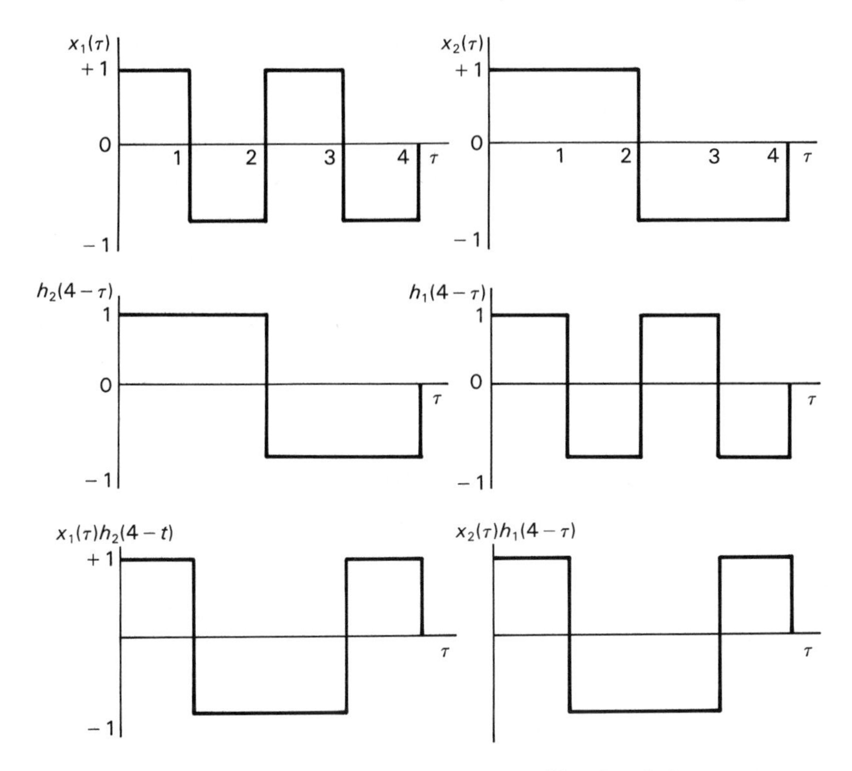

Figure 4.21 Application of signal $x_1(t)$ to filter 2 and vice versa.

corresponding products are also shown and the total area under each of these is zero. Hence the filters would reject the other signal (at $t = 4$) completely.

4.4 *Zero-input and zero-state responses*

In the previous section the concept of the impulse response as a memory was introduced. Generally the output of a system depends not only on its input at the present time instant but also on its inputs at previous time instants. Theoretically then the input to a system needs to be known back to $-\infty$ in order to obtain its output. This is unfortunate as a system is not accompanied by a 'pedigree' listing all the inputs applied in its past life. How then are these past (unknown) inputs to be taken into account?

The answer lies in the mechanism by which the system 'remembers' its previous inputs. In a continuous system it does so because it contains elements that store energy. In a discrete system there may be actual memory elements (as in a computer). A measure of this storage is used to replace the input prior to the instant that the input signal is applied (usually at $t=0$). This measure takes the form of a number of *initial conditions*, these initial conditions are not unique but an nth order system requires n initial conditions. Alternatively, as introduced in Section 3.10 these conditions can be thought of as giving a picture of the state of the system.

The system response to an input applied at $t=0$ can now be considered as the sum of two components:

1. The zero-state response. This is the response to the applied input when all the initial conditions (the system state) is zero.
2. The zero-input response. This is the system output due to the initial conditions only. The system input is taken as zero.

4.5 Continuous systems: solution of the differential equation

In Chapter 3 it was shown that a linear continuous system can in general be described by a differential equation relating the system output $y(t)$ to its input $x(t)$. The nth order equation can be written ($n \geq m$)

$$\frac{d^n y}{dt^n} + a_{n-1}\frac{d^{n-1}y}{dt^{n-1}} + \ldots\ldots a_0 y = b_m \frac{d^m x}{dt^m} + b_{m-1}\frac{d^{m-1}x}{dt^{m-1}} + \ldots\ldots b_0 x \qquad (4.5.1)$$

This section will be concerned with the solution of such an equation, i.e., given the system input $x(t)$ and any initial conditions obtain the resulting output $y(t)$. The solution will be in the time domain, later in the book in Chapter 7, it will be shown how this equation can be solved, usually with greater ease, by use of transform methods. If this is so the student may ask why does one bother with time domain solution? There are several answers:

1. It fits naturally into the development of the systems approach.
2. It gives greater appreciation of the transform methods when they are used.
3. It gives physical insight into the structure of the system.
4. It gives results that can be used in the next chapter, which in this development, is better covered before transform methods.

However because results can be obtained by transform methods the treatment is not given in very great depth and the interested student can expand upon it from numerous texts on differential equations that are available.

The form of the differential equation given as eqn (4.5.1) is rather cumbersome and a shorthand notation is useful. The symbol D is used to denote d/dt, D^2 to denote d^2/dt^2, etc. Equation (4.5.1) can then be written

$$(D^n + a_{n-1}D^{n-1} + \ldots\ldots a_0)y = (b_m D^m + b_{m-1}D^{m-1} + \ldots\ldots b_o)x$$

Assuming that D can be interpreted as obeying the normal rules of algebra, multiplying out the above equation and interpreting the operator D as $\mathrm{d}/\mathrm{d}t$ returns to the original form of the differential equation.

As introduced in Section 4.4 the response of a system can be considered as formed of two parts, the response due to initial conditions, which is the zero input response and the response to the input applied after $t = 0$, which is zero state response. These responses will be derived separately and the results added to give the complete response.

Other terms are often used to denote the two components of the response. For zero-input response one meets the terms, complementary function, free response, natural response, transient response. For the zero-state response one meets forced response, steady state response, particular solution. One has to be careful with these terms as often they will not be directly equivalent to the zero-input and zero-state responses as used here.

4.5.1 The zero-input response

The zero-input response is obtained by solving the differential equation of the system with the right-hand side set to zero. The equation in this form is known as an homogeneous equation. Using the operator D it can be written

$$(D^n + a_{n-1}D^{n-1} + \ldots\ldots a_0)y = 0 \tag{4.5.2}$$

Considering first the solution of a first order homogeneous equation

$$\frac{\mathrm{d}y}{\mathrm{d}t} + a_0 y = 0$$

This equation can be solved by the technique known as separation of variables. It can be expressed

$$\int \frac{1}{y}\,\mathrm{d}y = -\int a_0\,\mathrm{d}t$$

integrating

$$\log y = -a_0 t + \text{constant}$$

$$y = A\mathrm{e}^{-a_0 t} \tag{4.5.3}$$

Where A is an arbitrary constant. This constant can be evaluated by using information regarding the initial conditions. If at $t = 0$, $y = y_0$ then eqn (4.5.3) becomes

$$y = y_0\mathrm{e}^{-a_0 t}$$

If there is no initial condition then the output will be zero, this is not surprising as we are only solving for the zero-input component of the solution.

Taking now the nth order equation, this in its homogeneous form is

$$(D^n + a_{n-1}D^{n-1} + \ldots\ldots a_o)y = 0$$

As the solution to the first order equation was of exponential form this suggests a solution of the form $y = Ae^{\lambda t}$, where λ and A are constants. Substituting and performing the differentiation leads to the equation

$$A(\lambda^n + a_{n-1}\lambda^{n-1} + \ldots\ldots a_0)e^{\lambda t} = 0$$

As $e^{\lambda t} \neq 0$ and $A \neq 0$ (if there is to be a zero input solution) then

$$\lambda^n + a_{n-1}\lambda^{n-1} + \ldots\ldots a_0 = 0 \tag{4.5.4}$$

Equation (4.5.4) is known as the auxiliary equation and is easily obtained from the coefficients of the homogenous system equation, eqn (4.5.2). Solution of the auxiliary equation gives roots $\lambda_1, \lambda_2 \ldots \lambda_n$ and leads to a set of solutions $A_1e^{\lambda_1 t}$, $A_2e^{\lambda_2 t}, \ldots A_ne^{\lambda_n t}$.

However it is a feature of the homogeneous equation that the sum of solutions is also a solution, hence the solution becomes

$$y = A_1e^{\lambda_1 t} + A_2e^{\lambda_2 t} + \ldots\ldots A_ne^{\lambda_n t} \tag{4.5.5}$$

The solution contains n arbitrary constants that can be obtained from the initial conditions. This is not the easy process that it was for the first order equation and it involves the solution of n simultaneous equations. Each of the time responses of the form $e^{\lambda t}$ is known as a *mode* of the solution.

The roots of the auxiliary equation have been assumed to be distinct but in practice two or more equal roots can occur. This case will not be considered here but it will be examined by transform methods in Section 7.3.3.

The roots of the auxiliary equation need not be real. If they are complex however they will always occur in conjugate pairs (with conjugate arbitrary constants). This must be the case as the coefficients in the auxiliary equation are real and also from a practical viewpoint the system time response cannot be complex. The response can be expressed as the sum of complex exponentials but it is usually neater to express this sum as a trigonometric function. This is illustrated in the following example.

EXAMPLE 4.5.1

Obtain the zero-input response of the systems described by the following differential equations:

(a) $$\frac{d^2y}{dt^2} + 3\frac{dy}{dt} + 2y = \frac{dx}{dt} + 3x$$

(b) $$\frac{d^2y}{dt^2} + 2\frac{dy}{dt} + 2y = \frac{dx}{dt} + x$$

At $t = 0$ the initial conditions for both systems are $y(0) = 1$ and $dy/dt|_{t=0} = 2$

SOLUTION

(a) The auxiliary equation is

$$\lambda^2 + 3\lambda + 2 = 0$$

and this has roots $\lambda = -1$ and $\lambda = -2$. Hence the solution of the differential equation takes the form

$$y = A_1 e^{-t} + A_2 e^{-2t} \tag{4.5.6}$$

The constants A_1 and A_2 can be evaluated by use of the initial conditions. At $t = 0$, $y(0) = 1$ and

$$1 = A_1 + A_2 \tag{4.5.7}$$

Differentiation of eqn (4.5.6) gives

$$\frac{dy}{dt} = -A_1 e^{-t} - 2A_2 e^{-2t}$$

and using the initial condition that dy/dt at $t = 0$ is 2 then

$$2 = -A_1 - 2A_2 \tag{4.5.8}$$

Equations (4.5.7) and (4.5.8) are a pair of simultaneous equations in A_1 and A_2. They can easily be solved to give $A_1 = 4$ and $A_2 = -3$.
Hence the zero input solution is

$$y = 4e^{-t} - 3e^{-2t}$$

(b) The auxiliary equation is

$$\lambda^2 + 2\lambda + 2 = 0$$

this has roots

$$\lambda = \frac{-2 \pm \sqrt{4-8}}{2}$$
$$= -1 \pm j1$$

Hence the form of the solution is

$$y = Ae^{(-1+j)t} + A^* e^{(-1-j)t} \tag{4.5.9}$$

Note that the constants must be complex conjugates in order to give a real solution. Continuing as in (a) leads to the equations

$$1 = A + A^* \tag{4.5.10}$$
$$2 = (-1 + j1)A + (-1 - j1)A^*$$
$$2 = -(A + A^*) + j(A - A^*) \tag{4.5.11}$$

Substituting from eqn (4.5.10) into eqn (4.5.11) and remembering rules about the sum and difference of a complex number and its conjugate (see Appendix) gives

$$A = \tfrac{1}{2} - j\tfrac{3}{2}, \qquad A^* = \tfrac{1}{2} + j\tfrac{3}{2}$$

Substituting into eqn (4.5.9) gives the solution

$$y = \tfrac{1}{2}[(1 - j3)e^{(-1+j)t} + (1 + j3)e^{(-1-j)t}] \qquad (4.5.12)$$

Equation (4.5.12) does not represent a very convenient form for the solution as it is expressed in terms of complex exponentials.

Writing $(1 - j3) = Re^{j\varphi}$, $(1 + j3) = Re^{-j\varphi}$ where $R = \sqrt{10}$ and $\varphi = -\tan^{-1}3$, enables eqn (4.5.12) to be expressed

$$y = \frac{Re^{-t}}{2}(e^{j(t+\varphi)} + e^{-j(t+\varphi)})$$

$$= Re^{-t}\cos(t + \varphi)$$

and this is a more useful form of solution.

4.5.2 The zero-state response

The zero-state response is the solution to eqn (4.5.1) to a given input $x(t)$ when all the initial conditions are zero. As in the last section it is convenient to start with the solution of the first order equation, which in operator form is

$$(D + a_0)y = x \qquad (4.5.13)$$

The solution to the homogenous equation was of the form $y = Ae^{\lambda t}$ ($\lambda = -a_0$). Hence a solution of the form $y = \alpha(t)e^{\lambda t}$ where $\alpha(t)$ is a function to be determined, will be tried. Then

$$Dy = \frac{dy}{dt} = \frac{d\alpha}{dt}e^{\lambda t} + \lambda e^{\lambda t}\alpha$$

and substituting into eqn (4.5.13) gives

$$\frac{d\alpha}{dt}e^{\lambda t} + \lambda e^{\lambda t}\alpha + a_0 e^{\lambda t}\alpha = x$$

As $a_0 = -\lambda$ then

$$\frac{d\alpha}{dt}e^{\lambda t} = x(t)$$

$$\alpha = \int_0^t e^{-\lambda t}x(t)\,dt$$

$$= \int_0^t e^{-\lambda\tau}x(\tau)\,d\tau$$

where τ has been introduced as a dummy variable into the definite integral. Using

$y = \alpha e^{\lambda t}$ this gives a solution

$$y = e^{\lambda t} \int_0^t e^{-\lambda \tau} x(\tau)\, d\tau$$

and as the integral is with respect to τ the factor $e^{\lambda t}$ can be included under the integral sign giving

$$y = \int_0^t e^{\lambda(t-\tau)} x(\tau)\, d\tau$$

which can be written as

$$y = \int_0^t h(t-\tau) x(\tau)\, d\tau \qquad (4.5.14)$$

where $h(\tau) = e^{\lambda \tau}$. The student should recognise the above formula as the convolution integral. This is a point of great importance but it will be considered in more detail when the nth order system is investigated. Before doing this let us return to the operational form of the differential equation.

$$(D + a_0)y = x$$

If this had been a purely algebraic equation and we had wished to solve for y we would write

$$y = \frac{x}{(D + a_0)} \qquad (4.5.15)$$

Does this have a meaning when D is an operator ($D = d/dt$)? This question cannot be entered into in depth here and the interested student is referred to a mathematical text on the solution of differential equations by operational methods. As far as this book is concerned eqn (4.5.15) can be interpreted as the solution

$$y = \int_0^t h(t-\tau) x(\tau)\, d\tau$$

where

$$h(\tau) = e^{\lambda \tau} \text{ and } \lambda = -a_0$$

Turning now to the nth order equation which can be written in operational form

$$(D^n + a_{n-1}D^{n-1} + \ldots\ldots a_0)y = (b_m D^m + b_{m-1}D^{m-1} + \ldots\ldots b_0)x$$

This gives an operational form of solution as

$$y = \frac{b_m D^m + b_{m-1}D^{m-1} + \ldots\ldots b_0}{D^n + a_{n-1}D^{n-1} + \ldots\ldots a_0} x$$

The question now is, what is the interpretation of the right-hand side of this equation? It is going to be assumed, and this does really need proof, that the right-hand side can be expanded as a partial fraction expansion. This gives

$$y = \frac{k_1 x}{D - \lambda_1} + \frac{k_2 x}{D - \lambda_2} + \dots\dots + \frac{k_n x}{D - \lambda_n} \tag{4.5.16}$$

Here $\lambda_1, \lambda_2, \dots, \lambda_n$ are the roots of the auxiliary equation (as in the zero input solution) and $k_1, \dots, k_2$ are the coefficients of the partial fraction expansion. Again it is assumed that there are no repeated roots otherwise the expansion cannot be expressed in this form. Now each of the terms in this expansion is a first order term as in eqn (4.5.15) and a complete solution can be written

$$y = k_1 \int_0^t h_1(t - \tau)x(\tau)\, \mathrm{d}\tau + k_2 \int_0^t h_2(t - \tau)x(\tau)\, \mathrm{d}\tau + \dots\dots + k_n \int_0^t h_n(t - \tau)x(\tau)\, \mathrm{d}\tau$$

which can be written

$$y = \int_0^t h(t - \tau)x(\tau)\, \mathrm{d}\tau \tag{4.5.17}$$

where $h(t)$ is given by

$$h(t) = k_1 \mathrm{e}^{\lambda_1 t} + \dots\dots + k_n \mathrm{e}^{\lambda_n t} \tag{4.5.18}$$

As in the first order system, eqn (4.5.17) gives the output in the form of a convolution between the input and the function $h(t)$. However from Section 4.3 it is known that the output is the convolution of the impulse response and the system input. Hence $h(t)$ must represent the system impulse response. Alternatively setting $x(t) = \delta(t)$, i.e., putting a unit impulse as the system input, gives $h(t)$ as the corresponding output using the sifting property of the impulse function. Hence to recapitulate, the impulse response is of the form of eqn (4.5.18) (no repeated roots). The parameters $\lambda_1, \dots, \lambda_2$ can be obtained from the auxiliary equation and the constants $k_1, \dots, k_2$ from the partial fraction expansion of the operational form of the solution y. Having found the impulse response the response to any input can be found by convolution.

The complete response is obtained by adding the zero-input and zero-state responses to give

$$y(t) = A_1 \mathrm{e}^{\lambda_1 t} + A_2 \mathrm{e}^{\lambda_2 t} + \dots\dots A_n \mathrm{e}^{\lambda_n t} + \int_0^t h(\tau)x(t - \tau)\, \mathrm{d}\tau \tag{4.5.19}$$

The evaluation of these responses for a specific system is considered in the following example.

EXAMPLE 4.5.2

For the circuit shown in Figure 4.22 calculate the following:

(a) The response $v_o(t)$ for zero input if both capacitors are charged initially to 10 V.
(b) The response to an input step of magnitude 10 V applied at $t = 0$ if the capacitors are initially uncharged.

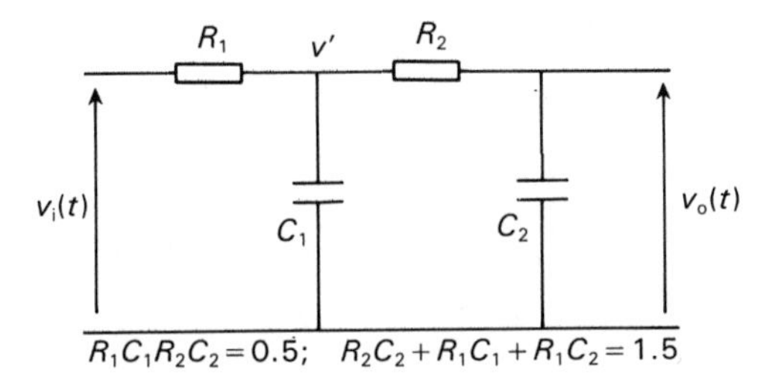

Figure 4.22 Circuit for Example 4.5.2.

(c) A combination of (a) and (b), the 10 V step is applied at $t = 0$ but the capacitors are initially charged to 10 V.

SOLUTION

Applying nodal analysis at node 1 (the voltage here is v' as in Figure 4.22)

$$\frac{v_i - v'}{R_1} + C_1 \frac{d(0 - v')}{dt} + \frac{v_o - v'}{R_2} = 0 \tag{4.5.20}$$

Also the currents in R_2 and C_2 are equal

$$\frac{v' - v_o}{R_2} = C_2 \frac{dv_o}{dt} \tag{4.5.21}$$

From eqn (4.5.21)

$$v' = R_2C_2 \frac{dv_o}{dt} + v_o \tag{4.5.22}$$

Differentiating

$$\frac{dv'}{dt} = R_2C_2 \frac{d^2v_o}{dt^2} + \frac{dv_o}{dt} \tag{4.5.23}$$

Substituting v' from eqn (4.5.22) and dv'/dt from eqn (4.5.23) in eqn (4.5.20) gives

$$R_1R_2C_1C_2 \frac{d^2v_o}{dt^2} + (R_1C_1 + R_2C_2 + R_1C_2) \frac{dv_o}{dt} + v_o = v_i$$

Inserting values and making the coefficients of d^2v_o/dt^2 unity gives

$$\frac{d^2v_o}{dt^2} + 3 \frac{dv_o}{dt} + 2v_o = 2v_i \tag{4.5.24}$$

(a) Zero-input response
The auxiliary equation corresponding to eqn (4.5.24) is

$$\lambda^2 + 3\lambda + 2 = 0$$

This has roots $\lambda = -1$ and $\lambda = -2$ giving the zero input response

$$v_o(t) = A_1 e^{-t} + A_2 e^{-2t} \tag{4.2.25}$$

The constants A_1 and A_2 are obtained from the initial conditions. At $t = 0$ both v_0 and v' are equal to 10 V, there is no potential difference across R_2 and no current through it. As this current is proportional to dv_o/dt, eqn (4.5.21) then dv_o/dt is initially zero.

Differentiating eqn (4.5.25) gives

$$\frac{dv_o}{dt} = -A_1 e^{-t} - 2A_2 e^{-2t} \tag{4.5.26}$$

and substituting initial values in eqns (4.5.25) and (4.5.26) gives

$$10 = A_1 + A_2$$
$$0 = -A_1 - 2A_2$$

These equations solve to give $A_1 = 20$ and $A_2 = -10$, hence the zero input response is given by

$$v_o = 20e^{-t} - 10e^{-2t} \tag{4.5.27}$$

(b) Zero-state response
Writing eqn (4.5.24) in operator form

$$(D^2 + 3D + 2)v_o = 2v_i$$

$$v_0 = \frac{2}{D^2 + 3D + 2} v_i$$

Expressing as partial fractions

$$\frac{2}{D^2 + 3D + 2} = \frac{2}{D+1} - \frac{2}{D+2}$$

Hence the impulse response (see Section 4.5.2) is

$$h(t) = 2e^{-t} - 2e^{-2t}$$

The step response can now be obtained by convolution where the input in $10u(t)$, $u(t)$ being the unit step.

$$v_o(t) = 10\int_{-\infty}^{+\infty} h(\tau)u(t - \tau)\,d\tau \tag{4.5.28}$$

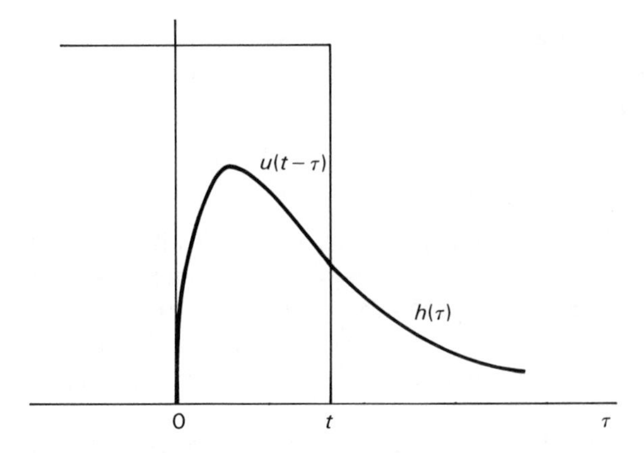

Figure 4.23 Convolution to obtain step response for Example 4.5.2.

Referring to Figure 4.23 it can be seen that the convolution integral given in eqn (4.5.28) can be evaluated as

$$v_o(t) = 10\int_0^t (2e^{-\tau} - 2e^{-2\tau})\,d\tau$$

$$= 10(1 + e^{-2t} - 2e^{-t}) \quad (4.5.29)$$

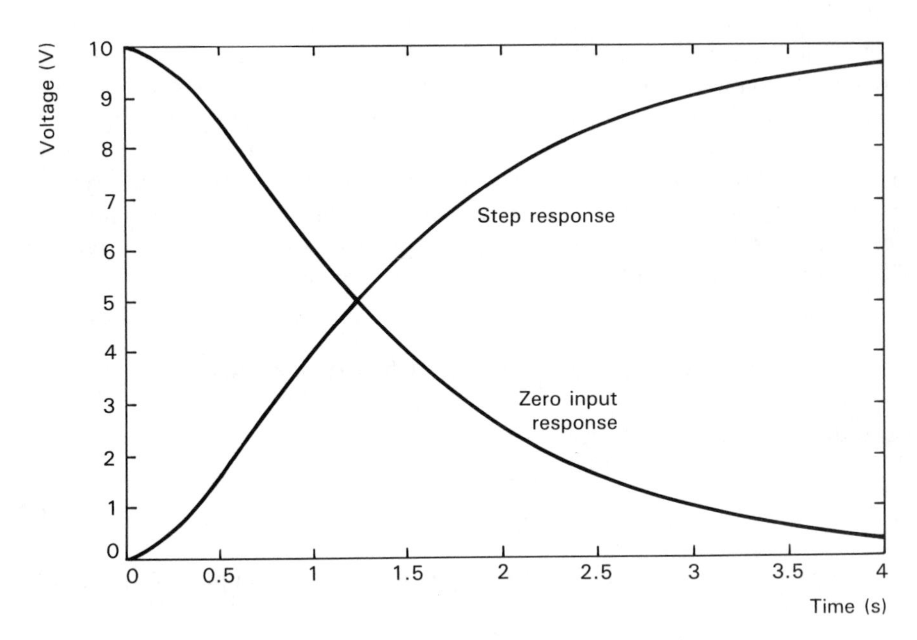

Figure 4.24 Zero input and step response for system of Example 4.5.2.

(c) Complete response
This is obtained by adding the zero-input and zero-state responses, i.e., the right-hand sides of eqns (4.5.27) and (4.5.29)

$$\begin{aligned} v_o(t) &= 20e^{-t} - 10e^{-2t} + 10 + 10e^{-2t} - 20e^{-t} \\ &= 10 \end{aligned}$$

The output remains constant at 10 V after the application of the step, this result can be easily verified by physical reasoning. At $t = 0$ the input is 10 V, the voltage v' is 10 V (due to the charge on C_2) and the voltage v_o is 10 V (due to the charge on C_2). Hence there is no potential difference across either of the resistors, no current flow and no discharge of the capacitors. Plots of the zero input response, and the step response are shown in Figure 4.24.

It will be seen later that problems of this nature are more easily solved by the use of operator methods and these will be considered in Chapter 7. The reasons for using the method of convolution considered here have already been discussed in the introduction to Section 4.5.

4.6 First and second order systems

These systems are rather special because it is relatively easy to obtain their step responses in terms of their system parameters. The response to a step input is a useful measure of a system's performance, giving a measure of the response to a rapidly changing input (the leading edge of the step) and then to a constant input. The step response can be more easily interpreted if the equations are written in a 'standard form' where the parameters have a direct physical interpretation.

The analysis of the first order system is straight forward using the methods developed in this chapter. Although the step response of the second order system can be obtained by these methods the transform methods of Chapter 7 offer a much easier route to the results. Hence only an outline of the results for the second order system will be given and these will be derived fully later (see Section 7.3.3).

4.6.1 Step response of a first order system

The differential equation describing the first order system is

$$\frac{dy}{dt} + a_0 y = b_0 x \tag{4.6.1}$$

where the coefficient of dy/dt has been taken as unity. The impulse response of this system is

$$h(t) = b_0 e^{-a_0 t}$$

The response to a unit step can be obtained by convolution

$$y(t) = \int_0^t u(t-\tau)h(\tau)\,\mathrm{d}\tau$$

where $u(t)$ is the unit step. Hence

$$\begin{aligned} y(t) &= \int_0^t b_0 \mathrm{e}^{-a_0\tau}\,\mathrm{d}\tau \\ &= b_0\left[-\frac{\mathrm{e}^{-a_0\tau}}{a_0}\right]_0^t \\ &= \frac{b_0}{a_0}[1-\mathrm{e}^{-a_0 t}] \end{aligned} \tag{4.6.2}$$

As $t \to \infty$ then $y(t) \to b_0/a_0$. This, for a unit step input, is the gain of the system when it reaches its steady state. It can be conveniently denoted by K, i.e. $K = b_0/a_0$. In theory the system never reaches steady state. However in practice the constant a_0 gives a measure of the time it takes the output to approach its final value. $1/a_0$ is defined as the system time constant T. Then from eqn (4.6.2) it is easily shown that the output reaches the following percentages of its final value at times T, $2T$, $3T$, $4T$, $5T$ respectively, 63.2%, 86.5%, 95.0%, 98.2%, 99.3%.

Using the parameters K and T eqn (4.6.1) can be re-written in a standard form

$$T\frac{\mathrm{d}y}{\mathrm{d}t} + y = Kx$$

and its solution, eqn (4.6.2) becomes

$$y(t) = K(1 - \mathrm{e}^{-t/T})$$

This response is plotted in Figure 4.25.

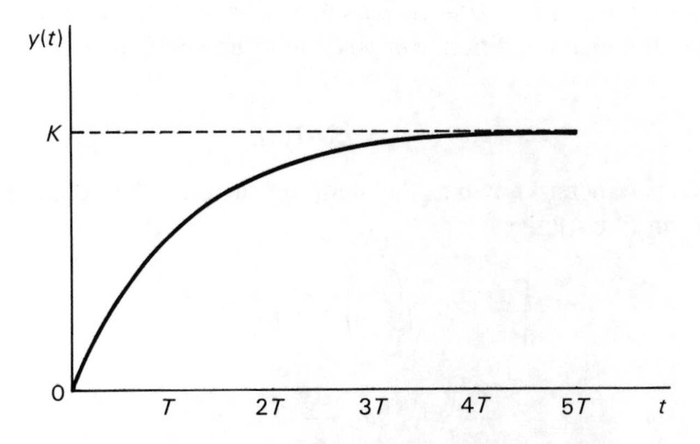

Figure 4.25 Step response of a first order system.

4.6.2 Step response of a second order system

The second order system is described by the differential equation

$$a_2 \frac{d^2y}{dt^2} + a_1 \frac{dy}{dt} + a_0 y = b_0 x \tag{4.6.3}$$

This equation can be solved for a unit step input by the method used for the first order equation, i.e., via the impulse response. This leads to modes in the solution of the form $e^{\lambda t}$ where λ is the solution of the equation

$$a_2\lambda^2 + a_1\lambda + a_0 = 0$$

which is

$$\lambda = \frac{-a_1 \pm \sqrt{(a_1^2 - 4a_0 a_2)}}{2a_2}$$

The form of the solution depends on whether the term $(a_1^2 - 4a_0a_2)$ is less than, equal to, or greater than zero.

As stated at the start of this section this approach will not be pursued here but the solution will be obtained in Chapter 7. However in order to obtain a standard form for the second order equation the following results will be stated without proof:

1. If $a_1 = 0$ then a component of the solution is a sinusoidal oscillation of angular frequency $\sqrt{(a_0/a_2)}$. This is known as the undamped natural frequency of the system and it is denoted by the symbol ω_n. Hence if eqn (4.6.3) is divided through by a_2 it can be written

$$\frac{d^2y}{dt^2} + \frac{a_1}{a_2}\frac{dy}{dt} + \omega_n^2 y = \frac{b_0}{a_0} x \tag{4.6.4}$$

2. If a_1 has a value such that $a_1^2 - 4a_0a_2 = 0$, i.e., $a_1 = 2\sqrt{(a_0a_2)}$ then this is termed the critically damped condition. Suppose now the coefficient a_1 is written in the form

$$a_1 = \zeta 2\sqrt{(a_0a_2)}$$

where ζ is a parameter known as the damping factor. Substituting this expression for a_1 into eqn (4.6.4) gives

$$\frac{d^2y}{dt^2} + 2\zeta\sqrt{\left(\frac{a_0}{a_2}\right)}\frac{dy}{dt} + \omega_n^2 y = \frac{b_0}{a_2} x$$

$$\frac{d^2y}{dt^2} + 2\zeta\omega_n \frac{dy}{dt} + \omega_n^2 y = \frac{b_0}{a_2} x \tag{4.6.5}$$

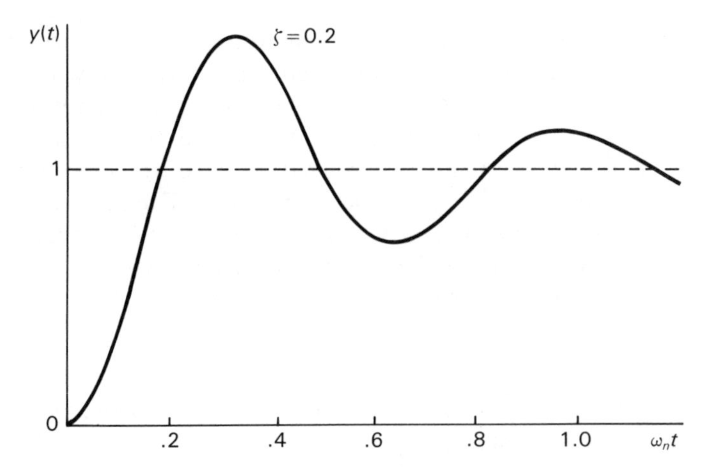

Figure 4.26 Step response of a second order system.

3. The left-hand side of eqn (4.6.5) is the standard form for the second order system. The right-hand side can be written

$$\frac{b_0}{a_2}x = \frac{b_0}{a_0}\frac{a_0}{a_2}x = K\omega_n^2 x$$

where $K = b_0/a_0$ is the gain factor. Often this is taken as unity and then taken into account by scaling the output response (system linearity). If this is the case then eqn (4.6.5) can be written

$$\frac{d^2y}{dt^2} + 2\zeta\omega_n\frac{dy}{dt} + \omega_n^2 y = \omega_n^2 x \qquad (4.6.6)$$

Equation (4.6.6) represents the standard form of the equation describing the second order system. The solution when obtained in Section 7.3.3 will be expressed in terms of the parameters ω_n and ζ. This is convenient as these parameters have a physical significance and salient features of the response can be expressed relatively simply in terms of them.

The form of the response to a unit step is shown in Figure 4.26 (for one value of ζ) and a more accurate plot will be given in Chapter 7 as Figure 7.11.

4.7 Solution of the difference equation

Linear difference equations can be solved formally using procedures that parallel those used for the solution of difference equations. A zero-input and a zero-state solution can be obtained as in the solution of differential equations. This approach will not be pursued here. One reason is, as with differential equations, the solutions are most easily obtained using transform methods and this topic will be covered in Chapter 7. Another reason is that if a closed form is not required then the solution can be obtained by numerical methods as covered in Section 3.9.1. Also if a recursive method of solution is used, starting from $n = 0$, for simpler systems, it is possible to obtain a closed form for the output response. Even if this is not possible the unit sample response can often be obtained by this method. The response to other inputs can then be obtained via the convolution summation. These methods are illustrated in the following example.

EXAMPLE 4.7.1

(a) In Chapter 3 the difference equation describing a discrete approximation to a first order system was derived in eqn (3.9.4)

$$v_o(n) - k_2 v_o(n-1) = k_1 v_i(n) \tag{4.7.1}$$

Obtain the solution to this equation if $v_i(n)$ is a discrete step of magnitude 10 V.

(b) A more accurate approximation to describe the first order system was derived in Example 3.9.1. This takes the form

$$v_o(n) = b_0 v_i(n) + b_1 v_i(n-1) + a_1 v_o(n-1) \tag{4.7.2}$$

Obtain the solution to this equation if $v_i(n)$ is the step input as in (a).

SOLUTION

(a) For convenience the input will be taken as a unit step (because of linearity the output can be scaled to give the result for a 10 V step). A table can be produced similar to Table 3.2, this is shown as Table 4.1.

If this table were continued the expression for $v_o(n)$ would be

$$v_o(n) = k_1(1 + k_2 + k_2^2 + \cdots\cdots + k_2^n)$$

The terms in the parentheses form a geometrical progression and its sum can be written in closed form (note the number of terms is $n + 1$) as

$$v_o(n) = k_1 \frac{(1 - k_2^{(n+1)})}{1 - k_2}$$

Table 4.1 Development of an expression for the step response of the system in Example 4.7.1 (a)

n	$v_o(n)$	$v_o(n-1)$	$v_o(n)$
0	1.0	0.0	k_1
1	1.0	k_1	$k_1 + k_2 k_1$
2	1.0	$k_1 + k_2 k_1$	$k_1 + k_2 k_1 + k_2^2 k_1$

Noting from Section 3.9.1 that $k_1 = 1 - k_2$, and remembering the scaling is 'times' 10, the expression for $v_o(n)$ can be written

$$v_o(n) = 10(1 - k_2^{(n+1)}) \tag{4.7.3}$$

Inserting the value for k_2 this equation will give the figures already obtained in Tables 3.2 and 3.3.

(b) Again a unit step input will be assumed and the corresponding scaling performed on the derived output expression. Following the approach of (a), Table 4.2. is constructed.

The output corresponding to the nth time instant is given by

$$\begin{aligned} v_o(n) &= (b_0 + b_1)(1 + a_1 + a_1^2 + \cdots\cdots + a_1^n) + b_0 a_1^n \\ &= b_0 \qquad\qquad n = 0 \\ &= (b_0 + b_1)\frac{(1 - a_1^n)}{(1 - a_1)} + b_0 a_1^n \qquad\qquad n > 0 \end{aligned}$$

From the definitions of the coefficients (see Example 3.9.1) it can be shown that $(b_0 + b_1)/(1 - a_1) = 1$. Again remembering the times 10 scaling, the expression for $v_o(n)$ becomes

$$\begin{aligned} v_o(n) &= 10 b_0 \qquad n = 0 \\ &= 1 - (1 - b_0) a_1^n \qquad n > 0 \end{aligned}$$

Substituting for b_0 and a_1 from Example 3.9.1 this equation will give the figures of Table 3.4.

An alternative method of deriving this result is to consider the unit sample response of the system described by eqn (4.7.2). Table 4.3 can be constructed to give the response to the unit sample.

The output is given by

$$\begin{aligned} h(n) &= b_0 \qquad n = 0 \\ &= (b_1 + a_1 b_0) a_1^{n-1} \qquad n > 0 \end{aligned}$$

Table 4.2 Development of an expression for the step response for the system of Example 4.7.1 (b)

n	$v_i(n)$	$v_i(n-1)$	$v_o(n-1)$	$v_o(n)$
0	1.0	0.0	0.0	b_0
1	1.0	1.0	b_0	$b_0 + b_1 + a_1 b_0$
2	1.0	1.0	$b_0 + b_1 + a_1 b_0$	$(b_0 + b_1)(1 + a_1) + a_1^2 b_0$
3	1.0	1.0	$b_0 + b_1(1 + a_1) + a_1^2 b_0$	$(b_0 + b_1)(1 + a_1 + a_1^2) + a_1^3 b_0$

Table 4.3 Development of an expression for the unit sample for the system of Example 4.7.1(b)

n	$v_i(n)$	$v_i(n-1)$	$h(n-1)$	$h(n)$
0	1.0	0.0	0.0	b_0
1	0.0	1.0	b_0	$b_1 + a_1 b_0$
2	0.0	0.0	$b_1 + a_1 b_0$	$a_1(b_1 + a_1 b_0)$
3	0.0	0.0	$a_1(b_1 + a_1 b_0)$	$a_1^2(b_1 + a_1 b_0)$

The response to a 10 V discrete input step can be obtained by convolution.

$$
\begin{aligned}
v_o(n) &= \sum_{k=0}^{n} u(n-k)h(k) \\
&= 10b_0 + 10\sum_{k=1}^{n}(b_1 + a_1 b_0)a_1^{k-1} \\
&= 10\left[b_0 + \frac{(b_1 + a_1 b_0)(1 - a_1^n)}{1 - a_1}\right]
\end{aligned}
$$

Which (after a little algebraic manipulation) is identical to the result derived earlier.

4.8 Summary

This chapter has been concerned with the derivation of the system output response. It has been shown that this response can be considered as comprising of two components, the zero-input response due to initial energy storage, and the zero-state response due to the input signal. These responses can be obtained directly from the system differential (or difference) equation. However such an approach does not give any 'feel' for the connection between the system description and the system response. An alternative method of obtaining the output is via the convolution integral (convolution summation). This uses the system impulse response as a model and it is the convolution of this impulse response with the system input that gives the required output signal.

The method of solution of the difference equation (for a discrete system) offers more freedom. Recursive methods can be used to obtain a numerical solution or can often lead to a closed form for the solution.

In Chapter 7 it will be shown that transform methods give yet another way of obtaining the system response. These methods are usually easier to use but do not give the insight offered by the more direct methods.

Problems

4.1 In Figure 4.27, $x(n)$ represents a system input and $h(n)$ represents the corresponding system impulse response. Obtain the system output by a convolution of $x(n)$ and $h(n)$. The signals are all zero outside the ranges indicated.

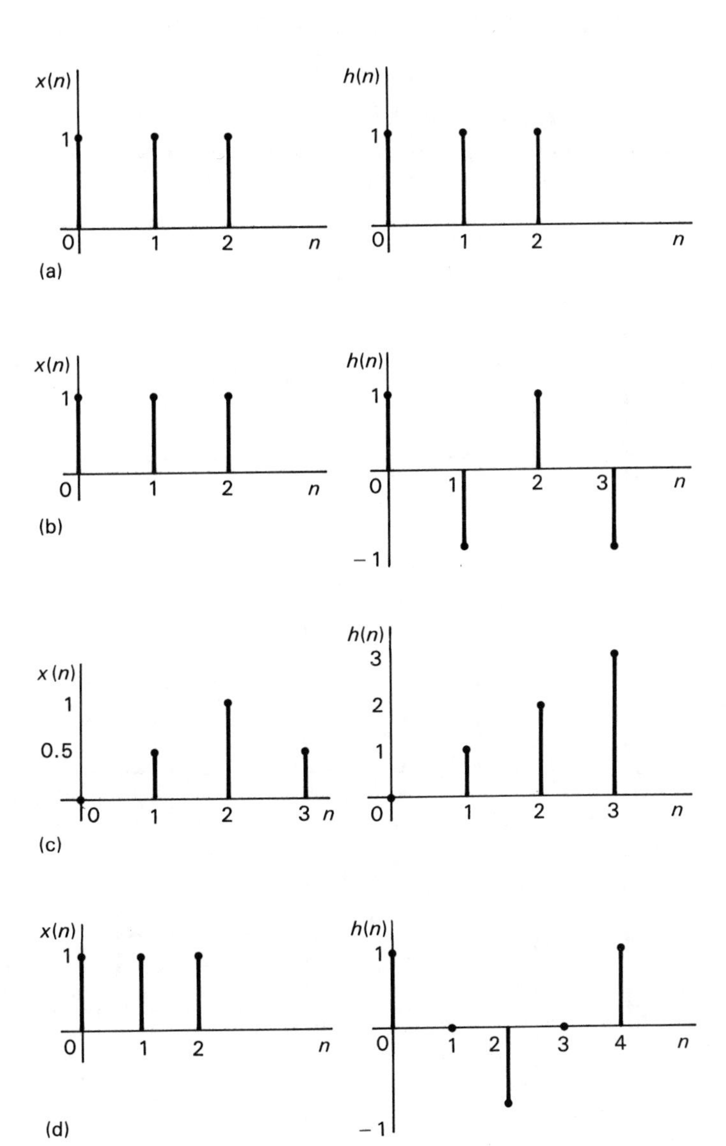

Figure 4.27

4.2 Obtain expressions for convolutions of the signals $x_1(n)$ and $x_2(n)$ as given:

(a) $x_1(n) = u(n)$ $\qquad x_2(n) = 2^n \quad n \geq 0$
$\qquad\qquad\qquad\qquad\qquad\qquad = 0 \quad n < 0$

(b) $x_1(n) = 2^n \quad n \geq 0 \qquad x_2(n) = 2^n \quad n \geq 0$
$\qquad = 0 \quad n < 0 \qquad\qquad = 0 \quad n < 0$

(c) $x_1(n) = 3(2)^n \quad n \geq 0 \qquad x_2(n) = 2(3)^n \quad n \geq 0$
$\qquad = 0 \quad n < 0 \qquad\qquad = 0 \quad n < 0$

4.3 In Figure 4.28, $x(t)$ represents the input to a system and $h(t)$ represents the corresponding impulse response. Use convolution to obtain the system output. Assume all initial energy storage in the systems is zero.

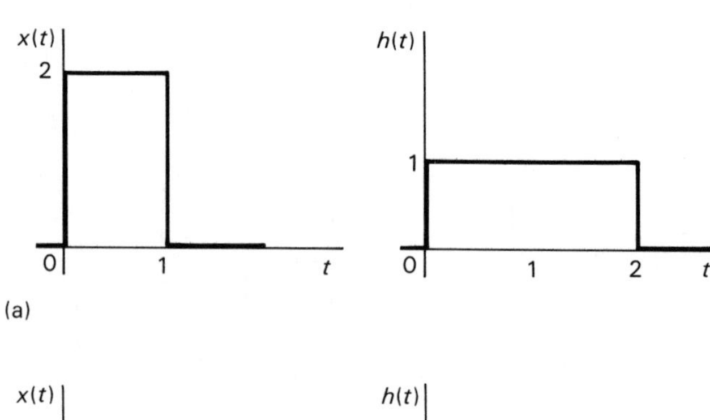

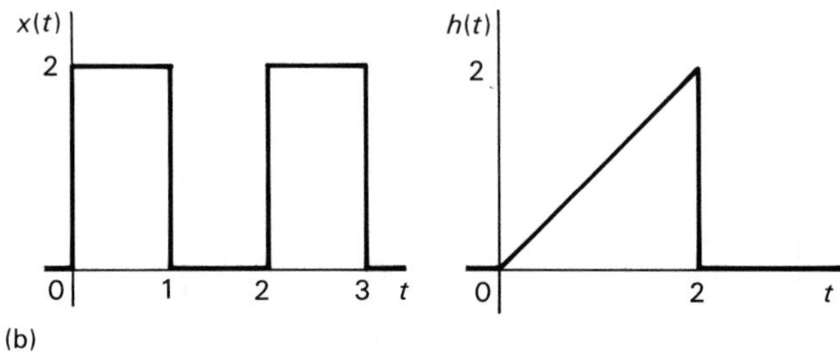

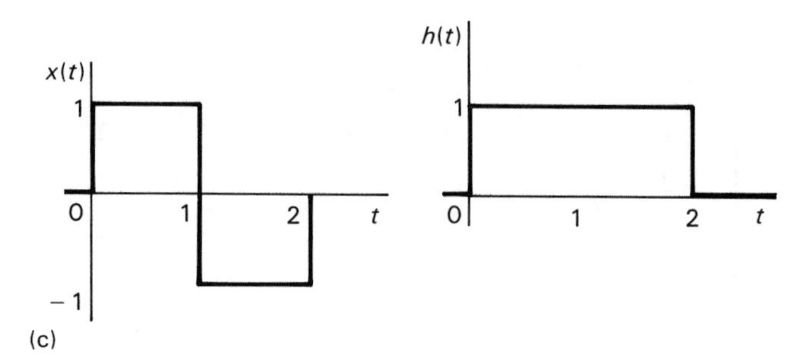

Figure 4.28

4.4 (a) Obtain the convolution of the following signals:

$$\begin{aligned} x_1(t) &= e^{-at} & t \geq 0 \\ &= 0 & t < 0 \\ x_2(t) &= e^{-bt} & t \geq 0 \\ &= 0 & t < 0 \end{aligned}$$

Obtain an answer for the case where $a = b$ and for case where $a \neq b$.

(b) Obtain the convolution of the following signals:

$$\begin{aligned} x_1(t) &= e^{-2t} & t \geq 0 \\ &= 0 & t < 0 \\ x_2(t) &= \sin 2\pi t & 0 \leq t \leq 1 \\ &= 0 & \text{otherwise} \end{aligned}$$

(Hint: Express $\sin 2\pi t$ in exponential form and use the result of (a).)

4.5 As an impulse signal cannot be realised in practice, a short rectangular pulse can be used as a test signal to approximate the impulse. Show that if the true impulse response shows little variation over the width of the test pulse then the approximation is valid provided a scaling factor is used. What is the scaling factor?

A system has a time impulse response

$$\begin{aligned} h(t) &= 10e^{-t} & t \geq 0 \\ &= 0 & t < 0 \end{aligned}$$

Use convolution to sketch the response to a finite width pulse. Using the appropriate scaling factors calculate the response to finite width pulse of 0.5, 0.1 and 0.01 seconds at: (a) a time equal to the pulse width; (b) time $t = 2$. Compare your answers with the true impulse response.

4.6 In Section 4.3.3 the commutative property of convolution was discussed. One result of this property is that the order of cascaded LTI systems can be interchanged.

Verify this result for the systems shown in Figure 4.29. Calculate the output signals $y(t)$, $y(n)$ for the systems as shown and then with the systems interchanged.

4.7 Deconvolution is the process of reforming one of the constituent signals in a convolution integral. The process is also known as inverse filtering and is illustrated for the discrete case in Figure 4.30(a), $g(n)$ in practice can represent some degrading effect on $x(n)$ and the inverse filter $h(n)$ restores the original signal.

The inverse filter must obey the relationship

$$x(n) * g(n) * h(n) = x(n)$$

$$h(n) * g(n) = \delta(n)$$

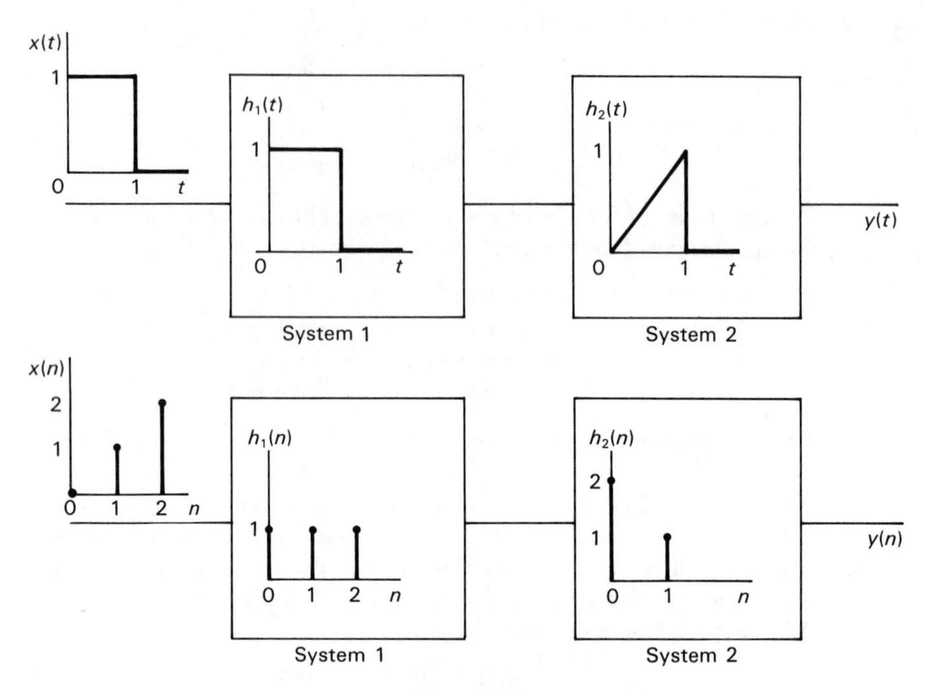

Figure 4.29

Given that the signal $g(n)$, as shown in Figure 4.30(b), is the impulse response of the degrading system, determine the impulse response, $h(n)$ of the inverse filter that will restore the original signal $x(n)$. (Hint: Write the impulse response $h(n)$ in the form $h(n) = h(0)\delta(n) + h(1)\delta(n-1) + h(2)\delta(n-2) + \cdots$ and use convolution to determine the coefficients $h(0)$, $h(1)\ldots$.)

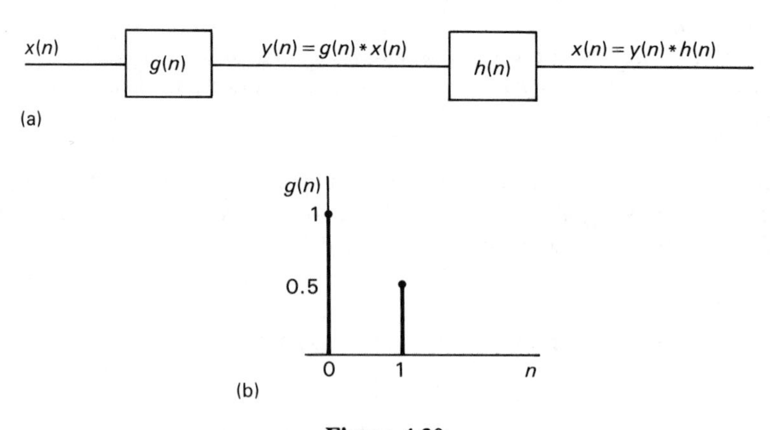

Figure 4.30

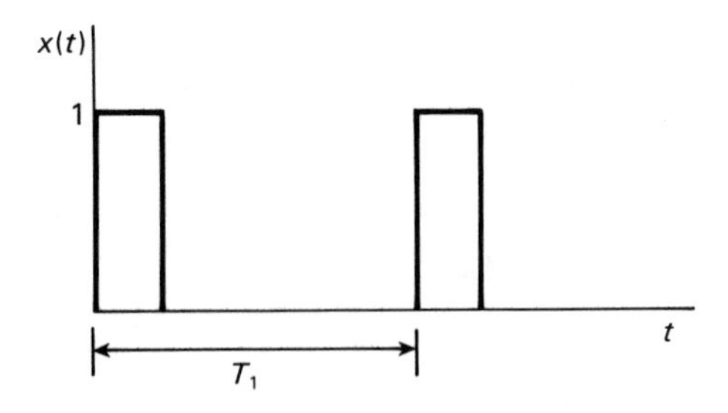

Figure 4.31

4.8 Figure 4.31 shows part of a radar signal $x(t)$ which consists of two rectangular pulses. The target range is obtained from the time difference T_1 as shown. Before the measurement can be made the pulses pass through a system whose output $y(t)$ is related to $x(t)$ by the differential equation.

$$T\frac{\mathrm{d}y(t)}{\mathrm{d}t} + y(t) = x(t)$$

Show the effect of increasing T on the system output $y(t)$ and hence comment on the difficulty in measuring T_1, especially if there is noise present in the system.

4.9 Figure 4.32 shows an armature controlled d.c. motor driving an inertial load. The motor torque T is related to the armature current by $T = K_T i_a$ and the back e.m.f., e, to the speed by $e = K_v \omega$ where K_v and K_T are constants. Determine the differential equation relating motor speed ω to the applied voltage v_i.

Given $K_v = K_T = 0.5$, $R_a = 10$, $J = 0.1$, in appropriate units, sketch the variation in motor speed following the application of 100 V step for v_i when the motor is at standstill. Obtain the final value of the speed and the value of the time constant associated with the variation. What is the motor speed 0.1 s after application of the step?

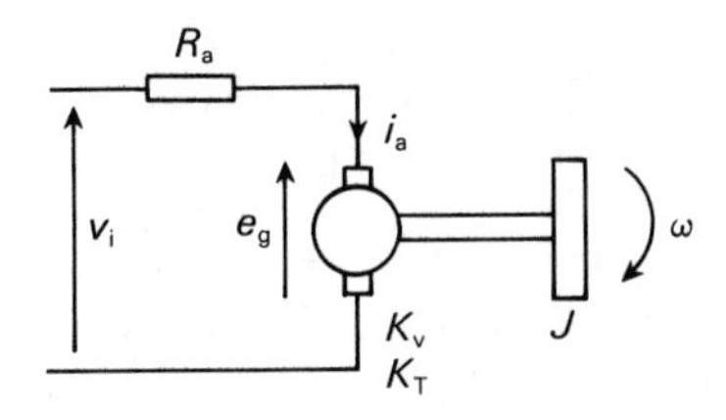

Figure 4.32

4.10 Obtain the response of the circuit shown in Figure 4.33 to the two input waveforms $x_1(t)$ and $x_2(t)$. Assume the capacitor is initially uncharged.

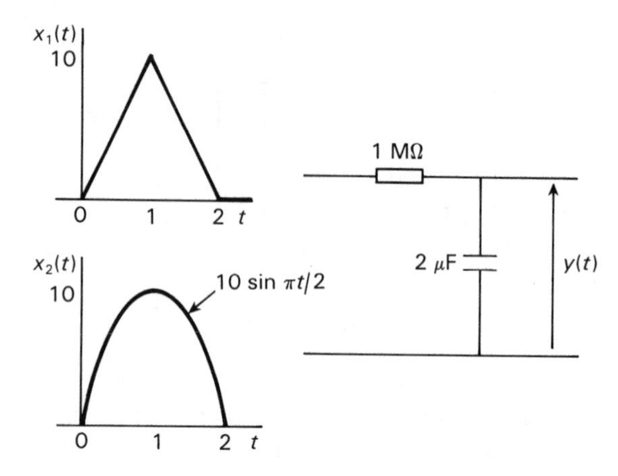

Figure 4.33

4.11 Figure 4.34 shows the basic construction of a moving-coil meter. Current flow in the coil causes a deflecting torque that is opposed by the restoring torque of the spiral springs. Damping is provided by eddy currents induced in the aluminium former supporting the coil.

In such a moving coil meter a steady current of 10 mA produces 90° deflection. The restoring torque is 3×10^{-5} N-m/rad, and the moment of inertia of the moving parts is 4×10^{-7} kg m^{-2}. Derive a differential equation relating the angle of deflection θ to the current i.

What is the undamped natural frequency of the system? What value of damping torque would be required to produce: (a) critical damping; (b) a damping factor of 0.6? Sketch the response to a step in current from zero to 5 mA for both these damping factors.

4.12 A small radar tracking aerial is operated by a control system. For the system the differential equation relating aerial angular position, θ_0, to demanded angle θ_i is given by

$$\frac{d^2\theta_0}{dt^2} + 25\frac{d\theta}{dt} + 126\theta_0 = 126\theta_i$$

At $t = 0$ the tracking demand is that $\theta_i = t$ (a unit ramp of demand). At $t = 0$ the aerial is stationary at position $\theta_0 = -5$ units. Calculate the aerial position at $t = 1, 2, 3$ s. Sketch the motion of the aerial and comment on the error between the demanded position and actual position as $t \to \infty$.

4.13 The following differential equations describe the relationship between the signals at the input and output of four systems. For each system determine: (i) the zero input response when the initial conditions are as given; (ii) the

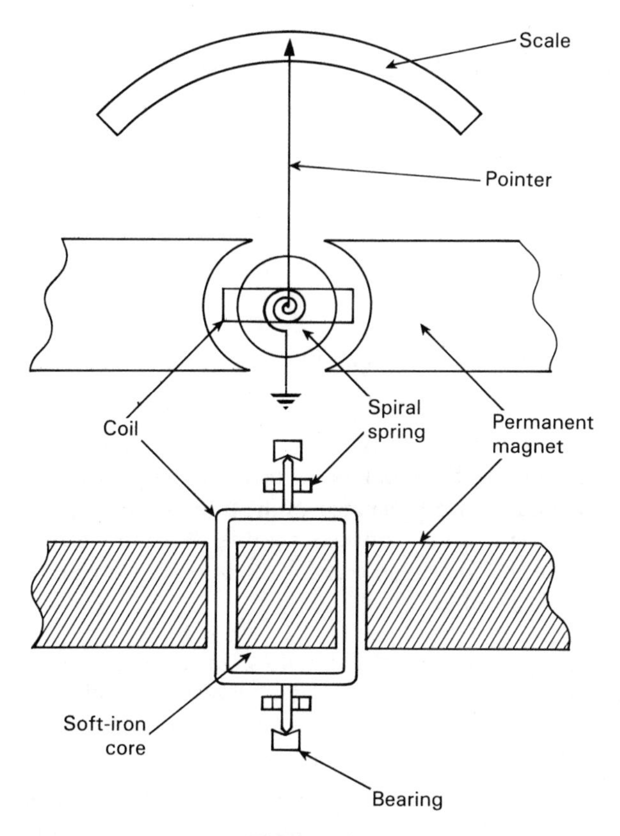

Figure 4.34

impulse response with zero initial conditions.

(a) $\dfrac{dy}{dt} + 2y = x \qquad y = 10 \text{ at } t = 0$

(b) $\dfrac{d^2y}{dt^2} + 5\dfrac{dy}{dt} + 6y = 2x \qquad \left.\begin{array}{r} y = 10 \\ \dfrac{dy}{dt} = -22 \end{array}\right] \text{at } t = 0$

(c) $\dfrac{d^2y}{dt^2} + 2\dfrac{dy}{dt} + 5y = x \qquad \left.\begin{array}{r} y = 4 \\ \dfrac{dy}{dt} = 6 \end{array}\right] \text{at } t = 0$

(d) $\frac{d^3y}{dt^3} + 7\frac{d^2y}{dt^2} + 12\frac{dy}{dt} = x$

$$\left.\begin{aligned} y &= 20 \\ \frac{dy}{dt} &= -12 \\ \frac{d^2y}{dt^2} &= 0 \end{aligned}\right] \text{at } t = 0$$

4.14 A digital filter is described by the following difference equation.

$$y(n) + 0.2y(n-1) - 0.5y(n-2) = x(n)$$

(a) If the input to the filter is a unit step $u(n)$, applied at $n = 0$, solve the equation numerically to obtain the resulting output for values of n, 0 to 10.
(b) Obtain the unit sample response of the filter for values of n, 0 to 10.
(c) By using convolution between the filter sample response and unit step input obtain the filter response for $n = 0$ to 10. Compare your answer with the result obtained in (a).

4.15 A person deposits an amount £A with a financial institution and this earns interest at a rate of r% per annum (compounded annually). However not fully trusting financial institutions he/she decided that each year he/she would take half the interest earned that month and keep it under his/her mattress.

(a) Derive a difference equation relating the amount in the account at the end of the nth year $P(n)$ (just after interest has been paid and half withdrawn) to the corresponding amount in the previous year $P(n-1)$. By writing out the first few terms for $P(n)$ obtain a general form for this quantity. Hence determine an expression for the amount $I(n)$ withdrawn from the account at the end of the nth year.
(b) Determine a difference equation relating the amount of money under the mattress at the end of the nth year to that at the end of the $(n-1)$th year. Solve this equation by the method used in (a).

4.16 A signal $y(n)$ from a sonar tracking system can be considered to consist of a range signal $x(n)$ plus a noise signal $w(n)$. Figure 4.35(a) shows the signal $x(n)$ for a particular target manoeuvre and Figure 4.35(b) shows a sample record of the noise signal $w(n)$. A measurement of the noise is its mean square value.

$$\text{Mean square value} = \frac{1}{N}\sum_{n=0}^{N}[w(n)]^2$$

In order to reduce the effects of the noise the signal $y(n)$ is filtered. Three possible filters are considered and these have impulse responses $h_1(n)$, $h_2(n)$, $h_3(n)$ as shown in Figure 4.35(c).

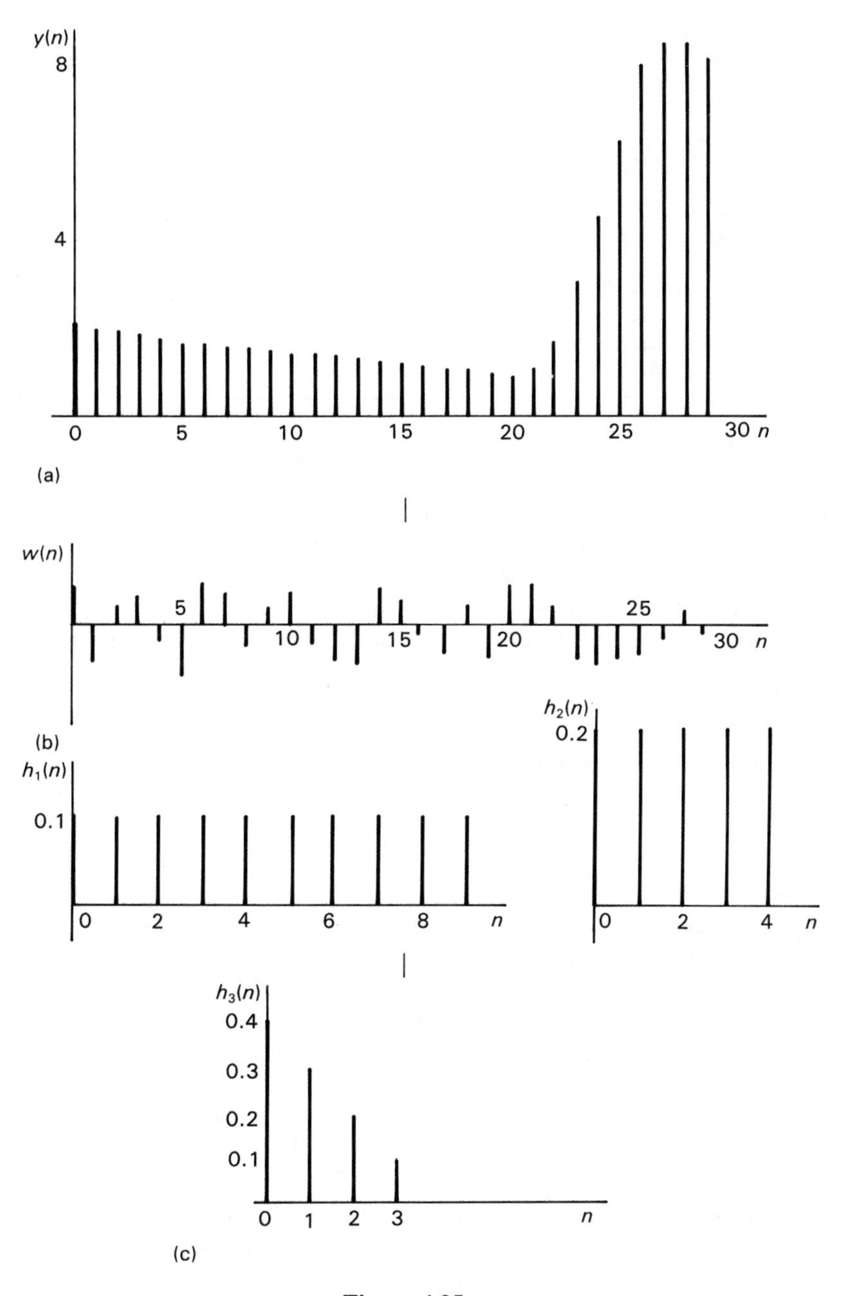
y(n)
8
4
0
5
10
15
20
25
30 n
(a)
w(n)
5
10
15
20
25
30 n
(b)
h₁(n)
0.1
0
2
4
6
8
n
h₂(n)
0.2
0
2
4
n
h₃(n)
0.4
0.3
0.2
0.1
0 1 2 3
n
(c)

Figure 4.35

Table 4.4

Sample	*Range*	*Noise*	*Sample*	*Range*	*Noise*	*Sample*	*Range*	*Noise*
0	2.00	+1.0	11	1.37	−0.4	22	1.70	+0.4
1	1.90	−0.8	12	1.35	−0.7	23	3.00	−0.7
2	1.83	+0.4	13	1.28	−0.9	24	4.50	−0.9
3	1.77	+0.6	14	1.20	+0.8	25	6.10	−0.7
4	1.72	−0.3	15	1.15	+0.5	26	7.90	−0.6
5	1.66	−1.1	16	1.10	−0.1	27	8.95	−0.2
6	1.62	+0.9	17	1.09	−0.5	28	8.50	+0.3
7	1.55	+0.7	18	1.05	+0.4	29	8.00	−0.1
8	1.52	−0.4	19	0.98	−0.6			
9	1.45	+0.4	20	0.90	+0.9			
10	1.40	+0.7	21	1.05	+0.9			

(a) Calculate the mean square value of the noise signal.
(b) Calculate the output of each of the filters due to the noise component and obtain its mean square value.
(c) Calculate the error caused by each of the filters in the signal component when the target is at its minimum and at its maximum range.

For convenience, values of the range and noise signals are shown in Table 4.4.

4.17 A continuous system is described by the following differential equation.

$$\frac{d^2y}{dt^2} + 2.4\frac{dy}{dt} + 4y = 4x$$

Use the following discrete approximations to obtain a difference equation that approximates this differential equation.

$$\frac{dy}{dt} \approx \frac{y(n) - y(n-1)}{T}$$

$$\frac{d^2y}{dt^2} \approx \frac{y(n) - 2y(n-1) + y(n-2)}{T^2}$$

Taking $T = 0.1$ obtain numerically the response of the discrete system to a unit step input. (Take values of n from 0 to 30 and assume zero initial conditions.)

Compare this response with the step response of the continuous system, this is given by

$$y(t) = 1 - 1.25e^{-1.2t}\sin(1.6t + 76.4°)$$

MATLAB exercises

4.18 Repeat Problem 4.1 using the `conv` command, plotting the system output as a discrete variable using `stem`.

Note Some care is needed to obtain the correct axis scaling for the output signal. Even more care is required if the signals being convolved exist for negative n. The student is encouraged to use `conv` for such signals and plot the result with the appropriate scaling.

4.19 Repeat Problem 4.3 using the `conv` command, plotting the system output as a continuous variable.

Note MATLAB can only perform convolution for discrete variables. Hence, the functions in Problem 4.3 must be approximated by discrete signals. The smaller the sampling interval chosen, the more accurate the approximation, but more time will be required for the convolution operation. The result must be scaled by the sampling interval to approximate continuous integration. The remarks made in Exercise 4.18 regarding axis scaling apply to this exercise also.

4.20 This exercise is based on the second half of Problem 4.5 and illustrates how an impulse may be approximated by a short pulse.

Plot $h(t)$ over the range 0 to 5 seconds (1000 points/second will give a good representation of the signal). Use `conv` to convolve with the rectangular pulses given and plot the results, in turn, on the same figure as $h(t)$. Don't forget to scale by the sampling time and the area of the finite width pulse.

Repeat using triangular pulses.

4.21 This exercise investigates the use of different integration algorithms for the discrete approximation of a first order differential equation.

A first order differential equation can be written in the form

$$\frac{\mathrm{d}y}{\mathrm{d}t} = f(t)$$

and its solution is

$$y = \int_0^t f(t)\,\mathrm{d}t$$

In discrete form this becomes

$$y(nT) = \int_0^{(n-1)T} f(t)\,\mathrm{d}t + \int_{(n-1)T}^{T} f(t)\,\mathrm{d}t$$

The first term on the right hand side is $y([n-1]T)$ and the second term can be approximated by the area of a rectangle, width T. Different integration algorithms can be used dependent upon the height chosen for this rectangle. (The dependence upon sampling time T is dropped in the following formulae.)

Forward difference, height of rectangle $f(n-1)$:

$$y(n) = y(n-1) + Tf(n-1)$$

Backward difference, height of rectangle $f(n)$:

$$y(n) = y(n-1) + Tf(n)$$

Trapezoidal approximation, height of rectangle $[f(n) + f(n-1)]/2$:

$$y(n) = y(n-1) + T[f(n) + f(n-1)]/2$$

Compare these three approximations when used for the solution of the differential equation describing the system in Problem 4.9. Use a range of sampling times and compare the responses obtained with the true step response of the continuous system.

4.22 The Student Edition of MATLAB contains the Symbolic Math Toolbox. This toolbox can be used to obtain the solution of differential equations.
Using this toolbox, repeat Exercise 4.13.

Note The zero input responses can be obtained directly using the command `dsolve`.
To derive the impulse response is a little more difficult. It can be obtained using `dsolve` as in the zero input case. However, all initial conditions are set to zero except that of the highest derivative involved. This is set at $1/a_n$, where a_n is the coefficient of the highest derivative in the differential equation. The general proof of this statement is easily shown by considering the initial conditions in the Laplace transform representation (see Section 7.2.2). For low order systems it can be shown by using the expressions for the zero input response and the impulse response given in Sections 4.5.1 and 4.5.2.

4.23 Use the Symbolic Math Toolbox to obtain the step response of the system given in problem 4.9. Obtain and plot responses for values of R_a of 5 Ω, 10 Ω and 20 Ω.

Note The step response can be obtained by making use of the property that the unit step is the integral of the unit impulse. Hence, the terms in the differential equation are differentiated with respect to time and the resulting equation solved for an impulse input as in Exercise 4.22. For comparison purposes, the three responses should be shown on the same display. This can be done by using `ezplot` and using the `hold on` command between plots.

4.24 Repeat Problem 4.11 using the method given in the previous exercise. Plot results for values of damping torque that give a wide range of damping factors.

4.25 Write a MATLAB program for the general numerical solution of difference equations. It should be able to solve equations with up to five terms on left and right hand sides and for input sequences of up to 20 samples. Use this program to repeat parts (a) and (b) of Problem 4.14. Verify that the impulse response can be obtained from the difference in step responses using the relationship

$$\delta(n) = h(n) - h(n-1)$$

Note Use the features of the MATLAB language to best advantage to write this program. The coefficients of the equation can be written as vectors and appropriate vector multiplications of input and output sequence vectors can be used.

4.26 Use the program developed in the last exercise to solve the difference equation obtained in Problem 4.17. Investigate the effect of varying the sampling time T, comparing the discrete responses with the given response of the continuous system.

5

Frequency Response Methods

5.1 Introduction

The methods used in the last chapter to obtain and describe the system response are known as time domain methods. The signals are described in terms of their variation in time and the system description takes the form of an nth order differential equation, n first order differential equations or the system impulse response. As most systems are subject to a wide variety of time varying inputs this is a method of description that has arisen naturally.

An alternative form of system description is formed by *frequency domain methods*. In these methods the system is described in terms of its response to one form of basic signal – the sinusoid. At first sight this may appear a very restrictive description and of limited use as the majority of input signals into working systems are not sinusoidal. However the use of frequency domain techniques forms a very important part of the analysis and design of systems. The reasons for this are as follows:

1. The use of the sinusoid as a basic signal is not as restrictive as first appears. By use of *Fourier methods* it is possible to construct a range of signals by linear combinations of sinusoids.

2. The mathematical methods used in the frequency domain are simpler to apply than the corresponding time domain methods. In particular, differential equations are replaced by algebraic equations (with the variable as a complex quantity).
3. Response in the time domain can be obtained from frequency domain results. If some degree of approximation is allowable this can be done by use of simple graphical methods.
4. For many applications, design in the frequency domain is conceptually easier and gives a greater physical insight into the system.
5. Sophisticated equipment exists for measuring frequency responses and thus enabling a frequency domain model to be obtained from practical measurement.

5.2 Response of a continuous system to a sinusoidal input

5.2.1 The frequency response function

Consider the block diagram of Figure 5.1, where an input $x(t) = A \cos \omega t$ is applied to a linear system at $t = 0$. Before that time the input is zero and there is no initial energy storage. It is required to obtain the resulting output $y(t)$.

The system is in general described by the differential equation

$$(D^n + a_{n-1}D^{n-1} + \ldots\ldots a_0)y = (b_m D^m + \ldots\ldots b_0)x$$

and has an impulse response given by

$$h(t) = k_1 e^{\lambda_1 t} + k_2 e^{\lambda_2 t} + \ldots\ldots k_n e^{\lambda_n t}$$

where

$$H(D) = \frac{b_m D^m + b_{m-1} D^{m-1} + \ldots\ldots b_0}{D^n + a_{n-1} D^{n-1} + \ldots\ldots a_0}$$

$$= \frac{k_1}{D - \lambda_1} + \frac{k_2}{D - \lambda_2} + \ldots\ldots + \frac{k_n}{D - \lambda_n} \qquad (5.2.1)$$

It is assumed that all the roots λ are distinct.

The output $y(t)$ is the convolution of the input $x(t)$ and the impulse response $h(t)$. The impulse response contains n terms and it is easier to perform the convolution for one term and then add the corresponding results for the other terms.

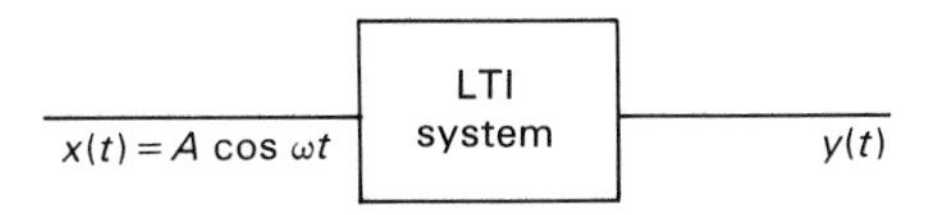

Figure 5.1 Frequency response of an LTI system.

If the input $x(t)$ is expressed in exponential form

$$A\cos\omega t = A\,\frac{(\mathrm{e}^{\mathrm{j}\omega t} + \mathrm{e}^{-\mathrm{j}\omega t})}{2}$$

then only one of the exponential terms needs to be considered initially. Denoting this restricted output as $y_1(t)$

$$y_1(t) = \frac{k_1 A}{2}\int_0^t \mathrm{e}^{\lambda_1(t-\tau)}\,\mathrm{e}^{\mathrm{j}\omega\tau}\,\mathrm{d}\tau$$

As the integral is with respect to τ the term $\mathrm{e}^{\lambda_1 t}$ can be taken outside the integral sign and the expression becomes

$$\begin{aligned} y_1(t) &= \frac{k_1 A}{2}\,\mathrm{e}^{\lambda_1 t}\int_0^t \mathrm{e}^{(\mathrm{j}\omega-\lambda_1)\tau}\,\mathrm{d}\tau \\ &= \frac{k_1 A}{2}\,\mathrm{e}^{\lambda_1 t}\left[\frac{\mathrm{e}^{(\mathrm{j}\omega-\lambda_1)\tau}}{\mathrm{j}\omega-\lambda_1}\right]_0^t \\ &= \frac{k_1 A}{2}\,\frac{(\mathrm{e}^{\mathrm{j}\omega t} - \mathrm{e}^{\lambda_1 t})}{\mathrm{j}\omega-\lambda_1} \end{aligned}$$

By now including the other terms and adding the complex conjugate the complete solution for $y(t)$ is obtained and this gives the output response for $t>0$. This response can be considerably simplified by considering its form as $t\rightarrow\infty$. Assuming the system is stable all the modes $\mathrm{e}^{\lambda_1 t}\ldots\mathrm{e}^{\lambda_n t}$ decay to zero then the total response becomes

$$y(t)\big|_{t\rightarrow\infty} = \frac{A}{2}\left[\frac{k_1}{\mathrm{j}\omega-\lambda_1} + \frac{k_2}{\mathrm{j}\omega-\lambda_2} + \ldots\ldots\frac{k_n}{\mathrm{j}\omega-\lambda_n}\right]\mathrm{e}^{\mathrm{j}\omega t} + \text{conjugate}$$

However comparing the terms in the bracket with those on the right-hand side of eqn (5.2.1) the output can be written

$$y(t)\big|_{t\rightarrow\infty} = \frac{A}{2}\,H(\mathrm{j}\omega)\mathrm{e}^{\mathrm{j}\omega t} + \frac{A}{2}\,H^*(\mathrm{j}\omega)\mathrm{e}^{-\mathrm{j}\omega t}$$

If $H(\mathrm{j}\omega)$ is written in modulus and angle form $|H(\mathrm{j}\omega)|$, $\underline{/\varphi}$ where

$$\varphi = \underline{/H(\mathrm{j}\omega)}$$

then

$$\begin{aligned} y(t)\big|_{t\rightarrow\infty} &= \frac{A|H(\mathrm{j}\omega)|}{2}\,[\mathrm{e}^{\mathrm{j}(\omega t+\varphi)} + \mathrm{e}^{-\mathrm{j}(\omega t+\varphi)}] \\ &= A|H(\mathrm{j}\omega)|\cos(\omega t+\varphi) \end{aligned} \qquad (5.2.2)$$

Equation (5.2.2) can be interpreted as follows. If a sinusoid is applied to an LTI system then, *after the transients have decayed*, the output is another sinusoid *of the same frequency*. However its magnitude and angle are different from those of the input sinusoid. Its magnitude is changed by a factor $|H(j\omega)|$ and its angle by an amount $\underline{/H(j\omega)}$ where

$$H(j\omega) = \frac{b_m(j\omega)^m + b_{m-1}(j\omega)^{m-1} + \ldots\ldots b_0}{(j\omega)^n + a_{n-1}(j\omega)^{n-1} + \ldots\ldots a_0}$$

$H(j\omega)$ is known as the *frequency response function* of the system and it can be obtained using the coefficients in the differential equation.

As the input and output signals are both sinusoidal in the steady state they can be completely specified by their frequency, amplitude and phase. This can be conveniently represented by writing these signals as $X(j\omega)$ and $Y(j\omega)$ where $x(t)$ and $y(t)$ are the corresponding time signals. The amplitude and phase of these signals are given by $|X(j\omega)|$, $|Y(j\omega)|$ and $\underline{/X(j\omega)}$, $\underline{/Y(j\omega)}$, respectively. Then

$$\frac{Y(j\omega)}{X(j\omega)} = H(j\omega)$$

and

$$\frac{|Y(j\omega)|}{|X(j\omega)|} = |H(j\omega)|, \underline{/Y(j\omega)} - \underline{/X(j\omega)} = \underline{/H(j\omega)}$$

This is an alternative method of stating the information given by eqn (5.2.2). The modulus of the frequency response functions gives the ratio of the amplitude output/input, its angle gives the phase difference between output and input.

Although strictly the signals at input and output should be written $X(j\omega)$ and $Y(j\omega)$ when frequency responses are considered, often for convenience, the dependence upon $j\omega$ is dropped and the signals are simply written X, Y. This representation of signals is known as a *frequency domain* representation.

The following example illustrates the process of obtaining the frequency response function for a simple first order system.

EXAMPLE 5.2.1

Obtain the frequency response function for the simple low-pass filter circuit shown in Figure 5.2.

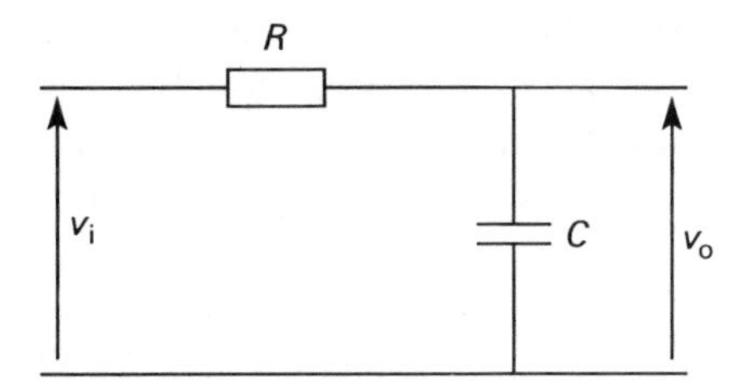

Figure 5.2 Low-pass filter of Example 5.2.1.

If $R = 5\ \text{k}\Omega$ and $C = 0.1\ \mu\text{F}$ obtain the magnitude change and phase shift through the filter at a frequency of 1 kHz.

SOLUTION

This circuit has been analysed as Example 3.4.1 (a) and it is described by the following differential equation

$$CR\frac{\mathrm{d}v_o}{\mathrm{d}t} + v_o = v_i$$

$$(CRD + 1)v_o = v_i$$

Substituting $\mathrm{j}\omega$ for D gives

$$H(\mathrm{j}\omega) = \frac{1}{1 + \mathrm{j}\omega CR}$$

and this is the the required frequency response function.

At $f = 1$ kHz ($\omega = 6.28 \times 10^3$), inserting values for C and R gives

$$H(\mathrm{j}\omega) = \frac{1}{1 + \mathrm{j}3.14} \qquad \text{at } f = 1\ \text{kHz}$$

This is a complex number and it is the magnitude and angle of this number that gives the magnitude and phase shift through the circuit. Taking the modulus and angle of $H(\mathrm{j}\omega)$ (see Appendix)

$$|H(\mathrm{j}\omega)| = \frac{1}{|(1 + \mathrm{j}3.14)|} = \frac{1}{\sqrt{(1^2 + (3.14)^2)}} = 0.303$$

$$\underline{/H(\mathrm{j}\omega)} = \tan^{-1}0 - \tan^{-1}\frac{3.14}{1} = -72.3°$$

5.2.2 *Use of reactance methods*

Students who have met the idea of reactance and the use of j notation will realise there is an easier method of obtaining the frequency response function derived in the last example. The circuit of Figure 5.2 can be considered as a potential divider with the impedance of the upper arm consisting of the resistance R and that of the lower arm the reactance $1/\mathrm{j}\omega C$. Then, the ratio of output to input is given by

$$\frac{V_o}{V_i}(\mathrm{j}\omega) = \frac{1/\mathrm{j}\omega C}{R + 1/\mathrm{j}\omega C} = \frac{1}{1 + \mathrm{j}\omega CR}$$

which is the frequency response function already derived.

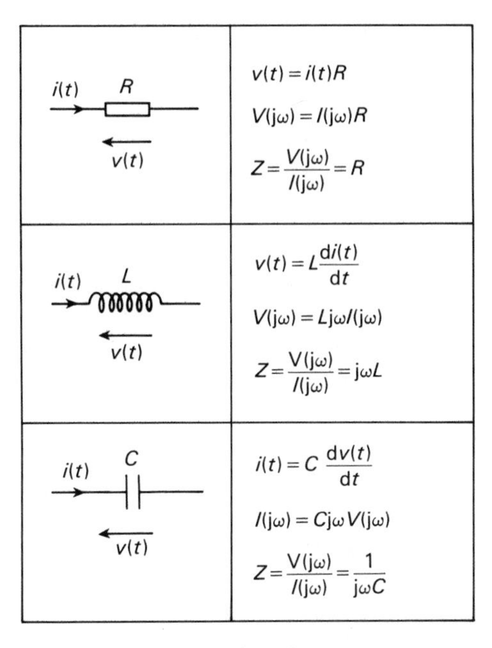

Figure 5.3 Impedance functions for electrical components.

The reactance method is of course not fundamentally different from the method considered in Section 5.2.1. If each component is described by the differential equation relating voltage and current at its terminals the corresponding frequency domain description is obtained by writing $d/dt \equiv j\omega$. The ratio $V(j\omega)/I(j\omega)$ then gives the impedance of the component. This is shown in Figure 5.3.

The concept of impedance has been derived with respect to electrical systems. A more generalised approach can be made and applied to systems other than electrical systems. Some care is required in these instances as there is more than one method of choosing the variables, and some ambiguity can arise. However when obtaining the frequency response of any type of system it is best to make the substitution $j\omega \equiv d/dt$ as early as possible. This avoids having to solve simultaneous differential equations, these being replaced by simultaneous algebraic equations.

5.2.3 General plot for a first order system

Example 5.1 considered a first order system and this led to the frequency response function

$$H(j\omega) = \frac{1}{1 + j\omega T}$$

Provided that there are no differential terms on the right-hand side of the equation describing the system then any first order system will lead to a frequency response of the form

$$H(j\omega) = \frac{K}{1 + j\omega T} \tag{5.2.3}$$

The constants K and T can be physically interpreted as the gain and time constants of the system (see Section 4.6.1).

In Example 5.1 the magnitude and angle of the response were calculated at one frequency only. In order to predict the time response from the frequency domain it is necessary to obtain values over a range of frequency and these are most useful if presented graphically. There is more than one method of doing this, the method considered in this chapter is to produce separate plots of $|H(j\omega)|$ and $\angle H(j\omega)$ against frequency. Taking the magnitude and angle of the frequency response function given by eqn (5.2.3) gives

$$|H(j\omega)| = \frac{K}{\sqrt{(1 + (\omega T)^2)}}$$

$$\angle H(j\omega) = -\tan^{-1}\omega T$$

These expressions give the magnitude and angle plots shown in Figure 5.4.

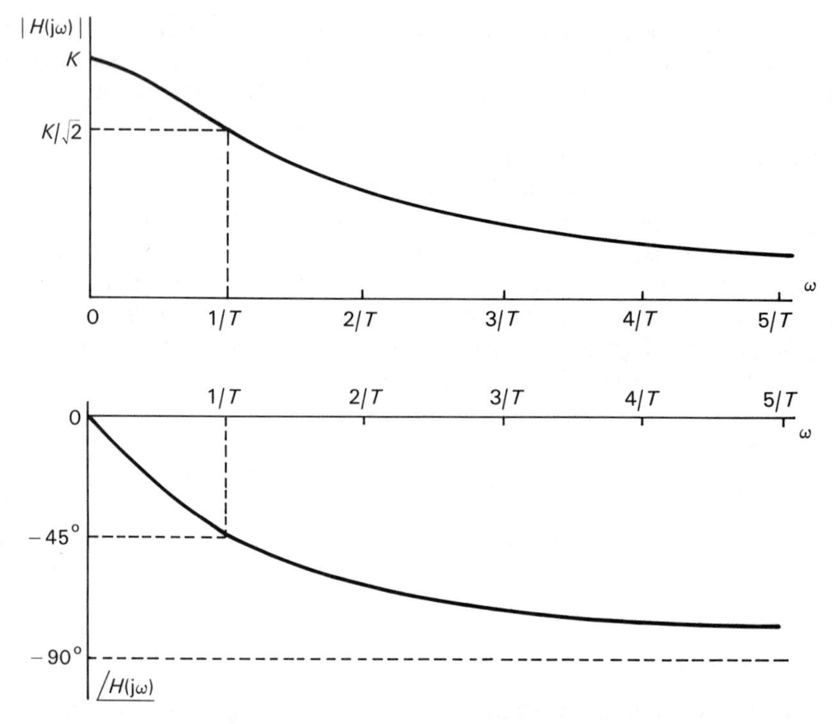

Figure 5.4 Magnitude and phase responses of a second order system.

Some of the general features of the first order response can be obtained from these plots:

1. The magnitude plot starts at K (the system gain factor) at $\omega = 0$ and falls to zero as $\omega \rightarrow \infty$.
2. The magnitude of the gain has fallen by a factor $1/\sqrt{2}$ from its low frequency gain when $\omega = 1/T$.
3. The phase plot starts at 0° when $\omega = 0$ and approaches −90° (90° lagging) as $\omega \rightarrow \infty$.
4. The phase shift is −45° when $\omega = 1/T$.

Hence, for the first order system, there is a complete correspondence between the frequency response and the time response. From the frequency response plots the constants T and K can be obtained. Hence from eqn (4.5.18) the impulse response can be obtained and then, via convolution, the response to any input.

Figure 5.4 shows the plots of magnitude and phase against frequency. The magnitude plot has been drawn with linear scales, there are however advantages in using logarithmic scales. Very large ranges in frequency and magnitude can be encompassed without losing the fine detail and as will be seen in Section 5.2.5 logarithmic scales lead to an easy method of obtaining approximate plots for more involved frequency response functions. To obtain the logarithmic scaling for the magnitude plot it is usual to express the magnitude in decibels. The decibel (dB) is obtained by taking twenty times the logarithm (to base 10) of the magnitude as a ratio. Frequency response plots when the magnitude is in dB and the frequency is plotted logarithmically are known as Bode plots. They will be considered in detail in Section 5.2.5.

5.2.4 General plot for a second order system

The differential equation describing the second order system (see Section 4.6.2) is given in standard form as

$$\frac{\mathrm{d}^2 y}{\mathrm{d}t^2} + 2\omega_n \zeta \frac{\mathrm{d}y}{\mathrm{d}t} + \omega_n^2 y = \omega_n^2 x$$

Writing $\mathrm{j}\omega$ for $\mathrm{d}/\mathrm{d}t$ and $(\mathrm{j}\omega)^2$ for $\mathrm{d}^2/\mathrm{d}t^2$ gives the frequency response function in standard form as

$$H(\mathrm{j}\omega) = \frac{Y(\mathrm{j}\omega)}{X(\mathrm{j}\omega)} = \frac{\omega_n^2}{(\mathrm{j}\omega)^2 + \mathrm{j}2\zeta\omega_n\omega + \omega_n^2} \tag{5.2.4}$$

To evaluate the magnitude and angle of $H(\mathrm{j}\omega)$ it is re-written

$$H(\mathrm{j}\omega) = \frac{\omega_n^2}{(\omega_n^2 - \omega^2) + \mathrm{j}2\zeta\omega_n\omega}$$

$$|H(\mathrm{j}\omega)| = \frac{\omega_n^2}{\sqrt{[(\omega_n^2 - \omega^2)^2 + (2\zeta\omega_n\omega)^2]}} \tag{5.2.5}$$

The following general features apply to the magnitude as given by eqn (5.2.5):

1. When $\omega = 0$, $|H(j\omega)| = 1$.
2. As $\omega \rightarrow \infty$, $|H(j\omega)| \rightarrow 0$.
3. When $\omega = \omega_n$, $|H(j\omega)| = 1/2\zeta$.
4. The maximum value of $|H(j\omega)|$ can be obtained by differentiation and setting the result to zero.

$$\frac{d|H(j\omega)|}{d\omega} = -\frac{1}{2}\frac{\omega_n^2[-2(\omega_n^2 - \omega^2)2\omega + 8\zeta^2\omega_n^2\omega]}{[(\omega_n^2 - \omega^2)^2 + 4\zeta^2\omega_n^2\omega^2]^{3/2}}$$

Setting this to zero gives the frequency where the peak occurs as

$$\omega_{\text{max}} = \omega_n\sqrt{(1 - 2\zeta^2)} \tag{5.2.6}$$

Substituting into the original expression for $|H(j\omega)|$, eqn (5.2.5) gives

$$|H(j\omega)|_{\text{max}} = \frac{1}{2\zeta\sqrt{(1 - \zeta^2)}} \tag{5.2.7}$$

It should be noted that if $\zeta > 1/\sqrt{2}$, i.e. $\zeta > 0.707$ then there is no peak, the maximum occurs at $\omega = 0$.

The corresponding phase shift is given by

$$\underline{/H(j\omega)} = -\tan^{-1}\frac{2\zeta\omega_n\omega}{\omega_n^2 - \omega^2} \tag{5.2.8}$$

The following features apply to this function:

1. When $\omega = 0$, $\underline{/H(j\omega)} = 0°$.
2. As $\omega \rightarrow \infty$, $\underline{/H(j\omega)} \rightarrow 180°$.
3. When $\omega = \omega_n$, $\underline{/H(j\omega)} = 90°$.

The magnitude and phase are plotted for various values, of ζ in Figure 5.5. These plots are not meant to be accurate, in fact it is quite difficult to show these plots for a reasonable range of ζ on linear scales. Hence they will be re-drawn using logarithmic scales in the next section.

The marked difference between the magnitude plot for this system and that of the first order system is that the second order system shows a peak in the response for values of $\zeta < 0.707$. This tends to associate with the overshoot shown in the time response (see Section 4.6) for values of $\zeta < 1$. The peak in the frequency response is often a useful feature of a system enabling it to 'tune' certain frequencies and reject others. For values of $\zeta > 0.707$ the response shows low-pass properties as does the first order system but its rate of cut-off is higher as will be shown in the next section.

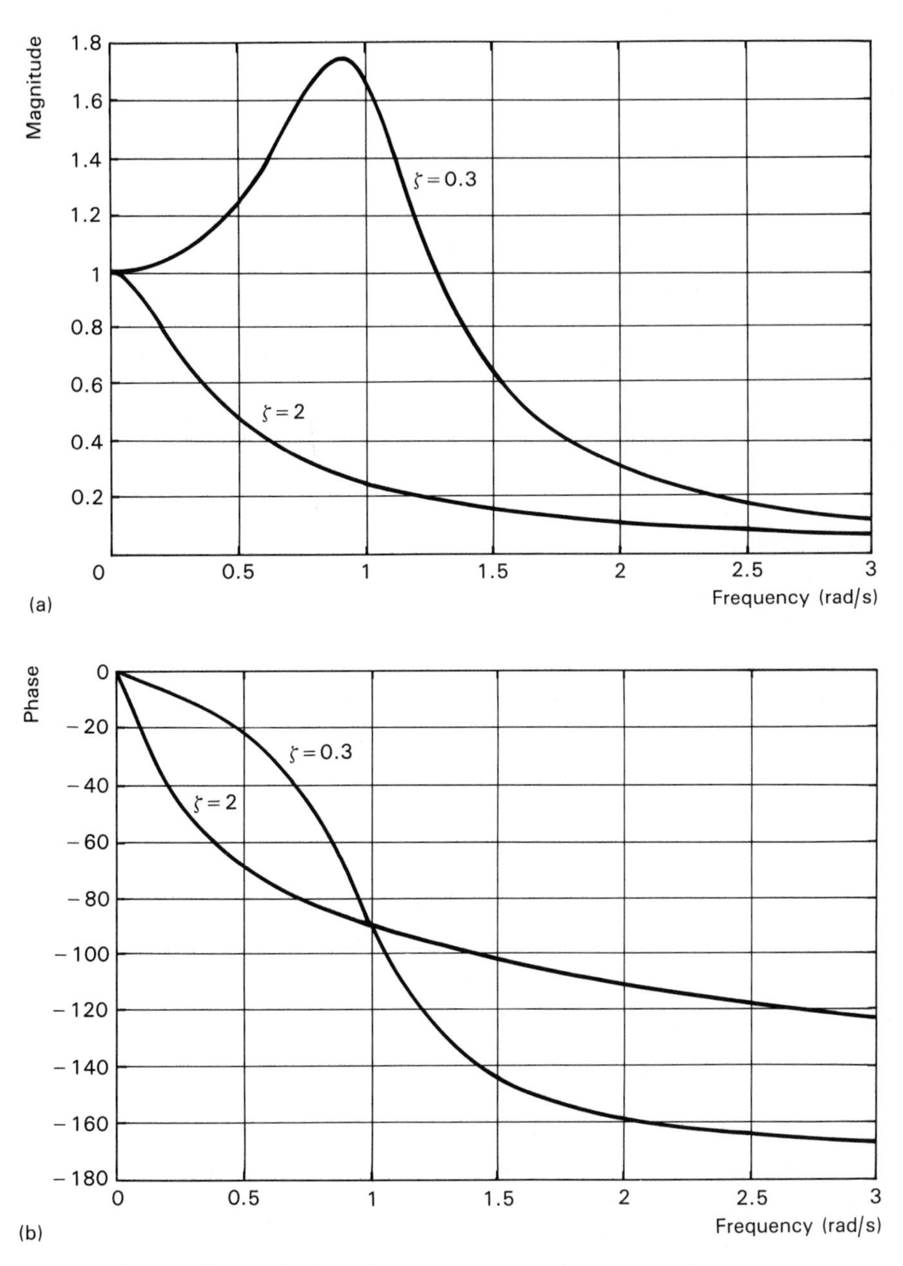

Figure 5.5 Magnitude and phase responses of a second order system.

EXAMPLE 5.2.2

The diagram shown in Figure 5.6 represents an electrically driven vibration table. The constants relating to the system are the following:

R and L = resistance and inductance of the vibrator coil.
K_v = back e.m.f. in coil, per unit of velocity (back e.m.f. = $K_v\, dx/dt$).
M = mass of table.
k = stiffness of suspension.
B = damper constant.
K_T = force on table per unit of current (force = $K_T i$).

Obtain the frequency response function relating the table displacement to the applied voltage. Show how this function can be simplified if $R \gg L$.

For the condition where $R \gg L$ the system parameters have the following values; $M = 1$ kg, $k = 2$ N/m, $B = 1$ N/m s^{-1}, $K_T = K_v = 2$ N/A, $R = 100\ \Omega$. Determine the values of ω_n and ζ for the system. Hence sketch the form of frequency response labelling any salient points.

SOLUTION

The current in the coil is given by

$$I = \frac{V_i - E}{R + j\omega L} \tag{5.2.9}$$

where E is the back e.m.f. The back e.m.f. (as a time signal) is given by

$$e = K_v \frac{dx}{dt}$$

which in the frequency domain becomes

$$E = K_v j\omega X \tag{5.2.10}$$

The force F is given by

$$F = K_T I \tag{5.2.11}$$

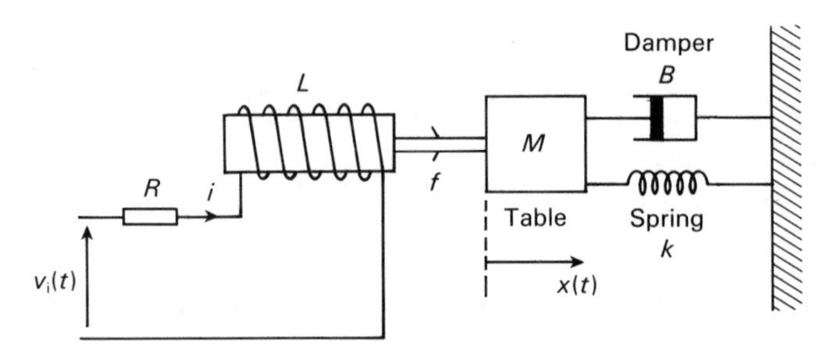

Figure 5.6 Vibration table for Example 5.2.2.

The mechanical system is described in time by the equation

$$f - kx - B\frac{\mathrm{d}x}{\mathrm{d}t} = M\frac{\mathrm{d}^2x}{\mathrm{d}t^2}$$

and in frequency by

$$F - kX - B\mathrm{j}\omega X = M(\mathrm{j}\omega)^2 X \tag{5.2.12}$$

Equations (5.2.9) to (5.2.12) form a set of simultaneous equations. Eliminating all the variables other than V_i and X leads to the frequency response function

$$\frac{X}{V_\mathrm{i}} = \frac{b_0}{a_3(\mathrm{j}\omega)^3 + a_2(\mathrm{j}\omega)^2 + a_1(\mathrm{j}\omega) + a_0}$$

where

$$b_0 = \frac{K_\mathrm{T}}{kR}, \qquad a_0 = 1, \qquad a_1 = \frac{B}{k} + \frac{L}{R} + \frac{K_\mathrm{T}K_\mathrm{v}}{kR},$$

$$a_2 = \frac{M}{k} + \frac{BL}{kR}, \qquad a_3 = \frac{ML}{kR}$$

If $R \geqslant \omega L$ then eqn (5.2.9) becomes

$$I = \frac{V_\mathrm{i} - E}{R}$$

and the frequency response function become

$$\frac{X}{V_\mathrm{i}} = \frac{K_\mathrm{T}/kR}{\frac{M}{k}(\mathrm{j}\omega)^2 + \left(\frac{B}{k} + \frac{K_\mathrm{T}K_\mathrm{v}}{kR}\right)(\mathrm{j}\omega) + 1}$$

Inserting values gives

$$\frac{X}{V_\mathrm{i}} = \frac{0.01}{0.5(\mathrm{j}\omega)^2 + 0.52(\mathrm{j}\omega) + 1} \tag{5.2.13}$$

The standard form for the frequency response function for a second order system is given by eqn (5.2.4)

$$\frac{X}{V_\mathrm{i}} = \frac{\omega_n^2}{(\mathrm{j}\omega)^2 + 2\zeta\omega_n(\mathrm{j}\omega) + \omega_n^2} \tag{5.2.14}$$

To fit into standard form eqn (5.2.13) is re-written as

$$\frac{X}{V_\mathrm{i}} = 0.01\left[\frac{2}{(\mathrm{j}\omega)^2 + 1.04\mathrm{j}\omega + 2}\right] \tag{5.2.15}$$

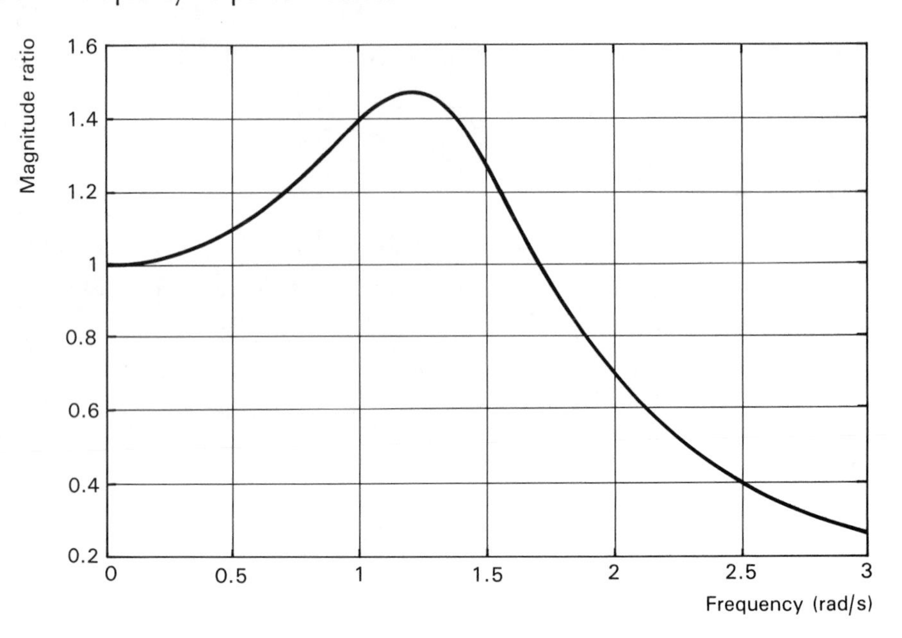

Figure 5.7 Frequency response of vibration table Example 5.2.2.

The term inside the square brackets corresponds to the standard form. The constant outside is a scaling factor that must be applied to any magnitude results derived.

By comparison of the terms in eqns (5.2.14) and (5.2.15) then

$$\omega_n^2 = 2, \qquad \omega_n = \sqrt{2};\ 2\zeta\omega_n = 1.04, \qquad \zeta = 0.367$$

Applying eqns (5.2.6) and (5.2.7) gives $\omega_{\max} = 1.21$ and $|H(\mathrm{j}\omega)|_{\max} = 1.464$.

The form of the response is shown in Figure 5.7, where the scaling factor has been accounted for by expressing the displacement of the vibration table in centimetres.

5.2.5 *Bode plots*

In the last two sections the frequency responses of first and second order systems have been considered. This section extends the idea of frequency response plots to higher order systems and also takes into consideration the effect of further terms on the right-hand side of the system differential equation. In general form, the frequency response function can be written

$$H(\mathrm{j}\omega) = \frac{b_m(\mathrm{j}\omega)^m + b_{m-1}(\mathrm{j}\omega)^{m-1} + \ldots\ldots b_0}{(\mathrm{j}\omega)^n + a_{n-1}(\mathrm{j}\omega)^{n-1} + \ldots\ldots a_0}$$

The magnitude and angle of this function can be calculated at any frequency. This can be quite a tedious operation and one seeks a method of obtaining approximate plots without excessive calculation.

The polynomials forming the numerator and denominator of the frequency response function can be factorised. Because the coefficients in the polynomials are real the factors will either be first order terms, or second order terms that cannot be further factorised into first order terms with real coefficients. Putting the first and second order terms into their standard forms the frequency response function can be written

$$H(j\omega)=\frac{K[1+j\omega T_1][(j\omega)^2+2\zeta_1\omega_{n1}(j\omega)+\omega_{n1}^2][......]}{[1+j\omega T_2][(j\omega)^2+2\zeta_2\omega_{n2}(j\omega)+\omega_{n2}^2][......]}$$

A numerical example will clarify the process.
Suppose

$$H(j\omega)=\frac{12(j\omega)^2+10(j\omega)+2}{(j\omega)^3+1.5(j\omega)^2+0.75(j\omega)+0.25}$$

The constant terms corresponding to a_0, b_0 are first made equal to unity

$$H(j\omega)=\frac{2[6(j\omega)^2+5(j\omega)+1]}{0.25[4(j\omega)^3+6(j\omega)^2+3(j\omega)+1]}$$

Factorising numerator and denominator

$$H(j\omega)=\frac{8(1+j\omega 2)(1+j\omega 3)}{(1+j\omega)(4(j\omega)^2+2j\omega+1)}$$

Converting the quadratic term to standard form gives

$$H(j\omega)=\frac{2(1+j\omega 2)(1+j\omega 3)}{(1+j\omega)((j\omega)^2+0.5j\omega+0.25)}$$

the quadratic term having $\omega_n=0.5$ rad/s, $\zeta=0.5$. In this example no details were given of the method by which the denominator cubic was factorised into a first order term and a quadratic. Many methods exist for root finding but for high order polynomials recourse to a suitable computer program is desirable. In practice the frequency response function is often obtained by considering cascaded subsystems, the numerator and denominator then occur naturally in factorised form.

Another type of factor can occur in numerator or denominator and this arises when either b_0 or a_0 is zero. In this case the term $j\omega$ can be removed as a factor leaving a reduced order polynomial. If this occurred in both numerator and denominator then the factor would of course cancel.

Hence the form of the frequency response function considered for approximate plots is

$$H(\mathrm{j}\omega) = \frac{K(\ )(\ )(\ )\ldots\ldots}{(\ )(\ )(\ )\ldots\ldots}$$

where the parentheses ( ) can indicate the following types of terms:

1. First order terms $(1 + \mathrm{j}\omega T)$
2. Second order terms $(\mathrm{j}\omega)^2 + 2\zeta\omega\omega_n + \omega_n^2$.
3. Terms $\mathrm{j}\omega$.

The magnitude and angle of the frequency response function is given by

$$|H(\mathrm{j}\omega)| = \frac{K|(\ )||(\ )|\ldots\ldots}{|(\ )||(\ )|\ldots\ldots} \tag{5.2.16}$$

$$\underline{/H(\mathrm{j}\omega)} = \underline{/(\ )} + \underline{/(\ )} + \ldots\ldots \text{ numerator terms}$$
$$- \underline{/(\ )} - \underline{/(\ )} + \ldots\ldots \text{ denominator terms}$$

As the forms of the terms that can appear are limited, the response of individual terms can be combined to obtain the overall response. With the angle response the individual responses add and subtract and it is not too difficult to combine them graphically. With the magnitude terms however the process is one of multiplication and division – much more difficult graphically. When considering the magnitude plots of the first and second order terms in Sections 5.2.3 and 5.2.4 it was stated that plotting using logarithmic scales often gives better proportioned plots. Logarithmic plotting will also change the multiplication and division of magnitude plots into additions and subtractions. Expressing the magnitude in eqn (5.2.16) in dB

$$\begin{aligned}\text{Magnitude in dB} &= 20 \log |H(\mathrm{j}\omega)| \\ &= 20 \log K + 20 \log |(\ )| + \ldots\ldots \text{ numerator terms} \\ &\quad -20 \log |(\ )| - \ldots\ldots \text{ denominator terms}\end{aligned}$$

Considering the terms individually the responses produced by the different types of term are as follows:

1. K in numerator. This is referred to as a gain term. Its magnitude in dB is obtained very easily as $20 \log_{10} K$, it does not depend upon frequency and has a constant response. Because it is a real term there is no associated phase shift. The rather trivial plots of magnitude and phase against frequency are shown in Figure 5.8.
2. Factor $\mathrm{j}\omega$ in numerator. This term arises from a simple differential equation expressing that the output is the rate of change of the input. For this reason it is called a *differential term*. Its magnitude in dB is given by

$$\begin{aligned}\text{Magnitude} &= 20 \log_{10} |\mathrm{j}\omega| \\ &= 20 \log_{10} \omega\end{aligned}$$

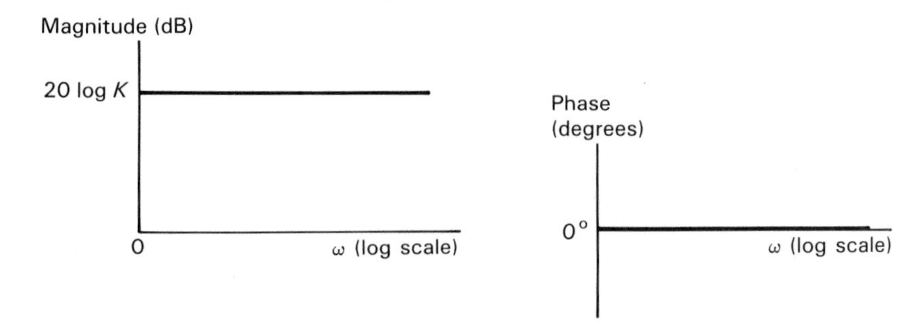

Figure 5.8 Magnitude and phase plots for the constant term.

If this is plotted on a linear scale it will be logarithmic in shape. If however it is plotted against $\log_{10} \omega$ (i.e., frequency on a log scale) it will be a straight line. In order to draw this line its slope and a point on it are required. The most convenient point is where the line crosses the 0 dB axis, when $0 = 20 \log \omega$, which occurs at $\omega = 1$.

One must be careful when specifying the slope using logarithmic scales. On a linear scale the slope would be specified by change of magnitude/change in frequency. However on a logarithmic scale a change in frequency, say 100 rad/s, would be represented by different distances at different points on the frequency scale. What are represented by equal distances are increases by equal *multiples* in frequency. The multiple chosen is a matter of convenience and two measures are in common use, the decade – a multiple of ten times, and the octave – a multiple of two times.

Returning now to the magnitude, magnitude $= 20 \log \omega$. Suppose the change in magnitude is required for a frequency change of a decade, i.e., from a frequency ω_1 to a frequency $10\omega_1$.

$$\begin{aligned}\text{Change in magnitude (dB)} &= 20 \log 10\omega_1 - 20 \log \omega_1 \\ &= 20 \log 10 + 20 \log \omega_1 - 20 \log \omega_1 \\ &= 20 \log 10 \\ &= 20\end{aligned}$$

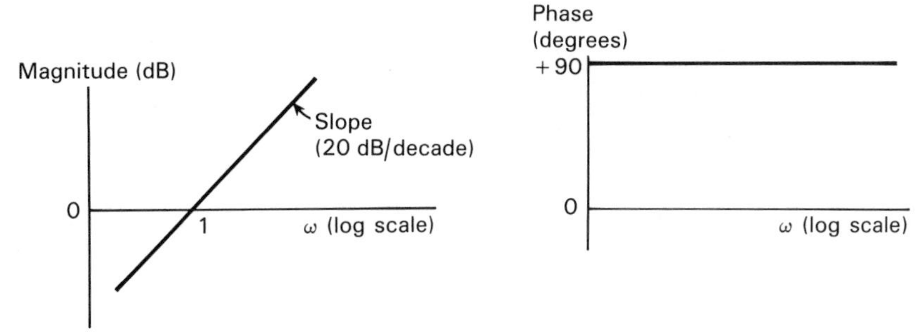

Figure 5.9 Magnitude and phase plots for the differential term.

The slope is 20 dB/decade. If a similar exercise is performed with a doubling of frequency this gives a slope of 6.02 dB/octave which is usually approximated as 6 dB/octave.

Hence the plot representing the magnitude of the term $j\omega$ when using logarithmic scales is a straight line of slope 20 dB/decade (6 dB/octave) passing through the 0 dB point when $\omega = 1$. This gives another reason for using logarithmic scales – simple shapes for the plots.

The phase associated with the differential term is given by

$$\begin{aligned} \text{Phase} &= \angle j\omega \\ &= \tan^{-1}\omega/0 \qquad \text{(there is no real part)} \\ &= \tan^{-1}\infty \\ &= +90^\circ \qquad (90^\circ \text{ leading}) \end{aligned}$$

Hence the phase is a constant at 90° throughout the frequency range. The magnitude and phase plots for this term are shown in Figure 5.9.

3. Factor $j\omega$ in the denominator. This term arises from a simple differential equation expressing that the input equals the rate of change of the output – the output is the integral of the input. For this reason it is called an *integral term.*

   This term can be treated in a similar manner to the differential term, however because it is in the denominator there is a negative sign in the magnitude expression

$$\text{Magnitude} = -20 \log_{10} \omega$$

   As with the differential term this gives a straight line plot on logarithmic scales, passing through the point $\omega = 1$. The slope however is −20 dB/decade (−6 dB/octave).

   The phase shift associated with this term is 90°, because it is in the denominator however it is −90° (90° lagging). Plots of magnitude and phase against frequency are shown in Figure 5.10.

4. Factor $(1 + j\omega T)$ in numerator. It will be shown that this term gives a leading phase shift throughout the frequency range. For this reason it is called a *lead term.*

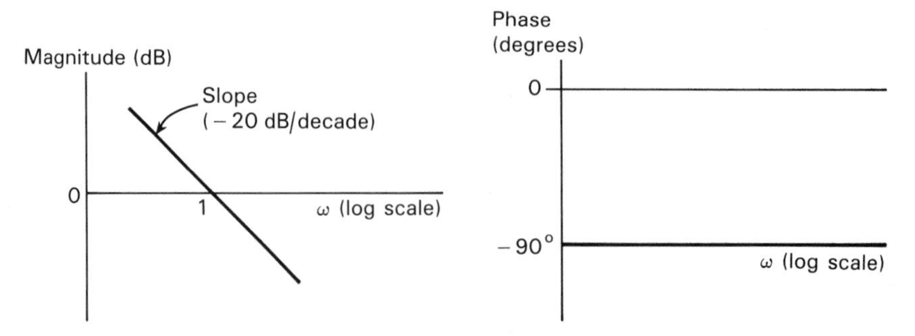

Figure 5.10 Magnitude and phase plots for the integral term.

The magnitude of this term is given by

$$\begin{aligned}\text{Magnitude} &= 20 \log |(1 + j\omega T)| \\ &= 20 \log \sqrt{(1 + \omega^2 T^2)} \qquad (5.2.17)\end{aligned}$$

This expression can be plotted for a given value of T but what is required is an approximate form that can be plotted with the ease of the integral and differential terms. This form can be obtained by considering the magnitude, eqn (5.2.17) at the extremes of the frequency range.

If ω is such that $\omega T \ll 1$ (this means $\omega \ll 1/T$) then $(1 + \omega^2 T^2) \approx 1$ and the magnitude becomes 20 log 1 which is 0 dB.

At the low frequencies (low here means frequencies such that $\omega \ll 1/T$) the magnitude plot approaches (is asymptotic to) the 0 dB axis. If now in eqn (5.2.17) ω is such that $\omega T \gg 1$ (this means $\omega \gg 1/T$) then the magnitude becomes

$$\begin{aligned}\text{Magnitude} &\approx 20 \log \sqrt{(\omega T)^2} \\ &\approx 20 \log(\omega T) \\ &\approx 20 \log \omega + 20 \log T\end{aligned}$$

Using logarithmic scales this represents a straight line slope 20 dB/decade and this line will intersect the 0 dB axis when

$$0 = 20 \log(\omega T), \qquad \omega T = 1, \qquad \omega = 1/T$$

The line is asymptotic to the plot at high frequencies and it is shown together with the asymptote at low frequencies in Figure 5.11.

The plot approaches these asymptotes at the extremes of the frequency range but as yet nothing has been established about its behaviour in between. An obvious point to check this behaviour is at $\omega = 1/T$. Putting $\omega = 1/T$ into the magnitude expression of eqn (5.2.17) gives

$$\begin{aligned}\text{Magnitude} &= 20 \log \sqrt{(1 + (T/T)^2)} \\ &= 3 \text{ dB}\end{aligned}$$

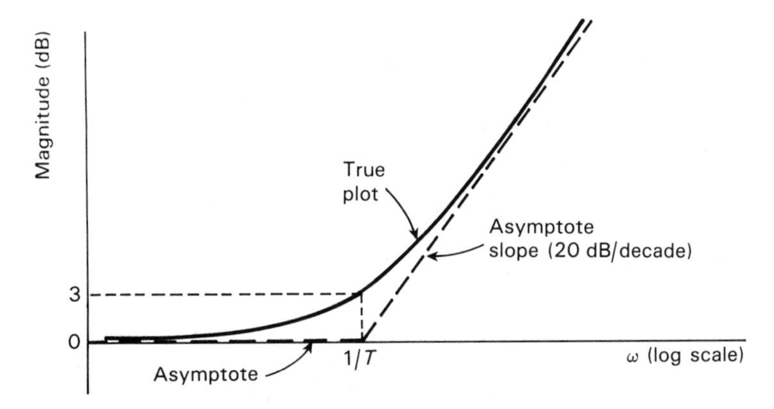

Figure 5.11 Magnitude plot for lead term.

The true plot passes through a magnitude of +3 dB at a frequency $\omega = 1/T$ and is shown in Figure 5.11. The range of frequencies used in the plots is often several decades giving a variation in magnitude of some 40 dB. Hence two methods are available to obtain the approximate plot:

(a) Use the straight line segments as an approximation to the plot. This is very easy to do but will produce an error of 3 dB at $\omega = 1/T$. This may not be significant over the range of magnitude involved.
(b) Outline the straight line segments and using these as asymptotes draw the plot to pass through 3 dB at $\omega = 1/T$. This produces a plot which is accurate enough for the majority of practical applications.

The frequency $\omega = 1/T$ goes under a variety of names, corner frequency, break point, 3 dB point. The phase of the lead term is given by

$$\text{phase angle} = \tan^{-1} \omega T/1$$

Its general shape is as follows:

$$\begin{aligned} &\text{As } \omega \to 0, && \text{phase angle} \to 0°. \\ &\text{As } \omega \to \infty, && \text{phase angle} \to +90°. \\ &\text{At } \omega = 1/T && \text{phase angle} = +45°. \end{aligned}$$

The plot of phase/frequency is shown in Figure 5.12 and takes the form of the tangent curve turned on its side.

The phase characteristic can be approximated by using three straight line segments (two break points) with break points $1/10T$ and $10/T$. This is shown in Figure 5.12. The error in this approximation is about 5° at the break points. Although this is easy to construct for one term, care is needed when several terms are combined as in Example 5.2.3. For this reason, together with the ease of modern computation, the phase is often calculated directly and this approximation is not used.

5. Factor $(1 + j\omega T)$ in denominator. As will be shown this term gives a lagging phase shift at all frequencies and for this reason it is known as a lag term.

   As this term is in the denominator of the frequency response function expressions for its magnitude and phase only differ by their sign from the corresponding expressions for the lead term.

$$\text{Magnitude} = -20 \log \sqrt{(1 + (\omega T)^2)}$$
$$\text{Phase} = -\tan^{-1} \omega T$$

   The magnitude and phase plots, together with the approximate plots are shown in Figure 5.13.

6. Factor $((j\omega)^2 + 2\zeta\omega_n\omega + \omega_n^2)$ in denominator. This is the quadratic term and has already been investigated using linear scales in Section 5.2.4. It is assumed that a factor ω_n^2 is present in the numerator (this can always be achieved by writing the

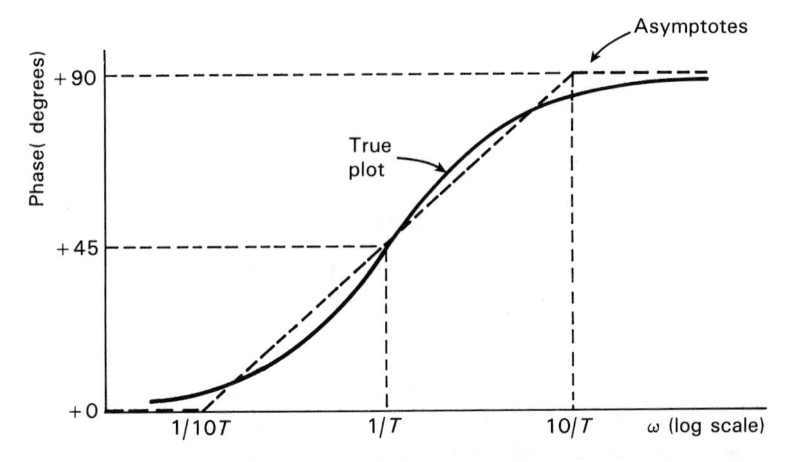

Figure 5.12 Phase plot for lead term.

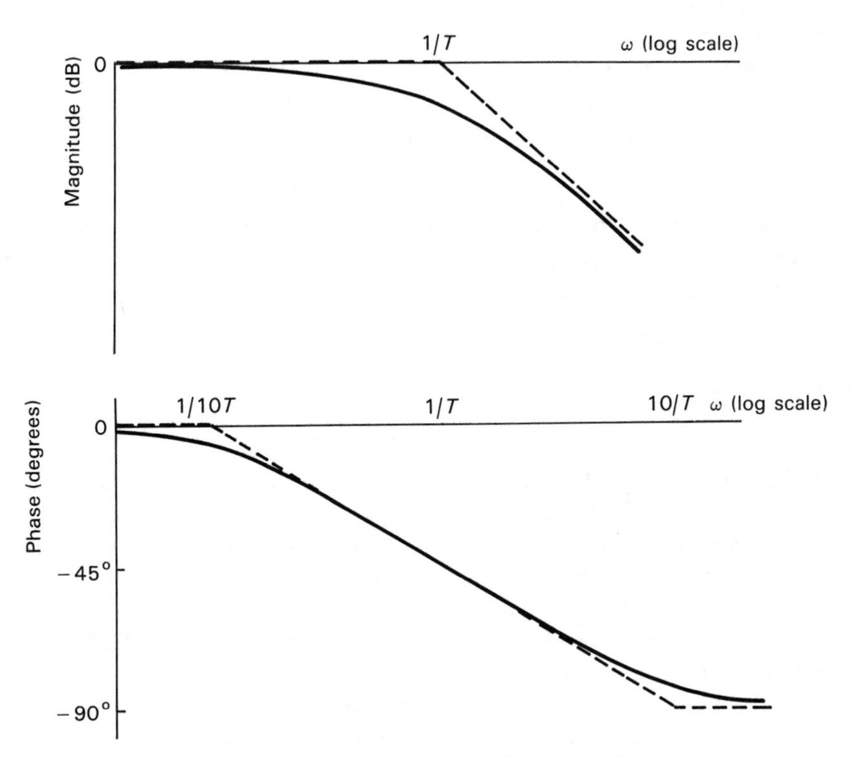

Figure 5.13 Magnitude and phase plots for the lag term.

numerator as a product of ω_n^2 and a constant factor). Hence the term in standard form is

$$H(j\omega) = \frac{\omega_n^2}{(j\omega)^2 + 2\zeta\omega\omega_n + \omega_n^2}$$

$$\text{Magnitude} = 20 \log \omega_n^2 - 20 \log |(j\omega)^2 + 2\zeta\omega_n\omega + \omega_n^2|$$

As $\omega \longrightarrow 0$, magnitude $\longrightarrow$ 0 dB

As $\omega \longrightarrow \infty$, magnitude is given by

$$\begin{aligned}\text{Magnitude} &= 20 \log \omega_n^2 - 20 \log \omega^2 \\ &= 20 \log \omega_n^2 - 40 \log \omega\end{aligned}$$

This is the equation of a straight line of slope -40 dB/decade (-12 dB/octave). This line would intersect the 0 dB axis when

$$20 \log \omega_n^2 - 20 \log \omega^2 = 0$$

$$\omega = \omega_n$$

Hence like the lead and lag terms the quadratic terms can be approximated at the extremes of the frequency range by two easily constructed asymptotes. Unlike the first order terms however the error at the break points can be considerable depending on the value ζ. The value of the magnitude at $\omega = \omega_n$ has already been calculated in Section 5.2.4 as $1/2\zeta$ and converting this to dB gives

$$\text{Magnitude at } \omega = \omega_n \text{ equals } -20 \log 2\zeta$$

More useful for constructing the plot is the maximum value of the magnitude and the frequency at which it occurs. These have been derived as eqns (5.2.6) and (5.2.7). Converting the maximum to dB gives

$$\text{Maximum magnitude} = -20 \log 2\zeta\sqrt{(1 - \zeta^2)} \quad \zeta < 0.707$$

and this occurs at

$$\omega = \omega_n\sqrt{(1 - 2\zeta^2)}$$

Hence the asymptotes and the maximum value can be easily established and these can form the basis of the plot. The result however will not be as accurate as the first order plots and extra points may require calculation in order to improve accuracy.

Plots for the magnitude of the quadratic term using logarithmic scales are shown in Figure 5.14. The phase of the quadratic term has already been derived as eqn (5.2.8). This is drawn using a logarithmic frequency scale in Figure 5.15.

The six terms considered are not the only terms that can appear in the frequency response function – they are however the most common. The quadratic term can appear in the numerator and this will of course alter the sign of the log magnitude and the associated phase.

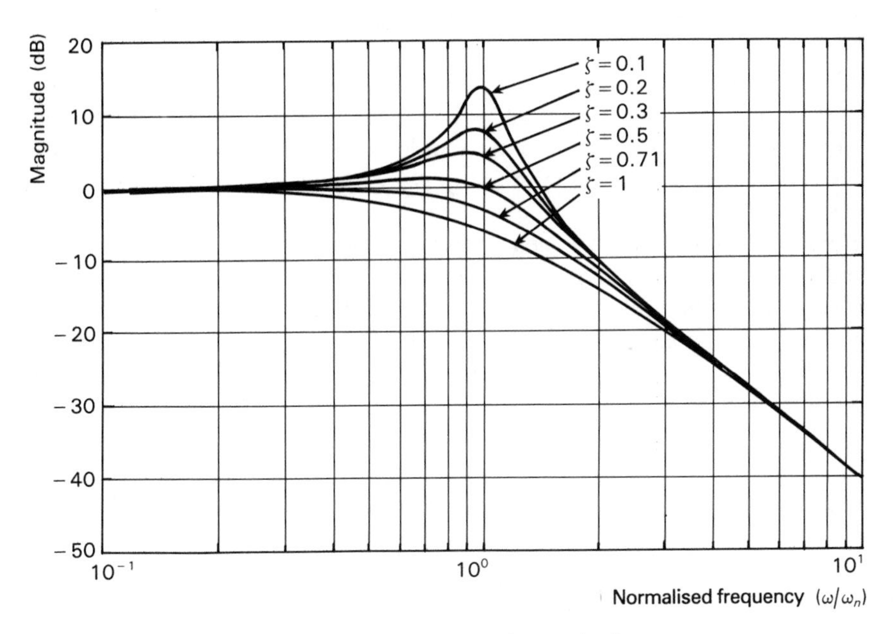

Figure 5.14 Magnitude plot for quadratic term.

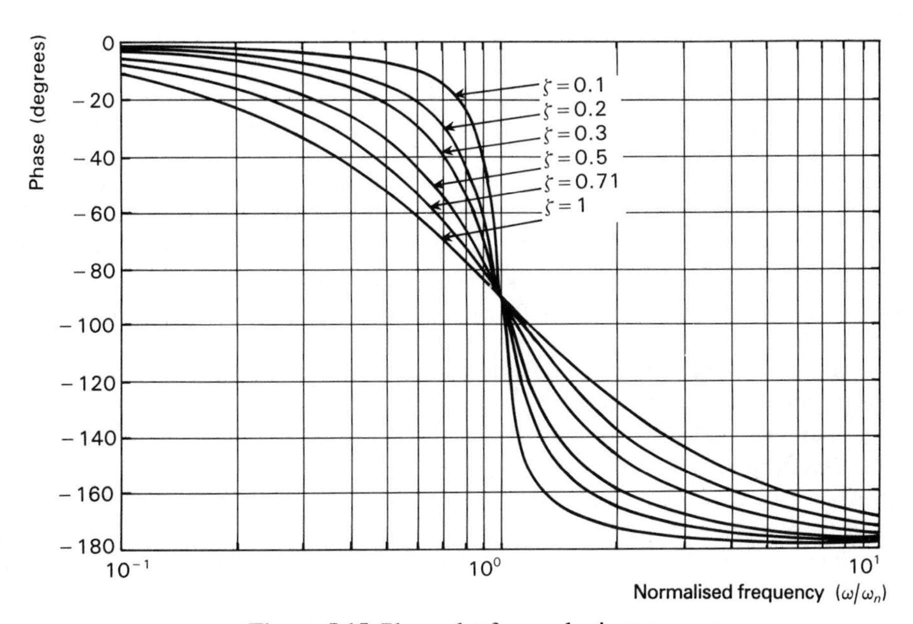

Figure 5.15 Phase plot for quadratic term.

When the frequency response polynomials are factorised terms of the form $(1 - j\omega T)$ can appear. It will be shown in Chapter 7 that such terms in the denominator indicate an unstable system and one is not usually interested in the frequency response of such a system. A term $(1 - j\omega T)$ can appear in the numerator and it is known as a non-minimum phase term. A little thought will show that its magnitude response is identical to the lead term $(1 + j\omega T)$. It has however the phase response of a lag term.

Having established the frequency response functions of individual terms these responses have to be combined to obtain the complete response. Because of the logarithmic scales this involves adding the individual plots of both magnitude and phase. This can be done on a point by point basis but advantage can be taken of the fact that the plots are constructed of straight line segments. The following example illustrates the method.

EXAMPLE 5.2.3

Obtain the Bode plots for the following frequency response function

$$H(j\omega) = \frac{20(1 + j\omega 0.5)}{j\omega(1 + j\omega 0.1)}$$

SOLUTION

The following terms are present in the frequency response function:

A constant term having value 20 log 20 = 26 dB.
An integral term.
A lead term having a corner frequency $1/0.5 = 2$ rad/s.
A lag term having a corner frequency $1/0.1 = 10$ rad/s.

As $\omega = 0$ cannot be represented on a logarithmic scale the frequency origin is chosen for convenience. As the integral term will cross the 0 dB axis at $\omega = 1$ then $\omega = 0.1$ is a convenient figure. Also it is reasonable to extend the plot to $\omega = 100$ a decade above the highest corner frequency. Hence 3 cycle logarithmic paper would be suitable. The magnitude plots for the individual terms are shown in Figure 5.16.

The complete plot is obtained by adding the plots for the individual factors. This could be done at selected frequency points throughout the range. However it is easier to use the fact that the plot will be comprised of straight line segments and to proceed as follows:

At $\omega = 0.1$ the magnitude is $20 + 26 = 46$ dB

From $\omega = 0.1$ to $\omega = 2$ the magnitude is falling by 20 dB/dec (6 db/octave) due to the integral term. By drawing a line with that slope (or by calculation at the point $\omega = 2$) the segment from $\omega = 0.1$ to $\omega = 2$ can be constructed.

From $\omega = 2$ to $\omega = 10$ the lead term increases the magnitude by the same amount as the lag term reduces it, the plot remains constant at 20 dB.

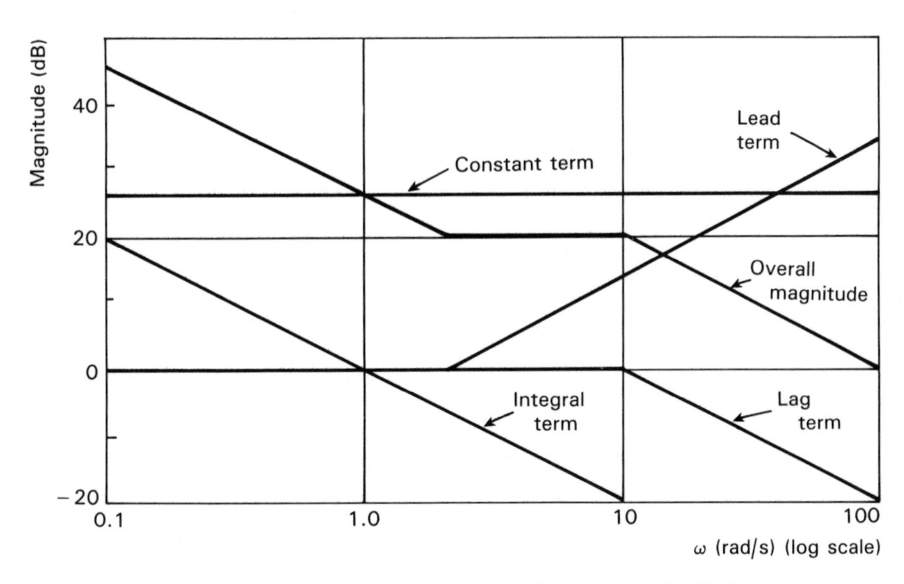

Figure 5.16 Bode plots of magnitude for Example 5.2.3.

From $\omega = 10$ to $\omega = 100$ both the lag and integral terms are decreasing the magnitude while the lead term is increasing it. The total magnitude must fall at 20 dB/decade until it reaches 0 dB at $\omega = 100$.

The total magnitude plot is also shown in Figure 5.16. By marking the 3 dB points at the corner frequencies a more accurate plot can be obtained if required.

Plotting the phase response of the individual terms does not help to obtain the overall phase response. It is easier to construct a table showing the phase shift of each term at selected frequencies. The overall phase shift can then be obtained by addition and subtraction of the phase shifts of the individual factors. This is shown in Table 5.1 and is plotted in Figure 5.17.

Table 5.1 Construction of phase plot for Example 5.2.3

ω *(rad/s)*	$\angle j\omega$ *(degrees)*	$\angle(1 + j\omega 0.5)$ *(degrees)*	$\angle(1 + j\omega 0.1)$ *(degrees)*	*Total phase (degrees)*
0.1	90.0	2.86	0.57	−87.71
0.2	90.0	5.71	1.15	−85.44
0.5	90.0	14.04	2.86	−78.82
1.0	90.0	26.57	5.71	−69.14
2.0	90.0	45.00	11.31	−56.31
5.0	90.0	68.20	26.57	−48.37
10.0	90.0	78.69	45.00	−56.31
20.0	90.0	84.29	63.43	−69.14
50.0	90.0	87.87	78.69	−80.98
100.0	90.0	88.85	84.29	−85.44

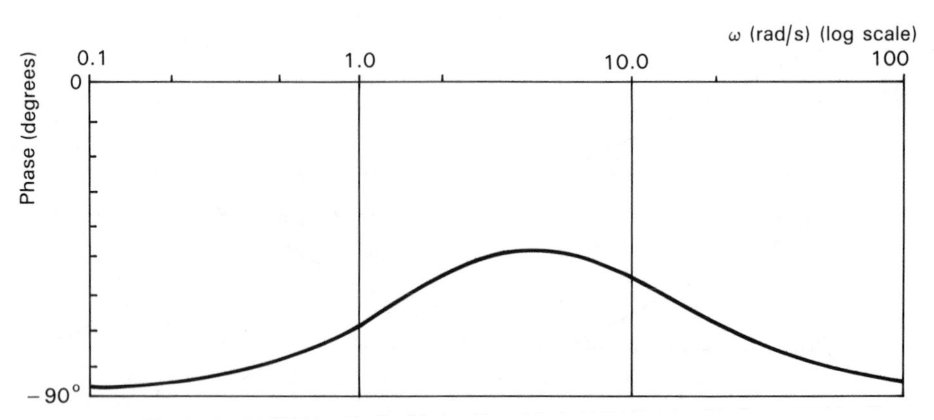

Figure 5.17 Phase plot for Example 5.2.3.

5.3 Frequency response of discrete systems

The frequency response of a discrete system can be investigated by methods similar to those used for continuous systems. This would show that the response consists of two components; a transient term that decays to zero for a stable system and the steady state sinusoidal response. Taking only the steady state component one can determine how the magnitude and phase of the response relate to the constants of the system.

Suppose the system is described by the difference equation

$$\begin{aligned} a_0 y(n) + a_1 y(n-1) + \ldots\ldots a_k y(n-k) \\ = b_0 x(n) + b_1 x(n-1) + \ldots\ldots b_l x(n-l) \end{aligned} \tag{5.3.1}$$

Following the analysis of the continuous system the input would be taken as a discrete cosine waveform

$$\begin{aligned} x(n) &= X \cos n\omega T \\ &= \frac{X}{2}\left[\mathrm{e}^{\mathrm{j}n\omega T} + \mathrm{e}^{-\mathrm{j}n\omega T}\right] \end{aligned}$$

However a slightly easier approach is to assume a steady state solution. A complex input signal $X\mathrm{e}^{\mathrm{j}n\omega T}$ will give a cosine term plus an imaginary sine term. Because of the linearity of the system these will produce a real and imaginary component in the steady state response. Hence the output $y(n)$ will take the form $Y\mathrm{e}^{\mathrm{j}(n\omega T+\varphi)}$. Using the normalised frequency $\theta = \omega T$ then

$$\begin{aligned} x(n) &= X\mathrm{e}^{\mathrm{j}n\theta} \\ x(n-1) &= X\mathrm{e}^{\mathrm{j}(n-1)\theta} = X\mathrm{e}^{\mathrm{j}n\theta}\mathrm{e}^{-\mathrm{j}\theta} \\ &\vdots \\ x(n-l) &= X\mathrm{e}^{\mathrm{j}(n-l)\theta} = X\mathrm{e}^{\mathrm{j}n\theta}\mathrm{e}^{-\mathrm{j}l\theta} \end{aligned}$$

$$
\begin{aligned}
y(n) &= Y\mathrm{e}^{\mathrm{j}(n\theta+\varphi)} = Y\mathrm{e}^{\mathrm{j}n\theta}\mathrm{e}^{\mathrm{j}\varphi} \\
y(n-1) &= Y\mathrm{e}^{\mathrm{j}((n-1)\theta+\varphi)} = Y\mathrm{e}^{\mathrm{j}n\theta}\mathrm{e}^{-\mathrm{j}\theta}\mathrm{e}^{\mathrm{j}\varphi} \\
&\vdots \\
y(n-k) &= Y\mathrm{e}^{\mathrm{j}((n-k)\theta-\varphi)} = Y\mathrm{e}^{\mathrm{j}n\theta}\mathrm{e}^{-\mathrm{j}k\theta}\mathrm{e}^{\mathrm{j}\varphi}
\end{aligned}
$$

Substituting into eqn (5.3.1) and dividing by $\mathrm{e}^{\mathrm{j}n\theta}$ gives

$$
\begin{aligned}
&Y\mathrm{e}^{\mathrm{j}\varphi}(a_0 + a_1\mathrm{e}^{-\mathrm{j}\theta} + \ldots\ldots a_k\mathrm{e}^{-\mathrm{j}k\theta}) \\
&\qquad = X(b_0 + b_1\mathrm{e}^{-\mathrm{j}\theta} + \ldots\ldots b_1\mathrm{e}^{-\mathrm{j}l\theta})
\end{aligned}
$$

$$
\begin{aligned}
H(\theta) &= \frac{Y}{X}\mathrm{e}^{\mathrm{j}\varphi} \\
&= \frac{b_0 + b_1\mathrm{e}^{-\mathrm{j}\theta} + \ldots\ldots b_l\mathrm{e}^{-\mathrm{j}l\theta}}{a_0 + a_1\mathrm{e}^{-\mathrm{j}\theta} + \ldots\ldots a_k\mathrm{e}^{-\mathrm{j}k\theta}} \qquad (5.3.2)
\end{aligned}
$$

$H(\theta)$ is the frequency response function of the discrete system. At any value of $\theta = \omega T$ it can be evaluated to give a magnitude and angle. Because numerator and denominator are not polynominal in θ they cannot be factored as in the continuous case and Bode plots cannot be constructed.

Details of the calculation for a specific frequency response function are given in Example 5.3.1. However three important general properties of the plot are as follows:

1. $\mathrm{e}^{\mathrm{j}\theta}$ is periodic with period 2π. Hence both the magnitude and the angle plots will be periodic with a period equal to the normalised frequency 2π.
2. At a normalised frequency $\theta = \pi$, $\mathrm{e}^{\mathrm{j}\pi} = -1$. Hence the frequency response function will be real at this point (there will be zero phase shift).
3. Consider two frequencies equidistant by frequency α from $\theta = \pi$. These can be written $(\pi + \alpha)$ and $(\pi - \alpha)$. $\mathrm{e}^{\mathrm{j}\theta}$ at these frequencies is

$$
\begin{aligned}
\mathrm{e}^{\mathrm{j}(\pi+\alpha)} &= \mathrm{e}^{\mathrm{j}\pi}\mathrm{e}^{\mathrm{j}\alpha} = -\mathrm{e}^{\mathrm{j}\alpha} \\
\mathrm{e}^{\mathrm{j}(\pi-\alpha)} &= \mathrm{e}^{\mathrm{j}\pi}\mathrm{e}^{-\mathrm{j}\alpha} = -\mathrm{e}^{-\mathrm{j}\alpha}
\end{aligned}
$$

The frequency response function $H(\mathrm{j}\theta)$ at frequency $(\pi + \alpha)$ must be the complex conjugate of that at $(\pi - \alpha)$. The magnitude plot will exhibit even symmetry about $\theta = \pi$, the phase plot odd symmetry.

These properties mean that any discrete frequency response plot can be presented on a normalised frequency scale using only the range $0 \longrightarrow \pi$.

EXAMPLE 5.3.1

In Section 3.9.1 the approximation of a differential equation by a discrete equation was considered. The equation was a first order equation describing a CR circuit

$$
\frac{1}{T_1}\frac{\mathrm{d}v}{\mathrm{d}t} + v_\mathrm{o} = v_\mathrm{i}
$$

The discrete approximation used led to the difference equation

$$v_o(n) - \frac{T_1}{T+T_1} v_o(n-1) = \frac{T}{T+T_1} v_i(n)$$

$$v_o(n) - kv_o(n-1) = (1-k)v_i(n) \tag{5.3.3}$$

where $k = T_1/(T+T_1)$ and T is the sampling time.

In Section 3.9.1 it was shown that as the sampling time T is made shorter the step response of the discrete approximation approached that of the continuous system. One would expect the same effect with frequency response as this is only an alternative description of the same system. Compare the frequency response of the continuous system with $T_1 = 1$ to that of the discrete system with sampling times $T = 1$ and $T = 0.5$.

SOLUTION

The frequency response of the continuous system has already been considered in Section 5.2.3 where it was shown that the frequency response function is given by

$$H(j\omega) = \frac{1}{1 + j\omega T_1}$$

To obtain the frequency response of the discrete system eqn (5.3.2) is used with the following coefficient values obtained from eqn (5.3.3), $b_0 = (1-k)$, $a_0 = 1$, $a_2 = -k$, giving

$$H(\theta) = \frac{(1-k)}{1 - ke^{-j\theta}} \tag{5.3.4}$$

where θ is the normalised frequency ωT.

Equation (5.3.4) can be re-written

$$H(\theta) = \frac{(1-k)}{1 - k\cos\theta + jk\sin\theta}$$

and the magnitude and angle of the response is

$$|H(\theta)| = \frac{(1-k)}{\sqrt{((1-k\cos\theta)^2 + k^2\sin^2\theta)}}$$

$$= \frac{(1-k)}{\sqrt{(1 - 2k\cos\theta + 2k^2)}}$$

$$\underline{/H(\theta)} = -\tan^{-1}\frac{k\sin\theta}{1 - k\cos\theta}$$

The magnitude responses of the discrete systems ($T = 1$, $T = 0.5$) are plotted in Figure 5.18 together with the response of the continuous system. Linear scales have been used

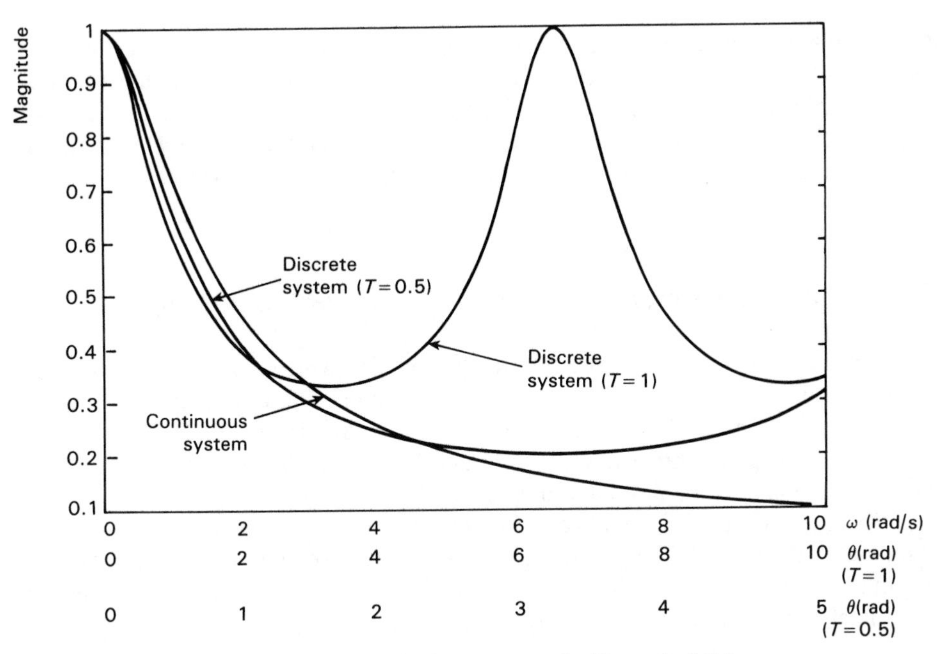

Figure 5.18 Magnitude responses for Example 5.3.1.

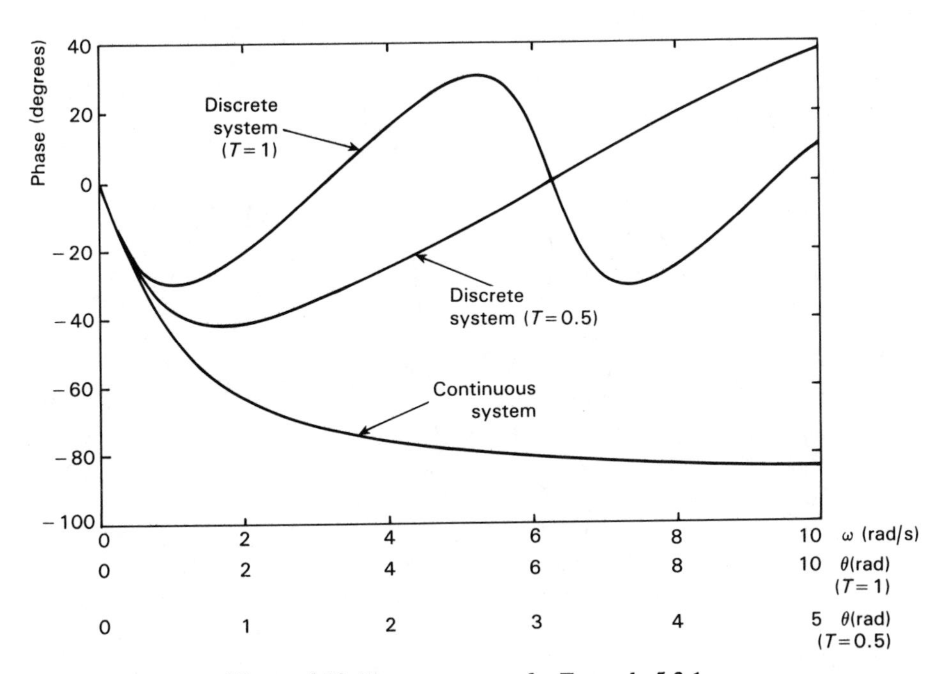

Figure 5.19 Phase responses for Example 5.3.1.

as they illustrate the periodic nature of the discrete response better. Frequency has been shown in Hz together with the normalised frequency for the two sampling times. As expected with the shorter sampling time the correspondence between the frequency response of the continuous and discrete systems extends over the greater frequency range. Figure 5.19 shows the corresponding phase response of the systems.

5.4 Fourier series representation of continuous signals

So far this chapter has considered the steady state frequency response of a system to a single sinusoidal input. The advantage of the method over time domain methods is that it does not require the solution of a differential equation or the evaluation of the convolution integral. The disadvantage is that it does appear somewhat restricted in so far as the input signal has to be sinusoidal.

This section extends the range of input signals that can be used in the frequency response method, from a single sinusoid, to signals consisting of linear combinations of sinusoids. Consider the signal shown in Figure 5.20, this does not appear very sinusoidal but it consists of a sine wave plus its third harmonic.

$$x(t) = 10 \sin \omega t + 3 \sin 3\omega t$$

Suppose it is required to determine the response of a system to such a signal. Taking a specific case suppose the frequency of the fundamental is 15 Hz ($\omega = 2 \times 15 \times \pi$)

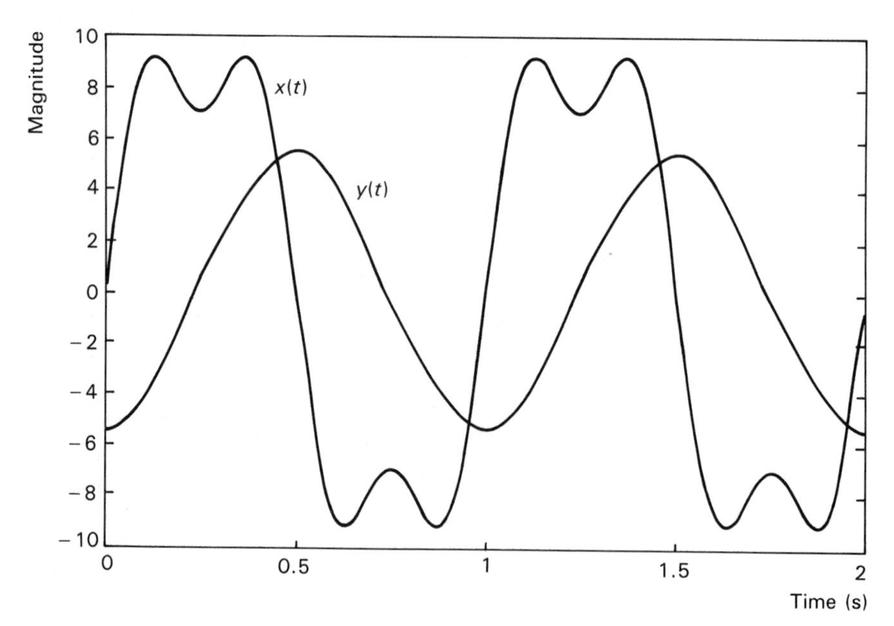

Figure 5.20 Response of a system to a non-sinusoidal input system.

and it is required to reduce the harmonic content by using a simple filter having a frequency response function

$$H(\mathrm{j}\omega) = \frac{1}{(1+\mathrm{j}\omega 0.01)^2}$$

The magnitude and phase response of the filter to the two components of the input signal can easily be calculated.

For the fundamental $\omega = 2 \times \pi \times 15$

$$|H(\mathrm{j}\omega)| = \frac{1}{1+(0.942)^2} = 0.529$$

$$\underline{/H(\mathrm{j}\omega)} = -2\tan^{-1} 0.942 = -86.6°$$

For the third harmonic $\omega = 2 \times \pi \times 3 \times 15$

$$|H(\mathrm{j}\omega)| = \frac{1}{1+(2.827)^2} = 0.111$$

$$\underline{/H(\mathrm{j}\omega)} = -2\tan^{-1} 2.827 = -141.04°$$

Hence the output of the filter $y(t)$ is (remember the angle is in radians)

$$y(t) = 5.29\,\sin(\omega t - 1.511) + 0.33\,\sin(\omega t - 2.462)$$

This waveform is also shown in Figure 5.20. It should be noted that the output still consists of the sum of a fundamental and a third harmonic. However the relative amplitudes and phase of the signals have been changed to produce a waveform different in shape from that at the input. (The relative amplitude of harmonic/fundamental has been reduced from 0.3 at the input to 0.06 at the output which indicates the filtering action.) The waveforms are both periodic with a period equal to that of the fundamental, $2\pi/\omega$.

This example has shown that it is possible to analyse systems subject to non-sinusoidal input signals by use of frequency response methods. However the signals were still restricted in as much as they were constructed from a linear combination of sinusoidal components and as such were periodic. As shown in Section 2.5.2 a signal of the form

$$x(t) = a_1\,\sin(\omega t + \varphi_1) + a_2\,\sin(2\omega t + \varphi_2) + \ldots\ldots\, a_n\,\sin(n\omega t + \varphi_n)$$

is periodic with period $\mathrm{T} = 2\pi/\omega$. The constants $a_1, a_2, \ldots, a_n, \varphi_1, \varphi_2, \ldots, \varphi_n$ indicate arbitrary amplitudes and phase angles. An important question is 'Is the reverse true: can any periodic waveform be represented as the sum of a fundamental plus harmonics?' The answer is yes (except for some mathematical oddities) and such a representation is known as a Fourier series. The following section considers such a representation in more detail.

5.4.1 Trigonometric form of the Fourier series

Fourier's theorem states that a periodic waveform $x(t)$ of period T can be represented by the series

$$x(t) = \frac{a_0}{2} + a_1 \cos \omega t + a_2 \cos 2\omega t + \ldots\ldots a_n \cos n\omega t$$
$$+ b_1 \sin \omega t + b_2 \sin 2\omega t + \ldots\ldots b_n \sin n\omega t \qquad (5.4.1)$$

The form of the series is slightly different from that given in the last section. The phase of the component has been indicated by including both a sine and a cosine component rather than using the form $\sin(\omega t + \varphi)$, this is only for convenience, both forms being mathematically identical. A term $a_0/2$ has been included, this represents a term of zero frequency, a d.c. term that may be present. The form of the term $a_0/2$, as opposed to a_0, will be explained later.

The series given by eqn (5.4.1) is the general form of the series, to represent a specific waveform values of a_n and b_n need to be calculated. To do this the evaluation of certain integrals is required. These will be quoted without proof as follows. In these integrals $\omega = 2\pi/T$, m and n are integers.

$$\int_0^T \sin(n\omega t)\, dt = 0 \qquad \int_0^T \cos(n\omega t)\, dt = 0$$

$$\int_0^T \sin(n\omega t)\sin(m\omega t)\, dt = 0 \qquad m \neq n$$
$$= T/2 \qquad m = n$$

$$\int_0^T \cos(n\omega t)\cos(m\omega t)\, dt = 0 \qquad m \neq n$$
$$= T/2 \qquad m = n$$

$$\int_0^T \sin(n\omega t)\cos(m\omega t) \quad = 0 \qquad \text{for all } m, n$$

Although no formal proof is given for the evaluation of these integrals the results are open to simple physical interpretation. The first two integrals state that the area under a complete number of cycles of a sine or cosine waveform is zero. In the remaining integrals the product can be expressed as sum and difference frequencies. For the case where $m \neq n$, again the area under a whole number of cycles is zero, when $m = n$ the difference frequency is a d.c. term whose integral gives the results quoted (in the last case however there is no difference term).

It should be noted that these results are true when the integration is over any complete period T, the range need not be 0 to T. In general it can be t_0 to $t_0 + T$ and great use will be made of the range $-T/2$ to $+T/2$.

The integrals can now be used to determine the coefficients in the Fourier series. If all the terms on both sides of eqn (5.4.1) are integrated over a complete period this gives

$$\int_{-T/2}^{+T/2} x(t)\,\mathrm{d}t = \int_{-T/2}^{+T/2} \frac{a_0}{2}\,\mathrm{d}t$$

$$= \frac{a_0 T}{2}$$

All other terms on the right-hand side give zero as a result of the integration. Hence

$$a_0 = \frac{2}{T}\int_{-T/2}^{+T/2} x(t)\,\mathrm{d}t \tag{5.4.2}$$

This expression gives the a_0 coefficient as twice the mean value of the waveform. The first term in the series, $a_0/2$, gives the mean value (it still has not explained why the factor 2 has been introduced!).

To now determine the coefficient, a_n, both sides of the series are multiplied by $\cos(n\omega t)$ and then integration performed over a complete period.

$$\int_{-T/2}^{+T/2} x(t)\cos n\omega t = \int_{-T/2}^{+T/2} a_n \cos n\omega t \cos n\omega t\,\mathrm{d}t$$

$$= \frac{a_n T}{2}$$

Again all other terms on the right-hand side give zero as a result of the integration. Hence

$$a_n = \frac{2}{T}\int_{-T/2}^{+T/2} x(t)\cos n\omega t\,\mathrm{d}t \tag{5.4.3}$$

It should be noted that if in this formula $n = 0$

$$a_0 = \frac{2}{T}\int_{-T/2}^{+T/2} x(t)\,\mathrm{d}t$$

This is identical to eqn (5.4.2) hence this equation can be used for all values of n, including $n = 0$. This would not have been possible if the first term of the Fourier series had not been written as $a_0/2$. However in some books the first term is written a_0 and an additional equation with the factor 2 missing, is required for its evaluation.

The method of obtaining b_n follows similar reasoning to that used to obtain a_n. This time both sides of the series are multiplied by sin $n\omega t$ and integration over a complete period leads to

$$b_n = \frac{2}{T}\int_{-T/2}^{+T/2} x(t)\sin n\omega t\,\mathrm{d}t \tag{5.4.4}$$

Using a summation notation, the formulae describing the Fourier series can be written

$$x(t) = \frac{a_0}{2} + \sum_{n=1}^{\infty} a_n \cos n\omega t + \sum_{n=1}^{\infty} b_n \sin n\omega t$$

where

$$a_n = \frac{2}{T}\int_{-T/2}^{+T/2} x(t)\cos n\omega t \, \mathrm{d}t$$

$$b_n = \frac{2}{T}\int_{-T/2}^{+T/2} x(t)\sin n\omega t \, \mathrm{d}t$$

and

$$\omega = 2\pi/T$$

The following example illustrates the evaluation of the Fourier series for a specific waveform.

EXAMPLE 5.4.1

Obtain the Fourier series for the square-wave signal shown in Figure 5.21.

SOLUTION

In order to obtain the Fourier coefficients the waveform must be represented by a mathematical function over a range of integration. Often the range has to be split into smaller intervals and different functions used in each interval. As stated earlier any range can be used provided it covers a complete period of the waveform. If the range $-T/2 \leqslant t \leqslant T/2$ is used then

$$x(t) = -V \qquad -T/2 \leqslant t < 0$$
$$x(t) = +V \qquad 0 \leqslant t < T/2$$

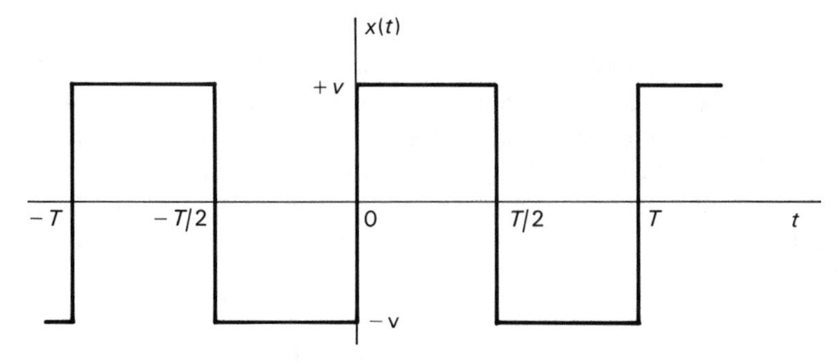

Figure 5.21 Square-wave signal for Example 5.4.1.

Then the coefficient a_n is given by (eqn 5.4.3)

$$a_n = \frac{2}{T}\int_{-T/2}^{+T/2} x(t)\cos n\omega t\,\mathrm{d}t$$

$$= \frac{2}{T}\int_{-T/2}^{0} (-V)\cos n\omega t\,\mathrm{d}t + \frac{2}{T}\int_{0}^{T/2} (+V)\cos n\omega t\,\mathrm{d}t$$

$$= \frac{2V}{n\omega T}\left[(-\sin n\omega t)_{-T/2}^{0} + (\sin n\omega t)_{0}^{T/2}\right]$$

$$= \frac{2V}{n\omega T}\left[0 + \sin\left(-\frac{n\omega T}{2}\right) + \sin\left(\frac{n\omega T}{2}\right) - 0\right]$$

Remembering that $\sin\theta = -\sin(-\theta)$, the right-hand side of this equation can be seen to be 0. (Care should be taken with the case where $n = 0$, the result $0/0$ then occurs – this case will be considered in more detail in Example 5.4.20)

The coefficient b_n is given by

$$b_n = \frac{2}{T}\int_{-T/2}^{+T/2} x(t)\sin n\omega t\,\mathrm{d}t$$

$$= \frac{2}{T}\int_{-T/2}^{0} (-V)\sin n\omega t\,\mathrm{d}t + \frac{2}{T}\int_{0}^{+T/2} (+V)\sin n\omega t\,\mathrm{d}t$$

$$= \frac{2V}{n\omega T}\left[(\cos n\omega t)_{-T/2}^{0} + (-\cos n\omega t)_{0}^{T/2}\right]$$

$$= \frac{2V}{n\omega T}\left[1 - \cos\left(-\frac{n\omega T}{2}\right) - \cos\left(\frac{n\omega T}{2}\right) + 1\right]$$

$$= \frac{2V}{n\omega T}\left[2 - 2\cos\left(\frac{n\omega T}{2}\right)\right]$$

This time, $\cos(-\theta) = \cos\theta$. Use is now made of the fact that $\omega = 2\pi/T$, giving $n\omega T/2 = n\pi$. Then $\cos(n\pi) = +1$ for n even and equals -1 for n odd. Hence

$$b_n = \frac{4V}{n\pi} \qquad \text{for } n \text{ odd } (n > 0)$$

$$= 0 \qquad \text{for } n \text{ even}$$

Hence the Fourier series can be written

$$x(t) = \frac{4V}{\pi}\left[\sin\omega t + \frac{1}{3}\sin 3\omega t + \frac{1}{5}\sin 5\omega t \ldots\ldots + \frac{1}{n}\sin n\omega t\right] \qquad (n \text{ odd})$$

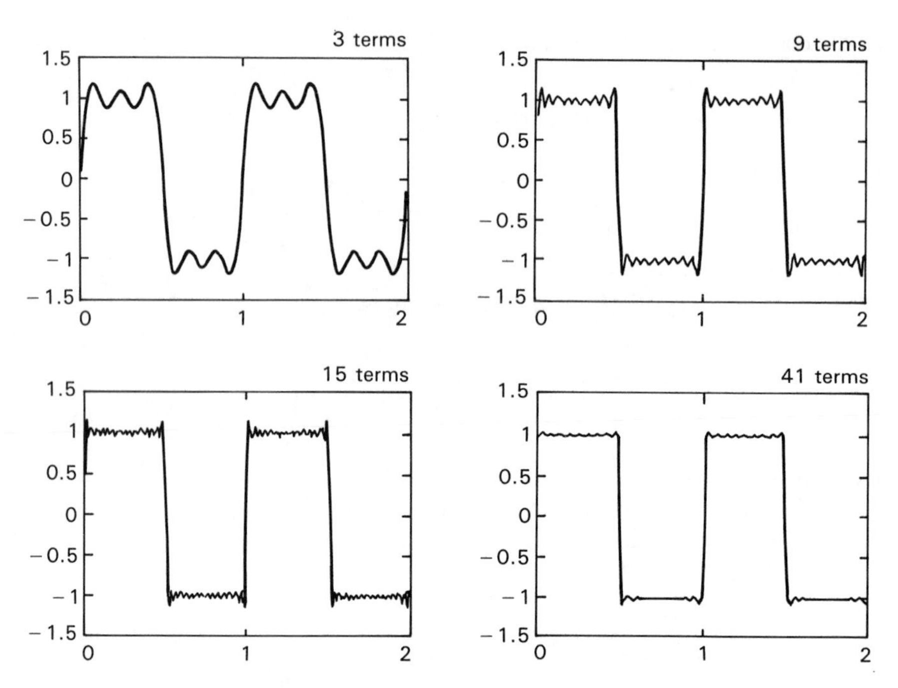

Figure 5.22 Effect of finite number of terms on Fourier series for a square wave.

It does seem remarkable that a function such as a square wave with discontinuities can be represented by the sum of smooth functions like sinusoids. However it must be remembered that an infinite number are required to obtain a time representation. Figure 5.22 shows the waveform obtained using only a finite number of harmonics.

5.4.2 Effect of symmetries

In the last example all the coefficients a_n were zero. If this would have been foretold it would have saved some effort in their evaluation. It could have in fact been predicted quite easily from the fact that $x(t)$ was an odd function.

Recalling Section 2.4.1 an even function is one where $x(-t) = x(t)$ and an odd function is one where $x(-t) = -x(t)$. Hence the cosine waveform $\cos \omega t$ and all its harmonics $\cos n\omega t$ are even functions, while the sine waveform $\sin \omega t$ and all its harmonics are odd functions. Figure 5.23 illustrates the effect of multiplying these functions by an odd (or even) function $x(t)$ and integrating over the range $-T/2$ to $+T/2$.

Figure 5.23(a) shows the case for the cosine components (only the fundamental is shown for clarity). If $x(t)$ is odd, multiplication by the even cosine component produces an odd function. Integration from $-T/2$ to $+T/2$ gives a net area of zero.

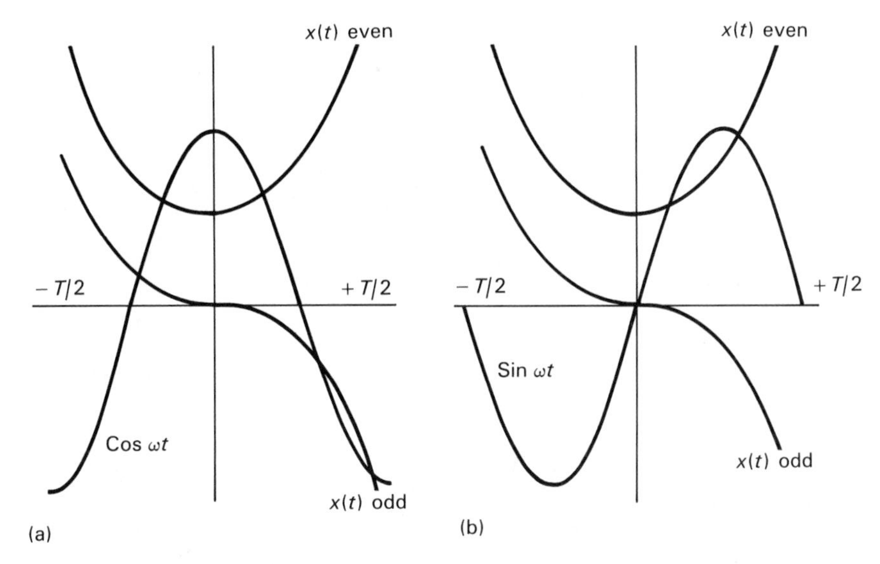

Figure 5.23 Effect of symmetries in $x(t)$ on Fourier coefficients.

This explains the disappearance of the cosine components in Example 5.9.1. If $x(t)$ is even, multiplication now produces an even function. The total integral over period T is now not zero, it is however equal to twice the integral over the range 0 to $T/2$.

A similar argument follows for the sine components as indicated in Figure 5.23(b). For this case when $x(t)$ is even all the sine components disappear and when $x(t)$ is odd the integral can be expressed as twice that over the range 0 to $T/2$.

The results could have been obtained more formally by splitting the ranges of integration ($-T/2$ to 0, and 0 to $T/2$) in the formulae for the evaluation of the Fourier coefficients (eqns (5.4.3) and (5.4.4)).

Half-wave symmetries

A form of symmetry that is applicable to periodic signals is half-wave symmetry. A periodic signal of period T shows even half-wave symmetry if it has the property that $x(t+T/2) = x(t)$. It shows odd half-wave symmetry if $x(t+T/2) = -x(t)$.

Both the sine and cosine signals have half-wave odd symmetry. Also even harmonics exhibit even half-wave symmetry and odd harmonics exhibit odd half wave symmetry (note T is the periodic time of the fundamental). This is illustrated in Figure 5.24 for second and third harmonics of a signal.

By using similar reasoning to that used when considering odd and even symmetry it follows that if a waveform has odd half-wave symmetry there will be no even harmonics in its Fourier series (as in Example 5.4.1). If it has even half-wave symmetry there will be no odd harmonics. This later case is rather unusual as it would imply that the waveform is periodic with period $T/2$ (not T).

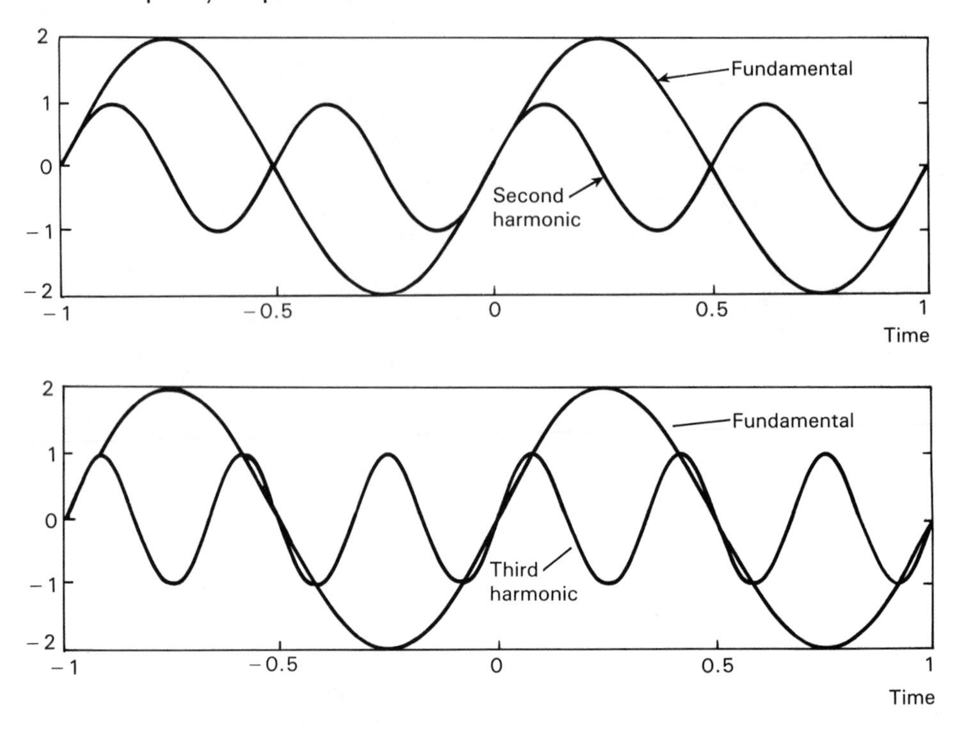

Figure 5.24 Illustration of odd and even half-wave symmetries.

5.4.3 Power representation in the Fourier series

The concept of power as represented by a signal is probably familiar to electrical engineering students. Consider a constant current I, the power dissipated in a resistor R is given by I^2R, the corresponding expression for a constant voltage is V^2/R. If the signal is not constant then the power varies at every instant and the expressions for instantaneous power become $i^2(t)R$ and $v^2(t)/R$. The concept of the power represented by a signal is very useful and the instantaneous power represented by *any* signal $x(t)$ is taken to be $x^2(t)$. More useful than instantaneous power is the mean power over a given time interval T and this is given by

$$\text{Mean power} = \frac{1}{T}\int_0^T x^2(t)\,dt$$

This quantity is the mean-square value (m.s. value) of the signal and its square root is the root-mean-square value (r.m.s.). The interval T to be chosen for an arbitrary signal can raise problems and these will be considered in Section 6.2.3. However for a periodic signal the interval can be taken as the periodic time. Hence for a single sine wave $x(t) = X \sin \omega t$

$$\text{m.s. value} = \frac{1}{T}\int_0^T X^2 \sin^2 \omega t \, \mathrm{d}t$$

$$= \frac{1}{T}\int_0^T \frac{X^2}{2}(1 - \cos 2\omega t)\, \mathrm{d}t$$

$$= \frac{X^2}{2}$$

$$\text{r.m.s. value} = \frac{X}{\sqrt{2}}$$

For the square wave analysed in Example 5.4.1 it is easily shown that its m.s. value is V^2. However by Fourier analysis it was shown that the waveform could be represented by the sum of sine waves and each of these must contribute to the total power. This leads to the more general question of attributing the power in any periodic signal to its harmonics. Taking $x(t)$ to be a periodic signal of period T

$$\text{Mean-square value} = \frac{1}{T}\int_0^T x^2(t)\, \mathrm{d}t$$

Expressing $x(t)$ as a Fourier series this can be written

$$\text{Mean-square value} = \frac{1}{T}\int_0^T \left(\frac{a_0}{2} + \sum_{n=1}^{\infty} a_n \cos n\omega t + \sum_{n=1}^{\infty} b_n \sin n\omega t\right)^2 \mathrm{d}t$$

This expression looks somewhat frightening but if the whole series were multiplied out term by term the following types of integral would emerge

$$\frac{1}{T}\int_0^T \left(\frac{a_0}{2}\right)^2 \mathrm{d}t = \frac{a_0^2}{4}$$

$$\frac{1}{T}\int_0^T a_n^2 \cos^2 n\omega t \, \mathrm{d}t = \frac{a_n^2}{2}$$

$$\frac{1}{T}\int_0^T b_n^2 \sin^2 n\omega t \, \mathrm{d}t = \frac{b_n^2}{2}$$

All other cross terms would integrate to zero over a complete period. These take the form

$$\frac{a_0}{2} a_n \cos(n\omega t) \qquad \frac{a_0}{2} b_n \sin(n\omega t)$$

$$a_n a_m \cos(n\omega t)\cos(m\omega t) \qquad b_n b_m \sin(n\omega t)\sin(m\omega t)$$

$$a_n b_m \cos(n\omega t)\sin(m\omega t)$$

These results have been obtained using the integrals of Section 5.4.1, hence

$$\frac{1}{T}\int_0^T x^2(t)\,dt = \frac{a_0^2}{4} + \sum_{n=1}^{\infty} \frac{a_n^2 + b_n^2}{2} \tag{5.4.5}$$

or

mean power in time = (d.c. term)^2
+ sum of mean squares values of cosine terms
+ sum of mean squares of sine terms

This relationship, eqn (5.4.5) between power in the time domain and power in the frequency domain is known as Parseval's theorem. Applying this result to the square wave analysed in Example 5.4.1

$$\text{Mean-square value from time} = V^2$$

$$\text{Mean-square value of fundamental} = \frac{(4V)^2}{\pi^2}\frac{1}{2} = 0.8106V^2$$

$$\text{Mean-square value of third harmonic} = \frac{(4V)^2}{\pi^2}\frac{1}{9}\frac{1}{2} = 0.0900V^2$$

Hence approximately 90% of the power contained in the waveform is contained in the fundamental and third harmonic.

5.4.4 *The exponential form of the Fourier series*

The trigonometric form considered so far provides a useful intuitive understanding of the Fourier series. One can imagine without too much difficulty a complex waveform constructed from combinations of sine and cosine waveforms. However, mathematically, the sine and cosine functions can be replaced by combinations of the complex exponential function producing a Fourier series in exponential form. Intuitively this is not as convenient as the trigonometric form. However it produces a much more compact form mathematically and it introduces concepts that lead into the Fourier transform of Chapter 6.

For convenience the trigonometric form of the Fourier series is repeated below

$$x(t) = \frac{a_0}{2} + \sum_{n=1}^{\infty} a_n \cos n\omega t + \sum_{n=1}^{\infty} b_n \sin n\omega t \tag{5.4.6}$$

where

$$a_n = \frac{2}{T}\int_{-T/2}^{+T/2} x(t) \cos n\omega t\,dt$$

$$b_n = \frac{2}{T}\int_{-T/2}^{+T/2} x(t) \sin n\omega t\,dt$$

the trigonometric functions in eqn (5.4.6) can be replaced by exponential functions

$$\cos n\omega t = \frac{e^{jn\omega t} + e^{-jn\omega t}}{2}$$

$$\sin n\omega t = \frac{e^{jn\omega t} - e^{-jn\omega t}}{2j}$$

giving

$$x(t) = \frac{a_0}{2} + \sum_{n=1}^{\infty} \frac{a_n(e^{jn\omega t} + e^{-jn\omega t})}{2} + \sum_{n=1}^{\infty} \frac{b_n(e^{jn\omega t} - e^{-jn\omega t})}{2j}$$

Collecting the exponential terms together according to the sign of the exponent and noting that $1/j = -j$

$$x(t) = \frac{a_0}{2} + \sum_{n=1}^{\infty} \frac{(a_n - jb_n)}{2} e^{jn\omega t} + \sum_{n=1}^{\infty} \frac{(a_n + jb_n)}{2} e^{-jn\omega t}$$

Remembering that a_n, b_n are coefficients in the Fourier expansion then $(a_n - jb_n)/2$ is also a constant coefficient although it is complex. Denoting this by $c_n = (a_n - jb_n)/2$ then $c_n^* = (a_n + jb_n)/2$ and $c_0 = a_0/2$ there being no b_0 coefficient. The series can now be written

$$x(t) = c_0 + \sum_{n=1}^{\infty} c_n e^{jn\omega t} + \sum_{n=1}^{\infty} c_n^* e^{-jn\omega t}$$

The c_0 term can be included into the first summation by making the lower limit $n = 0$ $(e^{j0} = 1)$. If the second summation were written $\sum_{n=-1}^{\infty} c_n e^{jn\omega t}$ this would be correct provided some interpretation can be given to the coefficient c_n, when n has a negative value, if this is taken as c_n^* then the complete series can be neatly expressed as

$$x(t) = \sum_{n=-\infty}^{+\infty} c_n e^{jn\omega t}$$

The coefficients c_n can be obtained by evaluating the expression for the coefficients a_n and b_n individually. However it is easier to proceed as follows

$$c_n = \frac{a_n - jb_n}{2}$$

$$= \frac{1}{2}\frac{2}{T}\int_{-T/2}^{+T/2} x(t)(\cos n\omega t - j \sin n\omega t)\, dt$$

$$= \frac{1}{T}\int_{-T/2}^{+T/2} x(t) e^{-jn\omega t}\, dt$$

Hence the Fourier series in exponential form can be expressed

$$x(t) = \sum_{n=-\infty}^{+\infty} c_n e^{jn\omega t} \tag{5.4.7}$$

$$c_n = \frac{1}{T}\int_{-T/2}^{+T/2} x(t) e^{-jn\omega t}\, dt \tag{5.4.8}$$

This form is much more compact than the trigonometric form. However it does raise questions of physical interpretation with regard to the complex coefficients and the negative frequencies that appear in the series. These questions are both answered by way of the following example.

EXAMPLE 5.4.2

Derive the Fourier series, in exponential form for the pulse train shown in Figure 5.25. The pulses are of width τ, amplitude V and are repeated with a periodic time T. Evaluate the series for the case where $T = 1$ s, $\tau = 0.2$ s, $V = 10$ V.

SOLUTION

As with the trigonometric form of the series the signal $x(t)$ has to be expressed in a form that is integrable over the range $-T/2$ to $T/2$ (or any range covering a full period). In this case

$$x(t) = 0 \qquad -\frac{T}{2} \leqslant t \leqslant -\frac{\tau}{2}$$

$$x(t) = +V \qquad -\frac{\tau}{2} \leqslant t \leqslant +\frac{\tau}{2}$$

$$x(t) = 0 \qquad +\frac{\tau}{2} \leqslant t \leqslant +\frac{T}{2}$$

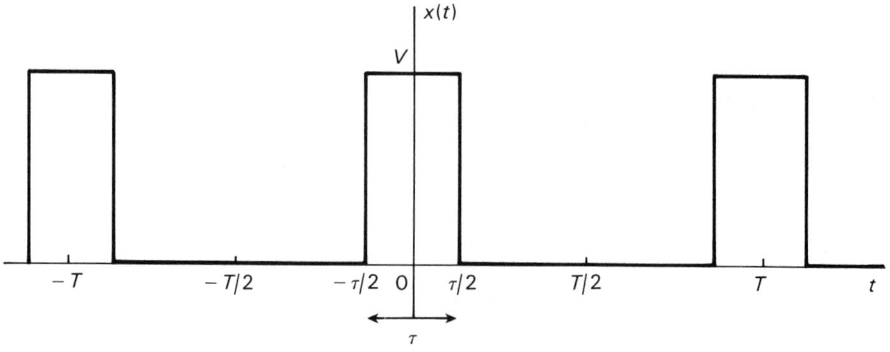

Figure 5.25 Pulse train for Example 5.4.3.

The integral can be divided into these ranges, however as the signal $x(t)$ is zero in two of these intervals the expression for c_n, eqn (5.4.8) becomes

$$c_n = \frac{1}{T}\int_{-\tau/2}^{+\tau/2} V\mathrm{e}^{-\mathrm{j}n\omega t}\,\mathrm{d}t$$

$$= \frac{V}{-\mathrm{j}n\omega T}[\mathrm{e}^{-\mathrm{j}n\omega t}]_{-\tau/2}^{+\tau/2}$$

$$= \frac{V}{-\mathrm{j}n\omega T}[\mathrm{e}^{-\mathrm{j}n\omega t/2} - \mathrm{e}^{+\mathrm{j}n\omega\tau/2}]$$

Remembering that the relationship $\omega = 2\pi/T$ still applies this expression can be written

$$c_n = \frac{V}{n\pi}\left[\frac{\mathrm{e}^{\mathrm{j}n\pi\tau/T} - \mathrm{e}^{-\mathrm{j}n\pi\tau/T}}{2\mathrm{j}}\right]$$

$$= \frac{V}{n\pi}\sin\left[\frac{n\pi\tau}{T}\right] \tag{5.4.9}$$

This has produced in this case a value of c_n that is real but in the more general case c_n will be complex. c_n can be evaluated using the figures given

$$c_n = \frac{10}{n\pi}\sin(n\,0.2\pi)$$

Values of c_n for $n = 1$ to 10 are given in Table 5.2.

One must be careful about evaluating the coefficient c_0. If n is made equal to zero in eqn (5.4.9) then the result is 0/0 which is indeterminate. The expression must be evaluated as a limit as $n \rightarrow 0$. Remembering that the limit of $\sin x/x$ as $x \rightarrow 0$ is unity, check on a calculator for decreasing x remembering to put it in radians. Then

$$c_n = \frac{V}{n\pi}\sin\left[\frac{n\pi\tau}{T}\right]$$

$$= \frac{V\tau}{T}\,\frac{\sin\left[\frac{n\pi\tau}{T}\right]}{\frac{n\pi\tau}{T}}$$

$$c_0 = \frac{V\tau}{T}$$

Table 5.2 Fourier series coefficients for Example 5.4.3

n	c_n	n	c_n
1	1.871	6	−0.312
2	1.514	7	−0.432
3	1.009	8	−0.378
4	0.468	9	−0.208
5	0.000	10	0.000

The alternative is to evaluate c_0 as a special case.

$$c_0 = \frac{1}{T}\int_{-\tau/2}^{+\tau/2} V e^2 \, dt = \frac{V\tau}{T}$$

Inserting the figures gives the numerical value of c_0 as 2.

To obtain the Fourier series the coefficients c_{-n} are required. These can be obtained by substituting negative values of n into eqn (5.4.9) and using the relationship $\sin(-\theta) = -\sin\theta$, these give the same coefficients as positive n. This result could have been foretold as $c_{-n} = c_n^*$ and in this example c_n is real.

Hence the Fourier series for the pulse train can be written

$$\begin{aligned} x(t) = 2 &+ 1.87e^{j\omega t} + 1.51e^{j2\omega t} + 1.11e^{j3\omega t} + \ldots\ldots \\ &+ 1.87e^{-j\omega t} + 1.51e^{-j2\omega t} + 1.11e^{-j3\omega t} + \ldots\ldots \end{aligned}$$

where $\omega = 2\pi$.

In this form the series looks rather unusual, it contains complex numbers and positive and negative frequencies. However by grouping the terms in pairs and replacing the exponentials by the cosine function the series can be written

$$x(t) = 2(1 + 1.87\cos\omega t + 1.51\cos 2\omega t + 1.11\cos 3\omega t + \ldots\ldots)$$

i.e., the trigonometric form of the series. Hence negative frequencies and complex terms are mathematical devices to enable the series to be expressed in exponential form. They will disappear if the series is converted to trigonometric form.

As stated earlier, in general c_n will be complex, it was real in this example because of the even symmetry of the signal being analysed. A waveform having even symmetry will have no sine terms, b_n will be zero, a waveform with odd symmetry will have no cosine terms a_n will be zero. As $c_n = (a_n + jb_n)/2$, c_n will be real for a signal with even symmetry and purely imaginary for one with odd symmetry.

5.4.5 The spectrum of a periodic signal

The waveform in the last example was a pulse train, the pulses were of width 0.2 s, amplitude 10 V and the repetition rate was 1 pulse/s. Together with information on the waveform's absolute position in time this gives a complete time description of the signal.

However the waveform was expressed as a Fourier series and again this was a complete description. Given the amplitude and phase of the frequency components present in the waveform it is possible to reconstitute the waveform. A useful way of presenting this information is by plotting the *spectrum* of the signal. The amplitude spectrum consists of a plot of $|c_n|$ against frequency, the phase spectrum a plot of $\angle c_n$ against frequency.

Because $|c_n| = |c_{-n}|$ the magnitude spectrum must be an even function and because $\angle c_n = -\angle c_{-n}$ the phase spectrum must be an odd function. This is shown in Figure 5.26 for the pulse train considered in the last example.

In the phase spectrum the angle has been represented as $+\pi$ and $-\pi$ in order to show it as an odd function. In this example as c_n is real it is possible to plot c_n

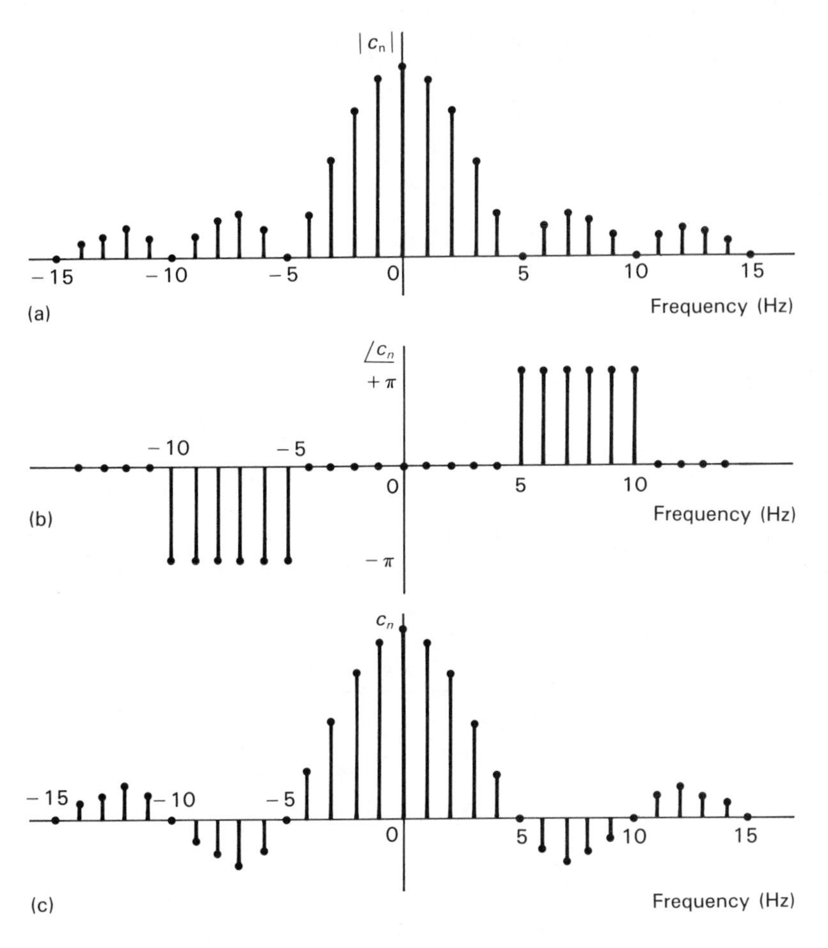

Figure 5.26 Spectrum of signal in Example 5.4.3.

against frequency as shown in Figure 5.26(c) and obtain both magnitude and phase information on a single plot.

5.4.6 Power and the power spectrum

The manner in which the Fourier series represents the power in a signal has already been discussed in Section 5.4.3. The power in a signal using the trigonometric form of the Fourier series was given by eqn (5.4.5).

$$\frac{1}{T}\int_0^T x^2(t)\,\mathrm{d}t = \frac{a_0^2}{4} + \sum_{n=1}^{\infty}\frac{a_n^2 + b_n^2}{2} \tag{5.4.10}$$

To obtain the power from the exponential form of the series, in terms of the coefficients c_n, the following relationships are used

$$c_n = \frac{a_n - \mathrm{j}b_n}{2}, \qquad c_n^* = \frac{a_n + \mathrm{j}b_n}{2}$$

From which

$$a_0 = 2c_0$$
$$a_n = c_n + c_{-n}$$
$$b_n = c_n - c_{-n}$$

Substituting into eqn (5.4.10)

$$\frac{1}{T}\int_0^T x^2(t)\,\mathrm{d}t = \frac{(2c_0)^2}{4} + \sum_{n=1}^{\infty}\frac{(c_n + c_{-n})^2 + (c_n - c_{-n})^2}{2} = \sum_{n=-\infty}^{+\infty}|c_n|^2$$

This is an alternative and more compact method of stating Parseval's theorem. The power associated with any frequency component is shown in Table 5.3.

Table 5.3 Power associated with Fourier series coefficients

	Trigonometric form	*Exponential form*
D.c. term	$\frac{a_0^2}{4}$	$\lvert c_0\rvert^2$
Others	$\frac{a_n^2 + b_n^2}{2}$	$2\lvert c_n\rvert^2 = \lvert c_n\rvert^2 + \lvert c_{-n}\rvert^2$

The power associated with the non-d.c. terms in the trigonometric series is the sum of the mean square values of the sine and cosine components. In the exponential form contributions from both positive and negative frequencies add to give the total power. A power spectrum can be constructed to show how the total power is distributed among the frequency components present.

5.5 Summary

This chapter has considered the response of a system to a specific form of input signal, the sinusoid. It has been shown that the response of a linear system in the steady state to such an input is another sinusoid. The output sinusoid is the same frequency as the input but in general differs in both magnitude and phase. These changes can be obtained from the system frequency response function which offers an alternative system description to that of the differential equation.

To use the frequency domain, description plots of magnitude/frequency and phase/frequency over a range of frequencies are often required. The Bode plot has been considered as an approximate method of plotting these responses. Although approximate this method can be produced quickly and it is often very useful for system investigation.

A disadvantage of the frequency response method is that it is restricted to an input signal that is sinusoidal in form. However it has been shown that Fourier analysis enables any periodic signal to be represented as a sum of sinusoids of harmonic frequencies. This widens the range of system input signals that can be covered using frequency response methods.

Two forms of the Fourier series have been investigated, the trigonometric form and the exponential form. Although the trigonometric form is most easily interpreted physically, the exponential form produces very compact mathematical expressions. It is this form that leads most easily to the Fourier transform that will be considered in the next chapter.

Problems

5.1 A system is described by the following differential equation

$$2\frac{d^2y}{dt^2} + 3\frac{dy}{dt} + y = x$$

Obtain the frequency response function of the system and evaluate its magnitude and angle at angular frequencies of $\omega = 0.1$ rad/s, $\omega = 1$ rad/s and $\omega = 10$ rad/s.

5.2 Figure 5.27 shows part of a speed control system involving a field fed d.c. motor. The power amplifier has a frequency response function given by

$$\frac{V_f}{V_i}(j\omega) = \frac{10}{1 + j0.01\omega}$$

The constants for the motor and load are the following:

R_f = motor field resistance
= 50 Ω
L_f = motor field inductance
= 25 H
K = torque per unit of field current
= 0.1 Nm/A
B = damper constant
= 10^{-3} Nm/rad s^{-1}
J = moment of inertia of motor and load
= 10^{-4} kgm^2

Derive an expression for the frequency response function relating motor speed to amplifier input voltage. Evaluate the magnitude and angle of this expression at angular frequencies of 0, 2 and 10 rad/s.

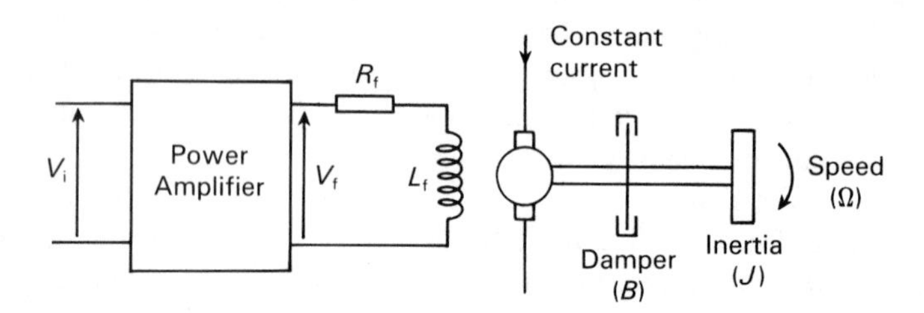

Figure 5.27

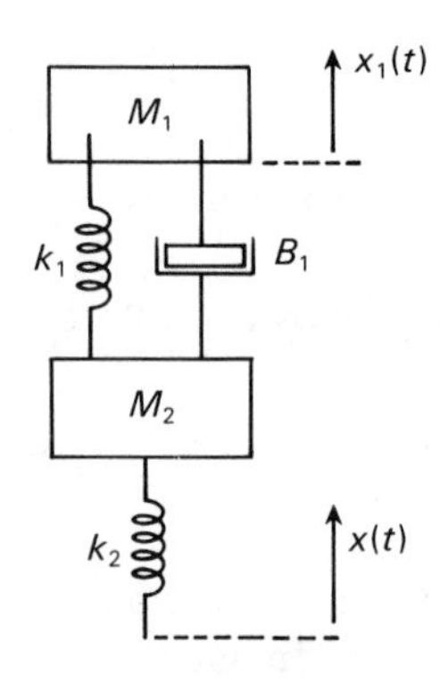

Figure 5.28

5.3 Figure 5.28 show a simplified version of a car suspension for one wheel only. Derive a frequency response function relating the body displacement $X_1(j\omega)$ to the road profile $X(j\omega)$.

5.4 Figure 5.29 shows a Wien bridge network that is used in oscillator circuits and measurement systems. Obtain an expression for the frequency response function $V_o(j\omega)/V_i(j\omega)$. At what frequency is the phase shift of the circuit 0° and what is the magnitude of the frequency response at this frequency?

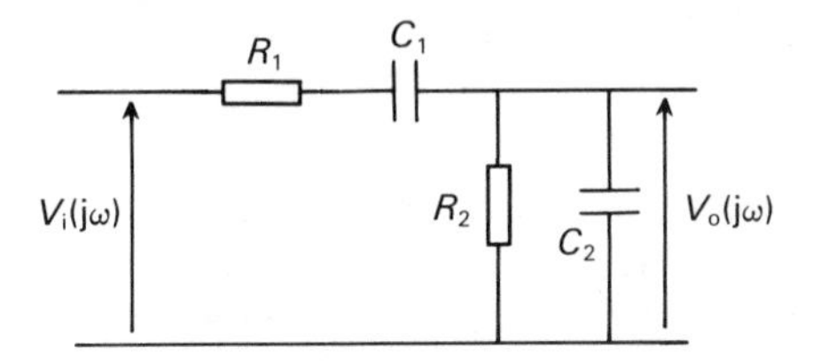

Figure 5.29

5.5 Figure 5.30 shows a ladder network used in R.C. oscillators. Derive the frequency response function $V_o(j\omega)/V_i(j\omega)$. At what frequency will the phase shift in the circuit be 180°? What is the magnitude of the frequency response function under these conditions?

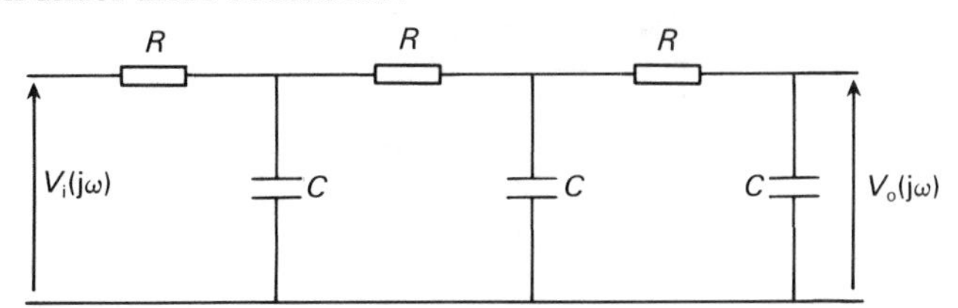

Figure 5.30

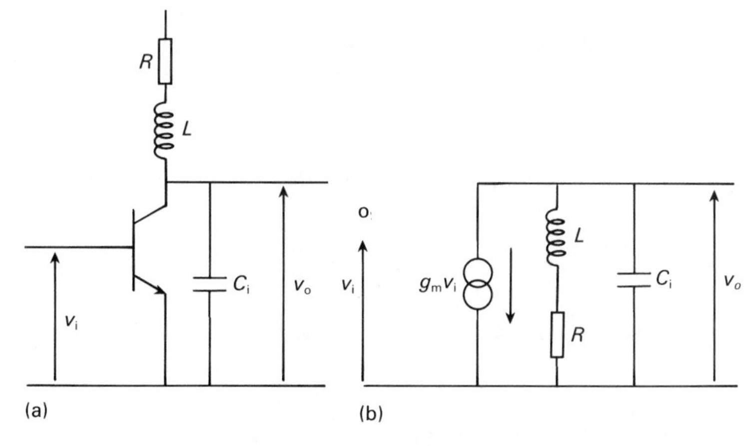

Figure 5.31

5.6 The circuit of Figure 5.31(a) is one stage of a video amplifier (the bias arrangements have been omitted). The small signal equivalent circuit is shown in Figure 5.31(b). The capacitance C_i represents the input capacitance of the following stages and this restricts the high frequency response of the amplifier. The fall-off in response can be compensated, to some extent, by inclusion of the inductor L.

Using the equivalent circuit obtain an expression for the frequency response function $V_o(j\omega)/V_i(j\omega)$. Given $g_m = 40$ mA/V, $R = 1$ kΩ, $C_i = 30$ pF and $L = 12$ μH evaluate the magnitude of this response at $f = 0$, $f = 5$ MHz, $f = 20$ MHz.

5.7 This problem considers the frequency response of the accelerometer introduced in Problem 3.7. The accelerometer is re-drawn in Figure 5.32. Derive a frequency response function relating accelerometer output z, $(z = y - x)$, to body acceleration.

Given $M = 0.1$ kg, $k = 1000$ N/m, $B = 12$ N/m s^{-1}, determine the maximum value of the magnitude of the frequency response function and the frequency at which it occurs.

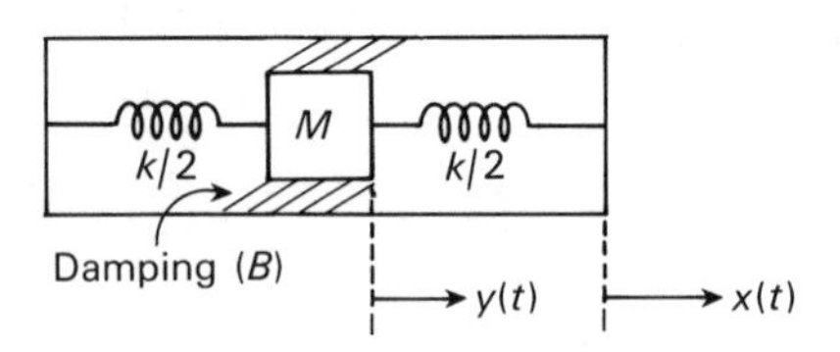

Figure 5.32

5.8 Figure 5.33 shows an active second order filter using operational amplifiers.

(a) Use the virtual earth principle to show that the following frequency response functions apply.

$$\frac{V_o}{V_2}(j\omega) = -\frac{1}{j\omega CR}$$

$$\frac{V_2}{V_1}(j\omega) = -\frac{1}{1 + j\omega CR}$$

(b) Obtain the overall frequency response function $V_o(j\omega)/V_i(j\omega)$ for the filter.

(c) If $R = 10$ kΩ and $C = 1$ μF determine the values ω_n and ζ for the filter. Sketch the magnitude of the frequency response (dB/log ω) showing approximate scaling.

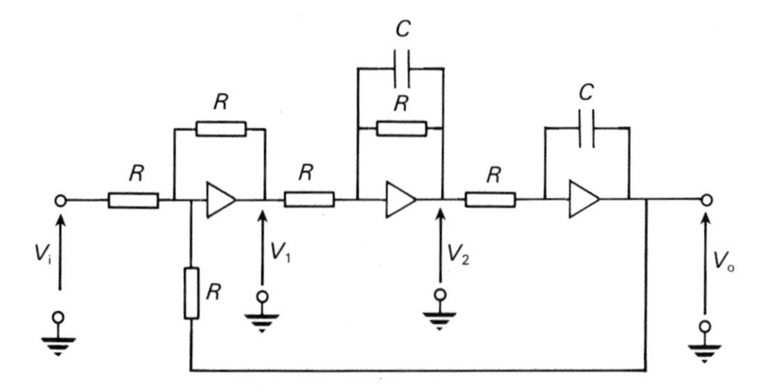

Figure 5.33

5.9 The circuit shown in Figure 5.34 is an active low-pass filter. Use the virtual earth principle ($V_1 = V_2$ for differential input) and nodal analysis at nodes X and Y to show that

$$\frac{V_o}{V_i}(j\omega) = \frac{2}{1 + j\omega CR + (j\omega CR)^2}$$

It can be assumed that the amplifier has infinite input impedance at both input terminals.

If it is required that the undamped natural frequency of the filter, ω_n shall be at 6280 rad/s determine the value for the time constant CR.

What is the low frequency gain of the circuit and what is the maximum gain under these conditions?

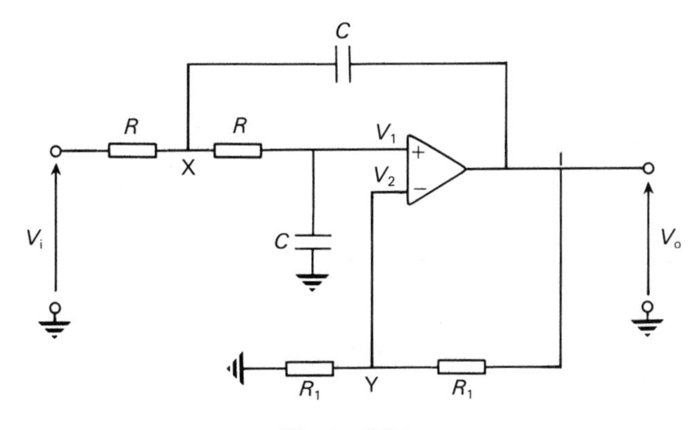

Figure 5.34

5.10 The circuit shown in Figure 5.35 is a cascade compensator for use in a control system.

(a) Obtain the frequency response function $V_o(j\omega)/V_i(j\omega)$ for the circuit.

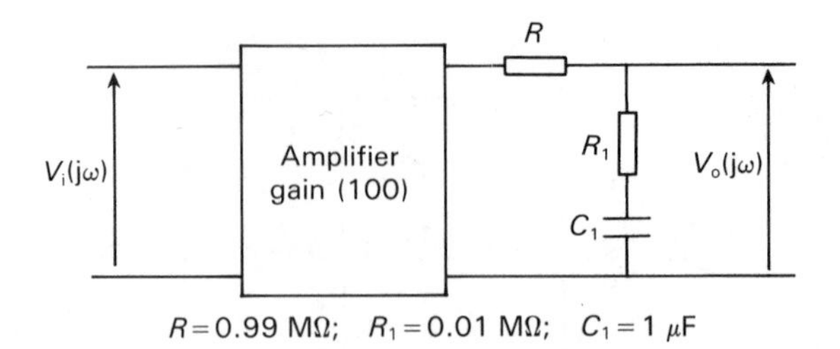

Figure 5.35

(b) Plot a Bode plot of the *magnitude* of the frequency response function. Use a range of ω from 0.1 rad/s to 1000 rad/s and a dB range of 0 to 45 dB.
(c) Sketch the form of the phase response $\underline{/V_o(j\omega)/V_i(j\omega)}$.

5.11 Figure 5.36(a) shows one stage in an audio amplifier and Figure 5.36(b) shows its low frequency, small signal equivalent circuit. Derive a frequency response function $V_o(j\omega)/V_i(j\omega)$ for the circuit. Plot the magnitude of this function as a Bode plot over a frequency range $f = 0.1$ Hz to $f = 100$ Hz. Calculate and plot the associated phase function and estimate the maximum phase lead and the frequency at which it occurs. (Do not take the phase reversal of the stage into account in the phase plot.)

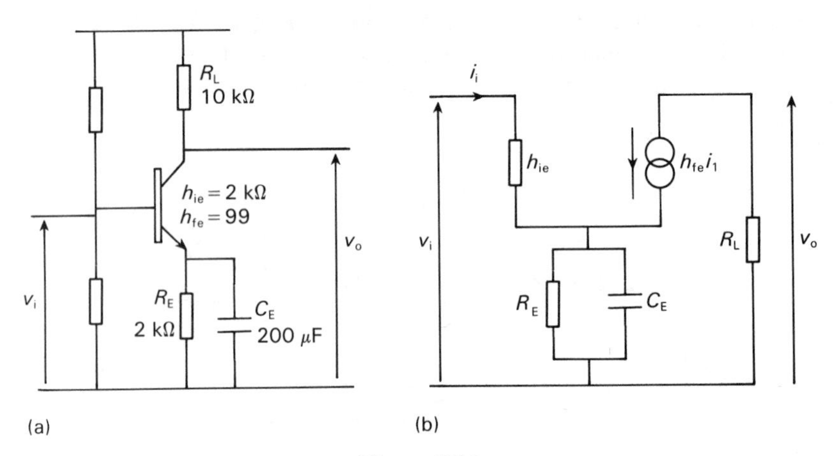

Figure 5.36

5.12 Bode plots provide a method of obtaining an analytical expression for a frequency response function that fits a set of values obtained by practical measurement on a control system.

Table 5.4 shows the results of such a test using a transfer function analyser. The investigation also suggests that the magnitude increases indefinitely as $\omega \rightarrow 0$. Plot the figures using logarithmic frequency scales and by drawing

Table 5.4

Frequency (rad/s)	0.1	0.3	1	3	10	30
Magnitude (dB)	41.6	32.1	21.7	13.0	7.3	5.8
Frequency (rad/s)	100	300	1000	3000	10000	
Magnitude (dB)	3.0	−4.1	−15.0	−28.6	−48.1	

appropriate straight line asymptotes estimates the frequency response function of the system.

5.13 A discrete system is described by the following difference equation:

$$y(n) - 0.8y(n-1) = x(n)$$

(a) If the input signal is given by

$$x(n) = \sin(0.1\, n\pi)$$

obtain numerically the values of $y(n)$ for values of n, 0 to 80. (Assume zero initial conditions.) Plot these results and estimate the amplitude and phase change between output and input signals after the decay of the transient.

(b) Obtain an expression for the frequency response function of the system. Evaluate the magnitude and angle of this function at $\omega = 2\pi$ (sampling time = 0.05) and compare with the result obtained in (a).

5.14 In Problem 4.16 the effect of three different filters on signal and noise signals was considered. The three filters can be expressed as the following difference equations

(a) $y(n) = 0.1 \sum_{k=0}^{9} x(n-k)$

(b) $y(n) = 0.2 \sum_{k=0}^{4} x(n-k)$

(c) $y(n) = 0.4x(n) + 0.3x(n-1) + 0.2x(n-2) + 0.1x(n-3)$

Obtain an expression for the frequency response function using normalised frequency of filters (a) and (b). Plot the magnitude and phase response over the range of normalised frequency 0 to π radians. (Hint: Obtain a closed form for the frequency response function by expressing it as a geometrical progression.)

Plot the magnitude and phase response of filter (c) by direct numerical calculation (do not attempt to obtain a general expression for the response).

5.15 Problem 4.17 was concerned with the approximation of a continuous system by a discrete equivalent. This problem considers the system using frequency response methods.

A continuous system is described by the following differential equation.

$$\frac{d^2y}{dt^2} + 2.4\frac{dy}{dx} + 4y = 4x$$

Determine the frequency response function of the system.

Using the following approximations, obtain a difference equation that approximates the differential equation.

$$\frac{dy}{dt} \simeq \frac{y(n) - y(n-1)}{T}$$

$$\frac{d^2y}{dt^2} \simeq \frac{y(n) - 2y(n-1) + y(n-2)}{T^2}$$

Taking $T = 0.1$ obtain an expression for the frequency response of the discrete system. Compare the magnitude and phase responses of the continuous and discrete systems over a frequency range 0 to 5 rad/s.

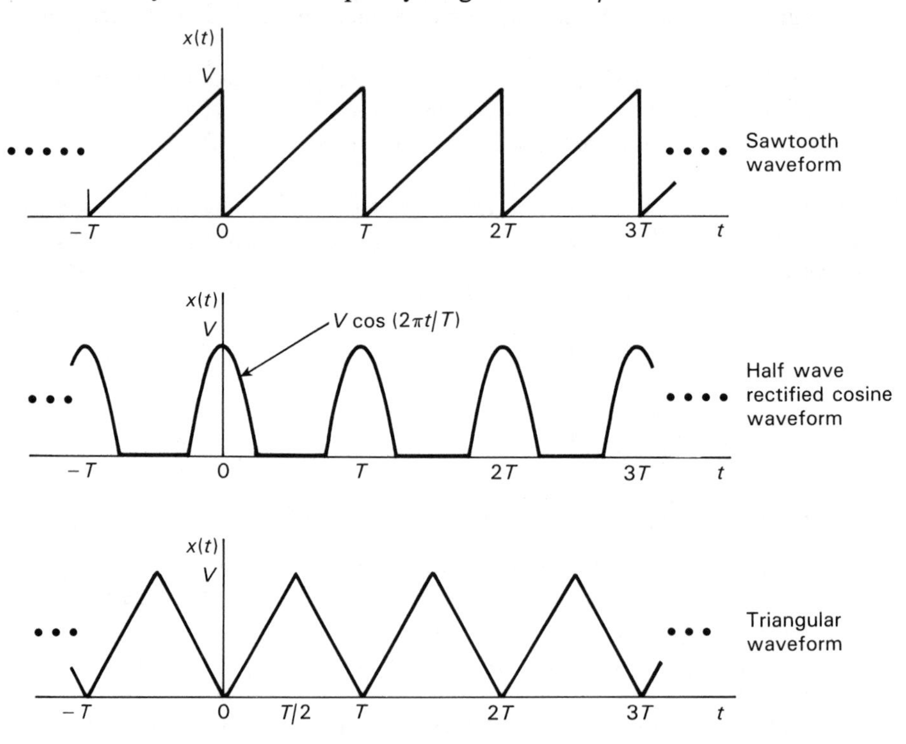

Figure 5.37

5.16 Obtain the Fourier series in both trigonometric and exponential forms for the signals shown in Figure 5.37. Sketch the magnitude and phase spectrum of these signals.

5.17 Figure 5.38 shows a square wave amplitude ± 5 V and having a periodic time $T = 1$ ms. Obtain the Fourier series for this waveform. If the waveform is applied to a 100 Ω resistor what would be the power dissipated in this resistor?

The waveform is applied to a perfect tuned filter which removes the fundamental completely but leaves the other components unaltered. If the output of the filter is now applied to a 100 Ω resistor what would now be the power dissipated?

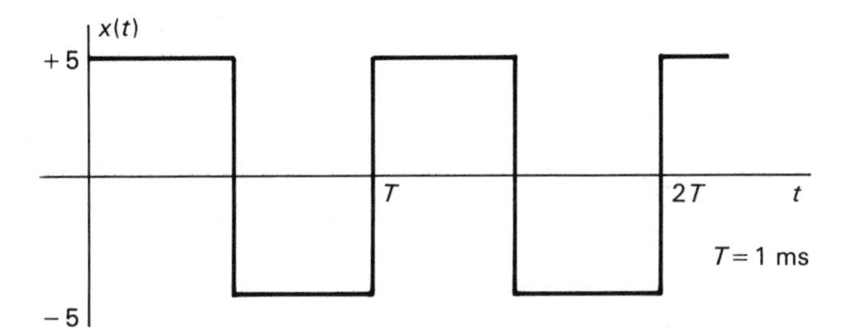

Figure 5.38

5.18 The repetitive signal shown in Figure 5.39 is defined by

$$x(t) = Ee^{-5t/T}$$

in the interval $0 \leqslant t \leqslant T$. Obtain the amplitudes of: (a) the d.c. component; (b) the fundamental in this waveform, in terms of E. In order to reduce the amplitude of the fundamental this waveform is fed to the simple low-pass filter shown. Obtain the amplitudes of the d.c. component and fundamental at the output of the filter.

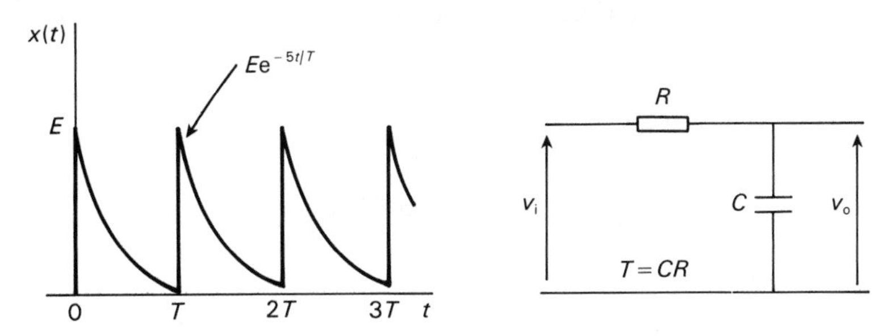

Figure 5.39

5.19 The waveform shown in Figure 5.40 is a repetitive waveform consisting of a pulse 0.2 s wide repeated every second. The height of the pulse is 1 V. It is fed

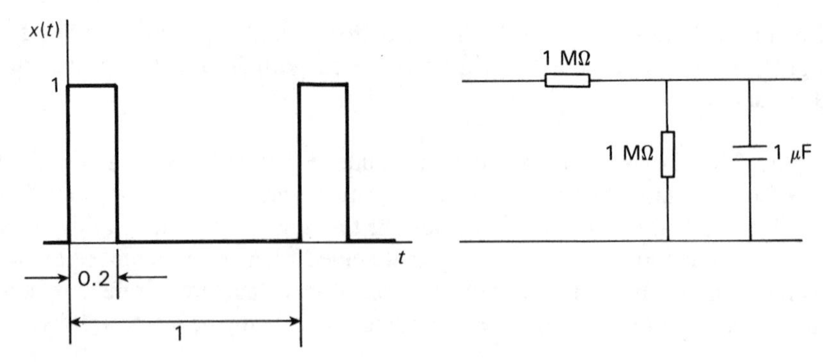

Figure 5.40

into the resistance capacitance network shown for a time long enough for all transients to subside. Calculate the ratios of the amplitudes of the third harmonic to the fundamental: (a) at the input to the network; (b) at the output of the network.

5.20 Determine the Fourier series for the fully rectified voltage waveform shown in Figure 5.41. (Use the trigonometrical form for the series.) Take the period of the waveform to be T as shown, not $T/2$ as might be expected. This is because the waveform has been derived by fully rectifying a sine wave of period T. This will result in no fundamental in the series.

In order to improve its d.c. power this waveform is passed through a low-pass filter having a frequency response function

$$H(\mathrm{j}\omega) = \frac{1}{(1 + \mathrm{j}\omega/\omega_0)^2}, \qquad \omega_0 = \frac{2\pi}{T}$$

Calculate the ratio d.c. power/total power and second harmonic power/total power for the waveform before and after the filter.

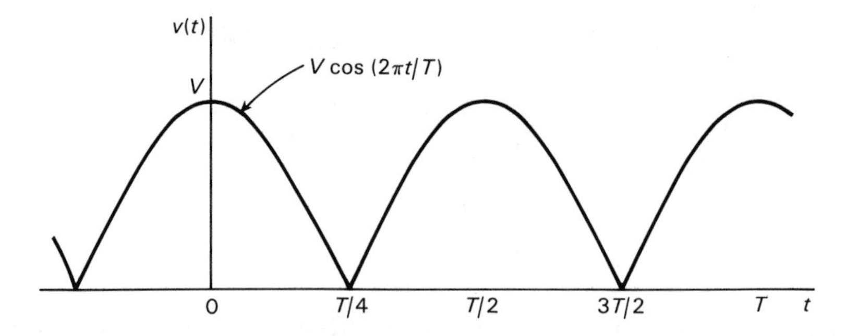

Figure 5.41

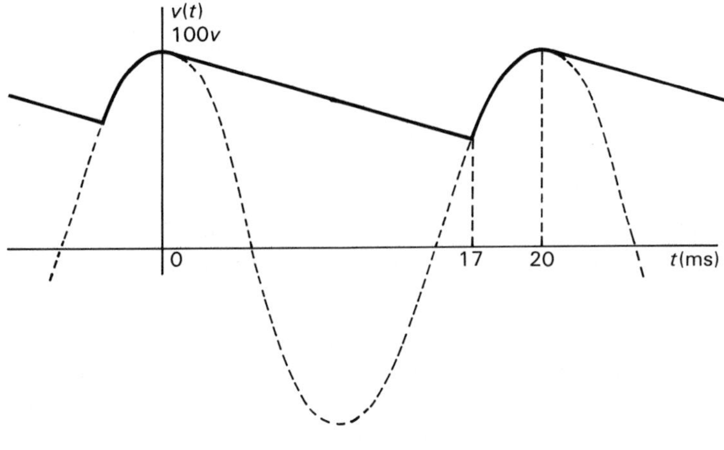

Figure 5.42

5.21 The waveform $v(t)$ shown in Figure 5.42 is a portion of the periodic voltage across the reservoir capacitor in a half-wave rectifier circuit. One cycle can be described by the following equations.

$$v(t) = 100e^{-t/T} \qquad 0 \leq t \leq 17 \text{ ms}$$

$$v(t) = 100 \cos \omega t \qquad 17 \text{ ms} < t < 20 \text{ ms}$$

where $T = 32$ ms and $\omega = 2\pi\ 50$.

The supply voltage to the rectifier is 100 V peak at a frequency of 50 Hz. Obtain the r.m.s. value of the 50 Hz component that is present in the waveform.

MATLAB exercises

5.22 Write a MATLAB program to evaluate a frequency response function when it is expressed as a ratio of polynomials in $j\omega$. Over a prescribed frequency range, plots of magnitude (dB) and phase (degrees), against frequency on a logarithmic scale, should be produced.

Note Use to best advantage the ability of MATLAB to operate on vectors and complex numbers. The polynomials can be expressed as

$$a_n(j\omega)^n + a_{n-1}(j\omega)^{n-1} + \cdots$$
$$= [a_n a_{n-1} \cdots] * [(j\omega)^n (j\omega)^{n-1} \cdots]^T$$

To obtain the phase angle in the correct quadrant, use the command `atan2`.

5.23 This exercise compares the true magnitude response of a system with that of a Bode plot approximation constructed from straight line segments.

Use the program developed in Exercise 5.22 to obtain a plot of magnitude (dB) against log frequency for the following frequency response function

$$G(j\omega) = \frac{20(1 + j\omega)}{j\omega(1 + j\omega T)}$$

when initially $T = 0.1$.

Obtain a straight line approximation to this function and plot on the same display. Compare these plots for a range of values of T.

Note The straight line approximation to the magnitude of a term $(1 + j\omega/\omega_1)$ can be obtained as follows. Construct a vector to represent the straight line approximation at higher frequencies, $20\log(\omega/\omega_1)$,over the full frequency range. Use the `find` command to determine the elements where $\omega < \omega_1$ and replace these with zero. The complete approximation can be obtained by adding or subtracting the individual terms.

5.24 Use the program written in Exercise 5.22 to obtain the magnitude and phase response of a second order system when expressed in standard form. Plot the results, for a range of ζ, against normalized frequency ω/ω_n and compare with Figures 5.14 and 5.15.

Use the program to check the answers to Problems 5.7 and 5.9.

5.25 Write a MATLAB program to obtain the frequency response of a discrete system when the system description takes the form of a difference equation. As in Exercise 5.22, use to best advantage the ability of MATLAB to operate on vectors and complex numbers.

$$\begin{aligned} &a_0 + a_1 e^{-j\theta} + a_2 e^{-j2\theta} + \cdots \\ &= [a_0 a_1 a_2 \cdots] * [1 e^{-j\theta} e^{-j2\theta} \cdots]^T \end{aligned}$$

Use the program to check the results of Problems 5.14 and 5.15.

5.26 This exercise investigates the Fourier series representation of a continuous waveform. The exercise is based on the periodic pulse waveform of Example 5.4.2, but, in order to illustrate the ability of MATLAB to handle complex numbers, the pulse is taken to exist between 0 and τ.

Evaluate the complex coefficients in the exponential Fourier series representation of the above waveform. Take $V = 10$, $\tau = 0.2$, $T = 1$. Hence, reconstruct the time waveform from its Fourier series over the time interval from -1.5 to $+1.5$ seconds, taking 3000 points in this interval.

Investigate the effect upon the time waveform of altering the number of harmonics in the representation.

Note The formula for c_n will now not be given by eqn (5.4.9). The limits for the integral are now 0 and τ and this leads to a complex expression for

c_n. However, this can be evaluated by MATLAB and it is easiest to leave it in exponential form. It is most convenient to evaluate the positive, negative and zero coefficients separately and then to add the resulting terms in the Fourier series.

5.27 This exercise uses the MATLAB command `fft` to obtain the Fourier series for the waveforms shown in Problem 5.16. The discrete Fourier transform (DFT) and the fast Fourier transform (FFT) are not fully covered until Chapter 6. However, the following definitions will suffice for this exercise.

The DFT of a discrete time series $x(n)$ is defined as

$$X(k) = \sum_{n=0}^{N-1} x(n)e^{-jnk2\pi/N}$$

Compare this definition to the expression for the kth coefficient in the continuous series

$$C_k = \frac{1}{T}\int_0^T x(t)e^{-jk\omega t}\,dt$$

With sampling time Δt, $T = N\Delta t$, $\omega = 2\pi/N\Delta t$, $t = n\Delta t$, the integral can be approximated by a summation and

$$c_k \approx \frac{1}{N}X(k)$$

$X(k)$ can be obtained from the command `fft(X)` where `X` is a vector representing the N sample points $x(n)$. If N is a power of 2, the fast Fourier transform algorithm is used, giving considerable savings in computing time. Use the `fft` command to obtain the Fourier series coefficients, in exponential form, for the waveforms shown in Problem 5.16. Compare the results with the analytic answers for a range of sample points N.

6

The Fourier Transform

6.1 Introduction

This chapter continues to develop frequency response methods. In the last chapter the Fourier series was introduced as a signal representation that enables frequency response methods to be used with non-sinusoidal signals. However, although the signals were not sinusoidal they were periodic. This chapter extends the representation to include non-periodic (aperiodic) signals.

One is immediately faced with a difficulty with such a representation – how can sinusoids that are infinite in duration be combined to produce a signal of finite duration? The approach taken will be to consider the aperiodic signal as the limit of a periodic signal when the periodic time approaches infinity. As will be seen the effect on the spectrum is to produce frequency components that are very closely spaced in the limit producing a continuous spectrum. However the amplitude of the components decreases and in the limit these become zero. To overcome this difficulty the concept of an amplitude density spectrum (amplitude/unit frequency) is introduced and this produces the physical interpretation of the Fourier transform.

6.2 The Fourier transform for continuous systems

6.2.1 From Fourier series to Fourier transform

The effect of letting the periodic time T, in the Fourier series, increase to infinity will be illustrated for the specific case of the pulse train analysed in Example 5.4.3. The waveform is shown again in Figure 6.1.

The expression for c_n the coefficient of the nth term in the exponential form of the Fourier series is

$$c_n = \frac{V}{n\pi} \sin \frac{(n\pi\tau)}{T} \tag{6.2.1}$$

Equation (6.2.1) can be re-written

$$c_n = \frac{V\tau}{T} \frac{\sin \dfrac{(n\pi\tau)}{T}}{\dfrac{(n\pi\tau)}{T}}$$

$$= \frac{V\tau}{T} \frac{\sin x}{x} \tag{6.2.2}$$

where

$$x = n\pi\tau/T$$

The function $\sin x/x$ defines the shape of the amplitude spectrum, the lines at frequencies $n2\pi/T$ being bounded by this function as shown in Figure 6.2.

The spectrum has the following properties:

The amplitude at $\omega = 0$ is $V\tau/T$.
The fundamental frequency (this will be denoted by ω_0) is $\omega_0 = 2\pi/T$.

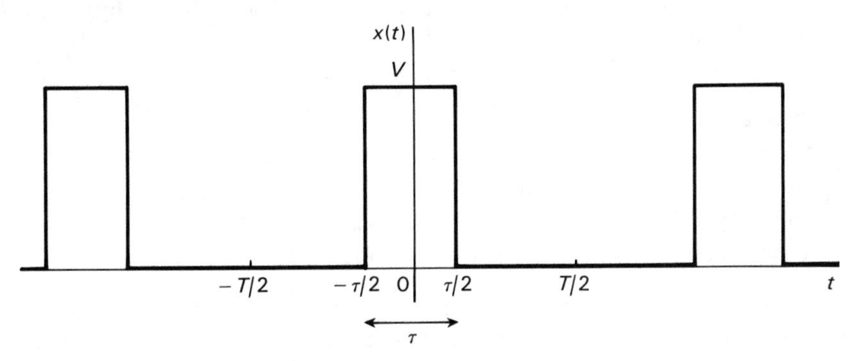

Figure 6.1 Periodic pulse train.

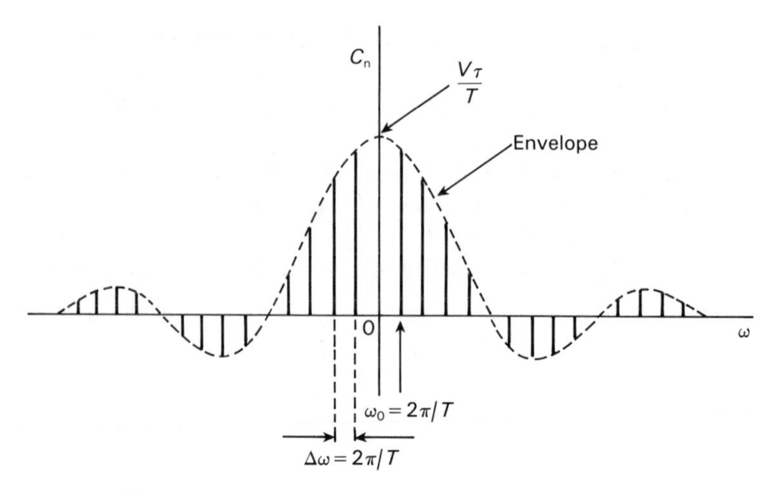

Figure 6.2 Spectrum of signal shown in Figure 6.1.

The spacing between the frequency components (this will be denoted by $\Delta\omega$) is $\Delta\omega = 2\pi/T$.

The first zero value of the envelope of the spectrum occurs when

$$\sin \frac{(n\pi\tau)}{T} = 0$$

$$n = T/\tau$$

(if this is not an integer then there will not be a component at this crossing frequency).

The frequency of the first crossing is

$$T\omega_0/\tau = 2\pi/\tau$$

Suppose now the periodic time T is increased. The case when $V = 10$ V, $\tau = 0.2$ and $T = 1$ has already been calculated in Example 5.4.3. The effect on the spectrum of making $T = 2$ and $T = 5$ is shown in Figure 6.3.

The following effects are noticed as T is increased:

Amplitude of the spectrum decreases.
Spacing between the components decreases ($2\pi/T$).
Number of components up to the first crossing increases (T/τ).
Frequency of first crossing is unchanged ($2\pi/\tau$).
General shape of spectrum ($\sin x/x$) is unchanged.

As $T \rightarrow \infty$ the effect will be to produce an infinite number of zero amplitude components with zero frequency spacing between them – an awful lot of nothing. This does not seem to be a very useful contribution towards obtaining the spectrum of a single pulse. However by redefining the spectrum it is possible to obtain a finite quantity from properties that disappear to zero.

Define a new coefficient

$$c(\omega) = \frac{c_n(n\omega_0)}{\Delta\omega} = \frac{1}{\Delta\omega T}\int_{-T/2}^{T/2} x(t)\mathrm{e}^{-jn\omega_0 t}\,\mathrm{d}t$$

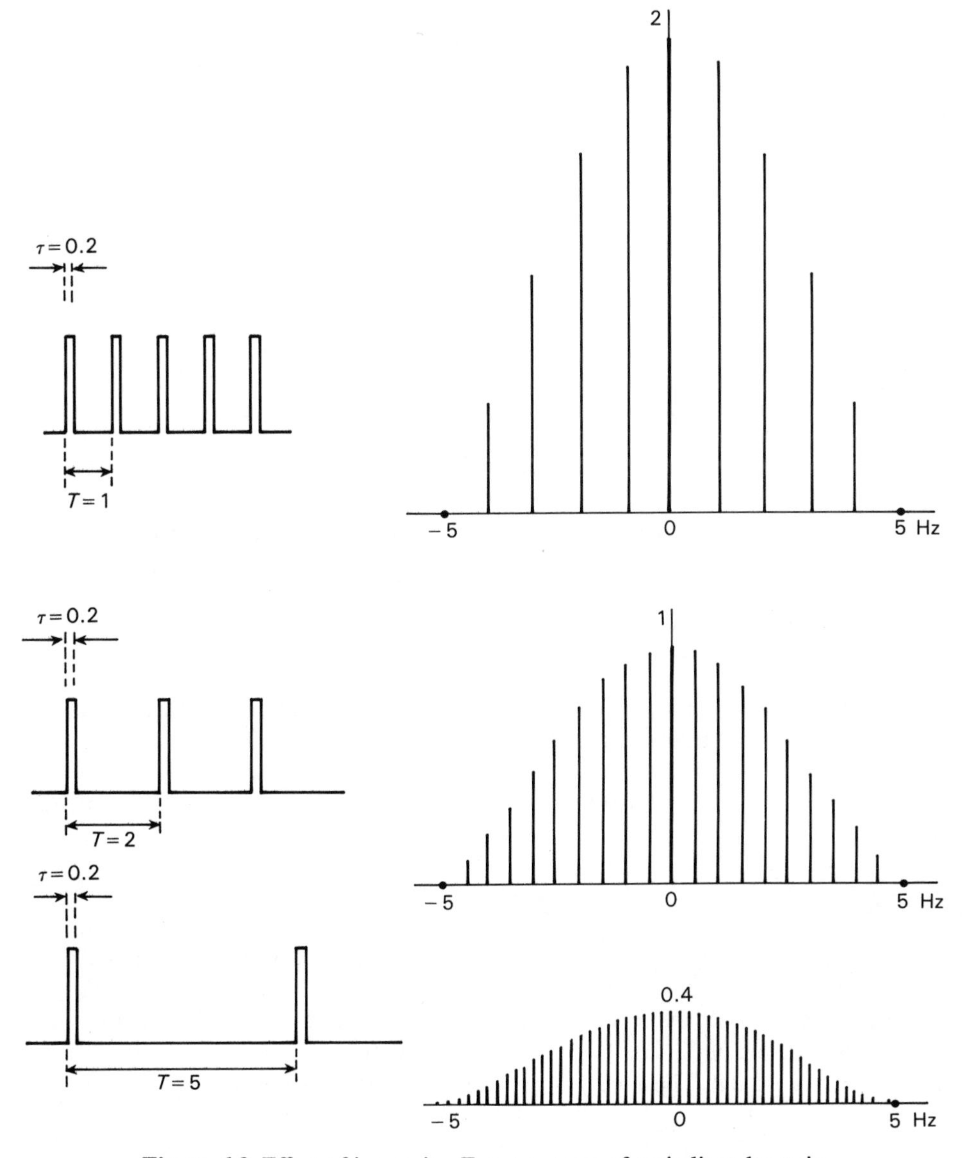

Figure 6.3 Effect of increasing T on spectrum of periodic pulse train.

as $T \rightarrow \infty$ then $n\omega_0$ approaches a continuous variable ω and

$$c(\omega) = \frac{1}{2\pi} \int_{-\infty}^{+\infty} x(t) \mathrm{e}^{-j\omega t} \, \mathrm{d}t$$

If this expression is multiplied by 2π the result is the *Fourier transform* $X(\omega)$ of the time signal $x(t)$.

$$X(\omega) = \int_{-\infty}^{+\infty} x(t) \mathrm{e}^{-j\omega t} \, \mathrm{d}t$$

The Fourier series is given by

$$x(t) = \sum_{n=-\infty}^{+\infty} c_n \mathrm{e}^{jn\omega_0 t}$$

and with

$$c_n = c(\omega)\Delta\omega$$

this becomes

$$x(t) = \sum_{n=-\infty}^{+\infty} c(\omega) \mathrm{e}^{jn\omega_0 t} \, \Delta\omega$$

As $T \rightarrow \infty$ then $c(\omega)$ becomes $X(\omega)/2\pi$, $n\omega_0$ becomes the continuous variable ω and the summation becomes an integral

$$x(t) = \frac{1}{2\pi} \int_{-\infty}^{+\infty} X(\omega) \mathrm{e}^{j\omega t} \, \mathrm{d}t$$

This is known as the inverse Fourier transform and together with the transform defining $X(\omega)$, forms a Fourier transform pair.

$$X(\omega) = \int_{-\infty}^{+\infty} x(t) \mathrm{e}^{-j\omega t} \, \mathrm{d}t \tag{6.2.3}$$

$$x(t) = \frac{1}{2\pi} \int_{-\infty}^{+\infty} X(\omega) \mathrm{e}^{j\omega t} \, \mathrm{d}\omega \tag{6.2.4}$$

The student may wonder why the factor 2π was introduced when going from $c(\omega)$ to $X(\omega)$. If it had not been, then the factor $1/2\pi$ would be present in the Fourier transform and it would be absent from the inverse transform. Such a definition is perfectly acceptable but in engineering texts the definitions given by eqns (6.2.3) and (6.2.4) are more usual. The fact that $x(t)$ and $X(\omega)$ form a Fourier transform pair is written

$$x(t) \leftrightarrow X(\omega)$$

The notation will also be used

$$X(\omega) = F\{x(t)\}$$

being interpreted as $X(\omega)$ is the Fourier transform of $x(t)$ and

$$x(t) = F^{-1}\{X(\omega)\}$$

being interpreted as $x(t)$ is the inverse transform of $X(\omega)$.

Returning to the example of the single pulse the calculation of the Fourier transform parallels closely the calculation of the exponential form of the Fourier series for the pulse train. As the function is zero for $t < -\tau/2$ and $t > \tau/2$ then the transform becomes

$$\begin{aligned} X(\omega) &= \int_{-\tau/2}^{+\tau/2} V\mathrm{e}^{-\mathrm{j}\omega t}\,\mathrm{d}t \\ &= \frac{V}{-\mathrm{j}\omega}(\mathrm{e}^{-\mathrm{j}\omega\tau/2} - \mathrm{e}^{+\mathrm{j}\omega\tau/2}) \\ &= V\tau\,\frac{\sin(\omega\tau/2)}{\omega\tau/2} \end{aligned}$$

As might be expected the transform has a sin x/x form and is plotted against frequency in Figure 6.4.

It has the same shape as the envelope of the amplitude spectrum shown in Figure 6.2. However unlike the spectrum for the pulse train it is a continuous function of frequency and $X(\omega)$ has the units of amplitude/frequency. It is termed the spectrum of the signal $x(t)$ but it is not an amplitude spectrum, it is an amplitude density spectrum. This concept of the amplitude of a component being spread across frequency is not easy to visualise. A very mundane analogy may help, consider three jars of jam on a table as shown in Figure 6.5(a). Where is the jam? It is concentrated in jars at points X, Y and Z.

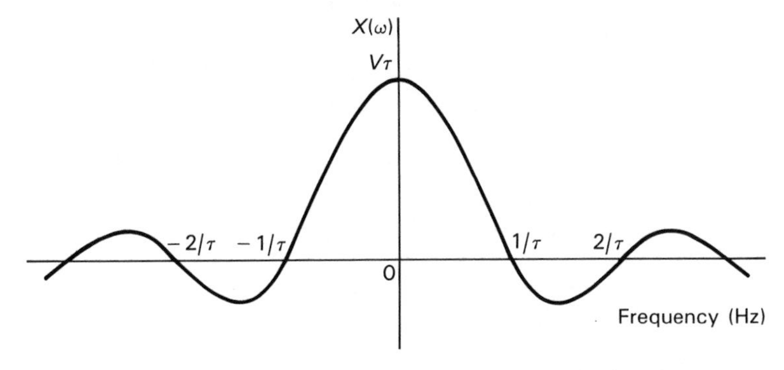

Figure 6.4 Amplitude density spectrum of a single pulse.

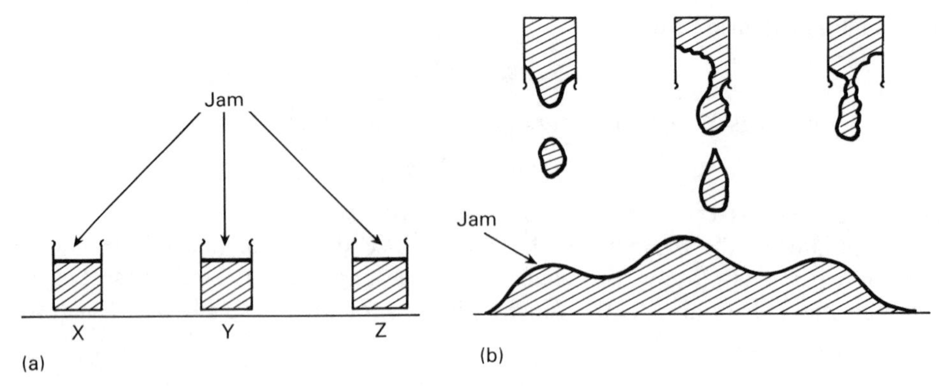

Figure 6.5 'Jam' analogy to amplitude density spectrum.

Suppose now the pots are turned upside down and the jam spread over the table as shown in Figure 6.5(b). Where is the jam now? It is no longer localised, one can no longer visualise it as being located at the specific points X, Y and Z. Instead one must consider a 'jam density'; so much jam per unit volume and this density varies over the table.

Like the amplitude spectrum for the Fourier series, in general the spectrum can be represented by a magnitude and phase spectrum. The following example illustrates the calculation and representation of the spectra.

EXAMPLE 6.2.1

Obtain the Fourier transform for the single-sided and double-sided exponential pulses as shown in Figure 6.6. Sketch the magnitude and phase spectrum in each case.

SOLUTION

For the single sided pulse the function is zero for $t<0$ and the transform becomes

$$X(\omega) = \int_0^\infty \mathrm{e}^{-at}\mathrm{e}^{-\mathrm{j}\omega t}\,\mathrm{d}t$$

$$= \int_0^\infty \mathrm{e}^{-(a+\mathrm{j}\omega)t}\,\mathrm{d}t$$

$$= \frac{1}{-(a+\mathrm{j}\omega)}\,[\mathrm{e}^{-(a+\mathrm{j}\omega)t}]_0^\infty$$

$$= \frac{1}{a+\mathrm{j}\omega}$$

$$|X(\omega)| = \frac{1}{\sqrt{(a^2+\omega^2)}}, \qquad \underline{/X(\omega)} = -\tan^{-1}(\omega/a)$$

The magnitude and phase spectra are plotted in Figure 6.7.

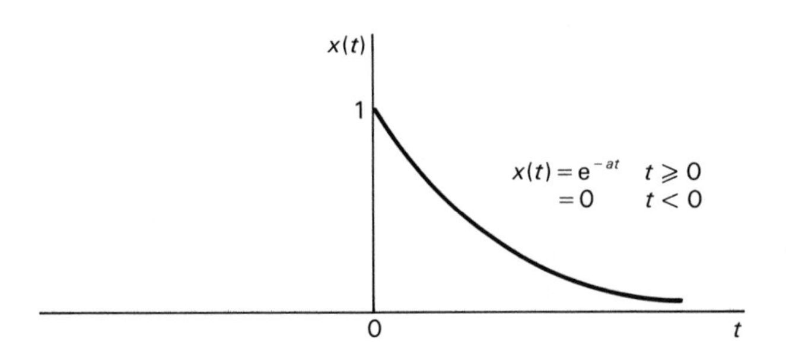

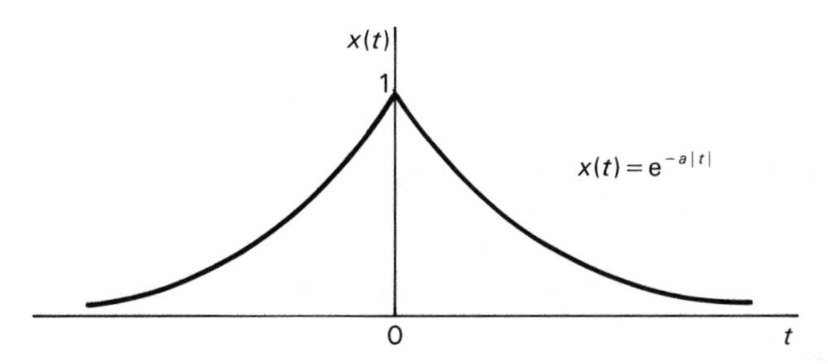

Figure 6.6 Exponential pulse for Example 6.2.1.

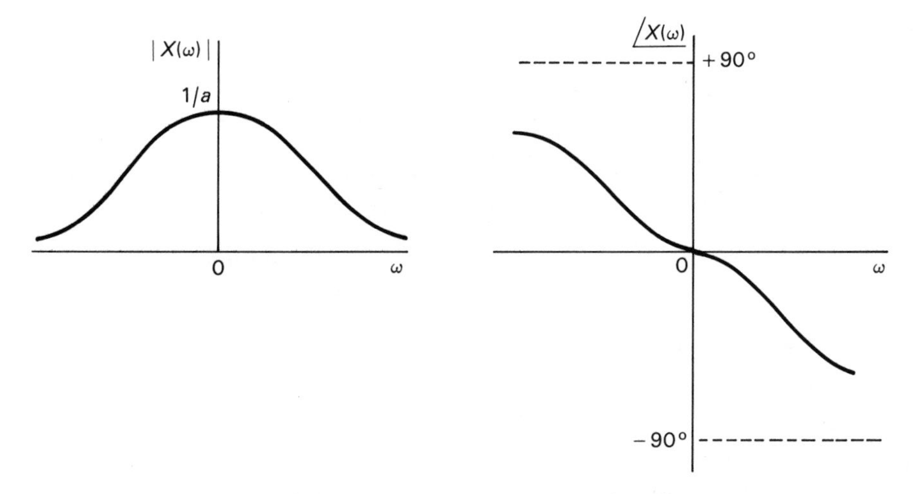

Figure 6.7 Magnitude and phase spectra for a single-sided exponential pulse.

For the double-sided pulse the integral can be split into two ranges $-\infty$ to 0 and 0 to $+\infty$

$$X(\omega) = \int_{-\infty}^{+\infty} e^{-a|t|} e^{-j\omega t}\, dt$$

$$= \int_{-\infty}^{0} e^{+at} e^{-j\omega t}\, dt + \int_{0}^{+\infty} e^{-at} e^{-j\omega t}\, dt$$

$$= \frac{1}{a - j\omega}\left[e^{(a-j\omega)t}\right]_{-\infty}^{0} + \frac{1}{-(a + j\omega)}\left[e^{-(a+j\omega)t}\right]_{0}^{+\infty}$$

$$= \frac{1}{a - j\omega} + \frac{1}{a + j\omega}$$

$$= \frac{2a}{a^2 + \omega^2}$$

The magnitude spectrum is identical in shape to that for the single-sided pulse, but it differs by the scaling factor $2a$. The transform is real, this might have been expected from the symmetry in the time waveform and the parallel with the exponential form of the Fourier series. This together with other properties of the Fourier transform will be proved in the next section.

Table 6.1 gives a short list of Fourier transforms. Many of the entries in this table are transforms of 'power signals'. The definition of this term and the derivation of the corresponding transforms will be covered in Section 6.2.3. However it is convenient to group all the transforms into one table at this point.

Table 6.1 A short table of Fourier transforms

Signal description	$x(t)$	$X(\omega)$
Unit impulse	$\delta(t)$	1
Unit step	$u(t)$	$\frac{1}{j\omega} + \pi\delta(\omega)$
Exponential pulse	$u(t)\, e^{-at}$	$\frac{1}{a + j\omega}$
Sine wave	$\sin \omega_0 t$	$-j\omega[\delta(\omega - \omega_0) + \delta(\omega + \omega_0)]$
Cosine wave	$\cos \omega_0 t$	$\pi[\delta(\omega - \omega_0) + \delta(\omega + \omega_0)]$
Complex exponential	$e^{j\omega_0 t}$	$2\pi\delta(\omega - \omega_0)$

$X(\omega) = \int_{-\infty}^{+\infty} x(t)\, e^{-j\omega t}\, dt$

The Fourier transform gives a frequency domain description of a time signal. Often it is useful to study the effect in frequency caused by an operation in time, or vice versa. In particular it is useful to know how some of the operations on a time signal that were introduced in Chapter 1 (scaling, reflection time shifting) affect the transform of that signal. The effect of these transformations and other properties of the Fourier transform are examined in the next section.

In deriving these results it will be taken that the function $x(t)$ and $X(\omega)$ form of Fourier transform pair

$$x(t) \leftrightarrow F(\omega)$$

6.2.2 *Elementary properties of the Fourier transform*

Even and odd symmetries

The Fourier transform can be easily expressed in trigonometric form

$$\begin{aligned} X(\omega) &= \int_{-\infty}^{+\infty} x(t)\mathrm{e}^{-\mathrm{j}\omega t}\,\mathrm{d}t \\ &= \int_{-\infty}^{+\infty} x(t)\cos(\omega t)\,\mathrm{d}t - \mathrm{j}\int_{-\infty}^{+\infty} x(t)\sin(\omega t)\,\mathrm{d}t \end{aligned} \tag{6.2.5}$$

Using the results of symmetry derived in Section 5.4.2, then if $x(t)$ is an even function the second integral is zero and $X(\omega)$ is real, the transform can then be expressed as

$$X(\omega) = 2\int_{0}^{\infty} x(t)\cos\omega t\,\mathrm{d}t$$

If $x(t)$ is an odd function then the transform is imaginary and can be expressed

$$X(\omega) = -\mathrm{j}2\int_{0}^{\infty} x(t)\sin\omega t\,\mathrm{d}t$$

Generally $x(t)$ is neither even nor odd but as shown in Section 2.4.1 it can be expressed as a combination of even and odd functions

$$x(t) = x_{\mathrm{e}}(t) + x_{\mathrm{o}}(t)$$

where

$$x_{\mathrm{e}}(t) = \frac{x(t) + x(-t)}{2}, \qquad x_{\mathrm{o}} = \frac{x(t) - x(-t)}{2}$$

If these expressions are substituted into eqn (6.2.5), as already seen some of the terms are zero giving

$$X(\omega) = \int_{-\infty}^{+\infty} x_{\mathrm{e}}(t)\cos\omega t\,\mathrm{d}t - \mathrm{j}\int_{-\infty}^{+\infty} x_{\mathrm{o}}(t)\sin\omega t\,\mathrm{d}t$$

The first integral is an even function in ω, the second an odd function. Hence the real part of the Fourier transform is even and the imaginary part is odd. The phase spectrum is given by

$$\begin{aligned}\text{phase} &= \angle X(\omega) \\ &= \tan^{-1}\left(\frac{\text{odd function}}{\text{even function}}\right)\end{aligned}$$

The ratio of an odd to an even function produces an odd function and as $\tan^{-1}$ is also an odd function the phase spectrum must be odd. The magnitude is given by

$$|X(\omega)| = \sqrt{[(\text{even function})^2 + (\text{odd function})^2]}$$

As a $(\text{function})^2$ is always even, $|X(\omega)|$ the magnitude spectrum is always even.

Hence

$$|X(\omega)| = |X(-\omega)| \qquad \text{and} \qquad X(\omega) = -X(-\omega)$$

i.e.,

$$X(-\omega) = X^*(\omega)$$

Linearity

If $X_1(\omega)$ is the Fourier transform of $x_1(t)$, and $X_2(\omega)$ is the Fourier transform of $x_2(t)$ $(x_1(t) \leftrightarrow X_1(\omega),\ x_2(t) \leftrightarrow X_2(\omega))$ then

$$\{ax_1(t) + bx_2(t)\} \leftrightarrow \{aX_1(\omega) + bX_2(\omega)\} \tag{6.2.6}$$

where a and b are constants. This follows directly from the definition of linearity.

Time scaling

Consider the Fourier transform of a single pulse as derived in the last section. Figure 6.8(a) shows the spectrum for a single pulse of width T, and Figure 6.8(b) shows the spectrum for a single pulse of width $2T$.

The effect of making the pulse more extended in time is to cause its spectrum to be more compressed in frequency. This is true generally and can be proved as follows.

The Fourier transform of $x(at)$, (a positive) is

$$F\{x(at)\} = \int_{-\infty}^{+\infty} x(at)e^{-j\omega t}\, dt$$

Making the substitution $\lambda = at$ then $t = \lambda/a$ and $dt = d\lambda/a$ gives

$$F\{x(at)\} = \frac{1}{a}\int_{-\infty}^{+\infty} x(\lambda)e^{-j(\omega/a)\lambda}\, d\lambda$$

A little thought will show that the integral on the right-hand side is the Fourier transform of $x(t)$ (the λ can be regarded as a dummy variable here) but with ω

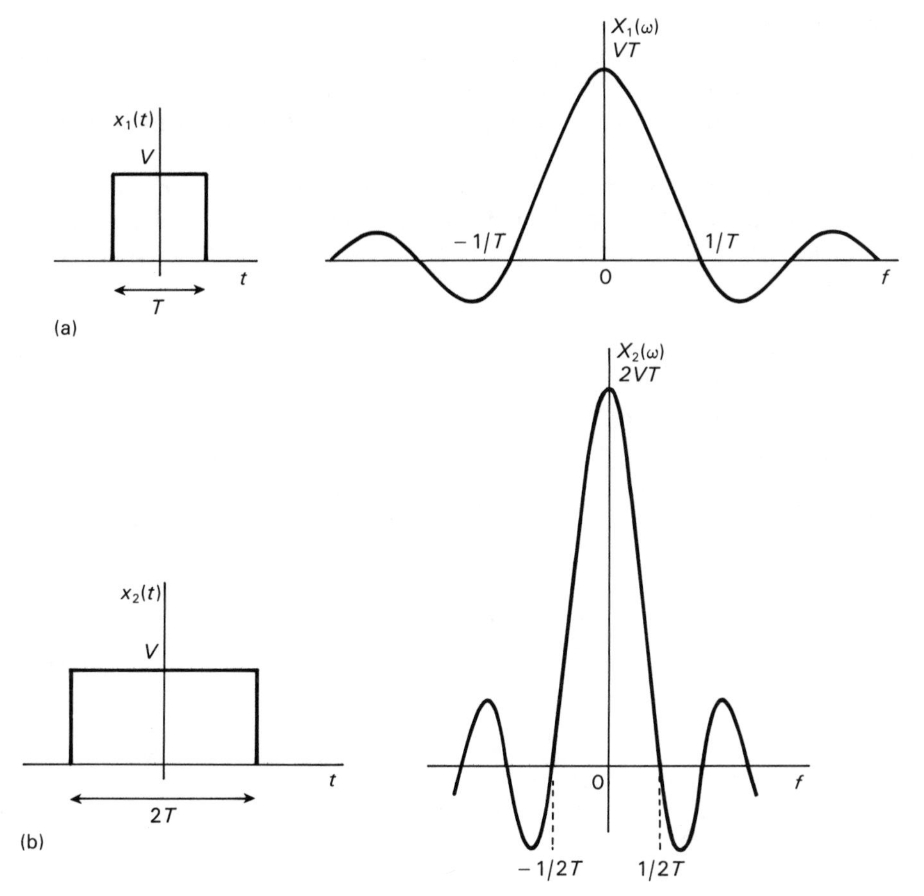

Figure 6.8 Effect of time scaling on a single pulse.

replaced by ω/a

$$F\{x(at)\} = \frac{1}{a} X\left(\frac{\omega}{a}\right)$$

$$x(at) \leftrightarrow \frac{1}{a} X\left(\frac{\omega}{a}\right) \tag{6.2.7}$$

This result is as would be expected intuitively. If the same shaped waveform is squashed in time then the signal has to change more rapidly and higher frequency components appear. The student will be familiar with this effect when a music tape is played at other than the recorded speed. The pitch of the music is shifted either up or down dependent on whether the playback speed is faster or slower than the recorded speed.

Time shifting property

Consider the Fourier transform of a pulse starting at $t=0$ and of duration T. The real and imaginary parts of the transform are found by multiplying the pulse waveform by the real and imaginary parts of $e^{-j\omega t}$ and integrating. This is shown in Figure 6.9.

Consider now the Fourier transform of the delayed pulse $x(t-\tau)$. Rather than use the delayed pulse this can be found by multiplying the original pulse by the real and imaginary parts of a time *advanced* function $e^{-j\omega(t+\tau)}$. Hence the transform of

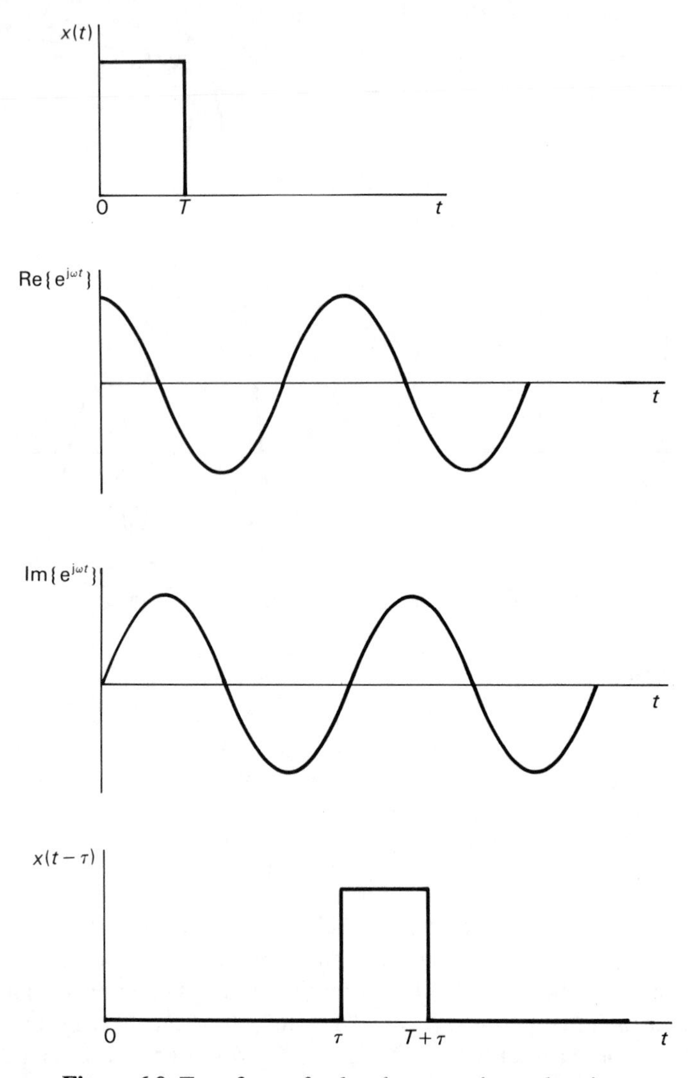

Figure 6.9 Transform of pulse shown as sine and cosine.

the delayed pulse can be written

$$F\{x(t-\tau)\} = \int_{-\infty}^{+\infty} x(t)\mathrm{e}^{-\mathrm{j}\omega(t+\tau)}\,\mathrm{d}t$$

$$= \mathrm{e}^{-\mathrm{j}\omega\tau}\int_{-\infty}^{+\infty} x(t)\mathrm{e}^{-\mathrm{j}\omega t}\,\mathrm{d}t$$

$$= \mathrm{e}^{-\mathrm{j}\omega\tau}X(\omega)$$

This can be written

if $\quad x(t) \leftrightarrow X(\omega)$

then $\quad x(t-\tau) \leftrightarrow \mathrm{e}^{-\mathrm{j}\omega\tau}X(\omega) \qquad (6.2.8)$

This result can be proved more formally by taking the transform of $x(t-\tau)$ and substituting a new variable for $t-\tau$. Hence delaying a signal by τ causes its transform to be modified by a factor $\mathrm{e}^{-\mathrm{j}\omega\tau}$. This factor has unit magnitude hence the magnitude spectrum of $x(t)$ is unaltered. The phase spectrum will be modified however by addition of an angle $\omega\tau$, representing a phase shift proportional to frequency. Such a characteristic is known as a linear phase characteristic and a system having such a characteristic will produce a time delay but will not otherwise distort the signal.

Frequency shift or modulation property

Suppose

$$x(t) \leftrightarrow X(\omega)$$

and it is required to obtain the Fourier transform of the signal $\mathrm{e}^{\mathrm{j}\omega_0 t}x(t)$. This is given by

$$F\{x(t)\mathrm{e}^{\mathrm{j}\omega_0 t}\} = \int_{-\infty}^{+\infty} x(t)\mathrm{e}^{\mathrm{j}\omega_0 t}\mathrm{e}^{-\mathrm{j}\omega t}\,\mathrm{d}t$$

$$= \int_{-\infty}^{+\infty} x(t)\mathrm{e}^{-\mathrm{j}(\omega-\omega_0)t}\,\mathrm{d}t$$

$$= X(\omega-\omega_0)$$

$$\mathrm{e}^{\mathrm{j}\omega_0 t}x(t) \leftrightarrow X(\omega-\omega_0) \qquad (6.2.9)$$

This property will be used in the section on modulation (Section 6.3).

Duality

If the time shift and modulation properties are written together

$$x(t-\tau) \leftrightarrow \mathrm{e}^{-\mathrm{j}\omega\tau}X(\omega)$$

$$\mathrm{e}^{\mathrm{j}\omega_0 t}x(t) \leftrightarrow X(\omega-\omega_0)$$

a certain symmetry or duality is apparent. This is a general property of the Fourier transform. The proof is as follows. Starting with the inverse transform

$$x(t) = \frac{1}{2\pi}\int_{-\infty}^{+\infty} X(\omega)e^{j\omega t}\,d\omega$$

then changing the sign of t and rearranging

$$2\pi x(-t) = \int_{-\infty}^{+\infty} X(\omega)e^{-j\omega t}\,d\omega$$

The variables ω and t can now be interchanged (strictly this should be done via dummy variables).

$$2\pi x(-\omega) = \int_{-\infty}^{+\infty} X(t)e^{-j\omega t}\,dt$$

$$2\pi x(-\omega) = F\{X(t)\} \tag{6.2.10}$$

This property can appear confusing and an example may help. Consider the Fourier transform of a rectangular pulse considered earlier, this was given by

$$X(\omega) = \frac{V\tau\sin(\omega\tau/2)}{\omega\tau/2}$$

Consider now that this represents a time function $X(t)$, i.e.,

$$X(t) = \frac{V\tau\sin(t\tau/2)}{t\tau/2}$$

Although there is nothing mathematically wrong with this expression it does not 'look right'. Remember τ is a constant and any symbol could be used here. However, one associates τ with a quantity having the units of time, however in this expression it has the units of frequency so let us use a different symbol ω_0. Then

$$X(t) = V\omega_0\,\frac{\sin(\omega_0 t/2)}{\omega_0 t/2}$$

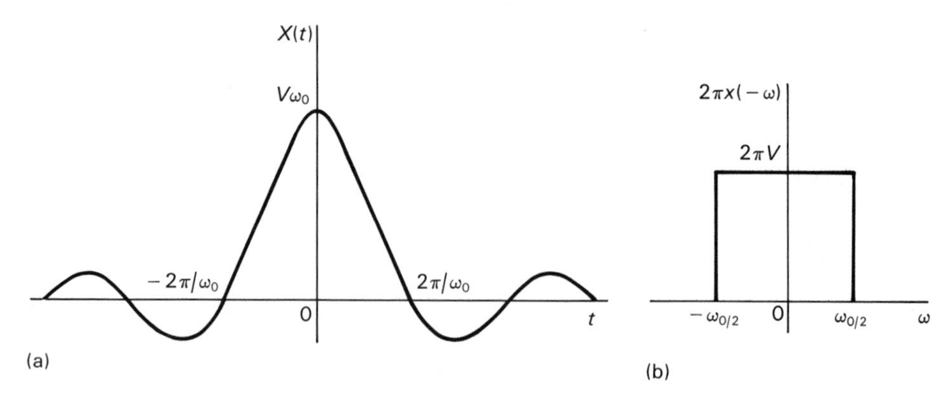

Figure 6.10 Use of pulse to illustrate duality principle.

Table 6.2 Some properties of the Fourier transform

Property	$x(t)$	$X(\omega)$
Linearity	$a_1x_1(t)+a_2x_2(t)$	$a_1X_1(\omega)+a_2X_2(\omega)$
Scaling $a>0$	$x(at)$	$\frac{1}{a}X\left(\frac{\omega}{a}\right)$
Time shift	$x(t-t_0)$	$e^{-j\omega t_0}X(\omega)$
Modulation	$e^{j\omega_0 t}x(t)$	$X(\omega-\omega_0)$
Duality	$X(t)$	$2\pi x(-\omega)$
Convolution	$x_1(t)*x_2(t)$	$X_1(\omega)X_2(\omega)$

Now using eqn (6.2.10) the Fourier transform is given by

$$F(x(t))=2\pi x(-\omega)$$

where $x(-\omega)$ is shown in Figure 6.10(b).

The duality principle has proved useful here as direct evaluation of the transform from $X(t)$ as shown in Figure 6.10(a) would have proved difficult. The properties of the transform described in this section are grouped together in Table 6.2.

The following example shows how the properties discussed in this section can be useful in the evaluation of transforms.

EXAMPLE 6.2.2

In Section 6.2.1 it was shown that the Fourier transform of the rectangular pulse shown in Figure 6.11(a) is given by

$$X(\omega)=\frac{V\tau\sin(\omega\tau/2)}{\omega\tau/2}$$

Using this result and the appropriate transform properties obtain the Fourier transform of the signal shown in Figure 6.11(b).

SOLUTION

The signal in Figure 6.11 can be regarded as the sum of three signals as shown in Figure 6.12

$$x(t)=x_1(t)+x_2(t)+x_3(t) \qquad (6.2.11)$$

The transform corresponding to the signal $x_2(t)$ is given. The signals $x_1(t)$ and $x_2(t)$ are scaled, time shifted versions of $x_2(t)$

$$x_1(t)=-\tfrac{1}{2}x_2\left[2\left(t+\frac{3\tau}{4}\right)\right]; \qquad x_3(t)=-\tfrac{1}{2}x_2\left[2\left(t-\frac{3\tau}{4}\right)\right]$$

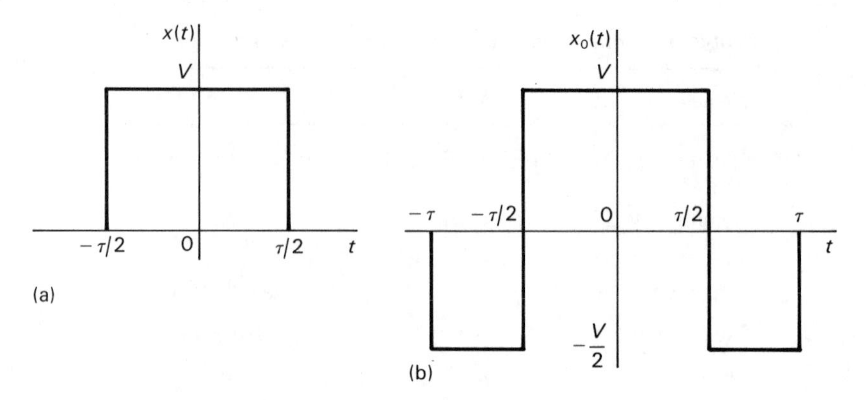

Figure 6.11 Signals for Example 6.2.2.

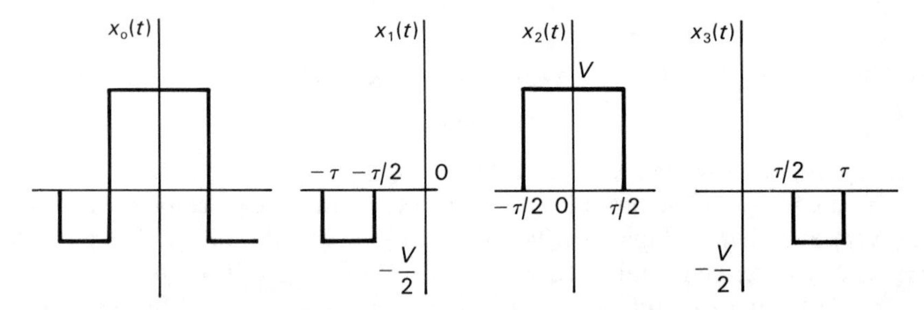

Figure 6.12 Decomposition of the signal shown in Figure 6.11(b).

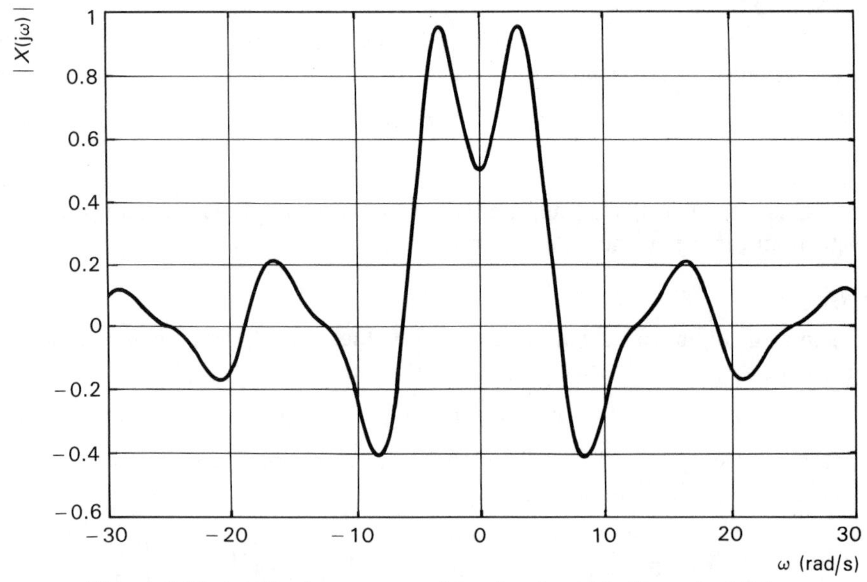

Figure 6.13 Amplitude spectrum of the signal used in Example 6.2.2.

Taking the Fourier transform of eqn (6.2.11) and using the scaling and shift properties of the transform gives

$$X(j\omega) = \frac{V\tau \sin(\omega\tau/2)}{(\omega\tau/2)} - e^{j3\omega\tau/4} \cdot \frac{1}{2} \cdot \frac{1}{2} \cdot \frac{V\tau \sin(\omega\tau/4)}{(\omega\tau/4)}$$

$$- e^{-j3\omega\tau/4} \cdot \frac{1}{2} \cdot \frac{1}{2} \cdot \frac{V\tau \sin(\omega\tau/4)}{(\omega\tau/4)}$$

$$= V\tau\left[\frac{\sin(\omega\tau/2)}{(\omega\tau/2)} - \frac{1}{2}\,\frac{\sin(\omega\tau/4)}{(\omega\tau/4)}\,\frac{(e^{j3\omega\tau/4} + e^{-j3\omega\tau/4})}{2}\right]$$

$$= V\tau\left[\frac{\sin(\omega\tau/2)}{(\omega\tau/2)} - \frac{1}{2}\,\frac{\sin(\omega\tau/4)}{(\omega\tau/4)} \cos 3\omega\tau/4\right]$$

It should be noted that the transform is real due to the symmetry in the waveform $x(t)$. The amplitude spectrum is shown in Figure 6.13.

6.2.3 *Signal energy: power signals*

In Section 5.4.3 the concept of signal power in relation to the Fourier series was introduced. For a periodic signal the power represented is the mean power obtained by averaging the instantaneous power over the periodic time. For a non-periodic signal the concept of power can cause difficulty, the mean power being dependent upon the time interval over which the average is taken. For finite duration signals the concept of energy is more useful where energy is represented by

$$E = \int_{-\infty}^{+\infty} [x(t)]^2 \, dt$$

Just as there was a relationship between power calculated in the time domain and power calculated in the frequency domain for the Fourier series there is a similar relationship for energy in terms of the Fourier transform.

Using the inverse Fourier transform for $x(t)$ the expression for energy can be written

$$E = \int_{-\infty}^{+\infty} x(t)\left[\frac{1}{2\pi}\int_{-\infty}^{+\infty} X(\omega)e^{+j\omega t}\, d\omega\right] dt$$

This is a double integral and one can in this case interchange the order of the integration. The inner integral is then with respect to time and one can take $X(\omega)$ outside the integral sign giving

$$E = \frac{1}{2\pi}\int_{-\infty}^{+\infty} X(\omega) \int_{-\infty}^{+\infty} x(t)e^{j\omega t}\, dt\, d\omega$$

The inner integral is almost a Fourier transform but the exponent is the wrong sign, a little thought will show it is $X(-\omega)$. Also $X(-\omega)=X^*(\omega)$ and $X(\omega)X^*(\omega)=|X(\omega)|^2$ hence

$$\int_{-\infty}^{+\infty} [x(t)]^2 \,\mathrm{d}t = \frac{1}{2\pi}\int_{-\infty}^{+\infty} |X(\omega)|^2 \,\mathrm{d}\omega \tag{6.2.12}$$

This relationship between the energy in the time and frequency domains is known as *Parseval's* theorem. A form of this relating to power for the Fourier series was given in Section 5.4.3.

Hence $|X(\omega)|^2$ when integrated with respect to frequency is proportional to the total energy in the signal, hence $|X(\omega)|^2$ can be interpreted as an energy spectral density or just a spectral density. Also

$$\int_{\omega_1}^{\omega_2} |X(\omega)|^2 \,\mathrm{d}\omega$$

gives the energy that is contained in the range of the spectrum between ω_1 and ω_2.

Returning to the calculation of the energy in the time domain this is no problem provided

$$0 < \int_{-\infty}^{+\infty} |x(t)|^2 \,\mathrm{d}t < \infty$$

Signals that meet this condition are known as energy signals. There are signals where these conditions are not satisfied. Unfortunately these are not 'weird' signals of only academic interest but are very useful signals like step functions and sinusoids.

For such signals the power P is defined

$$P = \lim_{T\to\infty} \frac{1}{2T}\int_{-T}^{+T} |x(t)|^2 \,\mathrm{d}t$$

If P satisfies the condition

$$0 < P < \infty$$

then the signal $x(t)$ is defined as a power signal.

There are two approaches by which the scope of the Fourier transform can be extended to include power signals. One is to introduce a convergence factor e^{-at} into the time signal, this leads to the Laplace transform which is the subject of the next chapter.

An alternative approach is to investigate signals that lead to the inclusion of the impulse function in their transforms. Consider first the Fourier transform of the impulse function $\delta(t)$

$$\begin{aligned} X(\omega) &= \int_{-\infty}^{+\infty} \delta(t)\mathrm{e}^{-j\omega t} \,\mathrm{d}t \\ &= 1 \end{aligned}$$

(by use of the sifting property of the impulse function).

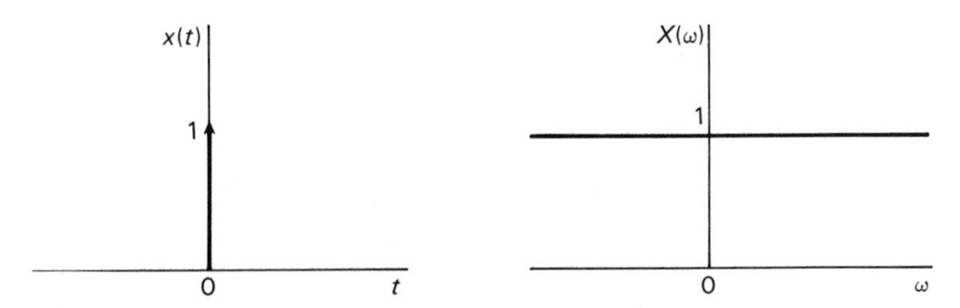

Figure 6.14 Transform of impulse.

The transform corresponding to the impulse function is illustrated in Figure 6.14.

This function illustrates the scaling property of the Fourier transform in an extreme case. The time signal involved is compressed into zero time and it produces a spectrum that extends uniformly to infinity. One would therefore expect the reverse, the spectrum of a constant d.c. signal would be an impulse. This can be verified by taking the inverse transform of an impulse in frequency $\delta(\omega)$

$$x(t) = \frac{1}{2\pi}\int_{-\infty}^{+\infty} \delta(\omega)e^{j\omega t}\,d\omega$$

$$= \frac{1}{2\pi}$$

This result is shown in Figure 6.15 for a constant level of 1. It could have been inferred directly from the duality property of the Fourier transform.

Considering a sinusoidal signal of frequency ω_0 one would intuitively expect the spectrum to be compressed and exist only at a frequency ω_0. However if a cosine signal is considered this is an even function and the spectrum must be real and possess even symmetry. Similarly the spectrum of a sine wave signal must be imaginary and possess odd symmetry.

These considerations lead to the spectra shown in Figure 6.16. The factor π must be included, as can easily be shown by consideration of the inverse transform.

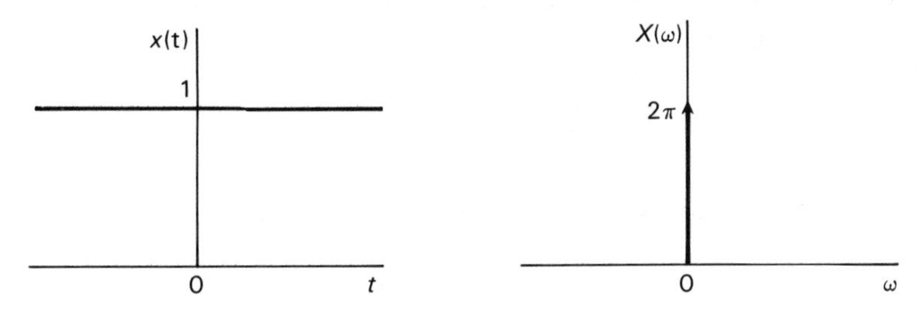

Figure 6.15 Fourier transform of a d.c. level.

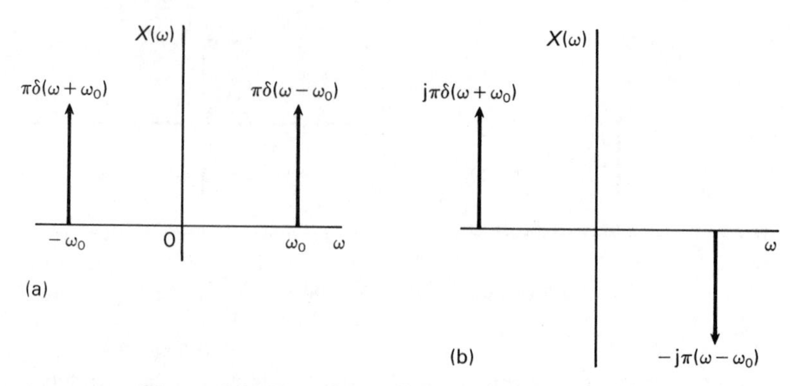

Figure 6.16 Transforms of sine and cosine signals.

For the cosine waveform, Figure 6.16(a)

$$\begin{aligned} x(t) &= \frac{1}{2\pi}\int_{-\infty}^{+\infty} \pi[\delta(\omega + \omega_0) + \delta(\omega - \omega_0)]e^{j\omega t}\,d\omega \\ &= \tfrac{1}{2}(e^{-j\omega_0 t} + e^{j\omega_0 t}) \\ &= \cos \omega_0 t \end{aligned}$$

Hence

$$\cos \omega_0 t \leftrightarrow \pi[\delta(\omega - \omega_0) + \delta(\omega + \omega_0)]$$

Similarly, with reference to Figure 6.16(b),

$$\sin \omega_0 t \leftrightarrow j\pi[\delta(\omega + \omega_0) - \delta(\omega - \omega_0)]$$

From these results the transform for $e^{j\omega_0 t}$ follows

$$\begin{aligned} e^{j\omega_0 t} &= \cos \omega_0 t + j \sin \omega_0 t \\ F\{e^{j\omega_0 t}\} &= \pi[\delta(\omega - \omega_0) + \delta(\omega + \omega_0)] + j^2\pi[\delta(\omega + \omega_0) - \delta(\omega - \omega_0)] \\ &= 2\pi\delta(\omega - \omega_0) \\ e^{j\omega_0 t} &\leftrightarrow 2\pi\delta(\omega - \omega_0) \end{aligned}$$

Having obtained the Fourier transform for the complex exponential it is now easy to obtain the Fourier transform for any periodic waveform.

As the waveform is periodic it can be expressed as a Fourier series.

$$x(t) = \sum_{n=-\infty}^{\infty} c_n e^{jn\omega_0 t}$$

Taking Fourier transforms and using the linearity property

$$X(\omega) = 2\pi \sum_{n=-\infty}^{\infty} c_n \delta(\omega - n\omega_0)$$

Hence the amplitude density spectrum of a periodic signal is similar to the amplitude spectrum of that signal. It is a discrete spectrum consisting of impulses each having strength $2\pi c_n$ where c_n is the coefficient in the Fourier series expansion.

6.2.4 System response in terms of Fourier transform

In the previous chapter the steady state response of a system to a single sinusoidal input was considered. It was shown that the magnitude of the sinusoid was multiplied by a factor $|H(\mathrm{j}\omega)|$ and the phase was shifted by an amount $\underline{/H(\mathrm{j}\omega)}$, where $H(\mathrm{j}\omega)$ is the system frequency response function. Because of the linear time invariant nature of the system this result could be extended to the component frequencies in the Fourier series. When expressed in exponential form a component having a complex coefficient c_n is modified by the factor $H(\mathrm{j}\omega)$ to produce the coefficient $c_n H(\mathrm{j}\omega)$ at the system output. Intuitively one would expect this result to carry over to the Fourier transform where the input is a continuum of sinusoids. One might then expect the Fourier transform of the output signal to be $X(\mathrm{j}\omega)/H(\mathrm{j}\omega)$ and this result can be proved formally as follows.

As usual the system input and output signals will be denoted by $x(t)$ and $y(t)$ and their respective Fourier transforms by $X(\omega)$ and $Y(\omega)$. The output time signal is related to the input time signal via convolution, if $h(t)$ is the system impulse response then

$$y(t) = \int_{-\infty}^{+\infty} h(t-\tau)x(\tau)\,\mathrm{d}\tau$$

Taking the Fourier transform of both sides gives

$$Y(\omega) = \int_{-\infty}^{+\infty}\int_{-\infty}^{+\infty} h(t-\tau)x(\tau)\,\mathrm{d}\tau \mathrm{e}^{-j\omega t}\,\mathrm{d}t$$

Interchanging the order of integration

$$Y(\omega) = \int_{-\infty}^{+\infty} x(t)\int_{-\infty}^{+\infty} h(t-\tau)\mathrm{e}^{-j\omega t}\,\mathrm{d}t\,\mathrm{d}\tau$$

The inner integral represents the Fourier transform of a delayed function and equals $\mathrm{e}^{-\mathrm{j}\omega\tau}H(\mathrm{j}\omega)$ hence

$$Y(\omega) = \int_{-\infty}^{+\infty} x(\tau)H(\omega)\mathrm{e}^{-j\omega\tau}\,\mathrm{d}\tau$$

taking $H(\mathrm{j}\omega)$ outside, the remaining integral represents the Fourier transform of $x(t)$, hence

$$Y(\omega) = X(\omega)H(\omega) \tag{6.2.13}$$

This is a useful result in general stating that convolution in time is equivalent to multiplication in frequency. From the viewpoint of the system response this is the result expected provided $H(\omega)$ (the Fourier transform of the impulse response) is the frequency response function defined in Section 5.2.1. This can be shown to be true by considering the input to a system to be a cosine wave and calculating the resulting output using eqn (6.2.13). If

$$x(t) = A \cos \omega t$$
$$X(\omega) = A\pi(\delta(\omega - \omega_0) + \delta(\omega + \omega_0))$$

and

$$Y(\omega) = A\pi H(\omega)[\delta(\omega - \omega_0) + \delta(\omega + \omega_0))$$

This is another cosine wave whose amplitude is modified by $|H(\omega)|$ and whose phase is modified by $\underline{/H(\omega)}$. This is precisely the definition of the frequency response function as in Section 5.2.1.

Taking the complex conjugate of both sides of eqn (6.2.13)

$$Y^*(\omega) = X^*(\omega)H^*(\omega)$$

and multiplying both sides by the original equation gives

$$|Y(\omega)|^2 = |X(\omega)|^2 |H(\omega)|^2 \tag{6.2.14}$$

As shown in Section 6.2.3 $|X(\omega)|^2$ and $|Y(\omega)|^2$ represent the spectral densities of the input and output signals respectively. Hence eqn (6.2.14) can be used to calculate how the energy in a signal is modified by a system. This will be illustrated by the following example.

EXAMPLE 6.2.3

An exponential pulse as shown in Figure 6.17 is fed into an ideal low-pass filter having a frequency response function as shown. Determine the filter cut-off frequency ω_c such that half the energy of the signal will appear at the filter output.

SOLUTION

The Fourier transform of a single-sided exponential pulse has already been calculated in Example 6.2.1. Applying the result here gives

$$X(\omega) = \frac{2}{(0.2) + j\omega}$$

$$|X(\omega)|^2 = \frac{4}{(0.2)^2 + \omega^2}$$

The total energy in the input signal is given by

$$E = \frac{1}{2\pi} \int_{-\infty}^{+\infty} \frac{4}{(0.2)^2 + \omega^2} \, d\omega$$

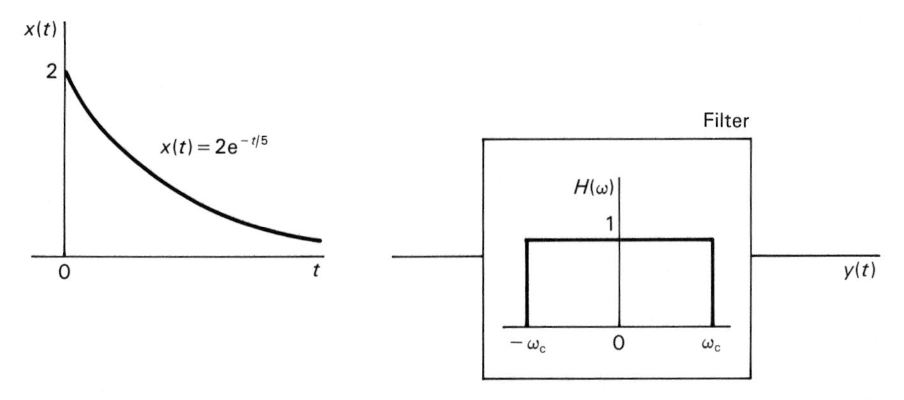

Figure 6.17 Pulse and filter for Example 6.2.3.

A standard integral is

$$\int \frac{1}{a^2 + x^2}\,\mathrm{d}x = \frac{1}{a}\tan^{-1}\frac{x}{a} + \text{constant}$$

Hence

$$\begin{aligned} E &= \frac{2}{2\pi}\int_0^{\infty} \frac{4}{(0.2)^2 + \omega^2}\,\mathrm{d}\omega \\ &= \frac{4}{\pi}\left[5\tan^{-1} 5\omega\right]_0^{\infty} \\ &= \frac{4}{\pi} \times 5 \times \frac{\pi}{2} \\ &= 10 \text{ units} \end{aligned}$$

The Fourier transform of the output signal is given as

$$\begin{aligned} Y(\omega) &= H(\omega)X(\omega) \\ &= \frac{2}{0.2 + \mathrm{j}\omega} \qquad -\omega_c \leqslant \omega \leqslant \omega_c \\ &= 0 \qquad \text{otherwise} \end{aligned}$$

Hence the energy at the output is given by

$$\frac{2}{2\pi}\int_0^{+\omega_c} \frac{4}{(0.2)^2 + \omega^2}\,\mathrm{d}\omega = \frac{4}{\pi} \times 5\tan^{-1} 5\omega_c$$

For this to be half the input energy

$$\frac{4}{\pi} \times 5 \tan^{-1} 5\omega_c = 5$$

$$\tan^{-1} 5\omega_c = \frac{\pi}{5}$$

$$\omega_c = 0.2 \text{ rad/s}$$

6.3 Amplitude modulation

Modulation is a process where the amplitude of one signal is used to control a parameter of another signal. Complete books are written on the subject and the object here is to introduce modulation as an example of the application of the Fourier transform. Hence only one form of modulation is investigated, amplitude modulation. In amplitude modulation the amplitude of the modulating signal controls the amplitude of a sinusoidal carrier as shown in Figure 6.18. Also shown is the output waveform for specific signals, $x_1(t)$, a square wave and $x_2(t)$ a sine wave. In order to appreciate the reasons why modulation is used the spectrum of the modulated carrier $y(t)$ will be obtained. It is instructive to obtain this by a number of different methods. The first being a restricted method applying only to the case where the input signal $x(t)$ is sinusoidal.

$$x(t) = X \cos \omega_s t$$

then

$$y(t) = X \cos \omega_s t \cos \omega_c t$$

$$= \frac{X}{2} [\cos(\omega_c + \omega_s)t + \cos(\omega_c - \omega_s)t]$$

The spectrum of the signal $x(t)$ is shown in Figure 6.19(a) and that of the modulated carrier $y(t)$ is shown in Figure 6.19(b) where it is assumed that $\omega_c \gg \omega_s$.

The process of modulation has caused two new frequency components to appear, one having frequency $(\omega_s + \omega_c)$, the other having frequency $(\omega_c - \omega_s)$, these are known as the upper and lower sidebands respectively. Note there is no carrier frequency present in the signal after modulation. Either of the sidebands contains information on the magnitude, frequency and phase of the signal and as will be shown later these can be extracted by a process known as demodulation. If $\omega_c \gg \omega_s$ then the information has been shifted from a frequency ω_s to a frequency close to ω_c. This is the principal use of modulation. Information in speech and music is contained in a frequency range in the region of 50 Hz – 20 kHz. To produce electromagnetic radiation in this frequency range would be very expensive, it would require huge amounts of power and the aerials needed would be extremely large.

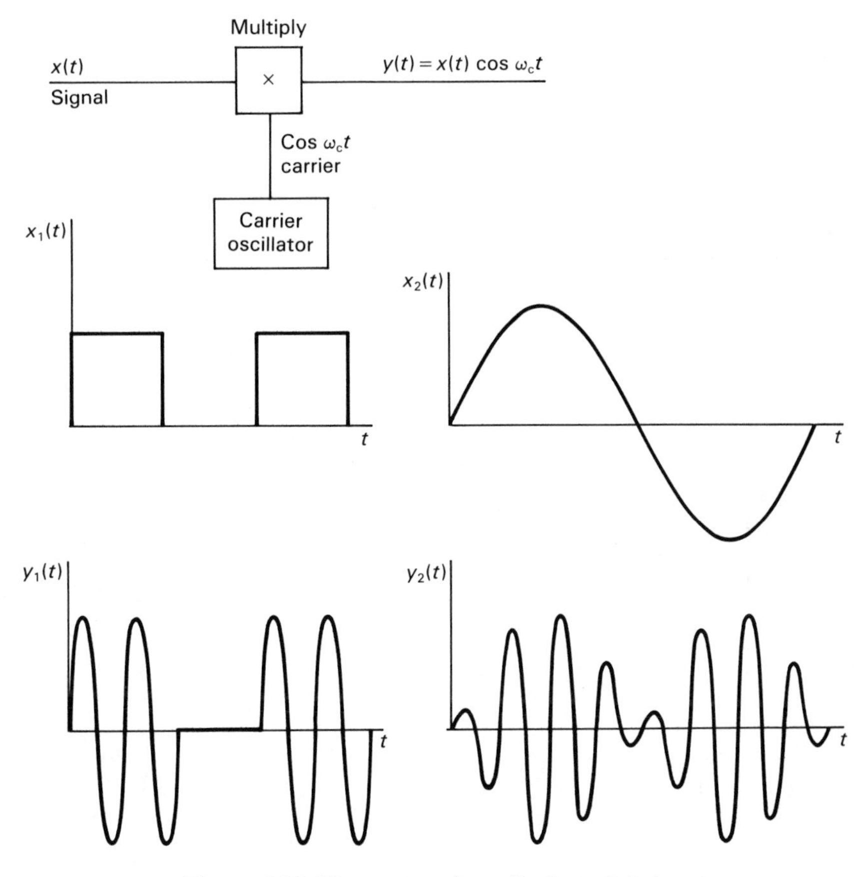

Figure 6.18 The process of amplitude modulation.

Hence the process of modulation is used to shift the information to a more manageable frequency range.

As stated in the last paragraph, usually the signal is not at a single frequency but covers a band of frequencies, the second method of considering modulation will take this into account.

The carrier can be expressed in exponential form

$$\cos \omega_c t = \frac{e^{j\omega_c t} + e^{-j\omega_c t}}{2}$$

Denoting the Fourier transform of the signal $x(t)$ by $X(\omega)$ then the Fourier transform of the modulated carrier can be obtained by using the modulation property of the Fourier transform eqn (6.2.9). Hence

$$Y(\omega) = \frac{X(\omega - \omega_c)}{2} + \frac{X(\omega + \omega_c)}{2} \tag{6.3.1}$$

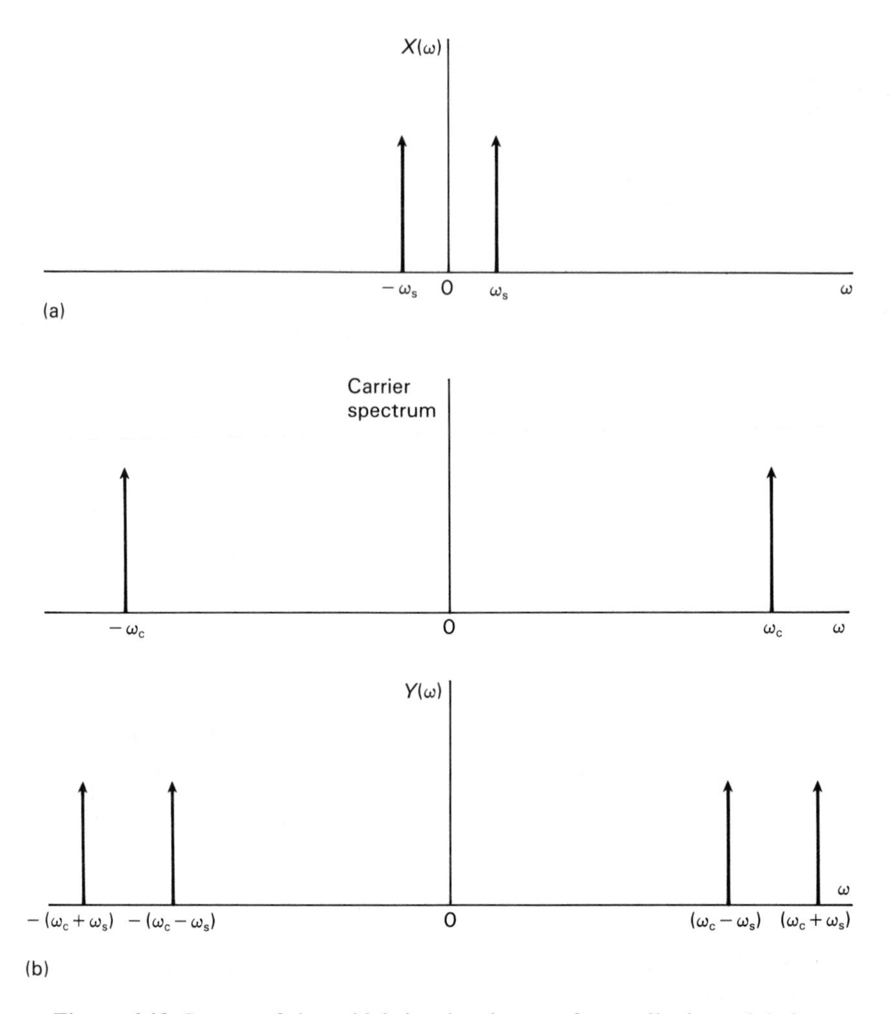

Figure 6.19 Spectra of sinusoidal signal and output for amplitude modulation.

The spectrum for the signal is shown in Figure 6.20(a) and that of the modulated signal is shown in Figure 6.20(b). The modulating process has now centred the signal spectrum on the carrier frequency ω_c.

The final method of investigating amplitude modulation is even more general. It is required to obtain the Fourier transform of the product of two time signals. It was shown in Section 6.2.4 that convolution in time is equivalent to a product in frequency. Using the duality principle it follows that a product in time is equivalent to a convolution in frequency

$$F(x_1(t)x_2(t)) = \frac{1}{2\pi} X_1(-\omega) * X_2(-\omega)$$

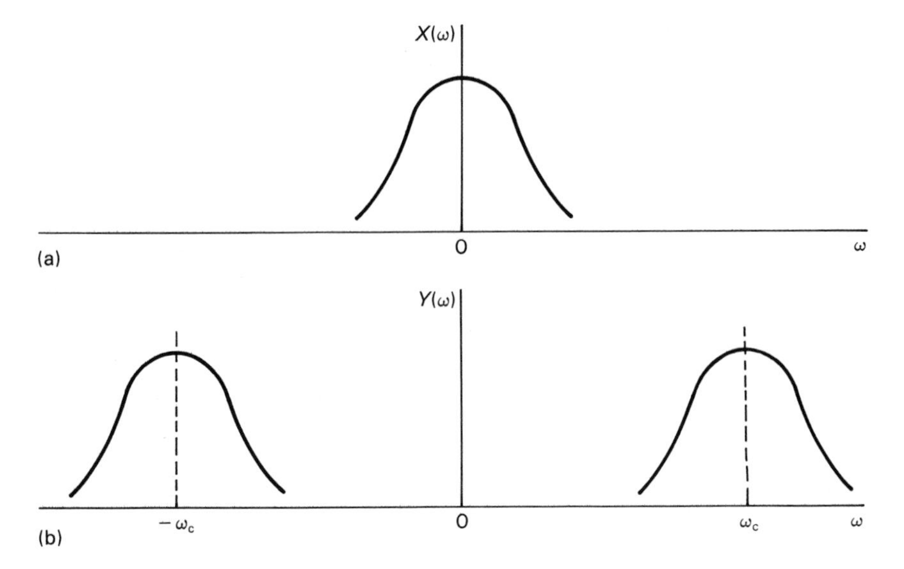

Figure 6.20 General spectrum for amplitude modulation.

Note the factor $1/2\pi$ arises because of the duality principle; this also accounts for the negative arguments in the transform. In this case the latter does not affect the result as $X_1(\omega)$ and $X_2(\omega)$ are even functions. The result of the convolution is indicated graphically in Figure 6.21.

Because of the sifting property of the impulse the result of the convolution is to shift the spectrum to around the carrier frequency. This is the same result as obtained previously but this method is more general as it can be applied to other than a sinusoidal carrier.

Demodulation

Demodulation is the process by which the original signal is recovered from the modulated carrier. A process is required that will shift the spectrum so that it is no longer centred at ω_c but is centred back at $\omega = 0$. However the modulation process as well as producing a shift to $+\omega_c$ also produced a shift to $-\omega_c$. If the modulation process were applied a second time one would expect this negative shift to reproduce the spectrum around $\omega = 0$.

The spectrum is illustrated in Figure 6.22. The original spectrum has been recovered (with a scaling factor) together with a spectrum centred on twice the carrier frequency. Because ω_c is usually much greater than the highest frequency component present in the signal the unwanted band of frequencies can be removed by filtering.

The form of amplitude modulation discussed so far has no carrier component present in the spectrum of the modulated carrier, it is known as 'suppressed carrier' modulation. This has the advantage from a broadcasting point of view that

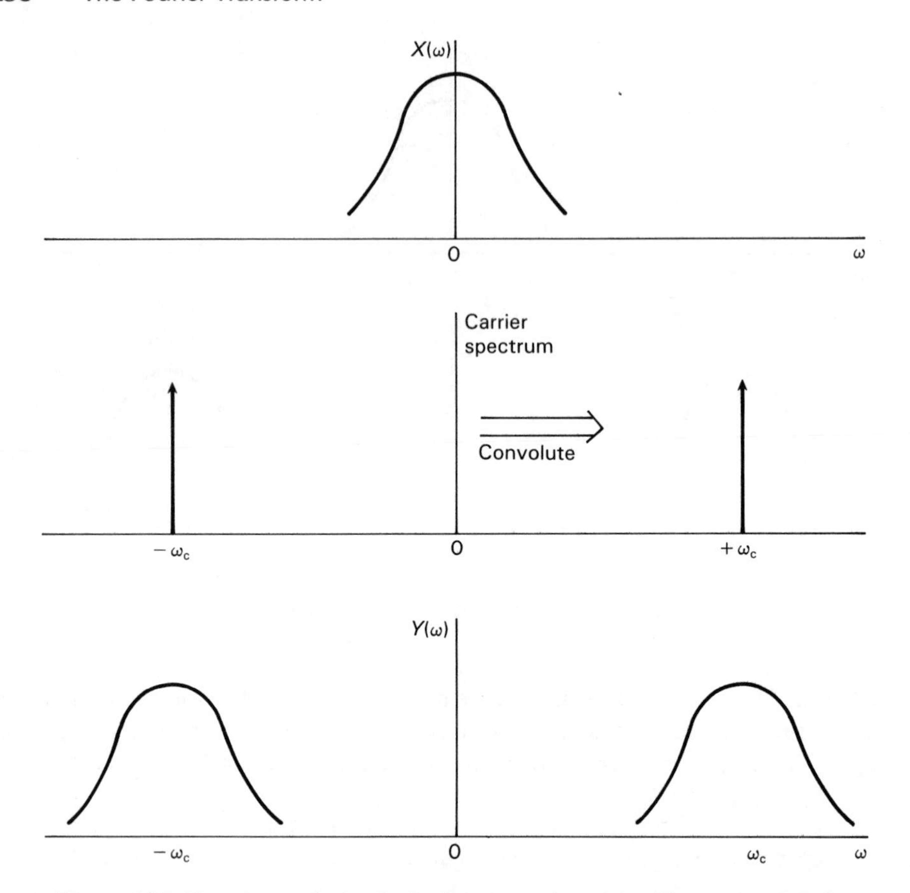

Figure 6.21 Use of convolution in the frequency domain to illustrate modulation.

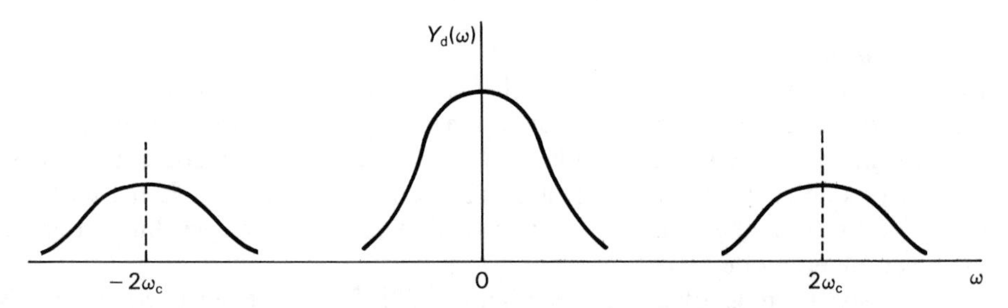

Figure 6.22 Demodulation as a result of further modulation.

transmitter power is saved by not having to radiate at carrier frequency. However the disadvantage lies at the receiving end. In order to demodulate the signal the carrier has to be reconstituted. Not only has the frequency of this reconstituted carrier to be correct but the phase relationship (which has not been considered in this analysis) has also to be correct. This leads to a complex and expensive receiver.

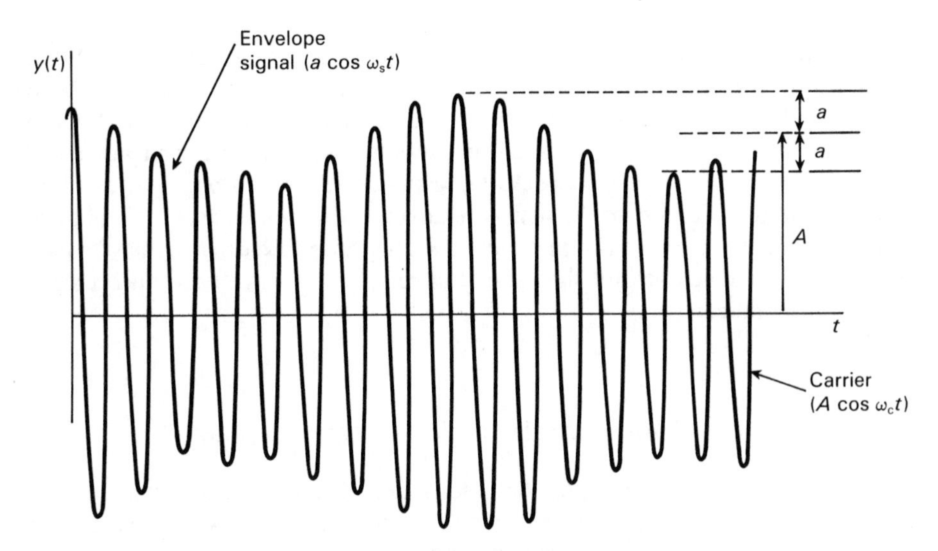

Figure 6.23 Waveform produced by amplitude modulation.

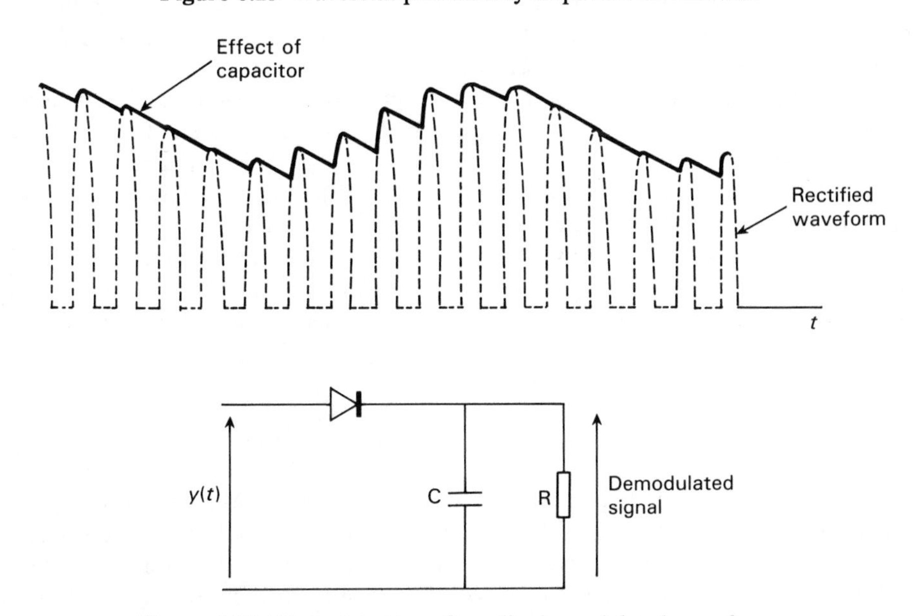

Figure 6.24 Diode detection of amplitude modulated waveform.

To avoid this problem a method is used in commercial broadcasting where the carrier is not suppressed, this is usually referred to as just amplitude modulation. Adding a carrier $A \cos \omega_c t$ to the modulated signal produces

$$\begin{aligned} y(t) &= x(t) \cos \omega_c t + A \cos \omega_c t \\ &= (A + x(t)) \cos \omega_c t \end{aligned}$$

The resulting waveform for a signal $x(t) = a \cos \omega_s t$ is shown in Figure 6.23.

Provided that $\omega_c \gg \omega_s$ and $a < A$ the envelope of this waveform represents the original signal. Reconstitution of the envelope is termed envelope detection. This can be done simply by use of a diode detector circuit as shown in Figure 6.24. The diode rectifies the waveform leaving only the positive half cycles. The capacitor charges through the diode on the positive peaks but cannot discharge through the rectifier. It discharges through the resistor R with a relatively long time constant. This gives a signal that approximates the original envelope (the ratio of carrier/signal frequency will generally be much greater than shown here giving a waveform much closer to the signal than indicated).

6.4 Fourier methods for discrete signals

Continuing the parallel between continuous and discrete signals that has been developed in previous chapters one would expect there to be a Fourier series representation of a discrete periodic waveform and a Fourier transform representation of a discrete aperiodic waveform. These transforms can be developed using methods that parallel those for the continuous case and they lead to the discrete time Fourier series and the discrete time Fourier transform.

From an engineering viewpoint one of the principal reasons for studying discrete signals is that they can be used to represent the samples values of a continuous signal. The Fourier transform of a continuous signal can then be approximated by a discrete equivalent using numerical computations on the sampled values. To this end a transform known as the discrete Fourier transform or the DFT is most useful. (This is especially so since an algorithm known as the fast Fourier transformer, the FFT, provides very efficient computation of the DFT.) This section will concentrate on the DFT covering the discrete time Fourier series and transform only briefly as a means towards this end.

6.4.1 Discrete time Fourier series

This is the discrete counterpart of the continuous time Fourier series and can be applied to discrete time periodic waveforms. As introduced in Section 2.5.2 a discrete signal $x(n)$ is periodic with period N if $x(n+N) = x(n)$ for all n. Figure 6.25 shows a discrete periodic signal with $N = 5$. As stated in the introduction the points forming $x(n)$ are usually samples of a continuous time waveform. However at this stage it leads to an easier development if the time dependence is not stressed.

The signal $x(n)$ is to be represented as a sum of discrete sinusoids these being harmonics of a fundamental having period N. Representing these directly in exponential form gives

$$x(n) = \sum_k c_k e^{+jnk2\pi/N}$$

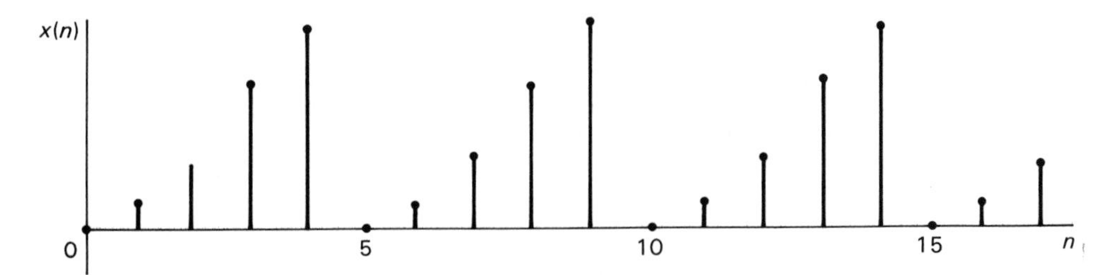

Figure 6.25 A periodic discrete waveform.

No limits have been placed on the summation in this formula. For the continuous case an infinite number of harmonics were required in the summation. However, as shown in Section 2.5.3 only N harmonics exist for a discrete periodic waveform and the summation need only cover these. Paralleling the continuous case any N successive values can be chosen, if the range 0 to $N-1$ is used then

$$x(n) = \sum_{k=0}^{N-1} c_k e^{+jnk2\pi/N} \tag{6.4.1}$$

Again drawing a parallel with the continuous case, one might expect the coefficient c_k to be given by

$$c_k = \frac{1}{N} \sum_{n=0}^{N-1} x(n) e^{-jnk2\pi/N} \tag{6.4.2}$$

Again the summation can be taken over any N successive values of n. The fact that this formula does give the coefficient c_n will not be proved, however the following example will illustrate the relationship.

EXAMPLE 6.4.1

Obtain the discrete time Fourier series for the waveform shown in Figure 6.26.

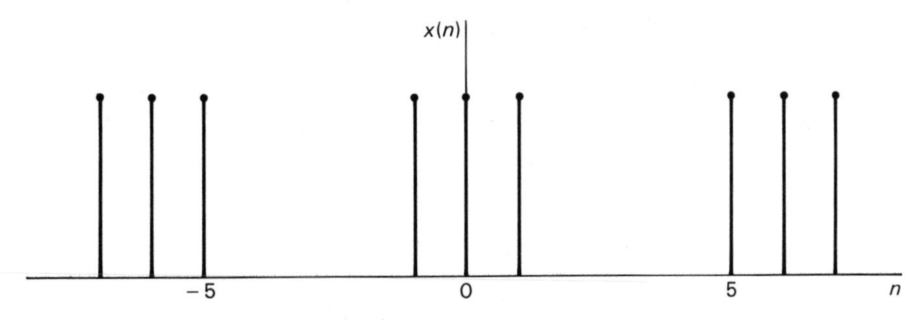

Figure 6.26 Discrete waveform for Example 6.4.1.

SOLUTION

This waveform has period $N = 6$.

This range of summation should be taken to include the points $n = -1, 0, +1$. As the other three points must have a value of zero the formula for c_k can be written

$$c_k = \frac{1}{N} \sum_{n=-1}^{+1} 1e^{-jnk2\pi/N}$$

This can be evaluated by recognising it as the sum of a geometric progression (a change of variable is required to give a range 0 to 2). However because there are so few terms involved it is easier to write

$$c_k = \frac{1}{N} (e^{jk2\pi/N} + 1 + e^{-jk2\pi/N})$$

$$= \frac{1}{N} (1 + 2 \cos k2\pi/N)$$

and for $N = 6$ this is

$$c_k = \tfrac{1}{6}(1 + 2 \cos k\pi/3)$$

For values $k = 0$ to 5 this gives

$$c_0 = 1/2 \qquad c_1 = 1/3 \qquad c_2 = 0$$
$$c_3 = -1/6 \qquad c_4 = 0 \qquad c_5 = 1/3$$

Table 6.3 Inversion of the discrete time Fourier series of Example 6.4.1

			$c_k \cos nk\, 60°$ $c_k \sin nk\, 60°$				
k	0	1	2	3	4	5	
n c_k	1/2	1/3	0	−1/6	0	1/3	*Sum*
0	1/2 0	1/3 0	0 0	−1/6 0	0 0	1/3 0	1 0
1	1/2 0	1/6 $\sqrt{3}/6$	0 0	1/6 0	0 0	1/6 $-\sqrt{3}/6$	1 0
2	1/2 0	−1/6 $\sqrt{3}/6$	0 0	−1/6 0	0 0	−1/6 $-\sqrt{3}/6$	0 0
3	1/2 0	−1/3 0	0 0	1/6 0	0 0	−1/3 0	0 0
4	1/2 0	−1/6 $-\sqrt{3}/6$	0 0	−1/6 0	0 0	−1/6 $\sqrt{3}/6$	0 0
5	1/2 0	1/6 $-\sqrt{3}/6$	0 0	1/6 0	0 0	1/6 $\sqrt{3}/6$	1 0

The inversion formula can be checked by using these coefficients.

$$x(n) = \sum_{k=0}^{N-1} c_k e^{jnk2\pi/N}$$
$$= \sum_{k=0}^{5} c_k(\cos nk2\pi/N + j \sin nk2\pi/N)$$

Using the fact that $2\pi/N = 60°$, Table 6.3 can be constructed to give the values of $x(n)$ in the range 0 to 5. As expected the imaginary component in the result is always zero. In constructing this table much effort is saved if the periodic properties of the functions cos nk 60° and sin nk 60° are taken into account. These properties lead to the construction of the efficient algorithm known as the fast Fourier transform which will be considered in Section 6.6.

6.4.2 *The discrete time Fourier transform*

Again this can be obtained by paralleling the development of the continuous time Fourier transform. A periodic discrete waveform is considered and the effect on the discrete time Fourier series investigated as the period $N \to \infty$.

Letting the periodic discrete series be $x_N(n)$ then

$$c_k = \frac{1}{N} \sum_{n=-N/2}^{N/2-1} x_N(n) e^{-jnk2\pi/N} \qquad (6.4.3)$$

$$x_N(n) = \sum_{k=0}^{N-1} c_k e^{jnk2\pi/N} \qquad (6.4.4)$$

The coefficient c_k can be regarded as a function of k as in the last example. However to aid in the development of the transform it is more convenient to take as a function of $k\theta$ where $\theta = 2\pi/N$. Paralleling the continuous case the spacing between the components in the transform will be represented by $\Delta\theta$ where $\Delta\theta = 2\pi/N$.

As $N \to \infty$ the coefficients c_k will tend to zero hence a new coefficient is defined.

$$c(k\Delta\theta) = 2\pi \frac{c_k}{\Delta\theta}$$
$$= 2\pi \frac{N}{2\pi} \frac{1}{N} \sum_{n=-N/2}^{N/2-1} x_N(n) e^{-jnk2\pi/N}$$

Then as $N \to \infty$

$k\Delta\theta$ becomes continuous variable θ.
The periodic sequence $x_N(n)$ becomes the aperiodic sequence $x(n)$.
The limits on the summation become $\pm\infty$.

$$c(\theta) = \sum_{n=-\infty}^{+\infty} x(n)e^{-jn\theta} \tag{6.4.5}$$

This formula represents the discrete time Fourier transform and it is a continuous function of the variable θ.

The inverse transform can be obtained from eqn (6.4.4). Substituting

$$c_k = \frac{\Delta\theta}{2\pi} c(k\Delta\theta)$$

gives

$$x_N(n) = \frac{1}{2\pi} \sum_{k=0}^{N-1} c(k\Delta\theta)e^{jnk\Delta\theta}\Delta\theta$$

As $N \to \infty$ this summation becomes an integral and putting $\theta = k\Delta\theta$ gives the limits as 0 and 2π. This gives

$$x(n) = \frac{1}{2\pi} \int_0^{2\pi} c(\theta)e^{jn\theta}\, d\theta \tag{6.4.6}$$

Although the limits have been shown as 0 and 2π any interval of 2π can be used, $\theta = \theta_0$ to $\theta_0 + 2\pi$.

EXAMPLE 6.4.2

Obtain the discrete time Fourier transform for the discrete pulse shown in Figure 6.27.

SOLUTION

As expected this example closely parallels Example 6.4.1.

$$\begin{aligned} c(\theta) &= \sum_{n=-\infty}^{\infty} x(n)e^{-jn\theta} \\ &= e^{jn\theta} + 1 + e^{-jn\theta} \\ &= 1 + 2\cos\theta \end{aligned}$$

This transform is plotted in Figure 6.28.

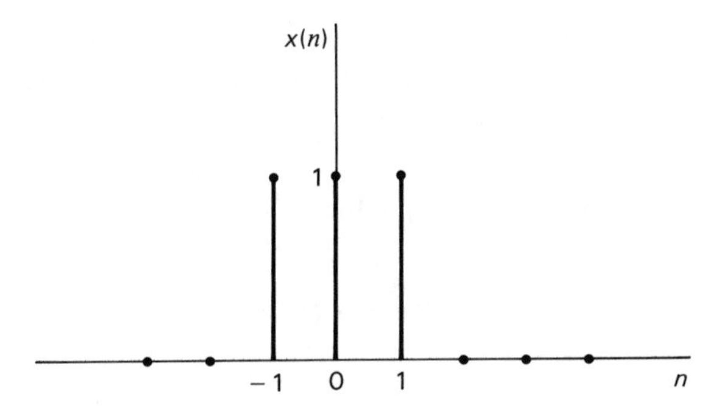

Figure 6.27 Discrete pulse for Example 6.4.2.

The inverse transform can be demonstrated by inverting this function

$$x(n) = \frac{1}{2\pi}\int_{-\pi}^{+\pi}(1 + 2\cos\theta)e^{jn\theta}\,d\theta$$

$$= \frac{1}{2\pi}\int_{-\pi}^{+\pi}(e^{jn\theta} + e^{j(n+1)\theta} + e^{j(n-1)\theta})\,d\theta$$

$$= \frac{1}{2\pi}\left[\frac{e^{jn\theta}}{jn} + \frac{e^{j(n+1)\theta}}{j(n+1)} + \frac{e^{j(n-1)\theta}}{j(n-1)}\right]_{-\pi}^{+\pi}$$

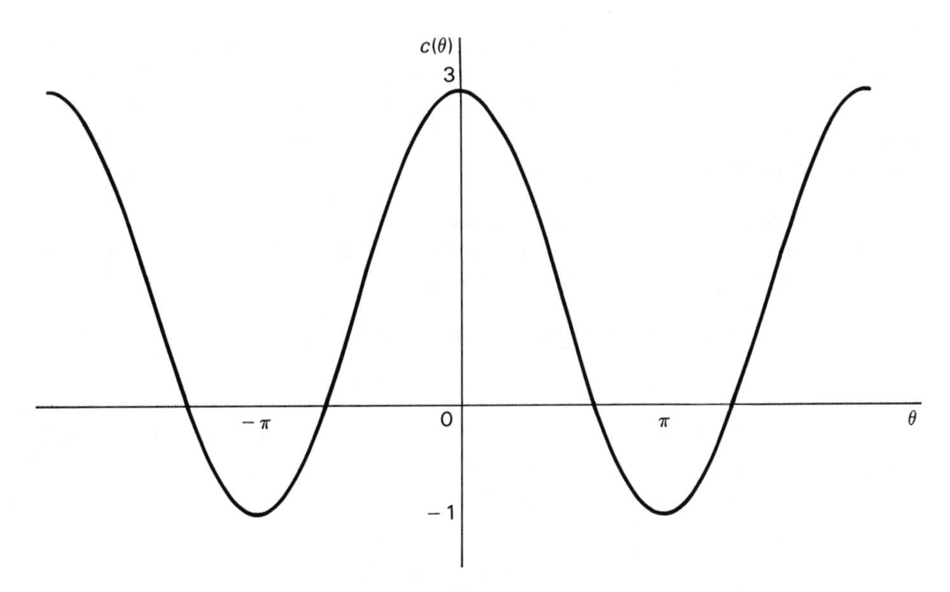

Figure 6.28 Discrete Fourier transform for Example 6.4.2.

The value of this expression is zero for all values of n except $n = -1, 0, 1$. Then some care must be taken in interpreting the terms on the right-hand side. Taking the first term and taking the limit as $n \to 0$

$$\lim_{n \to 0} x(n) = \lim_{n \to 0} \frac{1}{2\pi} \left[\frac{e^{jn\theta}}{n\theta} \right]_{-\pi}^{+\pi}$$

If the limits of integration are inserted this leads to the form $0/0$ as $n \to 0$. However, if the right-hand side is re-written

$$\begin{aligned} \lim_{n \to 0} x(n) &= \lim_{\substack{n \to 0 \\ a \to \pi}} \frac{1}{2\pi} \left[\frac{e^{jn\theta}}{n\theta} \right]_{-a}^{+a} \\ &= \lim_{\substack{n \to 0 \\ a \to \pi}} \frac{1}{2\pi} \frac{2a \sin na}{na} \\ &= \lim_{a \to \pi} \frac{1}{2\pi} 2a \\ &= 1 \end{aligned}$$

A similar result applies for the other limits as $n \to +1$ and $n \to -1$.

6.4.3 *The discrete Fourier transform (DFT)*

One of the principal reasons for investigating the Fourier transforms of a discrete waveform is to use the result to approximate the transform of a continuous signal. Assuming for the moment that a discrete transform calculated from sample points can form such an approximation, there are two difficulties in using the discrete time Fourier transform for this purpose:

1. The summation on the transform extends to infinity.
2. The transform is a continuous function of θ and this presents difficulty with the numerical evaluation of the inverse transform.

Considering the transform obtained in Example 6.4.2

$$c(\theta) = 1 + 2 \cos \theta$$

let us consider the effect of using an inversion formula based on a finite number of sample values. $c(\theta)$ is periodic and if N samples are taken, it seems reasonable to space these samples at intervals of $2\pi/N$ giving

$$c(k) = 1 + \cos \frac{2\pi k}{N}$$

Except for a factor $1/N$ this expression represents the coefficients in the discrete time Fourier series as obtained in Example 6.4.1. The signal represented there was a *periodic* pulse waveform of period N. This result is illustrated in more general terms in Figure 6.29. Figure 6.29(a) represents an aperiodic signal $x(n)$ which has a discrete time Fourier transform $X(\theta)$ as shown in Figure 6.29(b).

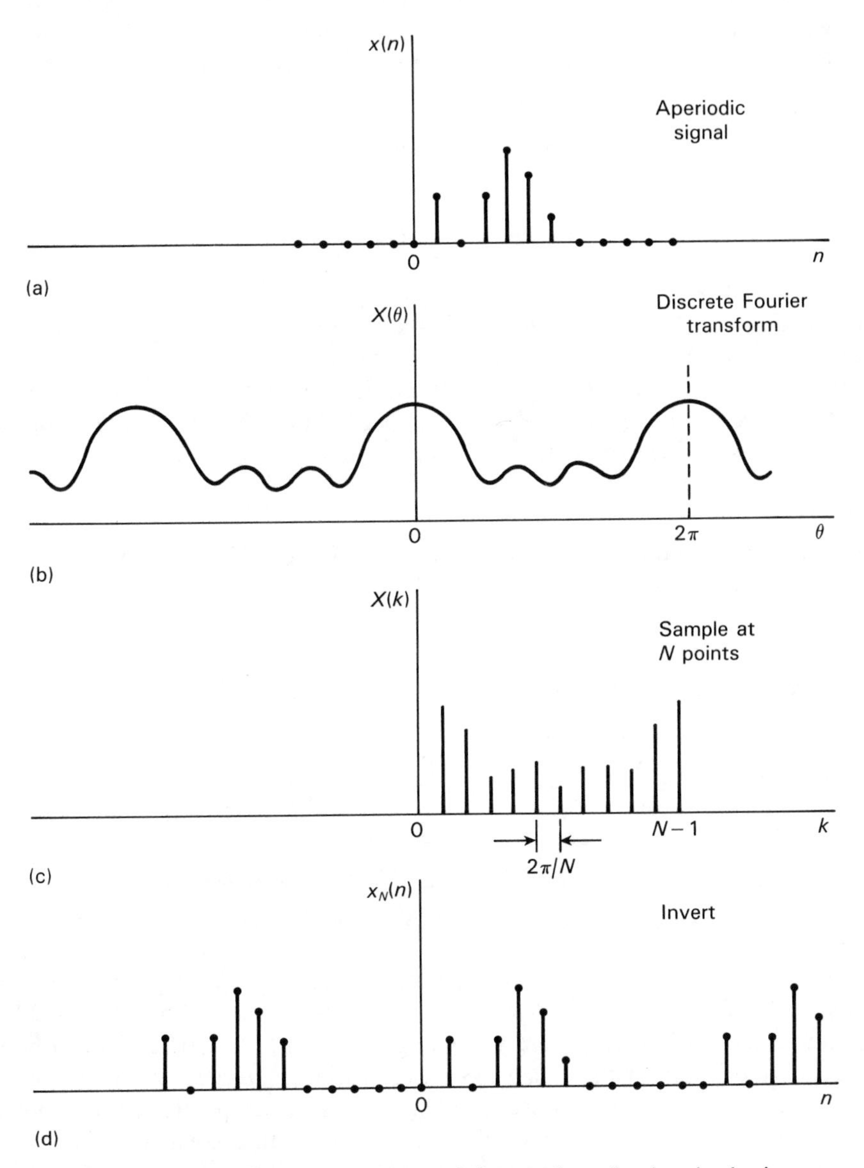

Figure 6.29 Illustration of how taking a finite number of points in the inverse transform leads to a periodic signal.

Sampling this function at N points produces the signal $X(k)$. Inversion of $X(k)$ gives the *periodic* signal $x_N(n)$, this has period N and one period consists of the signal $x(n)$.

This process indicates how the first of the problems posed earlier, the infinite summation on the transform, can be overcome. The summation is restricted to N points and the transform calculated as if it were a periodic signal period N. This produces N discrete frequencies in the spectrum the inversion of which will give a periodic signal. Hence the discrete Fourier transform (DFT) can be written

$$X(k) = \sum_{n=0}^{N-1} x(n)e^{-jnk2\pi/N}$$

$$x(n) = \frac{1}{N}\sum_{n=0}^{N-1} X(k)e^{+jnk2\pi/N}$$

When taking the DFT one has to decide the number of points N to be used in the transform. In the derivation of the DFT, N was introduced as the number of points in the spectrum – the resolution of the spectrum. Hence increasing N will not cause the transform at a given angle $k2\pi/N$ to change but it will increase the number of points available. This is illustrated in the following example.

EXAMPLE 6.4.3

Obtain the DFT for the discrete signal shown in Figure 6.30 for values of N equal to 5, 10, 20.

SOLUTION

As $x(n)$ only takes values at the points $n = 1, 2$ and 3 the DFT can be written

$$\begin{aligned} X(k) &= \sum_{n=0}^{2} x(n)e^{-jnk2\pi/N} \\ &= 1 + 2e^{-jk2\pi/N} + 3e^{-jk4\pi/N} \\ &= (1 + 2\cos k2\pi/N + 3\cos k4\pi/N) \\ &\quad -j(2\sin k2\pi/N + 3\sin k4\pi/N) \end{aligned}$$

Table 6.4 shows the numerical evaluation of the spectrum $X(k)$. By using $\theta = k2\pi/N$ it is seen that the results for $N = 5$ and $N = 10$ are included in those for $N = 20$. The magnitude and angle of $X(k)$ are plotted against k in Figure 6.31. As can be seen increasing the length of the signal by adding additional zeros gives more points in the transform. It does not however alter the value at points already obtained for specific values of θ. Adding additional zeros in this manner to improve the resolution of the spectrum is known as *zero padding*. It will be considered again in Example 6.5.1.

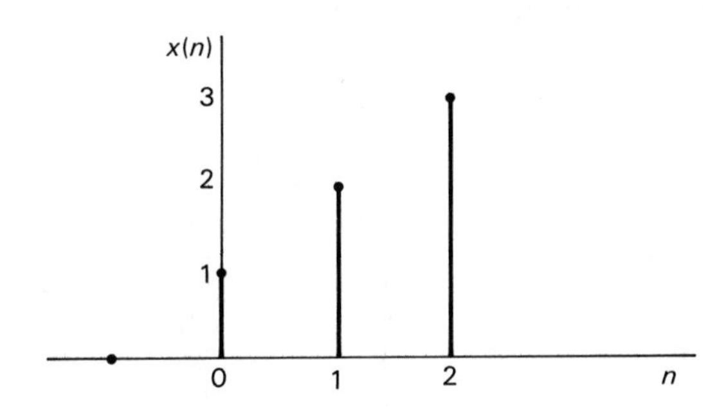

Figure 6.30 Discrete signal for Example 6.4.3.

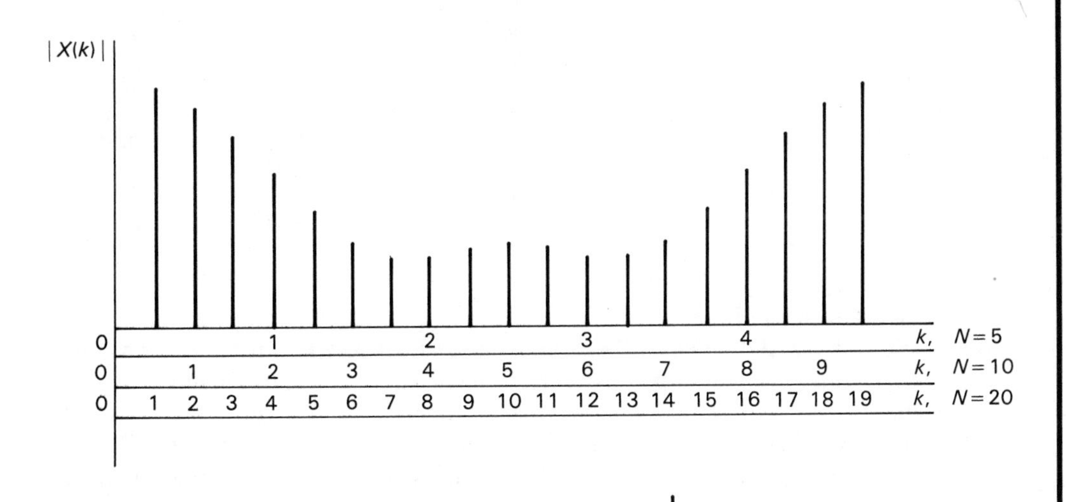

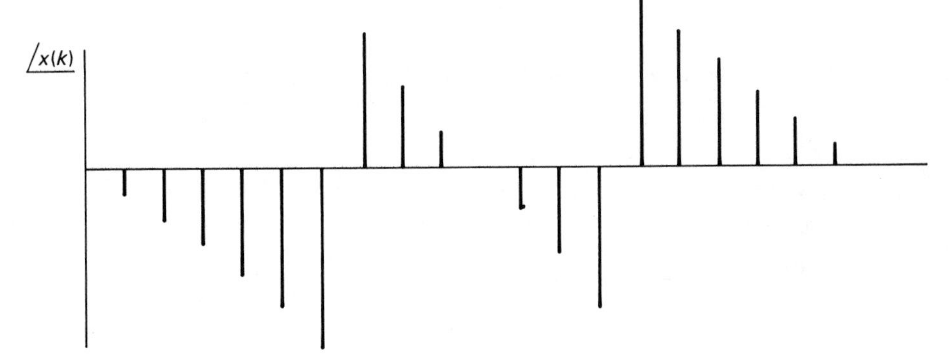

Figure 6.31 Magnitude and angle of the spectrum for Example 6.4.3.

Table 6.4 Numerical evaluation of the discrete Fourier transform for Example 6.4.3

k			$\theta = k2\pi/N$	$X(k)$		$X(k)$	
N=5	N = 10	N = 20	(deg)	Real	Imaginary	Magnitude	Angle (deg)
0	0	0	0.0	6.000	0.000	6.000	0.0
		1	18.0	5.329	−2.381	5.837	−24.1
	1	2	36.0	3.545	−4.029	5.366	−48.7
		3	54.0	1.249	−4.471	4.642	−74.4
1	2	4	72.0	−0.809	−3.666	3.754	−102.4
		5	90.0	−2.000	−2.000	2.828	−135.0
	3	6	108.0	−2.045	−0.139	2.050	−176.1
		7	126.0	−1.103	1.253	1.656	131.8
2	4	8	144.0	0.309	1.678	1.706	79.6
		9	162.0	1.525	1.145	1.907	37.0
	5	10	180.0	2.000	0.000	2.000	0.0
		11	198.0	1.525	−1.145	1.907	−37.0
3	6	12	216.0	0.309	−1.678	1.706	−79.6
		13	234.0	−1.103	−1.235	1.656	−131.8
	7	14	252.0	−2.045	0.139	2.050	176.1
		15	270.0	−2.000	2.000	2.828	135.0
4	8	16	288.0	−0.809	3.666	3.754	102.4
		17	306.0	1.248	4.471	4.624	74.4
	9	18	324.0	3.545	4.029	5.366	48.7
		19	342.0	4.029	5.329	5.837	24.1

Properties of the DFT

Because of the parallel between the continuous and discrete Fourier transforms one would expect many of the properties of the former to have discrete counterparts. This is generally true and Table 6.5 shows these properties which are quoted without proof. However care has to be taken with the interpretation of some of these properties. Although the signal under consideration may be aperiodic the DFT treats it as though it were a periodic signal, period N.

Table 6.5 Some properties of the discrete Fourier transform

Property	$x(n)$	$X(k)$
Linearity	$a_1x_1(n) + a_2x_2(n)$	$a_1X_1(k) + a_2X_2(k)$
Data shift	$x(n - n_0)$	$e^{-j2\pi kn_0/N}X(k)$
Modulation	$e^{j2\pi k_0 n/N}x(n)$	$X(k - k_0)$
Duality	$\frac{1}{N}X(n)$	$x(-k)$
Circular convolution	$x_1(n) * x_2(n)$	$X_1(k)X_2(k)$

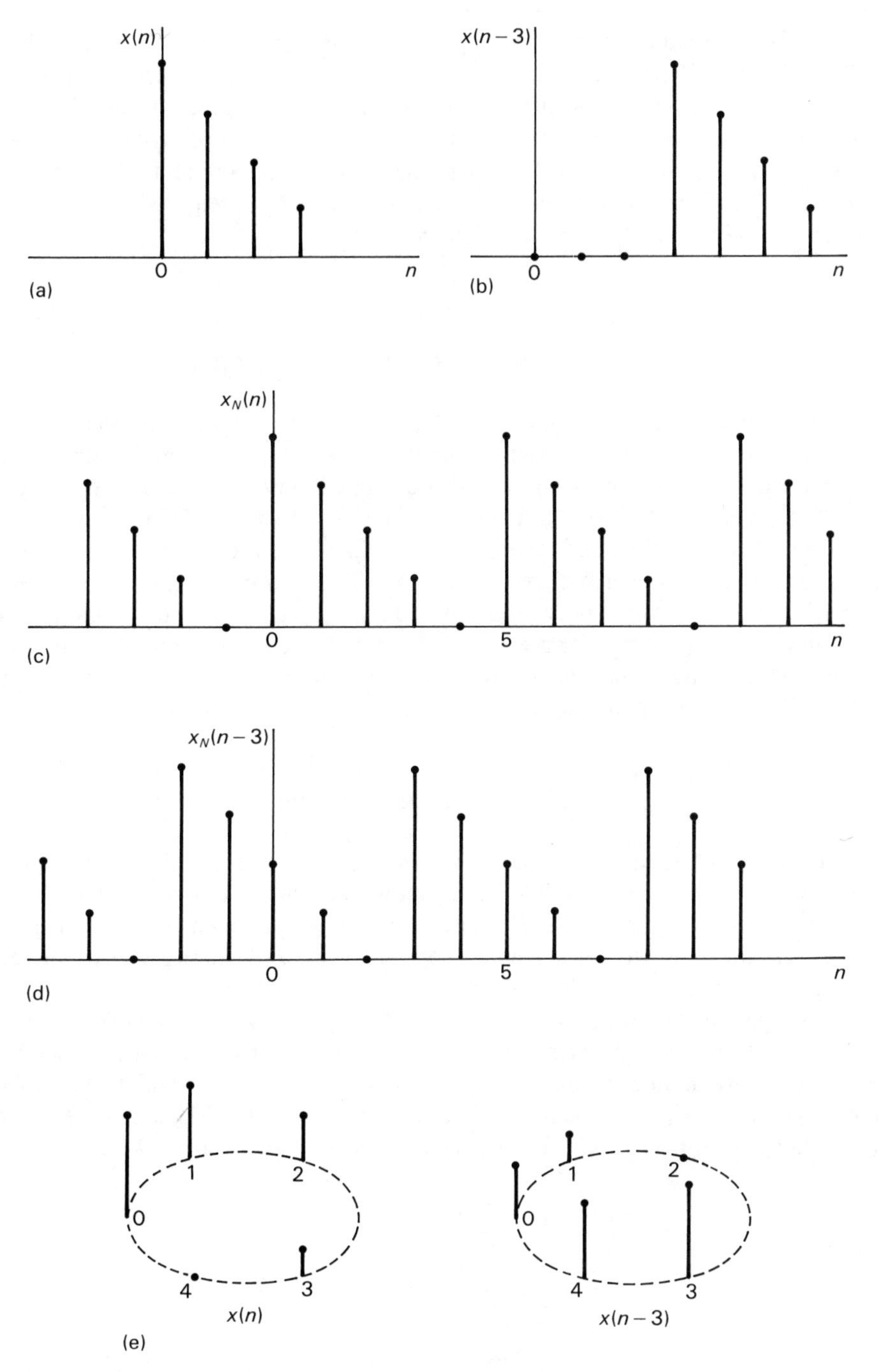

Figure 6.32 Shift property of the DFT.

Consider the data shift property applied to the signal $x(n)$ as shown in Figure 6.32(a). Figure 6.32(b) shows the signal $x(n-n_0)$ where $n_0=3$. However the shift property applies to periodic sequences as shown in Figures 6.32(c) and (d) ($N=5$). Hence the shift can be visualised as a 'circular shift' as shown in Figure 6.32(d).

The convolution property becomes one of circular convolution as the shifts involved are circular shifts. If an 'aperiodic convolution' is required the data can be 'padded out' with zeros to increase the value of N.

6.5 Sampling of continuous signals

As stated earlier one application of the DFT is to estimate the Fourier transform of a continuous signal by using numerical methods. At first sight one could envisage using the values of the continuous signal at discrete intervals as the data sequence for a DFT. However it is then difficult to relate the resulting DFT to the Fourier transform of the continuous waveform (there is no Fourier transform of a set of points as there is no area associated with them). This is important, accepting the fact there will be error in the representation via the DFT, one needs to be able to assess the magnitude of this error for a given representation. One such representation is by train of impulses, these have area hence give a continuous Fourier transform and also lead to a DFT representation as will be shown.

6.5.1 Impulse sampling

Consider the continuous signal as shown in Figure 6.33(a). This can be approximated by a set of finite pulses width T as shown in Figure 6.33(b) and then by a set of impulses as in Figure 6.33(c). (Note although T has been used in the past to denote periodic time it here denotes sampling time. If there is any danger of confusion then a subscript will be added.)

This representation is known as the impulse sampled representation. The strength of each impulse has been taken as the *area* of the associated rectangular pulse. (Note that all the impulses have infinite amplitude, they have however been drawn with finite amplitude in order to convey how they contain the information in $x(t)$.) This signal has been denoted $\hat{x}(t)$ meaning an estimate of $x(t)$ and its continuous Fourier transform is $\hat{X}(\omega)$

$$
\begin{aligned}
\hat{X}(\omega) &= \int_{-\infty}^{+\infty} \hat{x}(t)\mathrm{e}^{-j\omega t}\,\mathrm{d}t \\
&= \int_{-\infty}^{+\infty} \sum_{n=-\infty}^{+\infty} Tx(t)\delta(t-nT)\mathrm{e}^{-j\omega t}\,\mathrm{d}t \\
&= T\sum_{n=-\infty}^{+\infty} x(n)\mathrm{e}^{-jn\omega T} \qquad (6.5.1)
\end{aligned}
$$

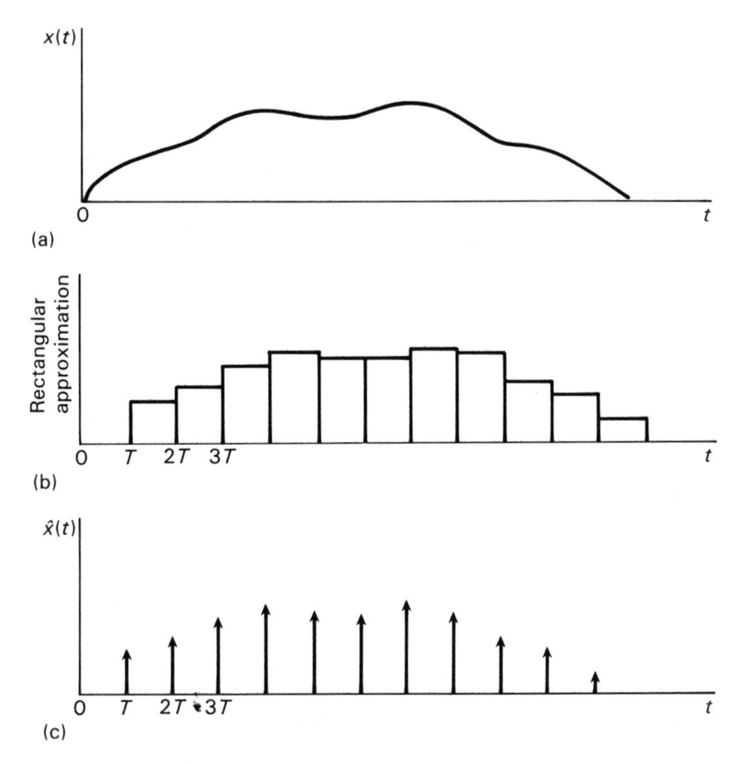

Figure 6.33 Representation of a continuous signal by a set of impulses.

$\hat{X}(\omega)$ is a continuous function of frequency, if however, it is evaluated at the discrete frequency points, $\omega = k2\pi/NT$, then

$$\hat{X}(\omega)\,|_{\omega = k2\pi/NT} = T\sum_{n=0}^{N-1} x(n)e^{-jnk2\pi/N}$$

$$= TX(k) \tag{6.5.2}$$

where $X(k)$ is the DFT of $x(n)$. Hence an estimate of the spectrum can be obtained by taking the DFT of sampled points of the continuous signal and multiplying by the sampling time T. The spectrum is only obtained at sample frequencies $\omega = k2\pi/NT$ and the resolution obtained will be determined by the number of sample points N. This method gives a result (except for the explanation of the factor T) that would have been obtained by the intuitive method rejected at the start of this section. However what can be obtained now is a measure of how accurately eqn (6.5.2) represents the continuous spectrum. Intuitively one would expect that the smaller the value of T, the sampling time, the nearer $\hat{X}(\omega)$ equals $X(\omega)$ at the sample frequencies. We can show this as follows.

In eqn (6.5.1) $\sum_{n=-\infty}^{+\infty} \delta(t - nT)$ represents an infinite train of unit impulses spaced T apart. This is shown in Figure 6.34 where the impulse train has been

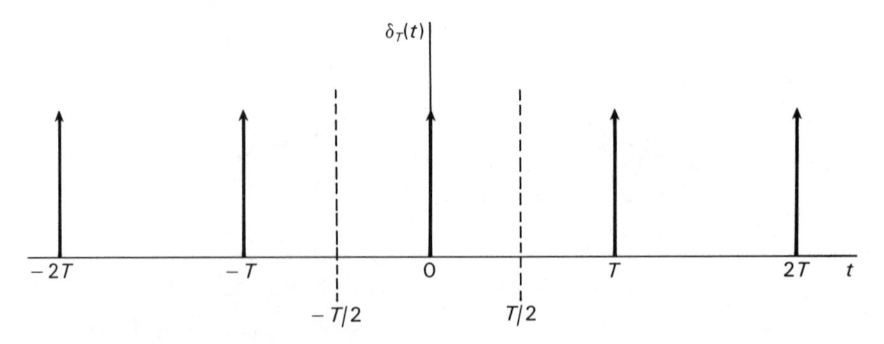

Figure 6.34 Train of unit impulses illustrated as a periodic signal.

denoted $\delta_T(t)$. $\delta_T(t)$ is periodic, period T and it can be represented as a Fourier series.

Taking the Fourier expansion over the range $-T/2$ to $T/2$

$$c_n = \frac{1}{T}\int_{-T/2}^{+T/2} \delta(t)\mathrm{e}^{-jn\omega t}\,\mathrm{d}t$$

$$= \frac{1}{T} \qquad \text{for all } n$$

Hence the Fourier series representation of the impulse train $\delta_T(t)$ is

$$\delta_T(t) = \frac{1}{T}\sum_{n=-\infty}^{+\infty} \mathrm{e}^{jn\omega_s t}$$

where

$$\omega_s = 2\pi/T$$

Substituting into eqn (6.5.1) gives

$$\hat{X}(\omega) = \int_{-\infty}^{+\infty} x(t) \sum_{n=-\infty}^{+\infty} \mathrm{e}^{jn\omega_s t} e^{-j\omega t}\,\mathrm{d}t \tag{6.5.3}$$

This equation looks a little frightening with the summation contained inside the integral. However the summation could be written out term by term and a separate integration performed on each term. Considering each term separately

$n = 0$

$$\hat{X}(\omega)\,|_{n=0} = \int_{-\infty}^{+\infty} x(t)\mathrm{e}^{-j\omega t}\,\mathrm{d}t$$

$$= X(\omega)$$

This is the true spectrum! Unfortunately the other terms in the summation must be taken into account and these will lead to errors. How significant these errors are can be seen by evaluating further terms

$n = 1$

$$\hat{X}(\omega)\big|_{n=1} = \int_{-\infty}^{+\infty} x(t)\mathrm{e}^{-\mathrm{j}(\omega-\omega_s)t}\,\mathrm{d}t$$
$$= X(\omega - \omega_s)$$

This is the true spectrum but displaced to a frequency $(\omega - \omega_s)$.

It is evident that other values of n will produce spectra $X(\omega + \omega_s)$, $X(\omega \pm 2\omega_s) \ldots X(\omega \pm n\omega_s)$. Hence eqn (6.5.3) can be written

$$\hat{X}(\omega) = \sum_{n=-\infty}^{+\infty} X(\omega - n\omega_s) \qquad (6.5.4)$$

$\hat{X}(\omega)$ is shown in Figure 6.35(a) and it is assumed for convenience that this spectrum is real. Samples from this at frequencies $\omega = k2\pi/NT$ equal the true spectrum $X(\omega)$ at these points.

However, in drawing this spectrum it has been assumed that the true spectrum $X(\omega)$, falls to zero within the frequency range $\omega = \pm\omega_s/2$. If this is not the case the spectrum $X(\omega)$ will be as shown in Figure 6.35(b).

If samples are now taken at frequencies $\omega = k2\pi/NT$ these will be in considerable error over part of the frequency range where the sideband overlaps.

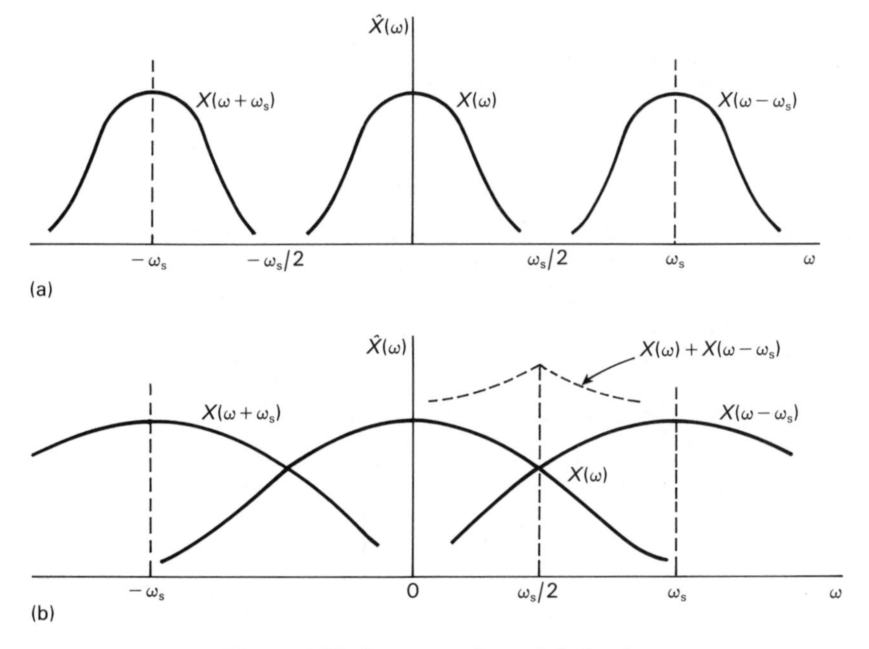

Figure 6.35 Spectrum of sampled signal.

This error is known as *aliasing* and has been introduced in Section 2.5.2. In order to avoid aliasing the highest frequency present in the unsampled signal must be less than half the sampling frequency. This result is known as the *sampling theorem* and is usually stated as follows:

> If a signal is band limited there is no loss of information if it is sampled at a frequency, at least twice as high as the highest frequency present in the signal.

In general the spectrum is complex, it has an amplitude and a phase component. The effect of aliasing is not easy to portray graphically under such conditions, however the sampling theorem still applies.

In practice a signal is often bandlimited by passing it through a low-pass filter before it is sampled. Although this procedure distorts the signal by removing high frequency components it is often preferable to do this rather than cause greater distortion due to aliasing effects. Such a filter is known as an anti-aliasing filter.

The number of points N used in the DFT depend on the resolution required. Resolution and the effect of aliasing will be illustrated by the following example.

EXAMPLE 6.5.1

It is required to obtain the Fourier transform of the rectangular pulse shown in Figure 6.36. This is to be done by sampling the pulse and using the DFT. Compare the resulting magnitude spectrum with the true transform of the pulse, investigating the effect of sampling time and number of points used.

SOLUTION

The continuous Fourier transform for this waveform has already been calculated in Section 6.2.1. For a pulse of unit width and unit amplitude this is given by

$$X(\omega) = \frac{\sin(\omega/2)}{(\omega/2)}$$

Because the signal is relatively simple an analytical expression can also be derived for the DFT. Referring to Figure 6.37 the pulse has been represented as having N_1 sample points while the total length of the record is N points. As the pulse is of one second duration this gives the sampling time $T = 1/N_1$. Although the original

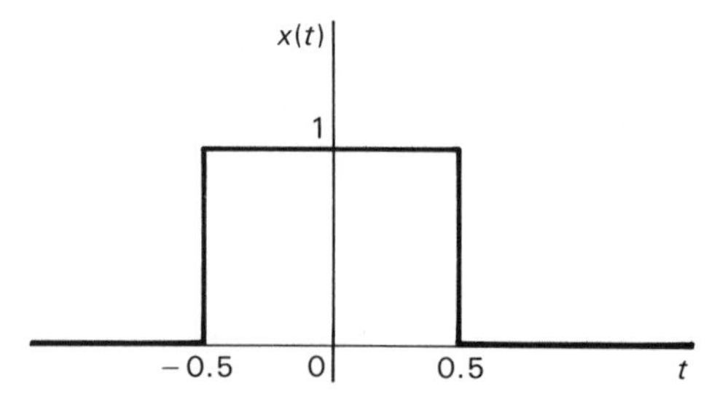

Figure 6.36 Rectangular pulse for Example 6.5.1.

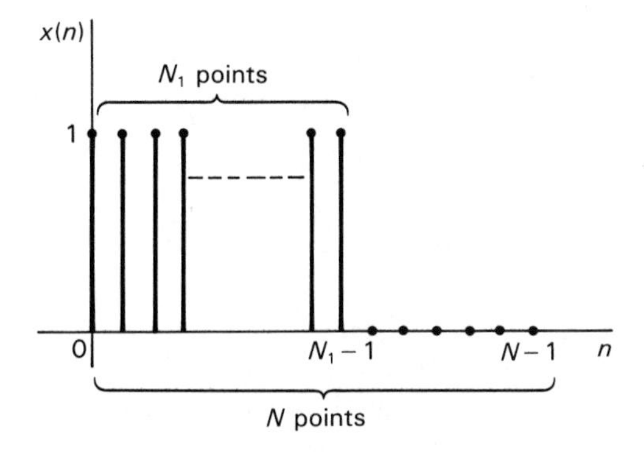

Figure 6.37 Discrete representation of the pulse for Example 6.5.1.

continuous pulse was symmetrical about $t = 0$ its discrete equivalent has been shown as starting at $n = 0$. This gives an easier calculation for the DFT and although this will alter the phase spectrum it will leave the required magnitude spectrum unaltered. The DFT, $X_D(k)$ is given by

$$X_D(k) = \sum_{n=0}^{N-1} x(n) e^{-jnk2\pi/N}$$

As $x(n) = 1$ for $n \leq (N_1 - 1)$ and is zero otherwise the expression $X_D(k)$ becomes

$$X_D(k) = \sum_{n=0}^{N_1-1} e^{-jnk2\pi/N}$$

This is in the form of a geometric progression and has been met in previous examples, summing the N_1 terms gives

$$X_D(k) = \frac{1 - e^{-jN_1k2\pi/N}}{1 - e^{-jk2\pi/N}}$$

Writing the exponentials in trigonometric form and taking the modulus gives

$$|X_D(k)|^2 = \frac{(1 - \cos N_1k2\pi/N)^2 + \sin^2 N_1k2\pi/N}{(1 - \cos k2\pi/N)^2 + \sin^2 k2\pi/N}$$

$$= \frac{1 - \cos N_1k2\pi/N}{1 - \cos k2\pi/N}$$

Remember that the estimate of the Fourier transform $\hat{X}(k\omega_0)$ is the DFT multiplied by the sampling time T then

$$\hat{X}(k\omega_0) = T\,\frac{1 - \cos N_1k2\pi/N}{1 - \cos k2\pi/N} \qquad (6.5.5)$$

where $\omega_0 = 2\pi/NT$. The effect of N and T on the estimate can now be investigated.

The first case considered is the case when $N_1 = N$. Intuitively this seems a ridiculous situation, remembering that the signal over N points must be considered as one period of a periodic waveform. This implies the identity of the signal as a pulse is lost, it appears as a constant level. However let us see, putting $N_1 = N$ into eqn (6.5.4) gives

$$\hat{X}(k\pi_0) = T\,\frac{1 - \cos k2\pi}{1 - \cos k2\pi/N}$$

For integer values of k, $\cos k2\pi = 1$ and the numerator is zero. This gives $\hat{X}(k\omega)$ zero for all values of k except $k = 0$, at this value the denominator is also zero and the limit as $k \longrightarrow 0$ must be obtained. This limit can be obtained by using the first two terms only of the power series expansion for $\cos x$.

$$\cos x \approx 1 - x^2/2$$

then

$$\begin{aligned}\lim_{k\to 0} \hat{X}(k\omega_0) &= \lim_{k\to 0} T\,\frac{1 - 1 + (k2\pi)}{1 - 1 + (k2\pi/N)^2}\\ &= TN\\ &= 1 \qquad (T = 1/N)\end{aligned}$$

Hence it appears that the result bears out that obtained by intuition, there is a component at $k = 0$, a d.c. component and no other. However if the result is plotted, together with the true magnitude spectrum the result is as in Figure 6.38. (Note: $\omega_0 = 2\pi/NT = 2\pi$.)

In fact at the sample frequency points the true spectrum is obtained. The resolution is however so poor that very little information is obtained about the true nature of the spectrum. One might expect the values at the sample frequencies to be incorrect due to aliasing. The sampling frequency however is $2\pi N$ and the zero values of the sidebands lie on the sample values of the spectrum thus causing no aliasing error at these points.

To improve the resolution the number of points must be extended beyond N_1, to take a specific case let $N_1 = 5$ and $N = 20$.

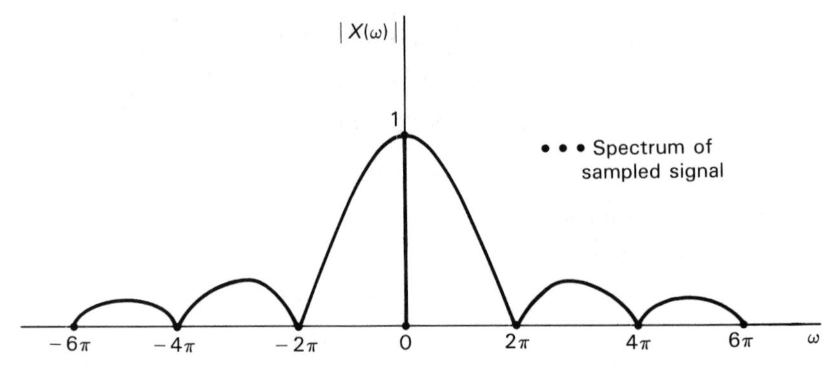

Figure 6.38 Discrete and true spectra for the case $N_1 = N$.

Then substituting these values into eqn (6.5.4)

$$|\hat{X}(k\omega_0)| = T\,\frac{1-\cos k\pi/2}{1-\cos 2\pi/20}$$

where $T=0.2$ and $\omega_0 = 2\pi/NT_s = \pi/2$.

Table 6.6 shows true magnitude and that obtained using the DFT approximation. The error is also shown at each frequency point. As shown in a previous example, increasing the number of zero points used to augment the record will not improve the accuracy but will give greater resolution. To improve the accuracy the effects of aliasing must be reduced and this can be done by decreasing the sampling time. If T is reduced to 0.1 second then $N_1 = 10$ (to maintain the one second pulse width) and to keep the same resolution $2\pi/NT_s = \pi/2$ giving $N = 40$.

Results for these values of N_1 and N are given in Table 6.6 and as expected the error has been reduced at all frequencies. The greatest error still occurs at the higher frequencies where the effects of aliasing are greatest. However there is no error at the frequencies where the contributions from the sidebands are zero, this includes $\omega = 0$.

The deviation of the DFT estimate from the true value is due to aliasing. This effect can be illustrated at one specific frequency. Taking $T = 0.2$ and considering an estimate at $\omega = 3\pi$, then using eqn (6.5.4)

$$\hat{X}(\omega) = \sum_{n=-\infty}^{+\infty} \frac{\sin(\omega - n\omega_s)/2}{(\omega - n\omega_s)/2}$$

$$\hat{X}(3\pi) = \sum_{n=-\infty}^{+\infty} \frac{\sin(3-n10)\pi/2}{(3-n10)\pi/2}$$

Table 6.6 Error between true and estimated spectra for Example 6.5.1

		Magnitude of extimated spectrum $|\hat{X}(\omega)|$ *and percentage error*			
		$T=0.2$ $N_1=5$ $N=20$		$T=0.1$ $N_1=10$ $N=40$	
ω *(rad/s)*	*Magnitude of true spectrum* $|X(\omega)|$	*Magnitude*	*Error %*	*Magnitude*	*Error %*
0	1.0000	1.0000	0.00	1.0000	0.00
$\pi/2$	0.9003	0.9040	0.41	0.9012	0.10
π	0.6366	0.6472	1.67	0.6392	0.41
$3\pi/2$	0.3001	0.3115	3.80	0.3029	0.93
2π	0.0000	0.0000	0.00	0.0000	0.00
$5\pi/2$	0.1801	0.2000	11.05	0.1847	2.55
3π	0.2122	0.2472	16.40	0.2203	3.82
$7\pi/2$	0.1286	0.1587	23.41	0.1353	5.23
4π	0.0000	0.0000	0.00	0.0000	0.00
$9\pi/2$	0.1000	0.1432	43.10	0.1088	8.80
5π	0.1273	0.2000	57.10	0.1414	11.07
$11\pi/2$	0.0818	0.1431	74.93	0.0929	13.56
6π	0.0000	0.0000	0.00	0.0000	0.00

Table 6.7 The effect of alias terms on the magnitude estimate for Example 6.5.1

N (limits of summation ±N)	*Sum*
0	−0.2122
2	−0.2444
4	−0.2463
6	−0.2467
8	−0.2469
10	−0.2470

Table 6.7 shows the result of this summation over a finite number of terms, $-N$ to $+N$. As can be seen when $N=0$ the true magnitude is obtained. For increasing N convergence is slow but the sum is converging on the value given by the DFT.

6.5.2 *Signal reconstitution*

The previous section has considered how a continuous signal can be represented as a set of sample points in order to use the DFT to estimate its Fourier transform. However another important reason for sampling a signal is for data transmission. The value of the signal at the sampling points can be converted into a suitable code (binary code) and the information transmitted via this code. Such an encoded signal is very much more immune to noise interference than a continuous signal and it also enables signals to be efficiently multiplexed (transmitting several signals over the same communication link). However at the receiving end the original signal has to be recovered or reconstituted. If the sampling theorem has not been violated all the original information is still present. The samples can be recovered by decoding the digital signal using a digital to analogue converter (DAC) but the continuous signal still needs to be recovered from the samples.

Consider the spectrum of a sampled signal as shown in Figure 6.39, the sampling theorem has not been violated so the signal suffers no aliasing. If a perfect low-pass filter could be used, as shown, this would remove the unwanted components and leave the original signal. As will be shown in Chapter 8 it is not possible to construct a perfect filter but it can be approximated (the greater the sampling frequency compared with the highest frequency present in the signal the less onerous the task of the filter).

It is useful to examine signal reconstitution from its interpretation in the time domain. The multiplication of the signal spectrum and the filter frequency response is equivalent to a convolution in the time domain. The signals convolved are the sampled time signal and the filter impulse response. The filter frequency response is shown in Figure 6.40(a) and its inverse transform gives its impulse response. This

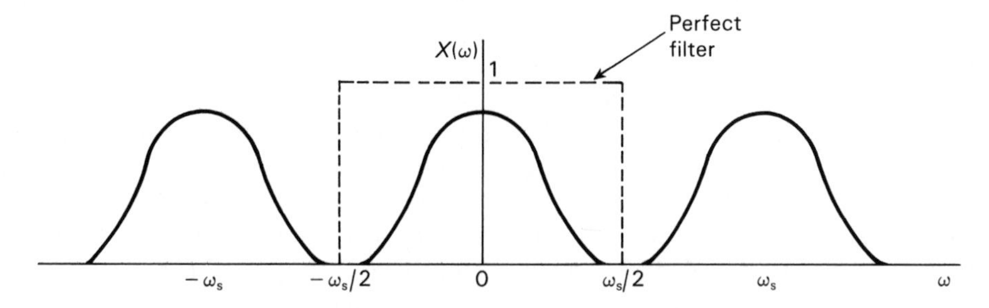

Figure 6.39 Spectrum of sampled signal.

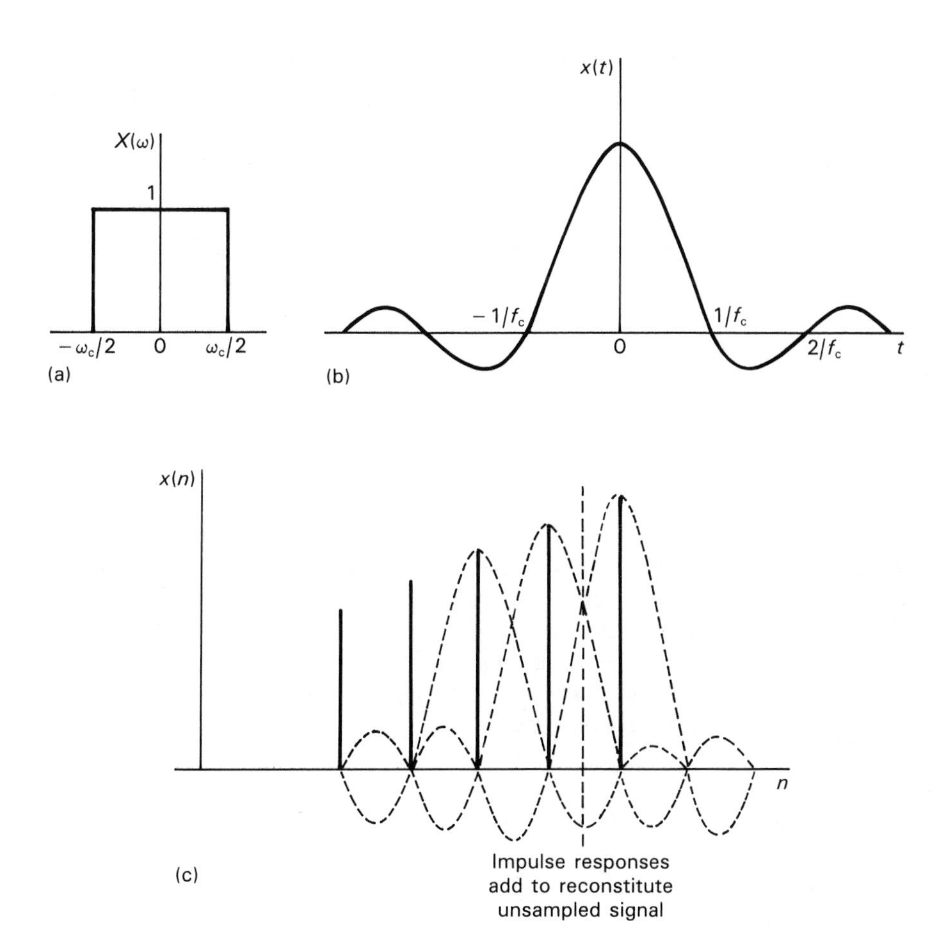

Figure 6.40 Frequency and impulse responses for interpolating filter.

can be obtained using the duality principle of the Fourier transform derived in Section 6.2.2 and is shown in Figure 6.40(b).

Note that the impulse response is non-causal (it gives a response before $t = 0$ to an impulse applied at $t = 0$) and for this reason the perfect filter cannot be implemented in practice. However remaining in the realms of theory for a while the time response of such a filter to the sampled signal is shown in Figure 6.40(c). The sampled signal consists of a train of impulses and each impulse produces a corresponding impulse response, these add to give a total response that is the original unsampled signal (note that any scaling factors have been ignored here).

The signal at the sampling instants is not modified as these lie on the zero points of all the component impulse responses.

Because this filter interpolates the signal giving values between the sampling instants it is known as an interpolating filter.

As already stated the response is non-causal but it can be approximated if some phase shift is allowed in the reconstruction. However an interesting application of the interpolating filter occurs in the process known as oversampling. In this process interpolation is used to recreate not the continuous signal but additional sample values between the ones available. Because the signal is in discrete form and only additional discrete points are required a digital interpolating filter can be used. This now produces a signal that 'appears' to have been sampled at a higher frequency (over sampled). Hence the sidebands due to aliasing are moved to higher frequencies and the task of the analogue filter required to reconstitute the continuous signal is less demanding. This technique is used in compact disc players when converting from the discrete recorded signal to the analogue output.

6.6 The fast Fourier transform (FFT)

From Examples 6.4.1, 6.4.3 and 6.5.1 the student will realise that the calculations involved to obtain the DFT can be shortened by making use of periodicities in the factor $e^{-jk2\pi/N}$. This together with suitable grouping of the terms in the transform leads to an algorithm for obtaining the DFT known as the fast Fourier transform. It should be emphasised that the result obtained by using this algorithm is identical to that which would be obtained by using the DFT, however the number of stages involved in the computation is drastically reduced.

6.6.1 Development of the FFT algorithm

Consider the expression defining the DFT and its inverse

$$X(k) = \sum_{n=0}^{N-1} x(n)e^{-jnk2\pi/N} \tag{6.6.1}$$

$$x(n) = \frac{1}{N}\sum_{n=0}^{N-1} X(k)e^{jnk2\pi/N} \qquad (6.6.2)$$

Ignoring the factor $1/N$ these two expressions differ only in the sign of the exponent. If one wrote a computer program to obtain the DFT, a relatively small change would be required in the program in order to obtain its inverse. Hence most computer programs are written with this facility, however it means that the multiplications required are multiplications of complex numbers and N^2 such multiplications are required. It is with the aim of reducing this number of multiplications that the FFT algorithm is formulated.

In order to provide easier manipulation of the expressions involved the following shorthand notation is used

$$W_N = e^{-j2\pi/N} \qquad \text{giving} \qquad W_N^{nk} = e^{-jnk2\pi/N}$$

The following properties of W will be used

$$W_N^{nk} = W_N^{(nk \pm mN)} \qquad m \text{ any integer}$$
$$W_N^{2M} = W_{N/2}^{M}$$
$$W_N^{(a+b)} = W_N^a W_N^b$$
$$W_N^{N/2} = -1$$
$$W_N^0 = 1$$

These properties are easily proved from the definition of W_N. Considering a two point transform W_N takes values according to Figure 6.41.

Writing $W_2^0 = 1$ the expression for $X(0)$ and $X(1)$ can be obtained from eqn (6.6.1)

$$X(0) = x(0) + x(1)$$
$$X(1) = x(0) + x(1)W_2^1$$

Noting that $W_2^1 = -1$ then the two point transforms can be represented by flow diagram of Figure 6.42.

k \ n	0	1
0	W_2^0	W_2^0
1	W_2^0	W_2^1

Figure 6.41 Construction of the two point FFT.

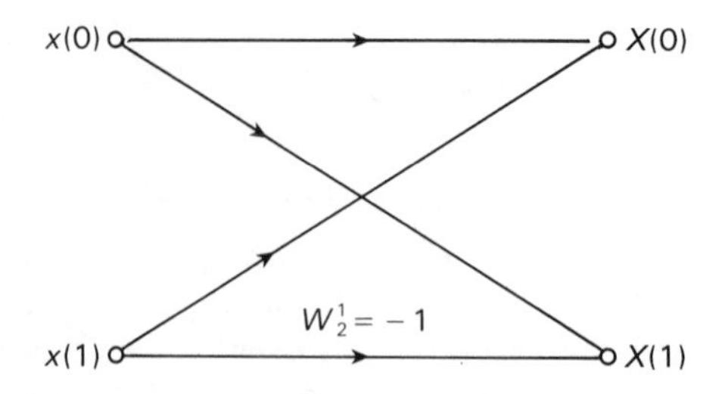

Figure 6.42 Flow diagram for two point transform.

The variable at each node is transmitted along the branch leading from the node and multiplied by the transmittance indicated (taken as unity if unmarked). Variables leading to a node are added at that node. Because of the shape of this flow diagram it is known as a 'butterfly representation'.

In this representation the four (2^2) multiplications of the transform have been replaced by one multiplication by -1. It could be argued that this is subtraction but it is better viewed as a multiplication to fit into the general development of the transform.

Considering now a four point transform Figure 6.43 gives the values W_4 where periodicity has been taken into account when $kn \geq 4$.

The expression for $X(0)$, $X(1)$, $X(2)$ and $X(3)$ can now be written out in terms of the appropriate W_4 multipliers. However for reasons that will become clearer the order of the terms has been changed, even values and odd values of n being grouped

k \ n	0	1	2	3
0	W_4^0	W_4^0	W_4^0	W_4^0
1	W_4^0	W_4^1	W_4^2	W_4^3
2	W_4^0	W_4^2	W_4^0	W_4^2
3	W_4^0	W_4^3	W_4^2	W_4^1

Figure 6.43 Construction of the four point FFT.

together,

$$X(0) = x(0)W_4^0 + x(2)W_4^0 + x(1)W_4^0 + x(3)W_4^0$$
$$X(1) = x(0)W_4^0 + x(2)W_4^2 + x(1)W_4^1 + x(3)W_4^3$$
$$X(2) = x(0)W_4^0 + x(2)W_4^0 + x(1)W_4^2 + x(3)W_4^2$$
$$X(3) = x(0)W_4^0 + x(2)W_4^2 + x(1)W_4^3 + x(3)W_4^1$$

Writing $W_4^0 = 1$, $W_4^2 = W_2^1$ and taking factors out of terms in the last two columns gives

$$X(0) = x(0) + x(2) \quad + W_4^0(x(1) + x(3))$$
$$X(1) = x(0) + x(2)W_2^1 + W_4^1(x(1) + x(3))W_2^1)$$
$$X(2) = x(0) + x(2) \quad + W_4^2(x(1) + x(3))$$
$$X(3) = x(0) + x(2)W_2^1 + W_4^3(x(1) + x(3)W_2^1)$$

These equations can be written

$$X(0) = G(0) + W_4^0 H(0)$$
$$X(1) = G(1) + W_4^1 H(1)$$
$$X(2) = G(0) - W_4^0 H(0)$$
$$X(3) = G(1) - W_4^1 H(1)$$

Where $G(0)$ and $G(1)$ are the outputs from a two point transform with inputs $x(0)$ and $x(2)$ and $H(0)$ and $H(1)$ are outputs from a two point transform with inputs $x(1)$ and $x(3)$. Use has been made of the property $W_4^2 = -1$ and although W_4^0 could be written as unity it helps develop a general pattern if left in this form. The process is shown diagrammatically in Figure 6.44.

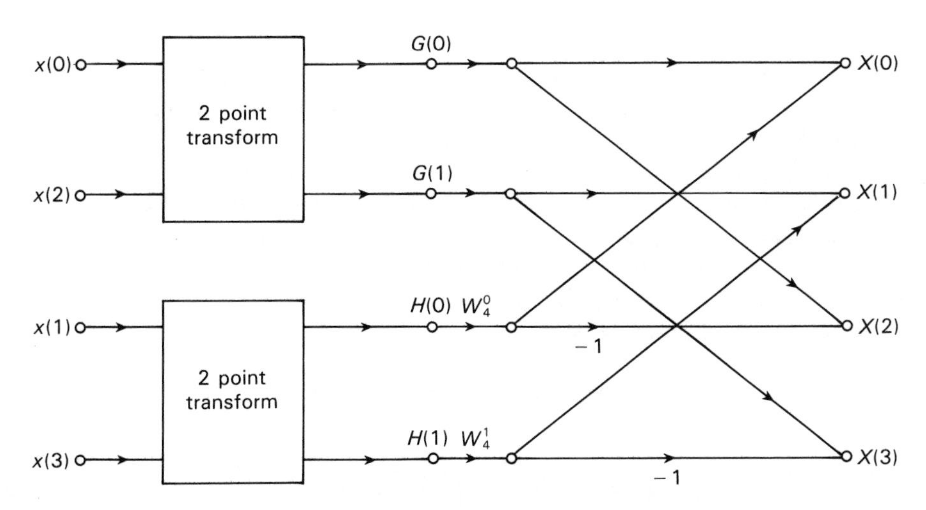

Figure 6.44 Flow diagram for four point transform.

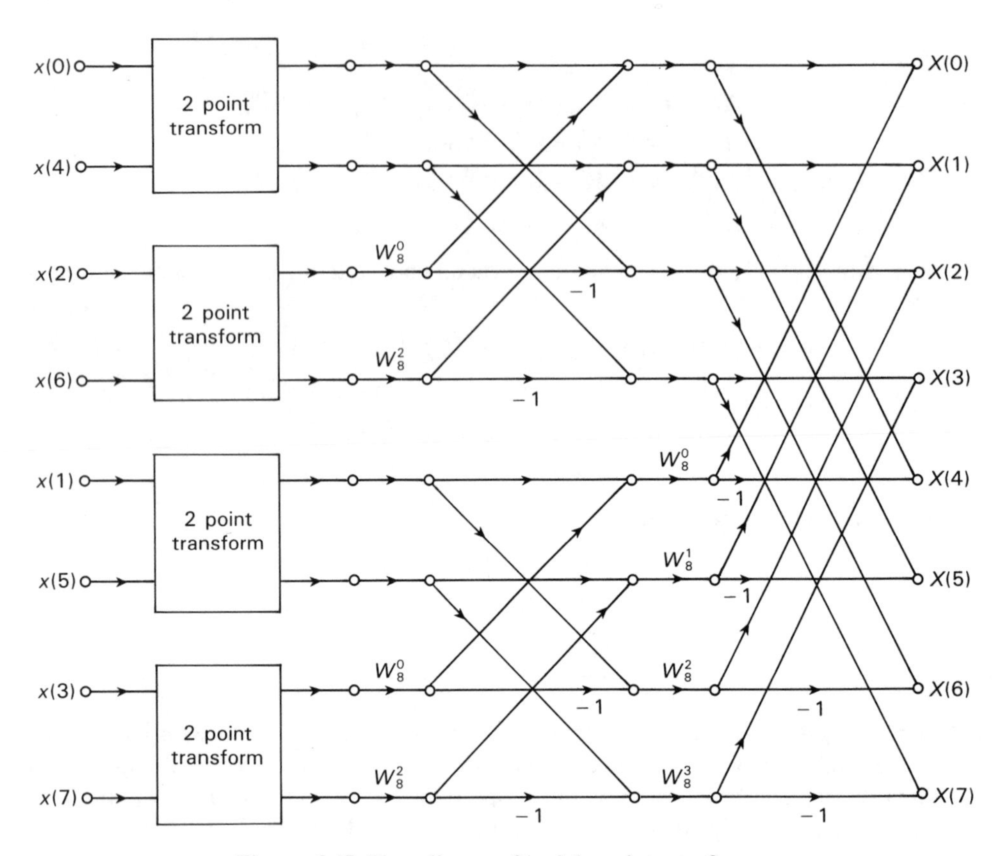

Figure 6.45 Flow diagram for eight point transform.

The number of complex multiplications required are two in the pair of two point transforms and two more when the terms are combined. For a two point transform the total number of complex multiplications is $2 \times (N/2) = 4$. This compares with 16 for the DFT calculations.

It is left as an exercise for the student to develop the eight point transform. The terms are again grouped odd and even to produce four point transforms and then these four point transforms realised by two two point transforms as just considered. The resulting chart is shown in Figure 6.45.

From this eight point transform it is hoped that the student can see the general pattern that is emerging. Four complex multiplications are required to combine the two four-point transforms, hence the total required is $3 \times (N/2) = 12$ compared with the N^2 or 64 for the DFT.

It is beyond the scope of this book to develop a general N point algorithm but it should be apparent that such an algorithm would require $p \times (N/2)$ complex multiplications where $N = 2p$. As many practical applications require transforms of

Table 6.8 Comparison of required multiplications FFT/DFT

N	p	$\dfrac{FFT\ multiplications}{DFT\ multiplications} \times 100\%$
8	3	18.750
128	7	2.734
1024	10	0.448
5096	12	0.118

the order $N = 2000$, a very large percentage saving in multiplications (and computer time) results. Table 6.8 shows the FFT multiplication/DFT multiplication as a percentage for different values of N.

Although the FFT provides the same result as the DFT, because of this marked reduction in computing time it has had a profound effect on all branches of signal processing. Many forms of the FFT algorithm are available and it has been implemented on mainframe and personal computers as well as via special purpose hardware designed to reduce the computing time even further. It has found application in almost every branch of engineering, e.g., speech and image processing and the processing of biological signals. The list is endless.

6.7 Summary

In this chapter the Fourier transform has been introduced as a limiting case of the Fourier series. The spectrum obtained shows the component frequencies that are present in the signal, however it must be interpreted as an amplitude density spectrum. The system frequency response function can still be used to show how the component frequencies of the input spectrum are modified by the system and it has been shown that the system frequency response function is the Fourier transform of the system impulse response.

The discrete Fourier transform (DFT) is the discrete counterpart of the continuous Fourier transform. This transform indicates the component frequencies present in discrete signals and it can be used in the analysis of discrete systems. It can also be used to analyse the sampled version of a continuous signal to approximate its continuous Fourier transform. This has the advantage that analysis of the signal can be performed by digital computer. Considerable reduction in processing time is possible if the fast Fourier transform (FFT) algorithm is used to evaluate the transform.

Problems

6.1 Determine the Fourier transforms of the signals shown in Figure 6.46. Sketch the associated amplitude and phase spectra.

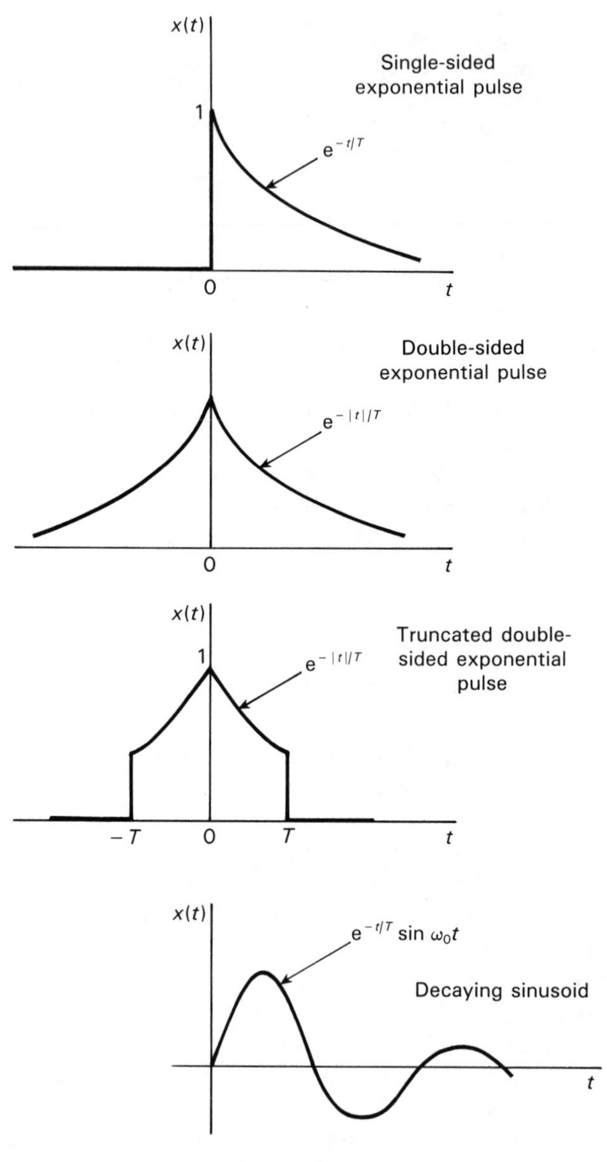

Figure 6.46

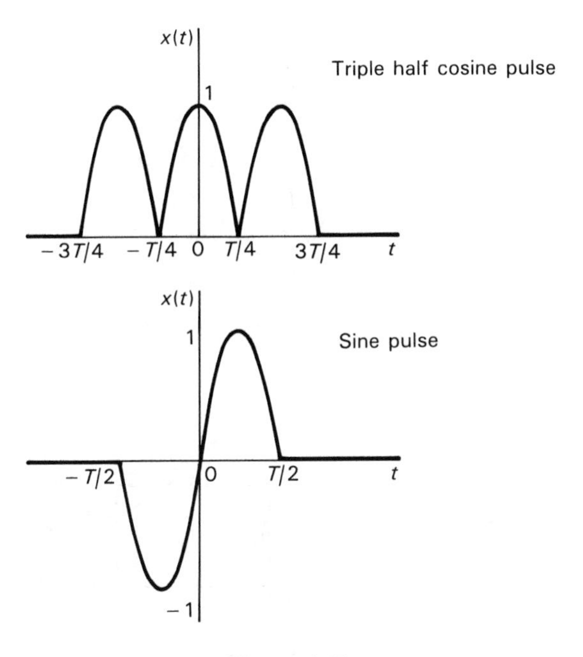

Figure 6.47

6.2 Determine the Fourier transform of the half cosine pulse given by

$$\begin{aligned} x(t) &= \cos(2\pi t/T) \qquad & -T/4 \leqslant t \leqslant +T/4 \\ &= 0 & \text{otherwise} \end{aligned}$$

By using the linearity and shift properties determine the transforms of the signals shown in Figure 6.47.

6.3 (a) The signal in Figure 6.48(a) is a 'sine-wave pulse' given by

$$\begin{aligned} x(t) &= \sin(2\pi t/T) \qquad & -T < t < +T \\ &= 0 & \text{otherwise} \end{aligned}$$

Determine directly the Fourier transform of this pulse. The pulse can be considered to be formed by the multiplication of an eternal sine wave and a unit amplitude rectangular pulse of width $2T$ as shown. Use the convolution property of the Fourier transform to obtain its Fourier transform.

(b) Determine directly the Fourier transform of the triangular pulse shown in Figure 6.48(b).

This pulse can be considered to be formed by the convolution of the two rectangular pulses shown. Use the convolution property of the Fourier transform to obtain its Fourier transform.

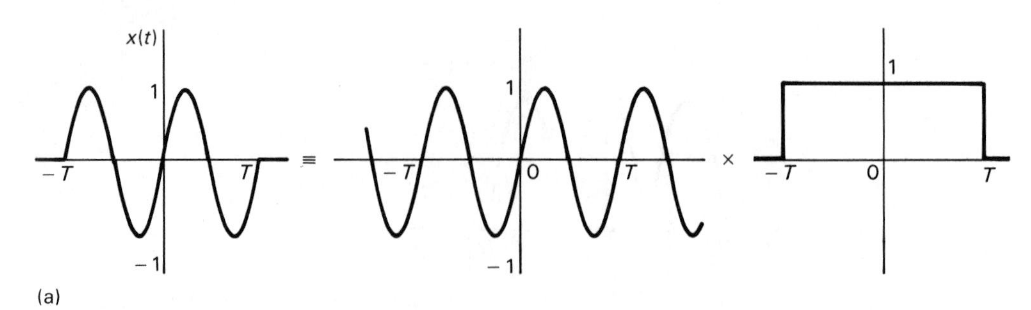

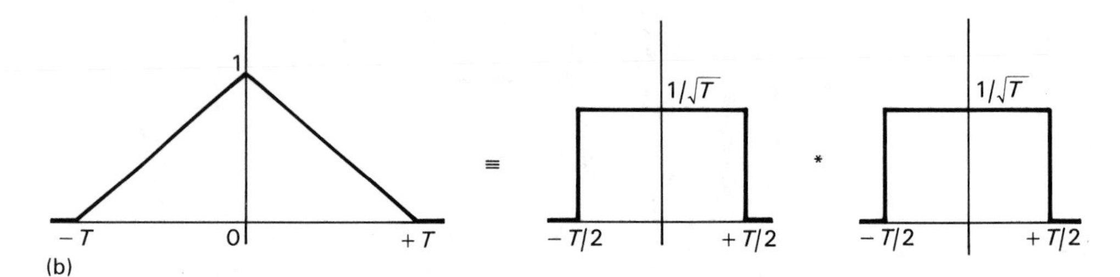

Figure 6.48

6.4. (a) Determine the impulse response of a band pass filter having magnitude and angle responses as shown in Figure 6.49(a).

(b) A filter having the frequency response shown in Figure 6.49(b) is known as a Hilbert transformer. Given that the Fourier transform corresponding to the unit step $u(t)$ is given by

$$F\{u(t)\} = \pi\delta(\omega) - \mathrm{j}\,\frac{1}{\omega}$$

use appropriate transform theorems to determine the impulse response of the Hilbert transformer.

6.5 The equivalent duration T_{eq} and equivalent bandwidth B_{eq} of a signal $x(t)$ are defined as follows

$$T_{eq} = \frac{\int_{-\infty}^{+\infty} x(t)\,\mathrm{d}t}{x(0)}$$

$$B_{eq} = \frac{\int_{-\infty}^{+\infty} X(\omega)\,\mathrm{d}\omega}{X(0)}$$

These quantities are only defined for the case where $x(0) \neq 0$, $X(0) \neq 0$.

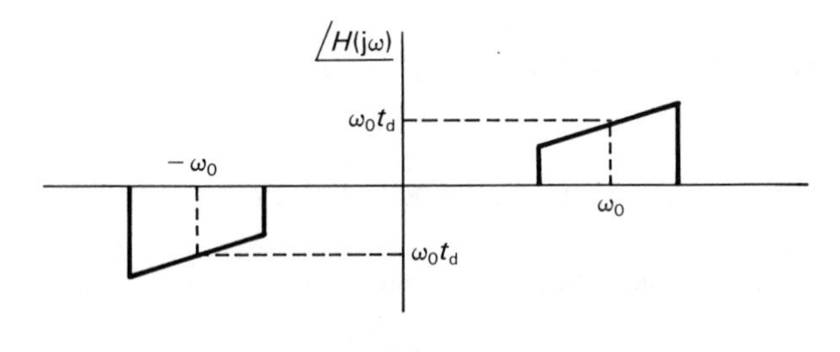

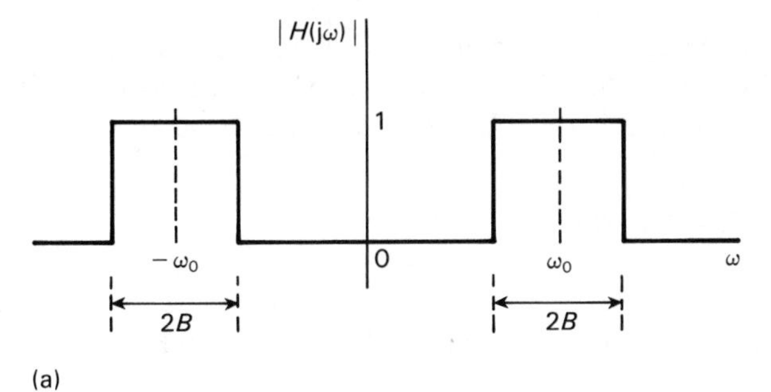

(a)

$H(j\omega)$

j1

ω

−j1

(b)

Figure 6.49

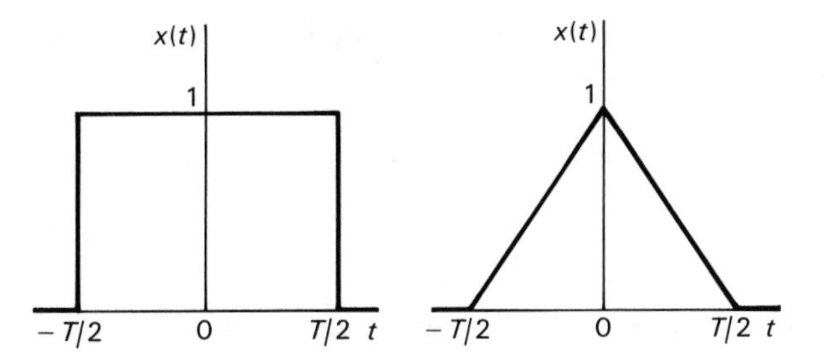

$x(t)$

1

$\cos \frac{\pi t}{T}$

$-T/2$ 0 $T/2$ t

Figure 6.50

Show the following relationship holds

$$T_{eq}B_{eq} = 2\pi$$

Hence determine the equivalent bandwidth of the signals shown in Figure 6.50.

6.6 Verify Parseval's theorem for the double-sided exponential pulse

$$x(t) = e^{-|t|T}$$

Using the definitions given in Problem 6.5 what is the effective bandwidth of this pulse? Denoting this bandwidth by $2B$ what percentage of the energy of the pulse is contained in the following frequency ranges?

(a) $-B \leqslant \omega \leqslant +B$
(b) $-B/2 \leqslant \omega \leqslant +B/2$
(c) $-2B \leqslant \omega \leqslant +2B$

6.7 A system has a frequency response function given by

$$H(j\omega) = \frac{1}{1 + j\omega T}$$

The input signal has a Fourier transform given by

$$X(\omega) = \frac{1}{1 + j\omega T_1}$$

Determine the ratio T_1/T such that 0.75 of the energy of the input signal will appear at the system output.

6.8 An instrumentation system makes use of an amplitude modulated suppressed carrier system. The carrier is given by

$$v_c(t) = 10\cos(2\pi f_c t) \qquad \text{where} \qquad f_c = 5\text{ kHz.}$$

If the signal is given by

$$v_s(t) = 2\cos(2\pi f_1 t) + 5\cos(2\pi f_2 t)$$

where $f_1 = 100$ Hz and $f_2 = 200$ Hz, determine the amplitude and frequency of the components of the modulated signal.

6.9 Figure 6.51 shows an a.c. strain gauge bridge where the supply voltage e_s is given by

$$e_s = 10\cos(2\pi \times 200t)$$

R_s represents the resistance of the strain gauge and this is given by

$$R_s = R + \lambda S$$

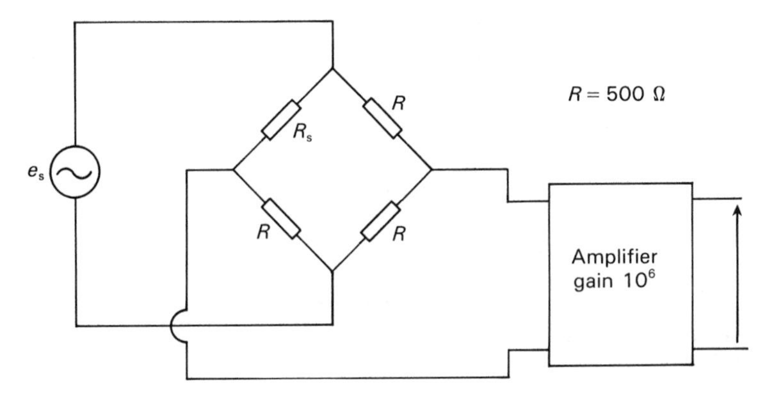

Figure 6.51

where λ is the gauge factor and S is the strain under measurement. For the particular gauge used $\lambda = 2$ and the gauge is subject to a sinusoidal strain given by

$$S = 10^{-3} \times \cos(2\pi \times 10t)$$

Obtain an expression for the amplifier output and determine the amplitude and frequency of the components present.

6.10 In Section 6.3 it was shown how demodulation can be achieved by use of a diode detector circuit. The rectifier action can be described as multiplication by a square wave having a fundamental frequency equal to that of the carrier frequency. The method is shown in Figure 6.52 and the waveform $x(t)$ is given for the case of full and half wave rectification.

Taking

$$y(t) = (A + a \sin \omega_s t) \sin \omega_c t$$

and expressing $x(t)$ as a Fourier series determine the amplitudes and frequencies of the components of the demodulated signal $v(t)$ in the range $\omega = 0$ to ω_c.

What is the advantage of using full wave rectification and what problems arise if ω_s is not very much less than ω_c?

6.11 Figure 6.53 shows two discrete periodic signals. Determine the magnitudes and angles of the coefficients in the discrete time Fourier series corresponding to each signal.

6.12 Determine the discrete time Fourier transform of the two signals shown in Figure 6.54.

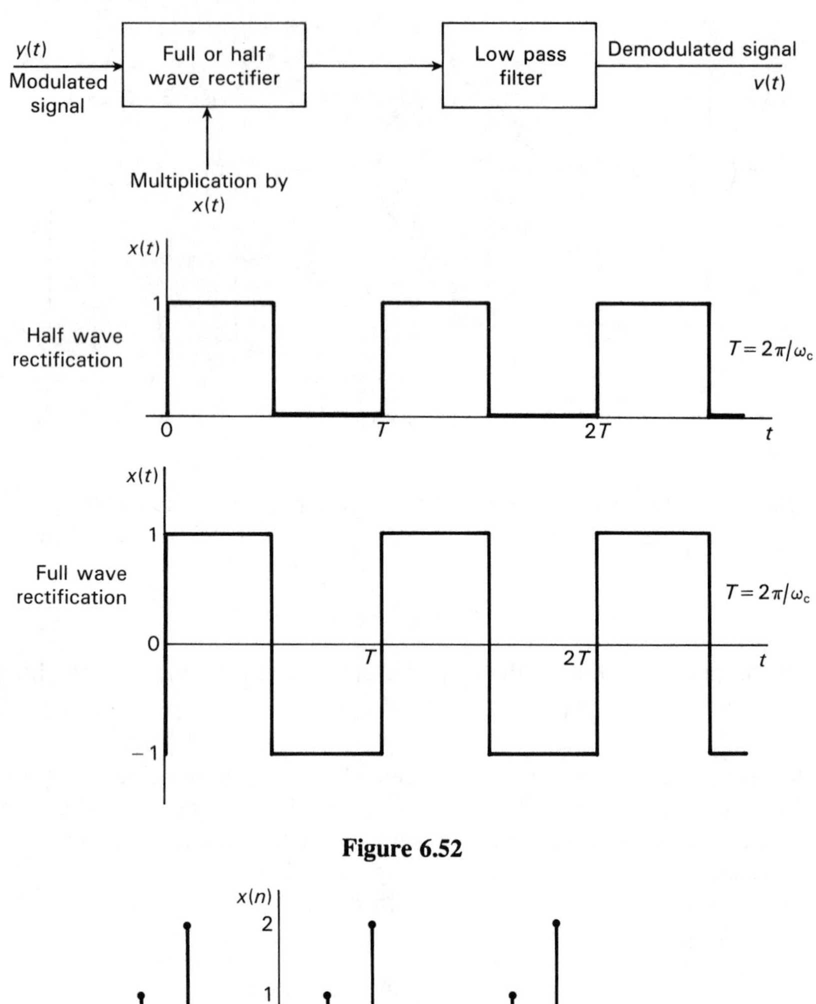

Figure 6.52

Figure 6.53

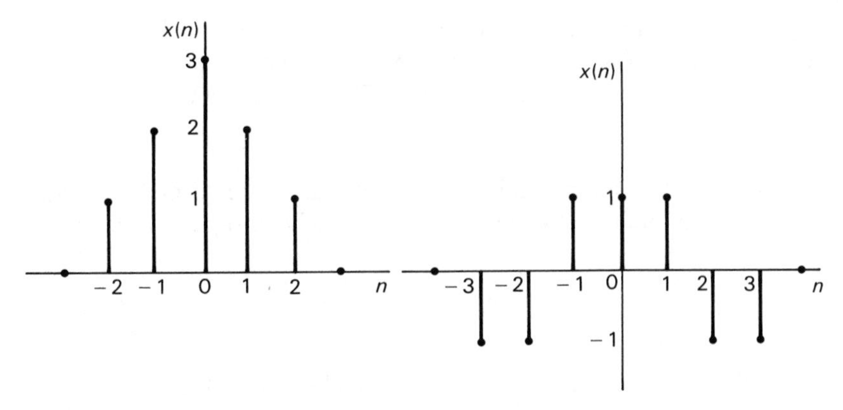

Figure 6.54

6.13 A discrete signal $x(n)$ has the following values $x(0) = 1$, $x(1) = 2$, $x(2) = 1$, $x(3) = 0$, $x(4) = -1$, $x(5) = 1$.

Determine the DFT of this signal for $N = 6$ and express the components in polar form.

Determine directly the DFT of the signal $x(n-2)$ where the delay is interpreted as a circular delay. Compare the result with that obtained using the delay theorem.

6.14 A discrete signal $x(n)$ takes the following values $x(0) = 3$, $x(1) = 2$, $x(2) = 1$, it has zero value otherwise.

(a) Determine numerically its discrete Fourier transform $X(k)$ over 4 points ($N = 4$).
(b) Determine the circular convolution of the signal with itself $x(n) * x(n)$.
(c) Compare the answer obtained in (b) with the inverse transform of $X^2(k)$.

6.15 In a biomedical experiment it is required to estimate the Fourier transform of a nerve potential pulse following excitation by an impulse. The pulse is 5 ms in duration, it is amplified and fed through an anti-aliasing filter that restricts the highest frequency present to 1 kHz. It is sampled and the discrete Fourier transform used to estimate its spectrum.

(a) Explain the purpose of the anti-aliasing filter and show that if the sampling time is 0.5 ms there will be no aliasing in the estimated spectrum.
(b) Explain what is meant by 'zero-padding' giving the reason why it is used. If the pulse sampled as in (a) is augmented by 30 zeros what resolution will be obtained in the estimated spectrum?
(c) The values of the amplified signal at the sampling points are as follows:

Sample	0	1	2	3	4	5	6	7	8	9
Voltage (V)	0	2	4	6	5	4	3	2	1	0

The sampling time is 0.5 ms as in (a) and the record is augmented by 30 zeros as in (b). Obtain by numerical evaluation of the discrete Fourier transform the magnitude of the 200 Hz components in the estimated spectrum.

6.16 Figure 6.55 shows a triangular pulse of duration one second and amplitude 1 unit. The magnitude of the Fourier transform of such a pulse is given by

$$|F(j\omega)| = \frac{1}{2}\left[\frac{\sin \omega/4}{\omega/4}\right]^2$$

The pulse is sampled at intervals of 0.1 s and the record augmented by zeros to give the 20 point discrete signal shown in Figure 6.55(b). Use the discrete Fourier transform (DFT) to estimate the magnitude of the frequency component at 0.5 Hz and calculate the error in the magnitude.

Explain qualitatively the reason for this error in terms of 'aliasing' and indicate how the aliasing effect could be reduced.

What is 'frequency resolution' when taking this DFT and indicate how this could also be improved?

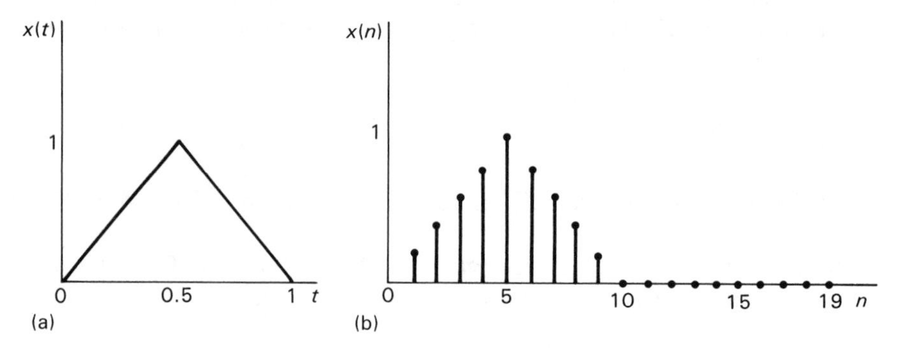

Figure 6.55

6.17 A sampling oscilloscope is used to display signals having bandwidths much larger than the bandwidth of a conventional oscilloscope amplifier. Its action depends upon the fact that the signal to be examined is periodic, it is illustrated in Figure 6.56.

The signal to be displayed $x(t)$ is sampled once per period but with a sampling time that is larger than the periodic time of the signal. The resulting signal $y(t)$ is as shown and passing this through a low-pass filter will produce an output signal proportional to $x(bt)$, where $b < 1$. The original signal is 'time stretched' and is now within the bandwidth of the oscilloscope amplifier.

Suppose the signal $x(t)$ consists of a d.c. level plus two sinusoids

$$x(t) = a_0 + a_1 \cos \omega t + a_x \cos 2\omega t, \qquad \omega = 2\pi/T$$

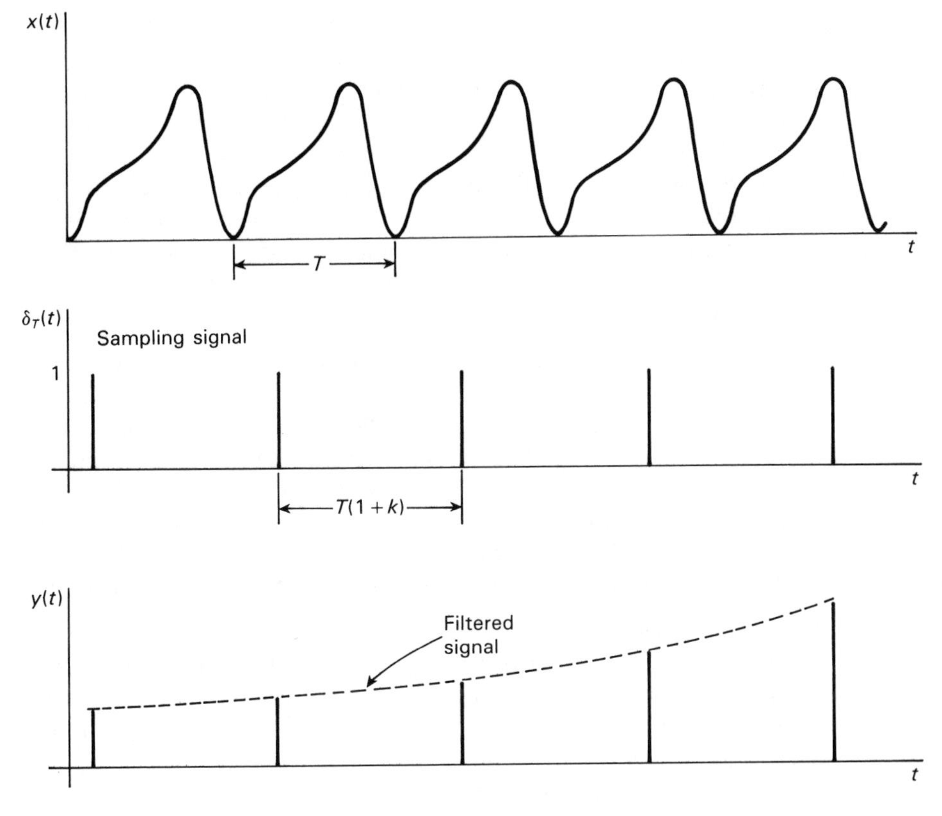

Figure 6.56

and it is sampled at a frequency ω_s given by

$$\omega_s = \frac{2\pi}{T(1+k)} \quad \text{where } k \text{ is a constant}$$

The low-pass filter is ideal and has a cut-off frequency ω_c equal to half the sampling frequency.

Show that the output signal $y(t)$ is proportional to $x(bt)$ and obtain an expression for the factor b.

MATLAB exercises

6.18 This exercise illustrates how the Fourier transform can be considered as a limiting case of the Fourier series. Using a sampling interval of 1 ms, create a vector representing a rectangular pulse of width 0.1 s. Determine the Fourier series for a periodic version of this pulse for periods 0.512, 1.024 and 2.048 s. Using `subplot` and `stem`, plot the magnitude of the Fourier series component against frequency for each of the periodic times. Note how the amplitudes of the components and the spacing between the components change with alteration of the periodic time.

Divide the amplitude of the Fourier coefficients derived above by the frequency spacing, $\Delta\omega$, for each of the periodic times used. Re-plot and compare amplitudes. Obtain a vector representing the magnitude of the Fourier transform of the pulse. Plot this on the same display as the above plots and compare results.

Note The Fourier series coefficient can be obtained by using the `fft` command and dividing by the number of points (see Exercise 5.27). To superimpose the discrete plots of the Fourier series components and the continuous plot of the Fourier transform, use the `hold on` command.

6.19 This exercise investigates some of the properties of the Fourier transform. In each case, time limited waveforms are used which are padded with zeros so that the `fft` command gives a good approximation to the Fourier transform.

(a) Time shift property.
Generate a vector representing a triangular pulse of amplitude 10 units. The rising and falling portions of the pulse should each consist of 100 sample points. Use the command `fft (X,n)` to obtain Fourier transform of the pulse with $n = 2048$.

Repeat this procedure for a pulse that has been delayed by 10 samples.

Plot, on the same display, the magnitudes of the transforms of the original and delayed pulses and, on a separate display, the angles of these transforms. Plot over a range of 100 frequency points and verify that these plots agree with theory.

(b) Frequency shift property.
Generate a vector representing signal samples

$$x(k) = \mathrm{e}^{-\mathrm{j}2\pi 100k/N}$$

where $N = 2048$ and k takes the range 0 to 2047. Form the product $x(n)x(k)$ where $x(n)$ is the undelayed triangular pulse of part (a) padded with zeros to make a vector of length 2048. Take the Fourier transform of this product and, by plotting its magnitude and angle, compare with the Fourier transform of $x(n)$. Verify that these plots agree with theory.

(c) Convolution property.
Generate a vector, $x_1(n)$, representing a unit amplitude rectangular pulse of width 200 samples and delayed by 200 samples from $n = 0$. Use the `conv` command to convolute this signal with the signal $x(n)$ generated in

part (a). Take the Fourier transform of the result over 2048 points and compare with the product of the Fourier transforms of $x(n)$ and $x_1(n)$.

Note In this exercise the signals can be treated as continuous signals by assuming a value for the sampling interval. The plots will then be against frequency. Alternatively, the signals can be treated as discrete signals. The plots will then be against harmonic number.

6.20 This exercise illustrates Parseval's theorem.

Produce a signal consisting of a vector of length 1024 where the elements consist of normal random variables having zero mean and unit variance. Choose a suitable sampling interval so that this vector represents a continuous random signal $x(t)$. Taking the Fourier transform of this signal, verify Parseval's theorem

$$\int_{-\infty}^{+\infty} x^2(t)\,\mathrm{d}t = \frac{1}{2\pi}\int_{-\infty}^{+\infty} |X(\omega)|^2\,\mathrm{d}\omega$$

Note To verify the theorem, a sampling time, Δt, has to be assumed. The time integral becomes a summation multiplied by the factor Δt. The discrete Fourier transform has to be multiplied by this factor and the frequency integral becomes a summation multiplied by a factor Δf where $\Delta f = 1/N\Delta t$.

The summations should be performed, without the use of 'for loops', by suitable vector multiplication.

The student should verify that the factor Δt cancels throughout, giving the discrete form of Parseval's theorem as

$$\sum_{n=0}^{N-1} x^2(n) = \frac{1}{N}\sum_{k=0}^{N-1} |X(k)|^2$$

6.21 This exercise uses MATLAB to investigate the spectrum of a modulated signal.

Generate a vector representing 1 cycle of a triangular modulating signal. This signal rises linearly from zero to a level of 1 unit over 256 sample points. It then falls linearly to a level of -1 over the following 512 samples. It then rises linearly to zero over the next 256 samples. The whole vector represents a signal of time duration 1 second.

Generate a vector representing a carrier signal. This is a sine wave having unit amplitude and frequency 20 Hz. The vector should consist of 1024 samples representing a time duration of 1 second.

Form a modulated signal from the product of the two signals described above. By use of the `fft` command, determine the amplitude spectra of both the modulating and modulated signal. Plot these as subplots against frequency over a range 0–30 Hz and compare with the theoretical spectra.

Note The modulating signal can be produced by forming separate vectors to represent each linear segment. These can then be combined to form a single vector.

Because of the discrete nature of the spectra, they are best displayed using the command `stem`. They can also be displayed as positive and negative frequency components by use of the command `fftshift`.

6.22 This exercise illustrates the effect of 'zero-padding'.

Generate a vector representing a triangular pulse. The pulse should rise linearly to unit amplitude in 100 samples and then fall linearly to zero in another 100 samples. The vector represents a pulse of time duration 0.2 seconds.

Increase the length of the vector representing the pulse to give a total length N by adding the appropriate number of zero elements. For values of N equal to 1024, 2048 and 4096, take Fourier transforms of the resulting signal and plot the magnitude of the transforms against frequency. Use a frequency scale 0–20 Hz and show the results as subplots.

Note Although the procedure described explains the mechanism of 'zero padding', the required transforms can be obtained more easily by use of the command `fft(X,n)`.

6.23 This exercise demonstrates some aspects of circular convolution.

One of the properties of the DFT is that the product of the transforms is the transform of the convolution of the data sequences. The convolution, however, is a circular convolution and this is not given by the command `conv(a,b)`. A program can be written to perform circular convolution, but an alternative method is to make one of the data sequences periodic when performing the convolution.

Generate two vectors, `x1` and `x2`, each of length 512 elements. These represent random signals and can be generated by the command `randn(1,512)`. From `x1`, generate a periodic signal, `xp=[x1 x1]` and use `conv` to obtain the convolution, `xc`, of `x2` and `xp`. Form the product of the DFTS of `x1` and `x2` and use the command `ifft` to obtain the inverse transform of this product. By plotting, compare the result with elements 512 to 1023 of `xc`. Because of the speed advantage offered by the FFT, it would be useful if the product of the transforms could be used to provide a non-circular convolution. Determine the FFTs of the signals `x1` and `x2` when they are both padded with 512 zeros and denote these transforms by $X_1(k)$ and $X_2(k)$. Determine the inverse transform of the product $X_1(k)X_2(k)$ and, by plotting, compare this with elements 1 to 1023 of the non-circular convolution of the padded sequences `x1` and `x2`.

6.24 This exercise examines the reduction in computing time that can be obtained by using the FFT algorithm.

By using the commands `tic` and `toc`, determine the time required to perform the DFT. Obtain times to execute the command `fft(X)` where the vector `X` is a random signal containing N elements. Compare times for the case where N is a power of 2, $N = 2^p$, to times when N has values $2^p + 1$, $2^p - 1$. Use a range of N from 128 to 4096.

Note The command `fft(X)` only performs the FFT if the length of `X` is a power of 2. If `X` is not a power of 2, but is not prime, a mixed radix transform is used which is slower than the FFT algorithm. If N is prime, then the slowest algorithm, the DFT, is used.

Lengths that are prime and are close to those which are a power of 2 are 127, 257, 511, 1021, 2053, 4093.

6.25 This exercise examines the reconstruction of a sine wave signal from its sampled values.

Construct a vector, containing 2048 points, representing 1 cycle of a sine wave of unit amplitude and frequency 1 Hz.

Construct a vector, containing 2048 points, representing a sampled version of the above signal where the sampling frequency is 8 Hz. Determine the amplitude spectrum of this sampled signal and plot it over the frequency range 0 to 30 Hz.

One method of signal reconstruction uses a zero order hold system. In this reconstruction the signal is held constant between sampling instants at a level equal to its sampled value at the last sampling instant.

Write a MATLAB program to reconstruct the unsampled signal by means of a zero order hold procedure. Determine the magnitude spectrum of the reconstructed signal and compare it with that of the sampled signal. Another method of signal reconstruction is to use an interpolating filter as described in Section 6.5.2. This method involves a convolution between the sampled signal and a sinc function. Write a MATLAB program to reconstruct the unsampled signal by this method and compare the reconstructed signal with the unsampled signal.

Note The sample values can easily be generated by using the formula

$$x(n) = \sin(2\pi f n / f_s)$$

where $f = 1$ Hz and $f_s = 8$ Hz. However, to represent the sampled signal on the same time scale as the unsampled signal, the appropriate number of zeros must be inserted between samples. The action of the sample and hold can be simulated in a similar manner. However, the points between the sample values must take on the value of the last sample.

To produce the impulse response of the interpolating filter, the command `sinc(x)` can be used, but it should be noted that this returns a value $\sin(\pi x)/\pi x$. Both the sinc function and the sampled response will have to be truncated before convolution. In order to reconstruct 1 cycle of the unsampled signal, more than 1 cycle of the sampled signal is required. Take care with the time scale for the sinc function: the zeros of this function must occur at intervals equal to the sampling interval.

7

The Laplace Transform

7.1 Introduction

The last chapter considered the use of the Fourier transform for system analysis. Although in theory it is possible to invert the Fourier transform at the system output in order to obtain the system time response, this step is rarely performed. The Fourier transform is used to interpret system performance in the frequency domain.

There are several reasons why the Fourier transform is rarely used to determine the system time response. The principal one is as shown in the last chapter, the Fourier transform does not exist for many useful engineering signals. Although it is possible to extend the Fourier transform to include some of these signals this involves the inclusion of the impulse function into the expression for the transform and the resulting transforms are somewhat intractable to use.

An alternative approach is to redefine the transform and include an exponential convergence factor within its integral. This leads to the concept of the Laplace transform and this is the subject matter of this chapter.

As well as existing for a larger range of signals than the Fourier transform the Laplace transform has other advantages for system analysis. The transforms themselves are simpler in form than the Fourier transform and there are standard methods that can be used for their inversion. Also initial conditions can be easily included into the calculation giving the system response.

However, as with the Fourier transform, considerable insight can be gained into the system performance without the need to invert the transform. This can be done by interpreting the transform in terms of the position of its poles and zeros in the complex plane. This often leads to a quick qualitative estimate of system performance that is very useful in engineering system design.

The equivalent of the Laplace transform when considering discrete systems is the z transform. This will be developed by the consideration of the Laplace transform of a sampled signal.

7.2 The one sided Laplace transform

As stated in the introduction, one of the difficulties with the Fourier transform is that the integral defining the transform will not converge for some commonly used signals. In the last chapter this difficulty was overcome at the expense of producing expressions for the Fourier transform that involved the impulse function. Such expressions are not very convenient to incorporate into general transform analysis.

7.2.1 Derivation of the transform

The alternative is to introduce a factor $e^{-\sigma t}$ into the transform where σ is a real number chosen to make the integral converge. The Fourier transform now becomes a function of ω and σ (it will be written as a function of $j\omega$ here)

$$X(j\omega, \sigma) = \int_{-\infty}^{+\infty} x(t)e^{-\sigma t}e^{-j\omega t}\,dt$$

$$= \int_{-\infty}^{+\infty} x(t)e^{-(\sigma + j\omega)t}\,dt$$

If evaluated for a given signal $x(t)$ this would be of a function of $\sigma + j\omega$.

Writing $s = \sigma + j\omega$ the transform can be written

$$X(s) = \int_{-\infty}^{+\infty} x(t)e^{-st}\,dt \qquad (7.2.1)$$

Equation (7.2.1) defines the *two sided or bilateral transform* of the function $x(t)$. In this context the complex variable s ($s = \sigma + j\omega$) is known as the Laplace operator. The student may be puzzled about σ, can any value of σ be chosen, will the integral always converge, does the σ chosen depend upon $x(t)$? It is not possible to give a full answer to these questions in a book of this nature. Briefly only a certain range of σ values will give convergence and the range chosen will depend on the function

$x(t)$. This can cause problems, different time functions can produce the same transform, however each function will have a different range of convergence for σ. These sorts of problems can be almost avoided by restricting the range of integration to between 0 and ∞. This is equivalent to saying that the signal $x(t)$ in the transform has the property that $x(t)=0$ for $t<0$. This then defines the unilateral or one sided Laplace transform.

$$X(s) = \int_0^{+\infty} x(t)e^{-st}\,dt \qquad (7.2.2)$$

It is only this version of the Laplace transform that will be considered for the remainder of this book and it will be referred to as just 'the Laplace transform'. As with the Fourier transform, different notations can be used to denote the transformation

$$X(s) \leftrightarrow x(t)$$
$$X(s) = \mathscr{L}\{(x(t)\}$$
$$x(t) = \mathscr{L}^{-1}\{(X(s)\}$$

These can be read, $X(s)$ and $x(t)$ are a Laplace transform pair, $X(s)$ is the Laplace transform of $x(t)$, $x(t)$ is the inverse transform of $X(s)$.

The lower limit on the transform in engineering texts is usually taken as 0^-. This allows functions such as the impulse function to be included under the integral sign. There is an inversion formula for the Laplace transform as with the Fourier transform, this is

$$x(t) = \frac{1}{2\pi j}\int_{\sigma - j\omega}^{\sigma + j\omega} X(s)e^{st}\,ds$$

This gives integration with respect to the complex variable s. Consideration of such an integral is beyond the scope of this book, however as will be seen it is not in practice necessary to evaluate this integral in order to invert a transform, there are alternative methods.

The actual evaluation of the transform for a given function is best shown by means of an example.

EXAMPLE 7.2.1

Obtain the Laplace transform of the unit step function $u(t)$ and the unit ramp defined as

$$x(t) = t \qquad t \geqslant 0$$
$$x(t) = 0 \qquad t < 0$$

SOLUTION

(a) The unit step.

As $x(t)=1$ for $t \geqslant 0$ then eqn (7.1.2) defining the transform becomes

$$X(s) = \int_0^\infty e^{-st}\,dt$$

$$= \left[\frac{-e^{-st}}{s}\right]_0^\infty$$

$$= -\left[\frac{e^{-\infty} - e^0}{s}\right]$$

$$= 1/s$$

(b) The unit ramp.
As $x(t) = t$ for $t \geqslant 0$

$$X(s) = \int_0^\infty te^{-st}\,dt$$

This can be integrated by parts giving

$$X(s) = \frac{1}{s}\,[-te^{-st}]_0^\infty + \frac{1}{s^2}\,[e^{-st}]_0^\infty$$

Inserting the limits gives the second term as $1/s^2$. The first term when evaluated at the lower limit gives zero, the upper limit depends on

$$\lim_{t\to\infty} te^{-st}$$

Table 7.1 A short table of Laplace transforms

Signal description	$x(t)$	$X(s)$
Unit impulse	$\delta(t)$	1
Unit step	$u(t)$	$1/s$
Unit ramp	t	$1/s^2$
Positive powers of t	t^n	$n!/s^{n+1}$
Exponential	e^{-at}	$\dfrac{1}{s+a}$
Sine wave	$\sin \omega t$	$\dfrac{\omega}{s^2+\omega^2}$
Cosine wave	$\cos \omega t$	$\dfrac{s}{s^2+\omega^2}$

$X(s) = \int_0^\infty x(t)\,e^{-st}\,dt$; all function assumed zero for $t < 0$

This limit can be shown to be zero. This will always be the case with functions that have a Laplace transform and it is consequent on the convergence of the defining integral. Hence for the unit ramp

$$X(s) = 1/s^2$$

Other functions can be evaluated in a similar manner to Example 7.1 and Table 7.1 lists the most common transforms used in engineering. Much more extensive lists of transforms are available in more specialist texts.

7.2.2 Properties of the Laplace transform

Because of the similarity of the integrals defining the Laplace and Fourier transforms many of the properties of the Fourier transform have their counterpart in the Laplace transform.

The proofs of some of these properties follow the corresponding proofs of properties of the Fourier transform. However stemming from the fact that the lower limit in the integral is zero there are some different and some additional properties.

Time differentiation

This is one of the most useful properties of the Laplace transform as it enables values of signals at $t = 0$ to be taken into consideration. This is very useful when the transform is applied to system analysis as it enables the effects of initial energy storage to be taken into account.

Consider the Laplace transform of the differential $\mathrm{d}x/\mathrm{d}t$. This is given by

$$\int_0^\infty \frac{\mathrm{d}x(t)}{\mathrm{d}t} \mathrm{e}^{-st} \,\mathrm{d}t = [x(t)\mathrm{e}^{-st}]_0^\infty + s\int_0^\infty x(t)\mathrm{e}^{-st} \,\mathrm{d}t$$

This expression has been obtained by integration by parts. The integral in the second term represents $X(s)$ the Laplace transform of $x(t)$. Inserting the limits in the first term gives

$$[x(t)\mathrm{e}^{-st}]_0^\infty = \lim_{t \to \infty} x(t)\mathrm{e}^{-st} - x(0)$$

If the function $x(t)$ has a Laplace transform then in the limit the first term becomes zero. Then

$$\mathcal{L}\left\{\frac{\mathrm{d}x(t)}{\mathrm{d}t}\right\} = sX(s) - x(0)$$

Where $X(s)$ is the transform of $x(t)$ and $x(0)$ is the value of $x(t)$ at $t = 0$.

Repeated application of this method leads to an expression for the Laplace transform of the nth differential $\mathrm{d}^n x(t)/\mathrm{d}t^n$

$$\mathcal{L}\left\{\frac{\mathrm{d}^n x(t)}{\mathrm{d}t^n}\right\} = s^n X(s) - s^{n-1}x(0) - s^{n-2}x_1(0) \ldots\ldots - x_{(n-1)}(0)$$

where $x^i(0)$ is the ith derivative of $x(t)$ at $t = 0$. To be consistent with the interpretation of the lower limit of the defining integral the initial values $x^i(0)$ in this formula should be interpreted as being at 0^-. However not all books follow this definition.

Initial and final value theorems

These again are useful properties that have no counterpart in the Fourier transform. They are useful because knowing the Laplace transform of a function they enable the initial and final values of the function to be obtained without inverting the transform.

Considering first the initial value theorem and using the time differential property

$$\int_0^\infty \frac{\mathrm{d}x(t)}{\mathrm{d}t} \mathrm{e}^{-st}\, \mathrm{d}t = sX(s) - x(0)$$

Taking the limit of both sides of this equation as $s \longrightarrow \infty$ and noting that

$$\lim_{s \longrightarrow \infty} \mathrm{e}^{-st} = 0$$

then

$$0 = \lim_{s \longrightarrow \infty} sX(s) - x(0)$$

or

$$x(0) = \lim_{s \longrightarrow \infty} sX(s)$$

This is the initial value theorem.

The proof of the final value theorem also starts with the time differentiation property.

$$\int_0^\infty \frac{\mathrm{d}x(t)}{\mathrm{d}t} \mathrm{e}^{-st}\, \mathrm{d}t = sX(s) - x(0)$$

Taking, this time, the limit of both sides as $s \longrightarrow 0$ gives

$$\int_0^\infty \frac{\mathrm{d}x(t)}{\mathrm{d}t}\, \mathrm{d}t = \lim_{s \longrightarrow 0} sX(s) - x(0)$$

$$x(\infty) - x(0) = \lim_{s \longrightarrow 0} sX(s) - x(0)$$

$$x(\infty) = \lim_{s \longrightarrow 0} sX(s)$$

This is the final value theorem.

Table 7.2 Some properties of the Laplace transform

Property	*Time signal*	*Laplace transform*
Linearity	$a_1x_1(t) + a_2x_2(t)$	$a_1X_1(s) + a_2X_2(s)$
Scaling $a > 0$	$x(at)$	$\frac{1}{a}X\left(\frac{s}{a}\right)$
Delay	$u(t-\tau)x(t-\tau)$	$e^{-s\tau}X(s)$
Multiplication by exponential	$e^{-at}x(t)$	$X(s+a)$
Differentiation	$\frac{d}{dt}x(t)$	$sX(s) - x(0)$
Initial value	$\lim_{t\to 0} x(t)$	$\lim_{s\to\infty} sX(s)$
Final value	$\lim_{t\to\infty} x(t)$	$\lim_{s\to 0} sX(s)$

Table 7.2 lists some of the most useful properties of the Laplace transform. The following example will illustrate some of these properties.

EXAMPLE 7.2.2

Use the Laplace transform to obtain the convolution of the two signals shown in Figure 7.1.

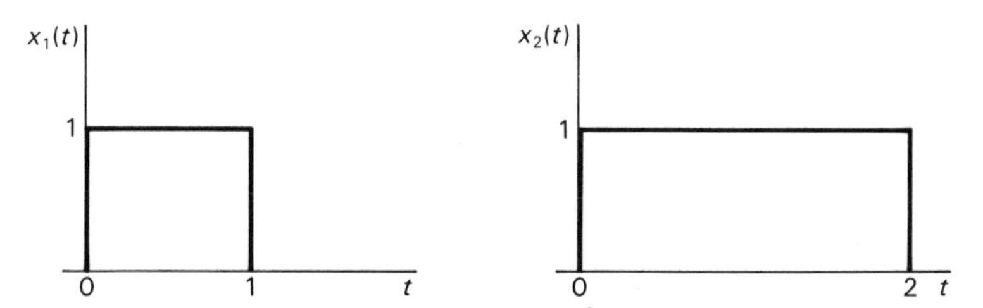

Figure 7.1 Signals for Example 7.2.2.

SOLUTION

The convolution of these functions is easily performed in the time domain and the resultant is shown in Figure 7.2.

It is instructive to obtain this result using the property that convolution in the time domain is equivalent to the multiplication of Laplace transforms. The transforms of the signal, $x_1(t)$ and $x_2(t)$ are required. These are not functions that appear in Table 7.1. However they can be constructed from standard forms with the aid of the delay property. Taking the signal $x_1(t)$, this can be constructed as shown in Figure 7.3.

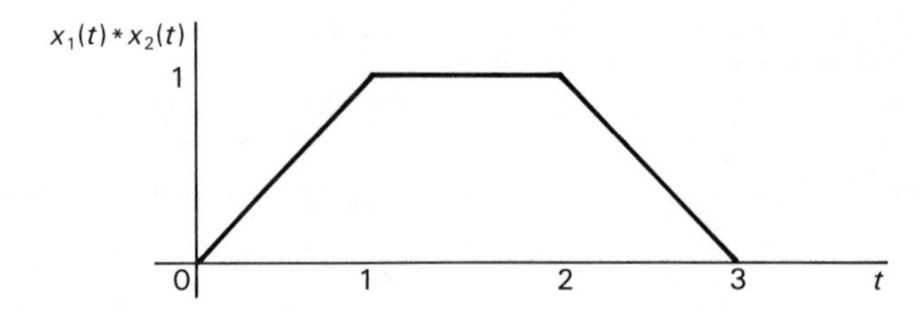

Figure 7.2 Convolution of signals for Example 7.2.2.

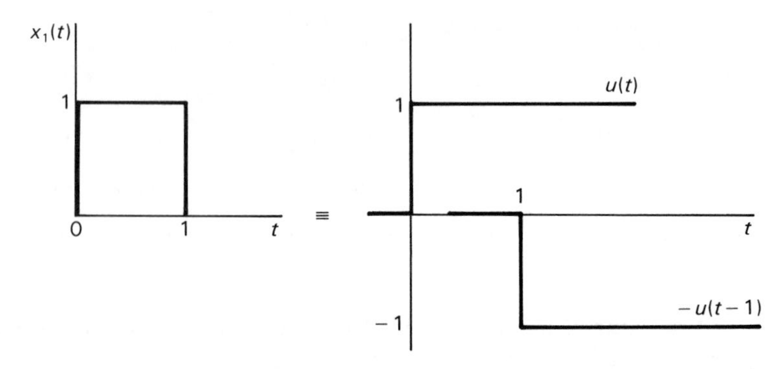

Figure 7.3 Construction of signal $x_1(t)$ from a combination of step functions.

Adding the two functions shown will produce $x_1(t)$. Because of the linearity property its Laplace transform is obtained by adding the transform of a unit step, $1/s$, and the transform of a delayed negative step, $-e^{-s}/s$. Hence

$$X_1(s) = \frac{1}{s} - \frac{e^{-s}}{s} = \frac{(1 - e^{-s})}{s}$$

Similarly

$$X_2(s) = \frac{1}{s} - \frac{e^{-2s}}{s} = \frac{(1 - e^{-2s})}{s}$$

Multiplication gives

$$\begin{aligned} X_1(s)X_2(s) &= \frac{(1 - e^{-s})(1 - e^{-2s})}{s^2} \\ &= \frac{1}{s^2} - \frac{e^{-s}}{s^2} - \frac{e^{-2s}}{s^2} + \frac{e^{-3s}}{s^2} \end{aligned}$$

The transform has been written out in this form to emphasise that is is formed of combinations of delayed ramp functions as shown in Figure 7.4.

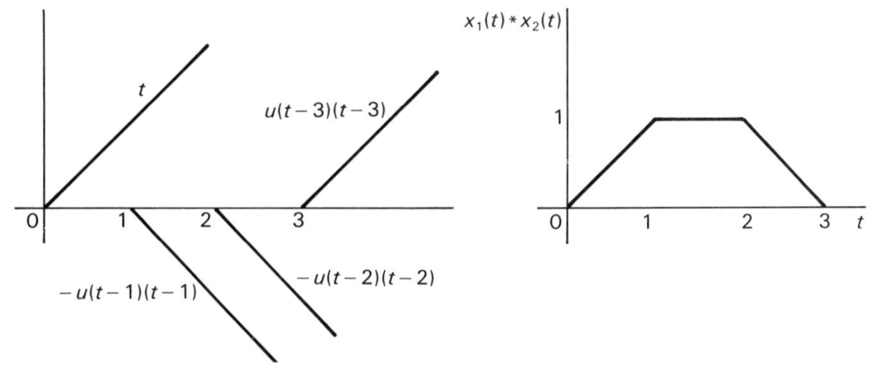

Figure 7.4 Convolution of signals $x_1(t)$ and $x_2(t)$ for Example 7.2.2.

It should be noted that if an inverse transform represents a delayed function, then this implies that the function was zero for $t<0$. Hence the inverse transform of e^{-s}/s^2 is not $(t-1)$ but is $u(t-1)(t-1)$, the delayed unit step ensuring that the time function takes zero value for $t<1$.

7.3 Application to system analysis

Chapter 4 considered the time response of systems to arbitrary input signals. The response could either be obtained by solution of the system differential equation or via the convolution integral and the system impulse response. Both of these methods are not very convenient from an engineering point of view.

The Laplace transform offers a much more convenient method of obtaining the system response. Two approaches are possible, by taking the Laplace transform of both sides of the system differential equation an algebraic equation can be obtained. Solving this equation gives an expression for the transform of the output signal and inversion then produces the time response.

The other approach is to use the convolution property of the Laplace transform. Convolution in time is equivalent to multiplication in the s domain. Hence the transform of the output signal is the multiplication of the transform of the input signal by the transform of the impulse response. The transform of the impulse response is known as the system transfer function and this is an important concept in system engineering.

Both of these approaches will be investigated in the following sections. However the student should bear in mind the comments made on time domain methods in Chapter 4. Although transform methods are easier they can very easily become 'turn the handle' methods and much physical insight into the problem can be lost.

7.3.1 Response via the differential equation

As shown in Chapter 3, continuous systems are in general described by a differential equation of the form

$$a_n \frac{\mathrm{d}^n y}{\mathrm{d}t^n} + a_{n-1} \frac{\mathrm{d}^{n-1} y}{\mathrm{d}t^{n-1}} + \ldots\ldots a_0 y = b_m \frac{\mathrm{d}^m x}{\mathrm{d}t^m} + \ldots\ldots b_0 x$$

Using the differential property of the Laplace transform $\mathrm{d}^n y/\mathrm{d}t^n$ can be replaced by $s^n Y(s)$, $\mathrm{d}^{n-1}y/\mathrm{d}t^{n-1}$ by $s^{n-1}Y(s)$, etc. This assumes there are no initial conditions on the variables, if initial conditions do exist they can be incorporated at this stage. Knowing the Laplace transform $X(s)$ of the input $x(t)$ an algebraic form for $Y(s)$ can be obtained, inversion of this transform produces $y(t)$.

This procedure is best illustrated by way of an example.

EXAMPLE 7.3.1

(a) Obtain the output response of the system shown in Figure 7.5 when the input is a 10 V step as shown, assume the capacitors are initially uncharged.
(b) Obtain the response to the step input as in (a) assuming that the capacitors are charged such that the voltage across C_2 is 10 V and that across C_1 is 5 V.
(c) Obtain the response to a half cycle sine wave pulse given by

$$\begin{aligned} v_i(t) &= 10 \sin \pi t \qquad 0 \geqslant t \geqslant 1 \\ &= 0 \qquad \text{otherwise} \end{aligned}$$

Assuming the capacitors are initially uncharged.

SOLUTION

All parts of this example require the differential equation relating $v_o(t)$ to $v_i(t)$. This can be obtained using nodal analysis, the voltage v_x is as shown in Figure 7.5.

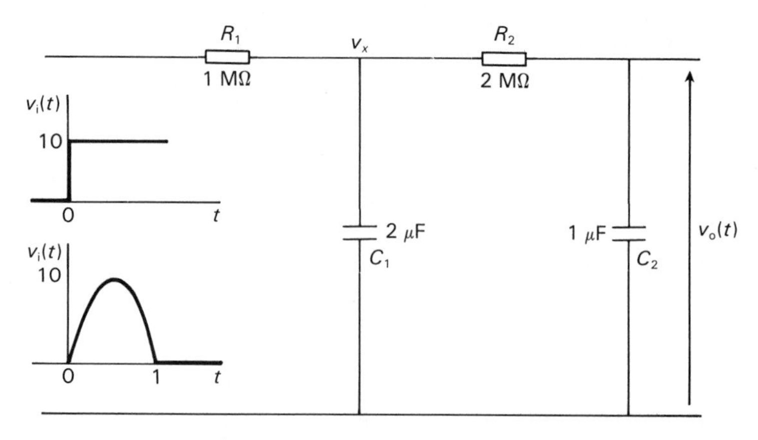

Figure 7.5 Circuit for Example 7.3.1.

$$\frac{v_i - v_x}{R_1} = C_1 \frac{dv_x}{dt} + \frac{v_x - v_o}{R_2} \tag{7.3.1}$$

$$\frac{v_x - v_o}{R_2} = C_2 \frac{dv_o}{dt} \tag{7.3.2}$$

Equation (7.3.1) can be re-written

$$v_i = C_1 R_1 \frac{dv_x}{dt} + v_x\left[1 + \frac{R_1}{R_2}\right] - \frac{R_1}{R_2} v_o \tag{7.3.3}$$

To obtain an equation relating input and output the variables v_x and dv_x/dt must be eliminated from the equation. From eqn (7.3.2)

$$v_x = C_2 R_2 \frac{dv_o}{dt} + v_o \tag{7.3.3}$$

and differentiating

$$\frac{dv_x}{dt} = C_2 R_2 \frac{d^2 v_o}{dt^2} + \frac{dv_o}{dt}$$

Substituting v_x and dv_x/dt into eqn (7.3.3) gives

$$v_i = C_1 R_1 C_2 R_2 \frac{d^2 v_o}{dt^2} + (C_1 R_1 + C_2 R_2 + C_2 R_1) \frac{dv_o}{dt} + v_o$$

Inserting component values

$$4 \frac{d^2 v_o}{dt^2} + 5 \frac{dv_o}{dt} + v_o = v_i \tag{7.3.4}$$

Taking the Laplace transform of both sides of eqn (7.3.4) gives

$$4(s^2 V_o(s) - s v_o(0) - v_o^1(0)) + 5(s V_o(s) - v_o(0)) + V_o(s) = V_i(s) \tag{7.3.5}$$

where $v_o(0)$ is the value of the output voltage at $t = 0$ and $v_o^1(0)$ is the value of its derivative at $t = 0$. From eqn (7.3.5) an expression for $V_o(s)$ can be obtained.

$$V_o(s) = \frac{V_i(s)}{4s^2 + 5s + 1} + \frac{(4s + 5)v_o(0) + 4v_o^1(0)}{4s^2 + 5s + 1} \tag{7.3.6}$$

Taking the inverse transform of eqn (7.3.6) will produce the time response $v_o(t)$. It can be seen that this response consists of two parts corresponding to the zero state and zero input responses.

(a) There are no initial conditions in this case, $v_o(0)=0$, $v_0^1(0)=0$. The input is a 10 V step giving $V_i(s)=10/s$. Hence from eqn (7.3.6)

$$V_o(s)=\frac{10}{s(4s^2+5s+1)} \tag{7.3.7}$$

To obtain $v_o(t)$ the inverse transform of the right hand side of eqn (7.3.7) is required. This expression is not a standard transform given in Table 7.1. However the denominator can be factorised giving

$$V_o(s)=\frac{2.5}{s(s+0.25)(s+1)} \tag{7.3.8}$$

This expression for $V_o(s)$, eqn (7.3.8), is in the form of a constant times the factors $1/s$, $1/(s+0.25)$, $1/(s+1)$. Each of these is a standard form and can be inverted individually by reference to Table 7.1. However multiplication of the corresponding time functions does *not* give the correct answer, multiplication of transforms corresponds to a *convolution* in time. If the expression for $V_o(s)$ could be rearranged to give a sum of these terms then in accordance with the linearity property of the transform the time function $v_o(t)$ would be the sum of the corresponding time functions. The process of representing $V_o(s)$ as the sum of the terms is that of partial fraction expansion. The general methods of partial fractions are given in books on higher algebra and only an outline will be given here.

It is assumed that the function can be written as a weighted sum of fractions whose denominators are the factors of the denominator of the original expression.

$$\frac{2.5}{s(s+0.25)(s+1)}=\frac{A}{s}+\frac{B}{(s+0.25)}+\frac{C}{(s+1)}$$

where A, B and C are constants that have to be determined. Multiplying through by the denominator of the left-hand side gives

$$2.5=A(s+0.25)(s+1)+Bs(s+1)+Cs(s+0.25)$$

This really should be written as an identity as it is true for all values of s. The trick is now to choose values of s that will give the required constants easily. If we chose $s=0$ the second two terms on the right-hand side go to zero leaving

$$\begin{aligned}2.5 &= A(0+0.25)(0+1)\\ A &= 10\end{aligned}$$

Similarly choosing $s=-1$ gives $C=10/3$ and choosing $s=-0.25$ gives $B=40/3$. Hence

$$V_o(s)=10\left[\frac{1}{s}-\frac{4}{3(s+0.25)}+\frac{1}{3(s+1)}\right] \tag{7.3.9}$$

It is always advisable to check at this point that combining the fractions on the right-hand side gives the original expression.

Each term on the right-hand side of eqn (7.3.9) can now be inverted as a standard form giving

$$v_o(t) = 10(1 - 1.33e^{-0.25t} + 0.33e^{-t}) \quad \text{for } t \geqslant 0$$
$$= 0 \quad \text{for } t < 0 \qquad (7.3.10)$$

This response is shown in Figure 7.6. As can be seen it starts at zero when $t = 0$ and approaches 10 V as $t \to \infty$. These limits agree with those that can be obtained from physical reasoning applied to the circuit of Figure 7.5.

(b) In this part the input is a 10 V step applied at $t = 0$ and this produces the response given by eqn (7.3.10). However the initial stored charge must be taken into account and the time response corresponding to the second term in eqn (7.3.6) must be evaluated. This requires values for $v_o(0)$ and $v_o^1(0)$. The value of $v_o(0)$ is straightforward, it is given as 10 V.

Although the value of $v_o^1(0) = dv_o/dt$ at $t = 0$ is not given directly, a second initial condition is given, the voltage across C_1 is 5 V. The current through R_2 at this instant is $(v_o - v_x)/R_2$ and this is 2.5 μA. However this current must equal $C_2 dv_o/dt$ giving $v_o(0)$ as -2.5 V/s. (The minus sign indicates the voltage v_o is decreasing at $t = 0$.)

Denoting the second term in eqn (7.3.6) as $V_{int}(s)$ and substituting the values for $v_0(0)$ and $v_o^1(0)$ gives

$$V_{int} = \frac{40(s+1)}{4s^2 + 5s + 1}$$

$$= \frac{10}{(s + 0.25)}$$

The inversion is straightforward giving a time response

$$v_{int}(t) = 10e^{-0.25t}$$

Combining this response with that due to the input step, eqn (7.3.10) gives a total response

$$v_{total} = 10 - 3.3e^{-0.25t} + 3.33e^{-t} \qquad t \geqslant 0$$
$$= 0 \qquad t < 0$$

This response is also plotted in Figure 7.6.

(c) In this part, the initial conditions as in (a) are zero. However obtaining the transform of the input function is not as straightforward as in (a).

The input signal $v_i(t)$ can be represented as in Figure 7.7. A sine wave 10 sin πt gives a half cycle until $t = 1$ then zero signal is produced for $t > 1$ by adding a delayed sine wave $u(t - 1)\sin \pi(t - 1)$. This gives the input $v_i(t)$ as

$$v_i(t) = 10u(t)\sin \pi t + 10u(t - 1)\sin \pi(t - 1).$$

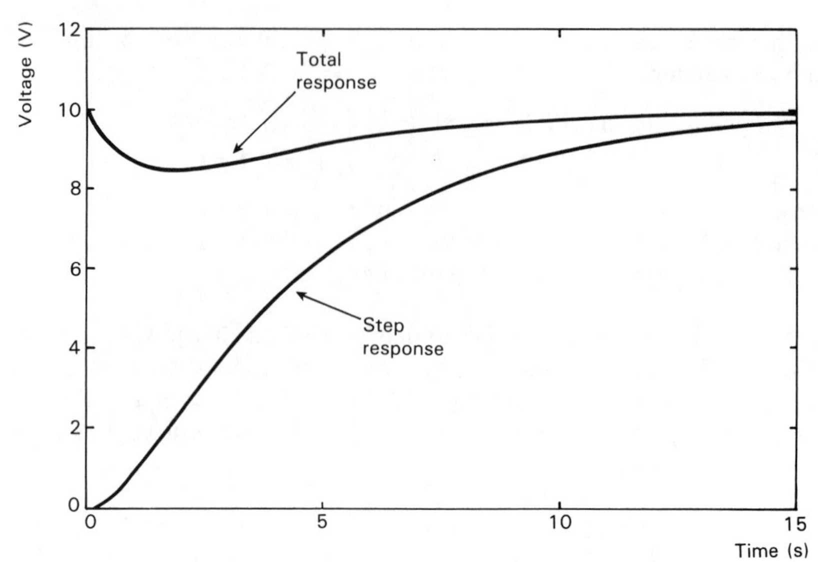

Figure 7.6 Responses of circuit in Example 7.3.1.

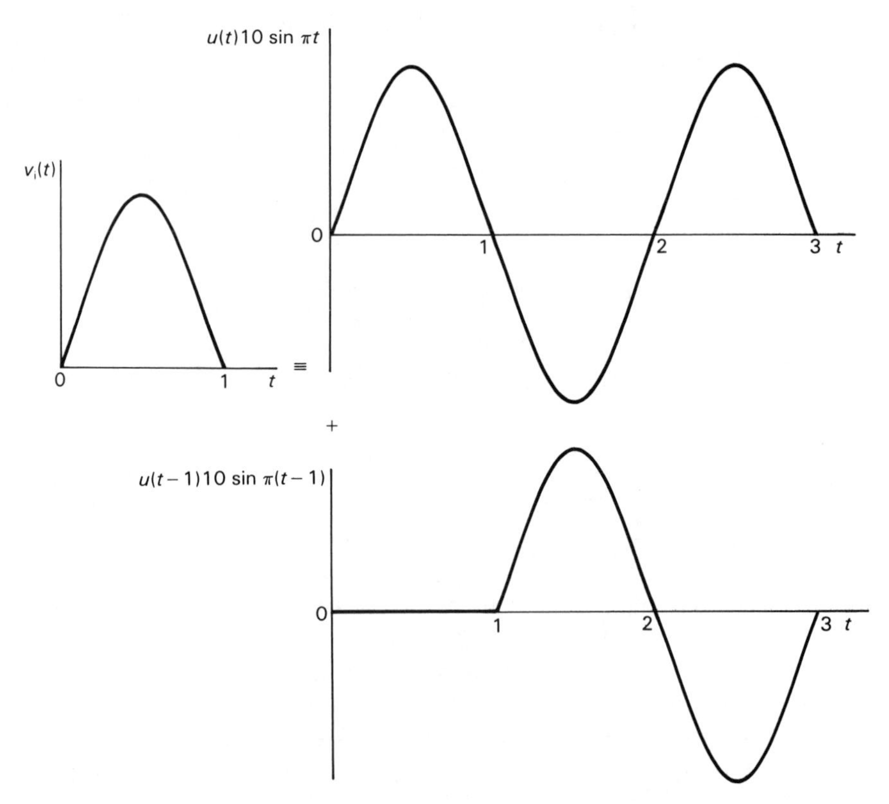

Figure 7.7 Construction of half sine wave pulse.

Taking Laplace transforms

$$V_i(s) = 10\left[\frac{\pi}{s^2+\pi^2}\right](1+e^{-s}) \tag{7.3.11}$$

Substituting into eqn (7.3.6) and taking into account the zero initial conditions gives $V_o(s)$ as

$$V_o(s) = \frac{2.5\pi(1+e^{-s})}{s^2+\pi^2}\cdot\frac{1}{(s+0.25)(s+1)} \tag{7.3.12}$$

where the denominator has been expressed in factored form. Inversion of eqn (7.3.12) will produce the required time response and this can be done via a partial fraction expansion. Care must be taken with the quadratic factor which will not split into factors with real coefficients. The form of the expansion is

$$\frac{1}{(s^2+\pi^2)(s+0.25)(s+1)} = \frac{As+B}{s^2+\pi^2}+\frac{C}{s+0.25}+\frac{D}{s+1} \tag{7.3.13}$$

where the multiplier $2.5\pi(1-e^{-s})$ has been omitted for the moment. Multiplying both sides of eqn (7.2.13) by the denominator of the left-hand side term gives

$$1 = (As+B)(s+0.25)(s+1)+C(s^2+\pi^2)(s+1)+D(s^2+\pi^2)(s+0.25) \tag{7.3.14}$$

The coefficients C and D can be evaluated by substitution of $s=-1$ and $s=-0.25$ into eqn (7.3.14). This gives

$$C = \frac{1}{0.75(0.0625+\pi^2)} = 0.1342$$

$$D = \frac{-1}{0.75(1+\pi^2)} = -0.1227$$

The coefficients A and B can be obtained by a similar evaluation at $s=j\pi$ and $s=-j\pi$. These lead to two complex simultaneous equations to solve for A and B. Although this is possible it is a rather tedious method. A better method is to arrange the right-hand side of eqn (7.3.14) as a polynomial in s. Equating the coefficients of corresponding terms on left- and right-hand sides now leads to the values of A and B. Equating the coefficient of the s^3 term

$$0 = A+C+D \text{ giving } A = -0.01158.$$

Equating the coefficients of the constant term

$$1 = 0.25B + C\pi^2 + 0.25D\pi^2 \text{ giving } B = -0.0891$$

Hence

$$\frac{1}{(s^2+\pi^2)(s+0.25)(s+1)} = \frac{-(0.0115s+0.0891)}{(s^2+\pi^2)}+\frac{0.1342}{(s+0.25)}-\frac{0.1227}{(s+1)}$$

The second and third terms on the right-hand side are standard transforms, however the first term on that side needs rearranging in order to express it as a combination of standard forms

$$-\frac{(0.0115s + 0.0897)}{(s^2 + \pi^2)} = \frac{-0.0115s}{(s^2 + \pi^2)} - \frac{0.0891}{\pi} \frac{\pi}{(s^2 + \pi^2)}$$

Expressing $V_o(s)$, eqn (7.3.12), in terms of these standard forms gives

$$V_o(s) = 2.5\pi(1 + e^{-s})\left[-0.0115 \frac{s}{s^2 + \pi^2} - \frac{0.0891}{\pi} \frac{\pi}{(s^2 + \pi^2)} + \frac{0.1342}{(s^2 + 0.25)} - \frac{0.1227}{(s + 1)}\right] \tag{7.3.15}$$

The effect of the term $(1 - e^{-s})$ can now be considered. Because of the linearity property of the transform each of the terms in this factor may be considered separately and the result combined. The term of unity leads to an output response that is obtained by inverting the terms inside the square brackets. Denoting this portion of the response as $v_{oi}(t)$ then

$$v_{oi}(t) = 2.5\pi[0.1342e^{-0.25t} - 0.1227e^{-t} - 0.0115 \cos \pi t - 0.0284 \sin \pi t] \tag{7.3.16}$$

It should be noted that this expression is true only for $t \geqslant 0$, for $t < 0$ the output is zero. This follows from the definition of the single sided Laplace transform.

The effect of the term e^{-s} must now be considered. From the delay property of the Laplace transform this will produce a time signal $v_{oi}(t - 1)$. However as $v_{oi}(t) = 0$ for $t < 0$ this should strictly be written $u(t - 1)v_{oi}(t - 1)$. Hence the complete response is given by

$$\begin{aligned} v_o(t) &= 0 && t < 0 \\ &= v_{oi}(t) && 0 \leqslant t \leqslant 1 \\ &= v_{oi}(t) + u(t-1)v_{oi}(t-1) && 1 < t \end{aligned} \tag{7.3.17}$$

where $v_{oi}(t - 1)$ is given by

$$\begin{aligned} v_{oi}(t-1) = 2.5\pi[&0.1342e^{-0.25(t-1)} - 0.1227e^{-(t-1)} \\ &-0.0115 \cos \pi(t-1) - 0.0284 \sin \pi(t-1)] \end{aligned} \tag{7.3.18}$$

However

$$e^{-0.25(t-1)} = e^{-0.25t}e^{0.25} = 1.284e^{-0.25t}$$

$$e^{-(t-1)} = e^{-t}e^{1} = 2.718e^{-t}$$

$$\cos \pi(t-1) = -\cos \pi t$$

$$\sin \pi(t-1) = -\sin \pi t$$

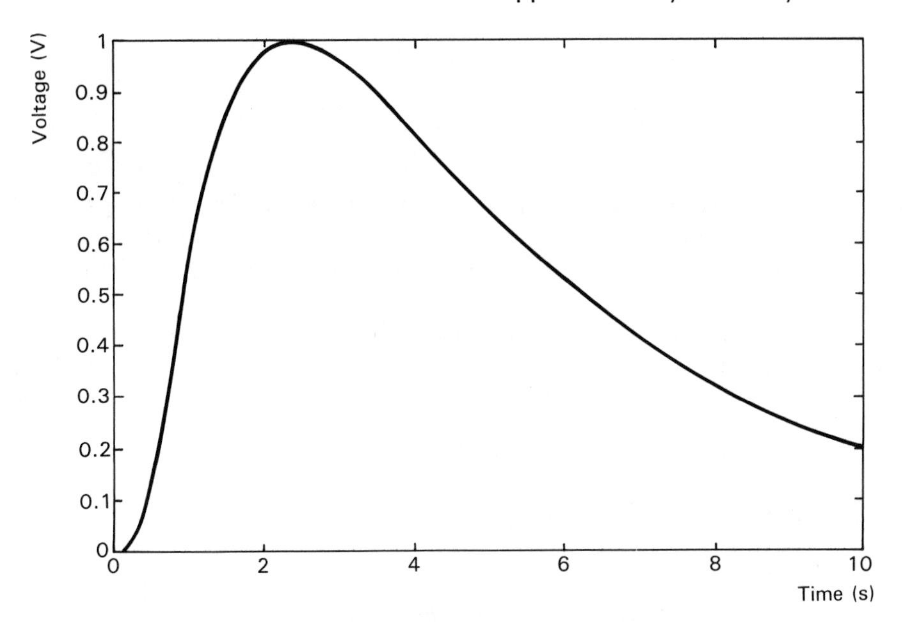

Figure 7.8 Response of circuit in Example 7.3.1 to half sine wave pulse.

Incorporating these results into eqn (7.3.18) and then substituting into eqn (7.3.17) gives the following expression for $v_o(t)$

$$v_o(t) = 0 \qquad t<0$$
$$v_o(t) = 2.5\pi[0.1342e^{-0.25t} - 0.1227e^{-t} - 0.0115 \cos \pi t - 0.0284 \sin \pi t] \qquad 0 \leqslant t \leqslant 1$$
$$v_o(t) = 2.5\pi[0.3065e^{-0.25t} - 0.4562e^{-t}] \qquad 1<t$$

This response is plotted in Figure 7.8.

7.3.2 The system transfer function

Returning to the differential equation of the system

$$a_n \frac{d^n y}{dt^n} + a_{n-1} \frac{d^{n-1} y}{dt^{n-1}} + \ldots\ldots a_0 y = b_m \frac{d^m x}{dt^m} + \ldots\ldots b_0 x$$

If the Laplace transform is taken of all the terms in this equation and *all the initial conditions are assumed to be zero* then the following equation is obtained.

$$(a_n s^n + a_{n-1} s^{n-1} + \cdots\cdots + a_0)Y(s) = (b_m s^m + b_{m-1} s^{m-1} + \ldots\ldots b_0)X(s)$$

Rearranging gives

$$\frac{Y(s)}{X(s)} = \frac{b_m s^m + b_{m-1} s^{m-1} + \ldots\ldots b_0}{a_n s^n + a_{n-1} s^{n-1} + \ldots\ldots a_0} \tag{7.3.19}$$

This ratio

$$\frac{\text{Laplace transform of output signal}}{\text{Laplace transform of input signal}}$$

with all initial conditions zero is known as the transfer function of the system. This function depends upon the coefficients in differential equations and hence on the parameters of the system. It does not depend on the input signal or the initial energy storage.

The expression for the transfer function $Y(s)/X(s)$ is that of the frequency response function $Y(j\omega)/X(j\omega)$ (see Section 5.2.1) with $j\omega$ replaced by s. As with the frequency response function it is not necessary to obtain the differential equation

$i(t)$ R $v(t)$

$$v(t) = i(t)R$$

$I(s)$ R $V(s)$

$$V(s) = I(s)R$$

$$Z(s) = \frac{V(s)}{I(s)} = R$$

$i(t)$ L $v(t)$

$$v(t) = L\frac{di(t)}{dt}$$

$I(s)$ $Z(s)$ $Li(0)$ $V(s)$

$$V(s) = LsI(s) - Li(0)$$

For zero energy storage

$$Z(s) = \frac{V(s)}{I(s)} = sL$$

$i(t)$ C $v(t)$

$$i = C\frac{dv(t)}{dt}$$

$I(s)$ $Z(s)$ $\frac{1}{S}v(0)$ $V(s)$

$$I(s) = CsV(s) - Cv(0)$$

$$V(s) = \frac{1}{sC}I(s) + \frac{1}{s}v(0)$$

For zero energy storage

$$Z(s) = \frac{V(s)}{I(s)} = \frac{1}{sC}$$

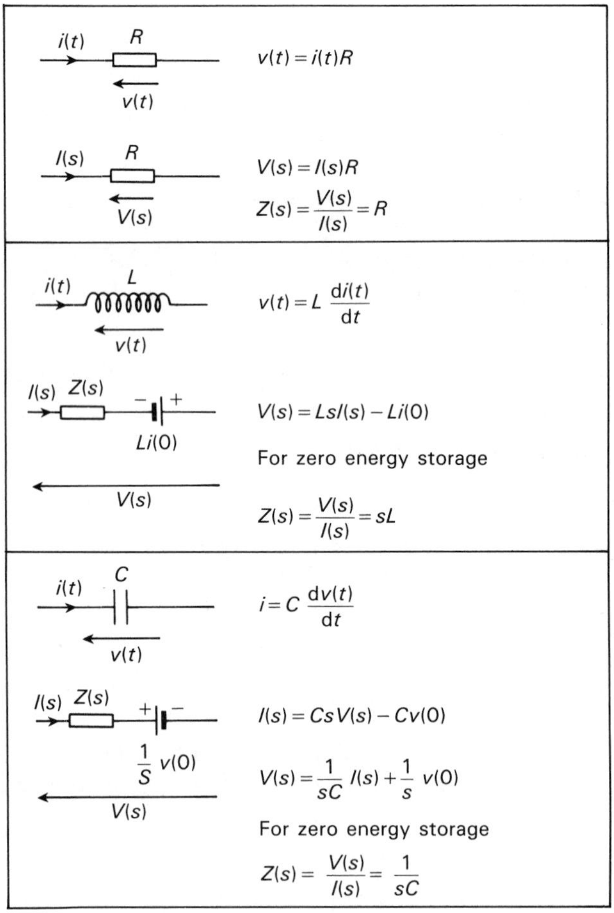

Figure 7.9 Operational impedance functions for electrical components.

of the system in order to obtain the function. With frequency response the concept of impedance is used, this can be extended by using the concept of *operational impedance* when using complex frequency s. Unlike the frequency domain it is possible to take into account any initial energy storage in the element at this stage. Figure 7.9 develops this idea where $Z(s)$ denotes operational impedance.

It should be noted that alternative forms of giving the initial conditions are possible and the student should consult a text book on circuit analysis for details.

Because of the similarity between the frequency response function and the transfer function, using Laplace transform methods is often termed, working in the s or complex frequency domain.

Using these concepts the last example can be approached as follows, referring to Figure 7.10.

Figure 7.10(a) shows the circuit with no initial conditions present. Marking the voltage $V_x(s)$ then nodal analysis gives

$$\frac{V_i(s) - V_x(s)}{R_1} = V_x(s)sC_1 + (V_x(s) - V_o(s))/R_2 \tag{7.3.20}$$

$V_o(s)$ and $V_x(s)$ are related by a potential divider with top arm R_2 and bottom arm $1/sC_2$

$$\frac{V_o(s)}{V_x(s)} = \frac{1/sC_2}{R_2 + 1/sC_2} \tag{7.3.21}$$

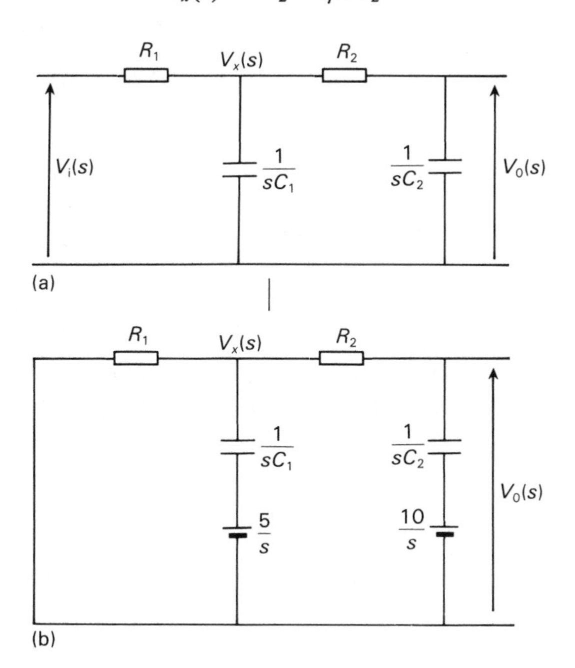

Figure 7.10 Circuit of Figure 7.5; use of operational impedance methods.

Substituting $V_x(s)$ from eqn (7.3.20) and rearranging gives the required transfer function.

$$\frac{V_o(s)}{V_i(s)} = \frac{1}{C_1R_1C_2R_2s^2 + (C_1R_1 + C_2R_2 + C_2R_1)s + 1}$$

Multiplying through by $V_i(s)$ and inserting component values from Example 7.3.1 gives the first term of eqn (7.3.6).

With the initial conditions present the circuit is as Figure 7.10(b). Again applying nodal analysis

$$\frac{-V_x(s)}{R_1} + \left[\frac{5}{s} - V_x(s)\right]sC_1 = \frac{V_x - V_o}{R_2} \qquad (7.3.22)$$

$$\frac{V_x(s) - V_o(s)}{R_2} + \left[\frac{10}{s} - V_o(s)\right]sC_2 = 0 \qquad (7.3.23)$$

It is left as an exercise for the student to verify that solution of eqns (7.3.22) and (7.3.23) will produce the time response obtained in Example 7.3.1 (b).

7.3.3 Step response of first and second order systems

In general the form of the transfer functions produced by continuous systems is

$$\frac{Y(s)}{X(s)} = \frac{b_m s^m + b_{m-1}s^{m-1} + \cdots\cdots + b_0}{a_n s^n + a_{n-1}s^{n-1} + \cdots\cdots + a_0} = \frac{B(s)}{A(s)}$$

where $B(s)$ and $A(s)$ represent polynomials in s. In physical systems $m \leqslant n$, the order of the numerator is less than or equal to the order of the denominator. The Laplace transform of the output signal is given by

$$Y(s) = \frac{B(s)}{A(s)}X(s)$$

In order to obtain the output time response this expression for $Y(s)$ is represented as the sum of standard terms using a partial fraction expansion. These terms can be first order terms

$$Y(s) = \frac{A}{s+a} + \frac{B}{s+b} + \cdots\cdots$$

or second order terms that cannot be factorised into *real* first order terms.

$$Y(s) = \frac{A_1s + B_1}{a_1s^2 + b_1s + c_1} + \frac{A_2s + B_2}{a_2s^2 + b_2s + c_2} + \cdots\cdots$$

or a combination of first and second order terms.

A complication may arise in that some of the terms in the expression for $Y(s)$ may be repeated, for example

$$Y(s) = \frac{B(s)}{(\ldots\ldots)(s+a)^3(a_1 s^2 + b_1 s + c_1)^2}$$

Leaving aside this complication for the moment, the total system response will consist of components due to first and second order terms only. It is therefore useful to be able to have the response of these terms to a standard input. Although the easiest input to deal with analytically is the unit impulse (it has a Laplace transform of unity) this is not very practical as a test signal. A more useful practical signal is the unit step (Laplace transform $1/s$) and the response of a first and second order system to this input will now be obtained. These transfer functions of the systems are expressed in the standard forms obtained from the differential equations given in Section 4.6.

First order response

The standard form of the differential equation describing the first order system is

$$T\frac{\mathrm{d}y}{\mathrm{d}t} + y = Kx$$

giving a system transfer function

$$\frac{Y(s)}{X(s)} = \frac{K}{1+sT}$$

As the input is unit step $X(s) = 1/s$

$$\begin{aligned} Y(s) &= \frac{K}{s(1+sT)} \\ &= K\left(\frac{1}{s} - \frac{T}{1+sT}\right) \\ &= K\left(\frac{1}{s} - \frac{1}{1/T+s}\right) \end{aligned}$$

Both of these terms are standard forms and inverting gives the output time response $y(t)$ as

$$y(t) = K(1 - \mathrm{e}^{-t/T}) \tag{7.3.24}$$

Second order response

The standard form of the differential equation describing a second order system is

$$\frac{\mathrm{d}^2 y}{\mathrm{d}t^2} + 2\zeta\omega_n\frac{\mathrm{d}y}{\mathrm{d}t} + \omega_n^2 y = \omega_n^2 x$$

where ω_n is the undamped natural frequency and ζ is the damping factor.

This gives a transfer function

$$\frac{Y(s)}{X(s)} = \frac{\omega_n^2}{s^2 + 2\zeta\omega_n s + \omega_n^2}$$

In order to obtain the partial fraction expansion the denominators will have to be factorised and this requires the roots of the equation

$$s^2 + 2\zeta\omega_n s + \omega_n^2 = 0$$

These are given by

$$s = \frac{-2\zeta\omega_n \pm \surd(4\zeta^2\omega_n^2 - 4\omega_n^2)}{2}$$

$$= -\omega_n\zeta \pm \omega_n\surd(\zeta^2 - 1)$$

The form of these roots will depend upon the value of ζ

$\zeta > 1$	two real unequal roots
$\zeta = 1$	a double real root
$\zeta < 1$	unequal complex roots

Each of these cases will be considered in turn.

Case 1

$\zeta > 1$ Overdamped response

Writing the roots as

$$s_1 = -\omega_n\zeta + \omega_n\surd(\zeta^2 - 1)$$

and

$$s_2 = -\omega_n\zeta - \omega_n\surd(\zeta^2 - 1)$$

gives an expression for $Y(s)$ when the input is a unit step, $X(s) = 1/s$, as

$$Y(s) = \frac{\omega_n^2}{s(s - s_1)(s - s_2)}$$

Using partial fractions this can be written

$$Y(s) = \omega_n^2\left[\frac{1}{s_1 s_2 s} + \frac{1}{s_1(s_1 - s_2)(s - s_1)} + \frac{1}{s_2(s_2 - s_1)(s - s_2)}\right]$$

and using the relationship $\omega_n^2 = s_1 s_2$ (product of roots) this can be inverted to give

$$y(t) = \left[1 + \frac{s_2}{(s_1 - s_2)}\,\mathrm{e}^{s_1 t} + \frac{s_1}{(s_2 - s_1)}\,\mathrm{e}^{s_2 t}\right] \qquad (7.3.25)$$

Although s_1, and s_2 can be expressed in terms of ω_n and ζ at this point, the resulting expression is quite involved. It is usually easier to substitute numerical values for s_1 and s_2 into eqn (7.3.25) when solving a particular example.

An alternative approach is to split the transfer function $Y(s)/X(s)$ into *two* terms by partial fractions. The standard form of the step response of each of these first order terms is then given by eqn (7.3.24). Combinations of these terms gives the required second order step response.

Case 2

$\zeta = 1$ Critical response

With $\zeta = 1$ the transfer function can be written

$$\frac{Y(s)}{X(s)} = \frac{\omega_n^2}{s^2 + 2\omega_n s + \omega_n^2} = \frac{\omega_n^2}{(s+\omega_n)^2}$$

giving the output transform for a step input as

$$Y(s) = \frac{\omega_n^2}{s(s+\omega_n)^2}$$

In the partial fraction expansion, for each individual term the degree of the numerator must be less than that of the denominator, this gives two alternative forms for the partial fraction expansion.

$$\frac{\omega_n^2}{s(s+\omega_n)^2} = \frac{A}{s} + \frac{B + Cs}{(s+\omega_n)^2}$$

or

$$\frac{\omega_n^2}{s(s+\omega_n)^2} = \frac{A}{s} + \frac{B}{(s+\omega_n)} + \frac{C}{(s+\omega_n)^2}$$

Taking the latter form and multiplying through by $s(s+\omega_n)^2$ gives

$$\omega_n^2 = A(s+\omega_n)^2 + Bs(s+\omega_n) + Cs \tag{7.3.26}$$

The coefficients A, B and C can easily be obtained from eqn (7.3.26). Substituting $s = 0$ and $s = -\omega_n$ gives $A = 1$ and $C = -\omega_n$. Equating coefficients of the s term gives $A + B = 0$ hence $B = -1$. The expression for $Y(s)$ becomes

$$Y(s) = \frac{1}{s} - \frac{1}{s+\omega_n} - \frac{\omega_n}{(s+\omega_n)^2}$$

The first two terms on the right-hand side are standard transforms and can be inverted directly. The last term can be inverted by using the shift property of the transform

$$X(s) \overset{\text{LT}}{\longleftrightarrow} x(t)$$
$$X(s+a) \overset{\text{LT}}{\longleftrightarrow} e^{-at}x(t)$$

Hence

$$\frac{1}{s^2} \overset{\text{LT}}{\longleftrightarrow} t$$

$$\frac{1}{(s+\omega_n)^2} \overset{\text{LT}}{\longleftrightarrow} te^{-\omega_n t}$$

This gives the output response $y(t)$

$$y(t) = 1 - e^{-\omega_n t} - \omega_n te^{-\omega_n t} \qquad t \geqslant 0 \tag{7.3.27}$$

Case 3

$$\zeta < 1 \qquad \text{Underdamped response}$$

The equation for the output transform $Y(s)$ is

$$Y(s) = \frac{\omega_n^2}{s(s^2 + 2\zeta\omega_n s + \omega_n^2)} \tag{7.3.28}$$

As $\zeta < 1$ some of the factors in the denominator will be complex. The partial fraction expansion can still be carried out as in Case 1, however the coefficients obtained will be complex. After inversion, conjugate time terms will be obtained and these can be combined to give a real time response as expected.

An alternative approach is to split $Y(s)$ in eqn (7.3.28) into partial fractions using a first order factor and a quadratic factor.

$$Y(s) = \frac{A}{s} + \frac{Bs + C}{s^2 + 2\zeta\omega_n s + \omega_n^2}$$

It is left to the student to show $A = 1$, $B = -1$, $C = -2\zeta\omega_n$ giving

$$Y(s) = \frac{1}{s} - \frac{s + 2\zeta\omega_n}{s^2 + 2\zeta\omega_n s + \omega_n^2} \tag{7.3.29}$$

The second term in the right-hand side of eqn (7.3.29) can be reduced to standard form. This can be done by use of the shift property in conjunction with the standard transforms for cosine and sine functions.

$$\frac{s+2\zeta\omega_n}{s^2+2\zeta\omega_n s+\omega_n^2}=\frac{s+\omega_n\zeta}{(s+\omega_n\zeta)^2+\omega_n^2(1-\zeta^2)}+\frac{\omega_n\zeta}{(s+\omega_n\zeta)^2+\omega_n^2(1-\zeta^2)}$$

$$=\frac{s+\omega_n\zeta}{(s+\omega_n\zeta)^2+\omega_n^2(1-\zeta^2)}+\frac{\zeta}{\sqrt{(1-\zeta^2)}}\frac{\omega_n\sqrt{(1-\zeta^2)}}{(s+\omega_n\zeta)^2+\omega_n^2(1-\zeta^2)}$$

Inverting $Y(s)$ gives

$$y(t)=1-e^{-\omega_n\zeta t}\left[\cos\omega_n\sqrt{(1-\zeta^2)}\cdot t+\frac{\zeta}{\sqrt{(1-\zeta^2)}}\sin\omega_n\sqrt{(1-\zeta^2)}\cdot t\right] \quad (7.3.30)$$

Equation (7.3.30) can be written more compactly

$$y(t)=1-\frac{e^{-\omega_n\zeta t}}{\sqrt{(1-\zeta^2)}}[\sin(\omega_n\sqrt{(1-\zeta^2)}\cdot t+\varphi)] \quad (7.3.31)$$

where

$$\varphi=\tan^{-1}\frac{\sqrt{(1-\zeta^2)}}{\zeta}$$

The step response for a range of values of damping factors is shown in Figure 7.11. A scale of non-dimensional time, $\omega_n t$, has been used for convenience.

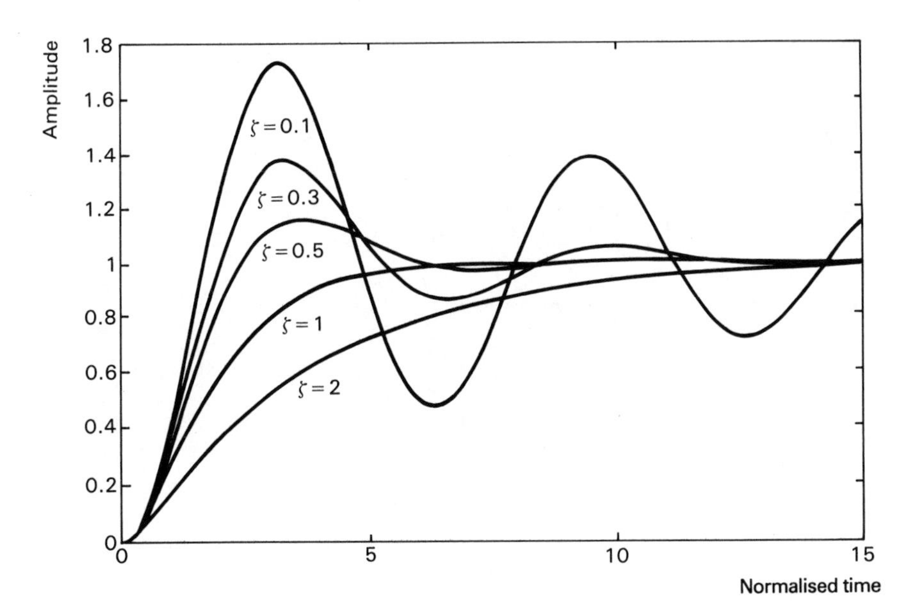

Figure 7.11 Step response of second order system.

As stated at the start of this section if the Laplace transform is given as a ratio of polynomials it can always be expanded as a combination of first and second order terms. However these terms will each have a weighting depending on the partial fraction expansion. This can make the responses of the terms differ from the step responses as shown in Figure 7.11. The following example will illustrate this point.

EXAMPLE 7.3.2

Figure 7.12 shows the construction of a typical moving-coil loudspeaker.

Current in the voice coil produces a force due to the interaction with the magnetic field. This force causes movement of the cone which is opposed by the stiffness of the suspension. Movement of the voice coil induces a back-e.m.f. which opposes the applied voltage. The following are the constants for a typical bass speaker:

R = resistance of voice coil = 8 Ω
L = inductance of voice coil = 1.0 mH
M = mass of cone and voice coil = 15 gm
λ = stiffness of suspension = 350 N/m
k = force produced for unit current = 5 N/A
(This is also equal to the back-e.m.f. produced per unit of velocity.)

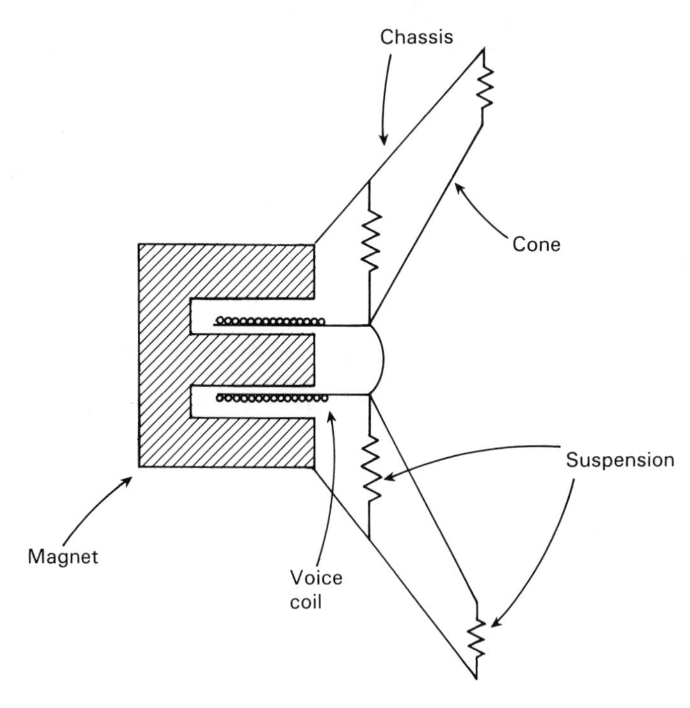

Figure 7.12 Loudspeaker construction for Example 7.3.2.

Determine the displacement of the cone following a 10 V step in applied voltage: (a) neglecting the inductance of the voice coil; (b) taking this into account.

SOLUTION

The current $I(s)$ in the voice coil due to applied voltage $V(s)$ is given by

$$I = \frac{V - E}{R + sL} \tag{7.3.32}$$

where E is the back-e.m.f.

$$E = ksX \tag{7.3.33}$$

The force on the coil is

$$F = kI \tag{7.3.34}$$

and the equation governing the mechanical motion is

$$F - \lambda X = Ms^2X \tag{7.3.35}$$

Combining eqns (7.3.32) to (7.3.35) gives the transfer function $X(s)/V(s)$ as

$$\frac{X}{V}(s) = \frac{k}{LM\left[s^3 + \frac{R}{L}s^2 + \frac{(L\lambda + k^2)}{LM}s + \frac{R\lambda}{LM}\right]} \tag{7.3.36}$$

(a) With $L = 0$ the transfer function becomes

$$\frac{X}{V}(s) = \frac{k}{MR\left[s^2 + \frac{k^2}{RM}s + \frac{\lambda}{M}\right]}$$

Inserting values

$$\frac{X}{V}(s) = \frac{41.66}{s^2 + 208.3s + 23333}$$

Comparing with the standard form

$$\frac{X}{V}(s) = K\frac{\omega_n^2}{s^2 + 2\zeta\omega_n s + \omega_n^2}$$

gives $\omega_n = \sqrt{23333} = 152.7$ rad/s (24.3 Hz)

$$2\zeta\omega_n = 208.3, \qquad \zeta = 0.68$$

$$K = 41.66/\omega_n^2 = 0.00178$$

Hence the step response can be written directly from eqn (7.3.31) noting that

$$\omega_n\sqrt{(1-\zeta^2)} = 112 \text{ rad/s}, \qquad \omega_n\zeta = 103.8$$

$$\tan^{-1}\frac{\sqrt{(1-\zeta^2)}}{\zeta} = 47.14°$$

The magnitude of the input step is 10 V and the equation describing the response is more conveniently written if the displacement is expressed in mm

$$x(t) = 17.8[1 - 1.364e^{-103t}\sin(112t + 47.14°)]$$

(b) Voice coil inductance included
The transfer function of eqn (7.3.36) now becomes

$$\frac{X}{V}(s) = \frac{0.333\times 10^6}{s^3 + 8\times 10^3 s^2 + 1.69\times 10^6 s + 0.186\times 10^9}$$

The denominator of this transfer function has to be factorised in order to obtain a partial fraction expansion. Equating the denominator to zero gives a cubic equation which cannot be solved analytically. A real root can be located by trial and error methods (for higher order equations recourse to a computer method of root finding may be necessary) and this is at $s = -7786$. This gives the transfer function with the denominator in factorised form as

$$\frac{X}{V} = \frac{0.333\times 10^6}{(s+7786)(s^2+214s+24\times 10^3)}$$

For the response to a 10 V input step $V(s) = 10/s$ and $X(s)$ can be expressed in partial fractions as

$$X(s) = 17.81\times 10^{-3}\left[\frac{1}{s} - \frac{0.4072\times 10^{-3}}{(s+7786)} - \frac{s+217}{s^2+214s+24\times 10^3}\right]$$

The first and second terms can be recognised as the Laplace transform of the step and exponential function. Inversion of the third term can be accomplished by re-writing the term as follows

$$\frac{s+217}{(s^2+214s+24\times 10^3)} = \frac{(s+107)}{(s+107)^2+(112)^2} + \frac{0.982\times 112}{(s+107)^2+(112)^2}$$

Inverting gives

$$\begin{aligned} x(t) &= 17.81[1 - 0.4072\times 10^{-3}e^{-7786t} - e^{-107t}(\cos 112t + 0.982\sin 112t)] \\ &= 17.81[1 - 0.4072\times 10^{-3}e^{-7786t} - 1.401e^{-107t}\cos(112t - 44.4°)] \end{aligned}$$

where the displacement is in mm.

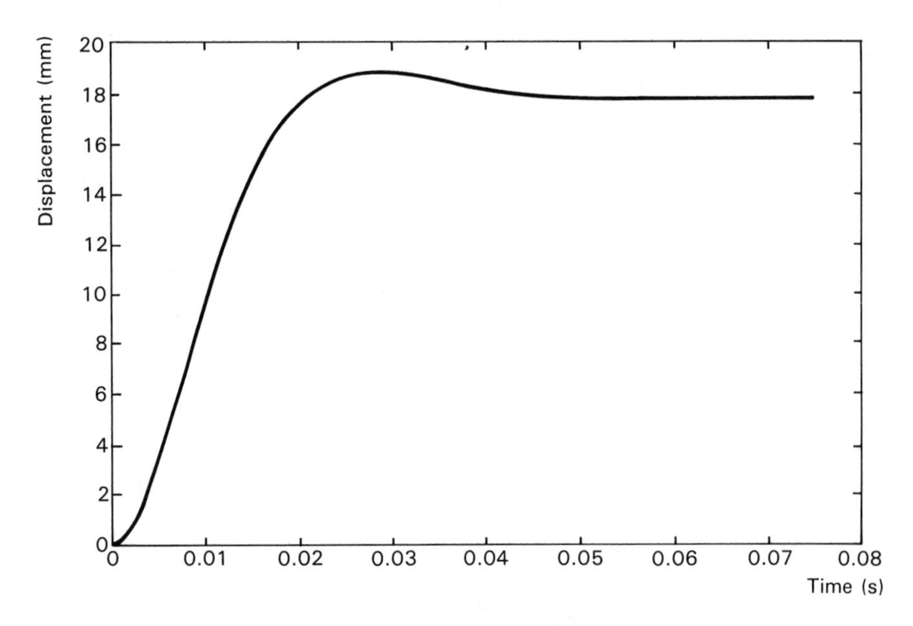

Figure 7.13 Loudspeaker response to step input.

If the responses, with and without voice coil inductance taken into account, were plotted they would be indistinguishable on the scales that could be used. Hence only one response is plotted in Figure 7.13 and it is applicable to either case.

7.4 Representation on the complex plane

In the introduction to this chapter it was stated that the Laplace transform was easier to invert than the Fourier transform when the time response of a system is required. So far this has been the application of the Laplace transform that has been investigated. However, like the Fourier transform considerable insight can be gained into system performance by direct interpretation of the transform without recourse to inversion. A helpful way of making this interpretation is via the complex plane and the poles and zeros of the transfer function.

7.4.1 Poles and zeros

The transfer function of a continuous system can be expressed in following form

$$G(s) = \frac{K(s^m + b_{m-1}s^{m-1} + \ldots\ldots b_0)}{s^n + a_{n-1}s^{n-1} + \ldots\ldots a_0} \tag{7.4.1}$$

(The coefficient of s^n can be reduced to unity by dividing through all the terms by a_n, the ratio b_m/a_n can then be taken out of all the numerator terms and written as K.)

In eqn (7.4.1), $G(s)$ is a function of s. If s were a real variable the relationship between $G(s)$ and s could be easily portrayed graphically. However s is complex and to interpret this transform graphically by plotting $G(s)$ against s would require four dimensions, two for the real and imaginary parts of s and two more for the real and imaginary parts of $G(s)$. The number of dimensions can be reduced to three by plotting only the magnitude of $G(s)$, $|G(s)|$ against s (a separate plot of $\angle G(s)$ could also be constructed). Take for example the transfer function

$$G(s) = \frac{(s+2)}{s^2+2s+2}$$

A representation of the plot of $|G(s)|$ against s is shown in Figure 7.14.

Even with the aid of computer graphics such a representation is not very convenient to produce and it is not very easy to interpret. However it does have some easily distinguishable features, there are two points where $|G(s)|$ is very large

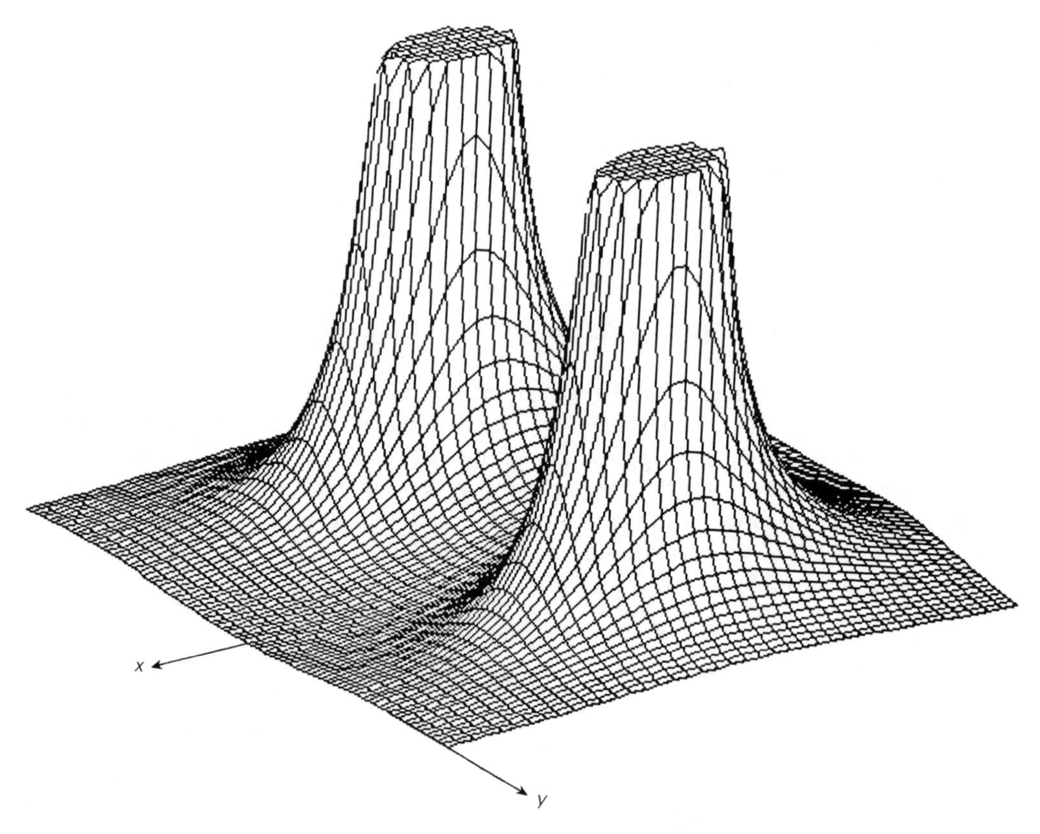

Figure 7.14 Three dimensional representation of the magnitude of a complex function.

(infinite in theory) and there is a point where the magnitude is zero. These points can be easily depicted in two dimensions as shown in Figure 7.15.

The axes are the real and imaginary parts of s (usually denoted by σ and $j\omega$) respectively and these axes define a point in the *complex plane*. The point of zero magnitude and those of infinite magnitude are shown by ⊙ and × respectively. The point of zero magnitude occurs when the numerator of the transfer function has zero value, i.e., when $s+2=0$, that is when $s=-2$, such a point is known as a *zero* of the function. The points of infinite magnitude occur when the denominator has zero value, that is when $s^2+2s+2=0$ when $s=-1\pm j1$, these points are known as the *poles* of the function.

More generally

$$G(s)=\frac{K(s^m+b_{m-1}s^{m-1}+\ldots\ldots b_0)}{s^n+a_{n-1}s^{n-1}+\ldots\ldots a_0)} \qquad (7.4.2)$$

can be written

$$G(s)=\frac{K(s-Z_1)(s-Z_2)\ldots\ldots(s-Z_m)}{(s-P_1)(s-P_2)\ldots\ldots(s-P_n)} \qquad (7.4.3)$$

$Z_1, Z_2, \ldots, Z_m$ are the roots of the equation

$$s^m+b_{m-1}s^{m-1}+\ldots\ldots b_0=0$$

and are the zeros of $G(s)$.

$P_1, P_2, \ldots\ldots P_n$ are the roots of the equation $s^n+a_{n-1}s^{n-1}+\ldots\ldots a_0=0$

and are the poles of $G(s)$.

The zeros and poles of a function can be real or complex but if complex they must occur in conjugate pairs (because the coefficients in the numerator and denominator polynomials are real). If the roots of the denominator polynomial are all distinct the poles are said to be simple poles, if a root is repeated r times then it is termed a multiple pole of order r. A similar terminology applies to the zeros.

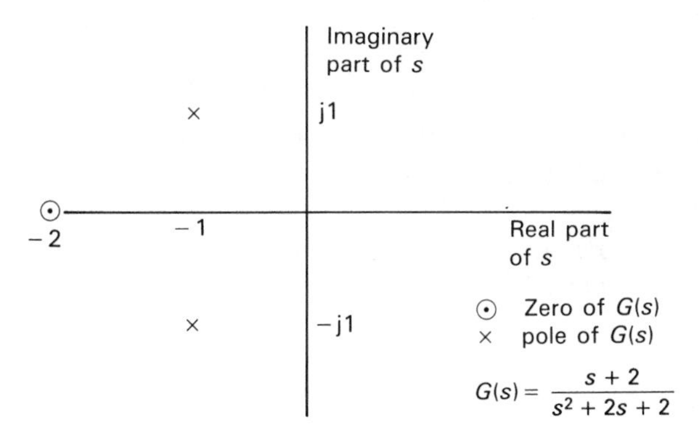

Figure 7.15 Depiction of poles and zeros on the complex plane.

EXAMPLE 7.4.1

Determine the poles and zeros of the following transfer functions

(a) $G(s) = \dfrac{20(s+1)}{s(s+2)(s+3)}$

(b) $G(s) = \dfrac{10(s+2)}{(s+1)^2(s^2+2s+3)}$

(c) $G(s) = \dfrac{4s+8}{2s^2+8s+6}$

SOLUTION

(a) Equating the numerator to zero and solving the resulting equation gives the zeros of the function

$$20(s+1) = 0 \qquad s = -1$$

there is a zero at $s = -1$.

Equating the denominator to zero and solving gives the poles of the function

$$s(s+2)(s+3) = 0 \qquad s = 0,\ s = -2,\ s = -3$$

there are poles at $s = 0$, $s = -2$, $s = -3$

(b) Proceeding as in (a) gives a zero at $s - 2$. The equation that results from setting the denominator to zero is

$$(s+1)^2(s^2+2s+3) = 0$$

This equation has solutions when

$$s^2+2s+2 = 0 \qquad s = -1 + j\sqrt{2},\ s = -1 - j\sqrt{2}$$

and when

$$(s+1)^2 = 0 \qquad s = -1$$

Hence there are complex poles at $s = -1 + j\sqrt{2}$ and $s = -1 - j\sqrt{2}$ (note complex conjugates) and a double pole at $s = -1$.

(c) There are two approaches that can be adopted here. If one proceeds as in (a) and (b) the zero occurs when

$$4s + 8 = 0 \qquad s = -2$$

and poles when

$$2s^2 + 8s + 6 = 0 \qquad s = -1 \text{ and } s = -3$$

Alternatively one can re-write the transfer function such that the highest power of s in numerator and denominator have coefficients of unity.

$$G(s) = \frac{4s+8}{2s^2+8s+6}$$

$$= \frac{4(s+2)}{2(s^2+4s+3)}$$

$$= \frac{2(s+2)}{s^2+4s+3}$$

It is easily confirmed that in this form there is a zero at $s = -2$ and poles at $s = -1$ and $s = -3$ as before. In order to locate the poles and zeros it does not matter which method is adopted. However the second method gives a transfer function that is of the form of eqn (7.4.3) and here the constant K can be determined. This is important when determining the system time response from the location of the poles and zeros as will be shown in Example 7.4.3.

7.4.2 Interpretation of system response via pole positions

So far this section has considered how the transfer function of a system can be characterised by its poles and zeros. We will now show how the positions of the poles in the complex plane indicate the system time response. The transfer function is the Laplace transform of the system impulse response. Hence to obtain this response the transfer function has to be expressed as a partial fraction expansion. Assuming all the poles are simple

$$G(s) = \frac{K(s^m + b_{m-1}s^{m-1} + \dots\dots b_0)}{(s-P_1)(s-P_2) - \dots\dots (s-P_n)}$$

$$= \frac{K_1}{s-P_1} + \frac{K_2}{s-P_2} + \dots\dots \frac{K_n}{s-P_n} \qquad (7.4.4)$$

where K_1, K_2 are constants determined by the partial fraction expansion.

Inverting eqn (7.4.4) gives the impulse response $g(t)$

$$g(t) = K_1 e^{P_1 t} + K_2 e^{P_2 t} + \dots\dots K_n e^{P_n t}$$

As previously stated in Section 4.5.1, terms of the form e^{Pt} are called the modes of the system. The form of these modes depends upon the position of the associated poles in the complex plane.

Considering first the case where the pole is on the real axis, $s = a$ (where a is real). This will give a mode e^{at} $(t \geqslant 0)$ and its form depends on whether the pole is to the left or right of the origin. To the left of the origin it represents an exponentially decaying response, to the right of the origin it represents a response that grows exponentially without limit. If the pole is at the origin it represents a response $e^{0t} = 1$, $(t \geqslant 0)$, a step response.

Considering now the case where the poles are complex, as stated earlier, these must occur in conjugate pairs. Also because the time response must be real the associated constants in the partial fraction expansion must also be complex conjugates. Hence the response $c(t)$ associated with a complex pole pair can be written

$$c(t) = Ke^{Pt} + K^{*}e^{P^{*}t}$$

Writing the pole position P in terms of its real and imaginary parts $P = \sigma + j\omega$ and writing K as $|K|e^{j\theta}$ where $\theta = \underline{/K}$ then the response becomes

$$\begin{aligned} c(t) &= |K|e^{j\theta}e^{(\sigma + j\omega)t} + |K|e^{-j\theta}e^{(\sigma - j\omega)t} \\ &= |K|e^{\sigma t}(e^{j(\omega t + \theta)} + e^{-j(\omega t + \theta)}) \\ &= 2|K|e^{\sigma t}\cos(\omega t + \theta) \end{aligned}$$

Thus these terms represent a sinusoidal response that either decays or grows without limit depending on the sign of σ (the real part of the pole position). If $\sigma = 0$ it represents a sinusoid.

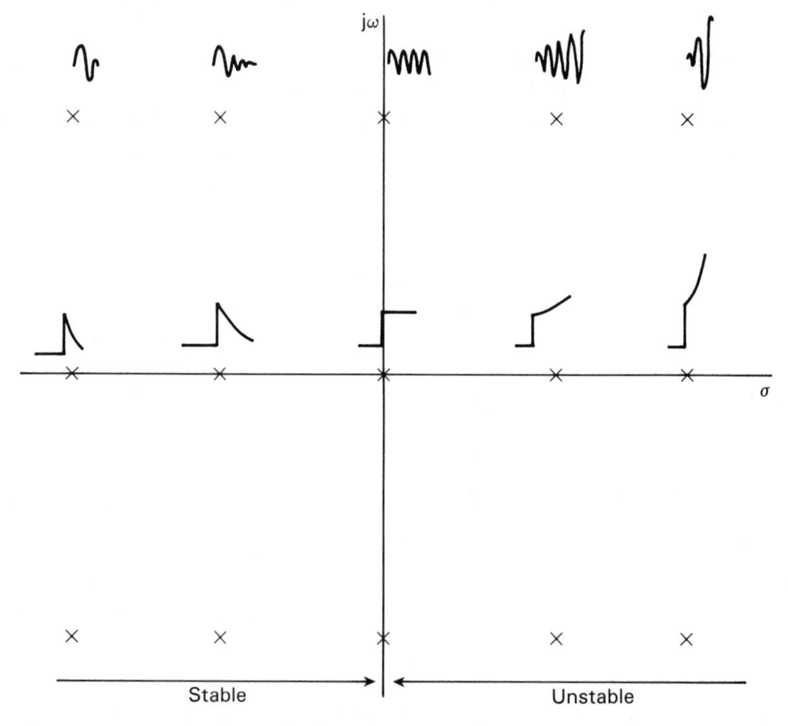

Figure 7.16 Responses associated with pole positions in the complex plane.

The responses associated with the position of the poles are shown in Figure 7.16. The response due to the complex poles is that due to the *pole pair* although for convenience it has been shown against the poles in the upper half plane only. One of the most important results that can be interpreted from these pole positions concerns stability. If the poles are in the left-hand half plane they represent an impulse response that decays to zero – a stable system. Poles in the right-hand half plane represent an impulse response that grows without limit – an unstable system. Poles on the imaginary axis give an impulse response that neither grows nor decays, this condition is referred to as marginal stability.

The closer the poles are to the imaginary axis the larger the time constant associated with the decay or growth of the mode. For poles on the real axis the reciprocal of the distance from the origin gives the time constant of the mode.

Earlier in the chapter the quadratic factor has been interpreted in terms of the undamped natural frequency ω_n and its damping factor ζ. These parameters can be easily obtained from the complex pole pairs representing this factor. Writing the factor $G(s)$ in standard form

$$G(s) = \frac{\omega_n^2}{s^2 + 2\zeta\omega_n s + \omega_n^2}$$

this has poles when

$$s^2 + 2\zeta\omega_n s + \omega_n^2 = 0$$

$$s = \frac{-2\zeta\omega_n \pm \surd(4\zeta^2\omega_n^2 - 4\omega_n^2)}{2}$$

$$= -\omega_n\zeta \pm j\omega_n\surd(1 - \zeta^2)$$

Hence the pole pair is located as shown in Figure 7.17.

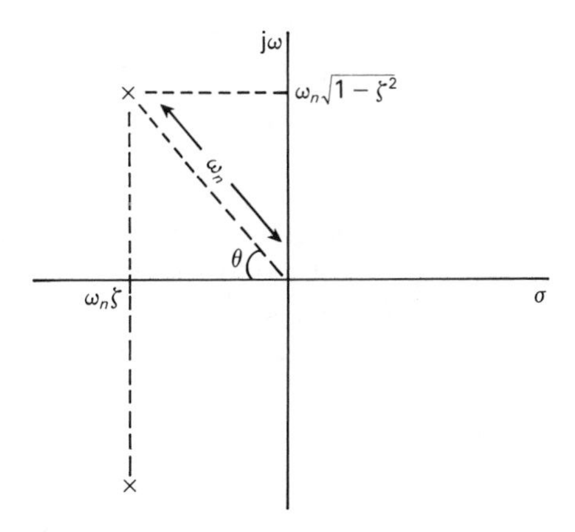

Figure 7.17 Interpretation of quadratic factor by pole positions.

The distance of either of the poles from the origin is

$$\sqrt{[(\omega_n\zeta)^2 + \omega_n^2(1 - \zeta^2)]} = \omega_n$$

The angle θ as shown is given by

$$\theta = \tan^{-1}\frac{\omega_n\zeta}{\omega_n} = \tan^{-1}\zeta$$

The imaginary part of the pole position, $\omega_n\sqrt{(1-\zeta^2)}$, is the damped frequency of oscillation ω_d and the real part of the pole position $\omega_n\zeta$ is reciprocal of the time constant associated with the decay of the transient.

7.4.3 Weighting of the modes

The student may wonder what part the zeros play in the response as the modes are determined by the pole positions only. However to obtain the complete response the modes are weighted by the coefficients in the partial fraction expansion. To obtain these coefficients both a knowledge of the poles and zeros of the functions are required. To understand how the pole zero positions can be used to determine this weighting it is first necessary to examine how the pole-zero positions can be used to evaluate the transfer function for any value of the complex variable s.

Consider the transfer function written in factorised form, eqn (7.4.3)

$$G(s) = \frac{K(s - Z_1)(s - Z_2)\ldots\ldots(s - Z_m)}{(s - P_1)(s - P_2)\ldots\ldots(s - P_n)}$$

and suppose it is required to evaluate $G(s)$ at point $s = x + \mathrm{j}y$. Taking just one of the factors $(s - Z_1)$ and writing Z_1 as, $Z_1 = x_1 + \mathrm{j}y_1$. Then

$$\begin{aligned} s - Z_1 &= x + \mathrm{j}y - x_1 - \mathrm{j}y_1 \\ &= (x - x_1) + \mathrm{j}(y - y_1) \\ |s - Z_1| &= \sqrt{[(x - x_1)^2 + (y - y_1)^2]} \\ \underline{/s - Z_1} &= \tan^{-1}\frac{(y - y_1)}{(x - x_1)} \end{aligned}$$

These operations are interpreted graphically in Figure 7.18. From this diagram it can be seen that $|(s - Z_1)|$ is L the length of the line joining s to Z_1 and $\underline{/(s - Z_1)}$ is the angle θ as shown. This process can be repeated for all the factors forming the numerator and denominator of the transfer function. Then

$$|(G(s)| = \frac{K|(s - Z_1)||(s - Z_2)|\ldots\ldots|(s - Z_m)|}{|(s - P_1)||(s - P_2)|\ldots\ldots|(s - P_n)|}$$

$$\begin{aligned} \underline{/G(s)} = {} & \underline{/(s - Z_1)} + \underline{/(s - Z_2)} + \cdots\cdots + \underline{/(s - Z_m)} \\ & - \underline{/(s - P_1)} + \underline{/(s - P_2)} + \cdots\cdots + \underline{/(s - P_n)} \end{aligned}$$

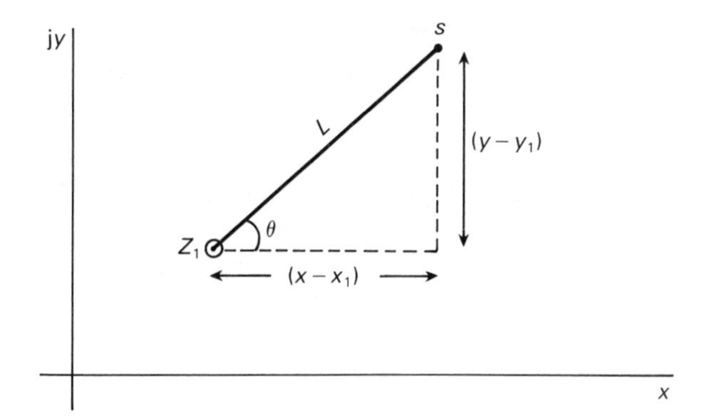

Figure 7.18 Interpretation of the factor $s - Z_1$.

Hence at any point s the magnitude of $G(s)$ (except for the constant K) can be determined as

$$\frac{\text{Product of distances from zeros}}{\text{Product of distances from poles}}$$

Knowledge of the pole zero positions does not allow evaluation of the constant K and this has to be determined separately.

The angle of $G(s)$ can be determined as

(Sum of the angles of the zeros) − (Sum of the angles from the poles).

This procedure is illustrated in the following example.

EXAMPLE 7.4.2

A system has a transfer function given by

$$G(s) = \frac{4s + 8}{2s^2 + 4s + 10}$$

From the pole-zero diagram determine the magnitude and angle of $G(s)$ at:
(a) $s = -1.5$; (b) $s = -0.5 + j2$.

SOLUTION

Before determining the positions of the poles and zeros the transfer function must be re-written in a form where the coefficients of the highest powers of s in both numerator and denominator are unity. If this is not done the constant K in the magnitude evaluation will not be correct (see Example 7.4.1)

$$G(s) = \frac{4(s + 2)}{2(s^2 + 2s + 5)}$$

$$= \frac{2(s + 2)}{s^2 + 2s + 5}$$

The positions of the poles and zeros are obtained by equating the numerator and denominator polynomials to zero and solving for s. This gives a zero at $s = -2$ and poles at $s = -1 + j2$ and $s = -1 - j2$.

(a) Evaluation at $s = -1.5$
The poles and zero are shown in Figure 7.19(a). Dotted lines are shown from these to the point $s = -1.5$. Direct measurement gives the distances from the poles and zero to the point $s = -1.5$. These are shown in the diagram, hence

$$|G(s)| = \frac{2 \times 0.5}{2.06 \times 2.06} = 0.235$$

The factor of 2 in the numerator appears because of the constant in the expression for $G(s)$, this factor cannot be determined from the pole-zero diagram.

The angle at the point $s = -1.5$ is

$$0° - (\theta - \theta) = 0°$$

The angle θ has not been measured as because of the symmetry of the diagram it will always partner its negative value due to the complex pole pair. As the angle from any

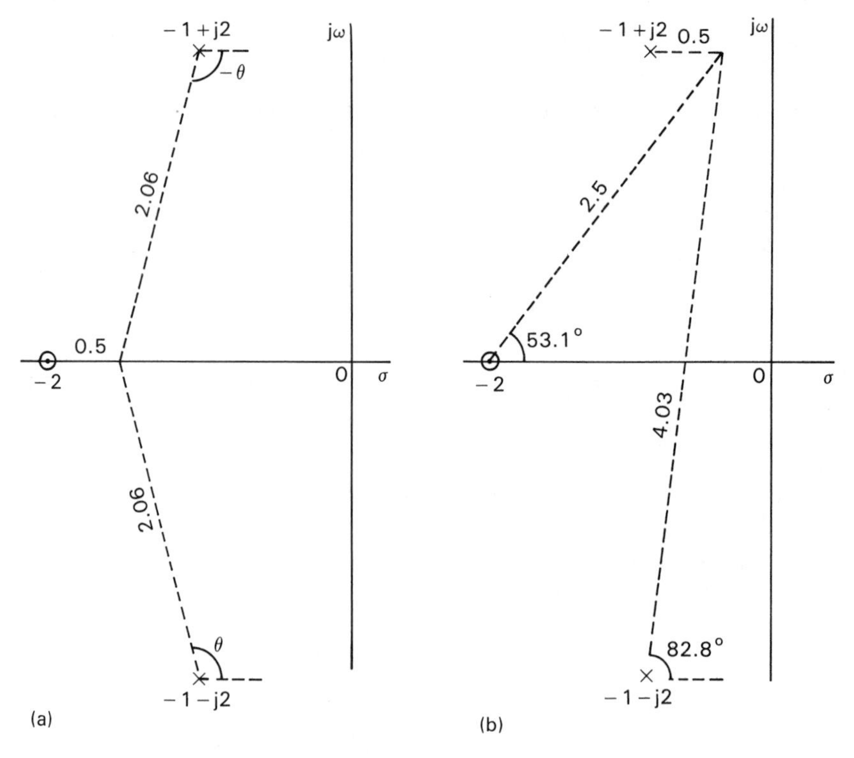

Figure 7.19 Pole zero diagram for Example 7.4.2.

poles or zeros on the real axis will always be 0° or 180°, then the angle at a point on the real axis will always be 0° or 180°.

(b) Evaluation at $s = -0.5 + j2$
Figure 7.19(b) shows this point with the distances and angles from the other poles and zeros marked.
Hence

$$|G(s)| = \frac{2 \times 2.5}{0.5 \times 4.03} = 2.48$$

$$\angle G(s) = 53.1° - (0° + 82.8) = -29.7°$$

Returning now to the evaluation of the weighting of the modes in the system time response. The partial fraction representation of $G(s)$, eqn (7.4.4) is

$$G(s) = \frac{K_1}{s - P_1} + \frac{K_2}{s - P_2} + \ldots\ldots \frac{K_n}{s - P_n}$$

the coefficients $K_1, K_2, \ldots, K_n$ give the weighting of the associated modes in the time response. To enable the coefficient K_1 to be determined each side of eqn (7.4.4) is multiplied by $(s - P_1)$ giving

$$(s - P_1)G(s) = K_1 + \frac{K_2(s - P_1)}{(s - P_2)} + \ldots\ldots \frac{K_n(s - P_1)}{(s - P_n)}$$

Evaluation at $s = P_1$ gives

$$K_1 = (s - P_1)G(s)\big|_{\text{evaluated at } s = P_1}$$

$$= \left.\frac{(s - P_1)K(s - Z_1)(z - Z_2)\ldots\ldots(s - Z_m)}{(s - P_1)(s - P_2)\ldots\ldots(s - P_n)}\right|_{\text{evaluated at } s = P_1}$$

Hence K_1 can be obtained by evaluating the transfer function $G(s)$ at the point $s = P_1$ on the complex plane *assuming the pole at this point was not there*. This can be done graphically at all the poles to obtain the weighting of the modes. It should be noted that this method is applicable to simple poles only.

To illustrate this method Example 7.3.2 will be reworked using pole-zero methods.

EXAMPLE 7.4.3

The loudspeaker of Example 7.3.2 has a transfer function between applied voltage and cone displacement given by

$$\frac{X}{V}(s) = \frac{0.333 \times 10^6}{s^3 + 8 \times 10^3 s^2 + 1.69 \times 10^6 s + 0.186 \times 10^9}$$

Use pole-zero methods to obtain the response of the speaker to an input step of 10 V magnitude.

SOLUTION

To determine the poles of the speaker transfer function the denominator has to be obtained in factored form. The real factor has already been obtained in Example 7.3.2 as $(s + 7786)$ and the roots of the quadratic factor can be obtained by solving the equation

$$s^2 + 214s + 24 \times 10^3 = 0$$

giving

$$s = -106 \pm j112$$

The response of the system to a 10 V step is required $V(s) = 10/s$. This has the effect of giving a pole at the origin in the transform $X(s)$ and including a factor of 10 in the numerator.

Hence $X(s)$ has poles at

$$s = 0$$
$$s = -7786$$
$$s = -106 \pm j112$$

there are no zeros and the constant in the numerator is 3.33×10^6. These poles are shown on the s plane in Figure 7.20. The associated modes are as follows.

Pole at $s = 0$	gives mode $e^{0t} = 1$
Pole at $s = -7786$	gives mode e^{-7786t}
Pole at $s = -107 + j112$ $= e^{-107t}e^{+j112t}$	gives mode $e^{(-107 + j112)t}$
Pole at $s = -107 - j112$ $= e^{-107t}e^{-j112t}$	gives mode $e^{(-107 - j112)t}$

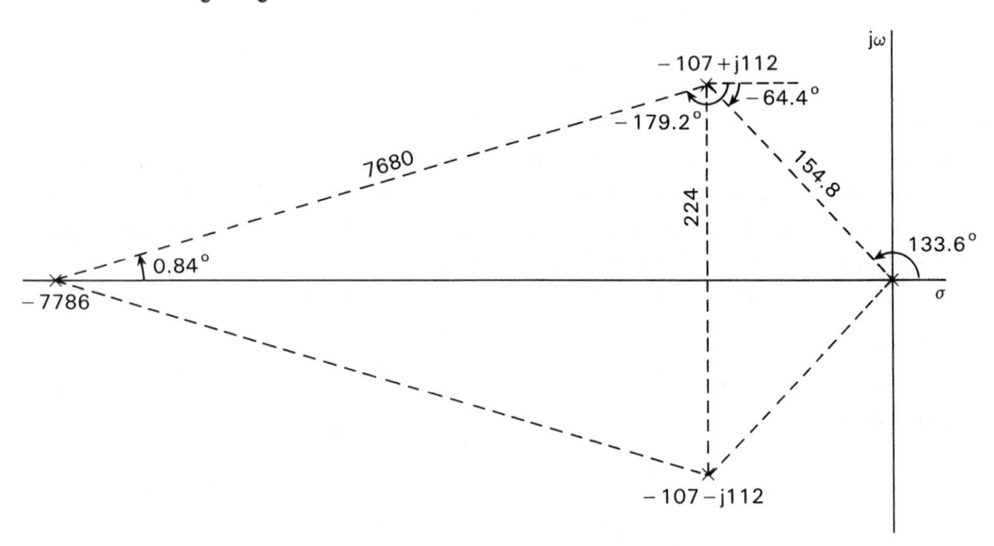

Figure 7.20 Pole positions for Example 7.4.3 (Note: Pole at $s = -7786$ is not shown to scale).

Response to an input step is given by

$$x(t) = K(K_1 + K_2 e^{-7786t} + K_3 e^{-107t} e^{j112t} + K_4 e^{-107t} e^{j112t})$$

With reference to Figure 7.20 the weighting K_1, K_2, K_3, K_4 can be obtained as

$$K_1 = \frac{1}{7786\underline{/0^\circ} \times 154.8\underline{/-66.4^\circ} \times 154.8\underline{/+66.4^\circ}}$$

$$= 0.00535 \times 10^{-6} \qquad \text{a real coefficient}$$

$$K_2 = \frac{1}{7786\underline{/180^\circ} \times 7680\underline{/-179.2^\circ} \times 7680\underline{/179.2^\circ}}$$

$$= -0.0021776 \times 10^{9} \qquad \text{a real coefficient but note the negative sign}$$

$$K_3 = \frac{1}{154.8\underline{/133.6^\circ} \times 7680\underline{/0.8^\circ} \times 224\underline{/90^\circ}}$$

$$= 0.00375 \times 10^{-6}\underline{/135.6^\circ} \qquad \text{a complex coefficient}$$

$$K_4 = \frac{1}{154.8\underline{/-133.6^\circ} \times 7860\underline{/-0.8^\circ} \times 224\underline{/-90^\circ}}$$

$$= 0.00375 \times 10^{-6}\underline{/-135.6^\circ} \qquad \text{the complex conjugate of } K_3$$

The constant K (which cannot be obtained from the pole zero diagram) is 3.33×10^9 if the displacement is in mm. Hence the response $x(t)$ is given by

$$\begin{aligned} x(t) = 3.33 \times 10^9 (&0.00535 \times 10^{-6} - 0.0021775 \times 10^{-9} e^{-7786t} \\ &+ 0.00375 \times 10^{-6} e^{-107} (1\underline{/135.6^\circ} e^{j112t} \\ &+ 1\underline{/-135.6^\circ} e^{-j112t})) \end{aligned}$$

The complex terms are in the form of a complex number plus its conjugate – this can be written as twice the real part giving

$$x(t) = 17.81(1 - 0.4070 \times 10^{-3}\, e^{-7786t} + 1.40 e^{-107t} \cos(112t + 135.6))$$

Using the relationship $\cos(\theta - 180^\circ) = -\cos\,\theta$ this is the same as the expression for the response obtained in Example 7.3.2.

In this example, because of the large relative distances from the origin to the real and complex poles direct measurements on the pole-zero diagram would be difficult. As shown in Example 7.3.2 the pole at $s = -7886$ has a very small weighting and could be ignored for most practical purposes.

7.4.4 Frequency response from pole-zero diagram

The frequency response function can be obtained by making the substitution $s = j\omega$ in the system transfer function. This can be interpreted on the pole-zero diagram as an evaluation of $G(s)$ along the imaginary axis, i.e., for points where $s = j\omega$. This will be investigated for the case of first and second order systems.

The transfer function for a first order system (ignoring the gain constant) can be written

$$G(s) = \frac{1}{1 + sT}$$

For interpretation on the pole-zero diagram the coefficient of the s term should be unity.

$$G(s) = \frac{1}{T(s + 1/T)}$$

There is a pole at $s = -1/T$ as shown in Figure 7.21(a). The magnitude of the response is given by $1/L$ where L is the distance from the pole to point s as shown. As ω increases, L also increases to give a response that falls monotonically as shown in Figure 7.21(b). The phase shift is given by φ and this angle increases from 0 to 90° as ω varies from 0 to ∞, this is shown in Figure 7.21(c).

From the pole-zero diagram it can be seen that

$$L = \sqrt{(\omega^2 + 1/T^2)}$$
$$\varphi = \tan^{-1} \omega T$$

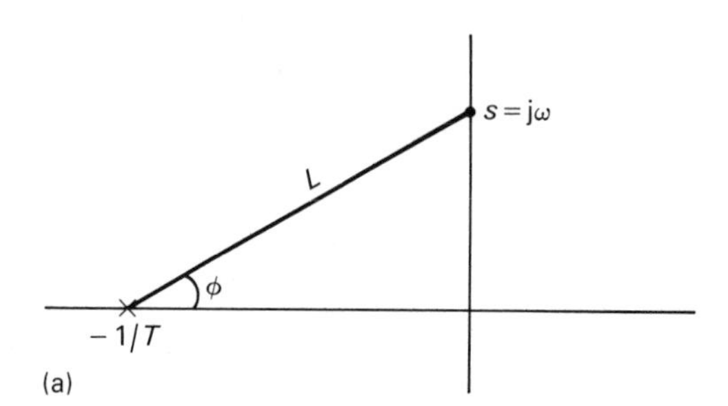

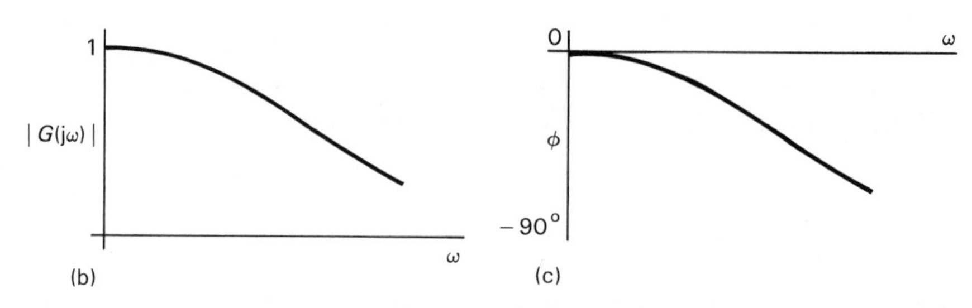

Figure 7.21 Frequency response of first order system from pole positions.

hence

$$|G(j\omega)| = \frac{1}{T}\frac{1}{L} = \frac{1}{\sqrt{(1+\omega^2T^2)}}$$

$$\underline{/G(j\omega)} = -\tan^{-1}\varphi$$

The constant $1/T$ must be included in the expression for magnitude, this constant cannot be obtained from the pole-zero diagram.

The transfer function of a second order system can be written

$$G(s) = \frac{\omega_n^2}{s^2 + 2\zeta\omega_n s + \omega_n^2}$$

this has poles at

$$s = -\omega_n\zeta \pm \omega_n\sqrt{(\zeta^2 - 1)} \qquad \text{for } \zeta > 1$$

and

$$s = -\omega_n\zeta \pm j\omega_n\sqrt{(1 - \zeta^2)} \qquad \text{for } \zeta < 1$$

These are shown in Figure 7.22 together with lines drawn from the poles to the point $s = j\omega$.

Taking first the case where $\zeta > 1$, shown in Figure 7.22(a) as ω increases both L_1 and L_2 increase monotonically. Also the angles φ_1 and φ_2 increase from 0 when $\omega = 0$ to $+90°$ as $\omega \rightarrow \infty$. Hence

$$|G(j\omega)| = \frac{1}{L_1L_2}$$ falls monotonically from unity at $\omega = 0$ to zero as $\omega \rightarrow \infty$

$$\underline{/G(j\omega)} = -(\varphi_1 + \varphi_2)$$ falls monotically from zero at $\omega = 0$ to $-180°$ as $\omega \rightarrow \infty$

The situation is more complicated when $\zeta < 1$, see Figure 7.22(b). Again L_1 increases monotonically as ω increases from $0 \rightarrow \infty$. However L_2 decreases as ω varies from zero to $\omega = \omega_n\sqrt{(1 - \zeta^2)}$ it then increases monotonically as ω varies from $\omega = \omega_n\sqrt{(1 - \zeta^2)}$ to infinity. The effect of this is to give a peak in $|G(j\omega)|$ in the vicinity of $\omega = \omega_n\sqrt{(1 - \zeta^2)}$. The smaller the damping factor ζ the larger this peak and the closer is the frequency at which the peak occurs to $\omega = \omega_n\sqrt{(1 - \zeta^2)}$, (which is equal to ω_n when $\zeta = 0$).

The phase shift $-(\varphi_1 + \varphi_2)$ decreases monotonically from zero to $-180°$ as ω varies $0 \rightarrow \infty$. However the angle φ_2 changes very rapidly with frequency in the vicinity of $\omega = \omega_n\sqrt{(1 - \zeta^2)}$ giving very rapid changes in phase shift. Plots for these cases have been shown in Chapter 5 and will not be repeated here.

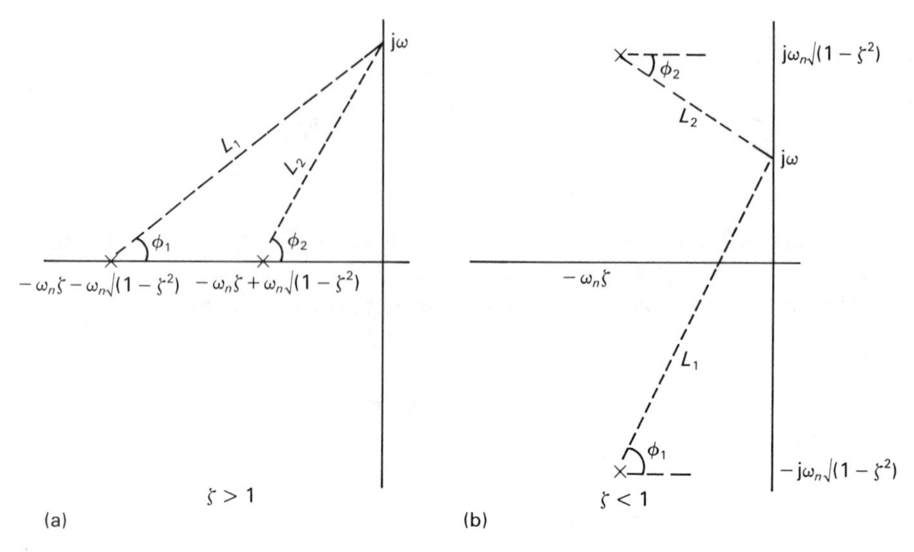

Figure 7.22 Frequency response of second order system from pole positions.

7.5 The z transform

As might be expected from the development of transform theory in this book there should be an equivalent of the Laplace transform that can be applied to discrete systems. The equivalent is the z transform. As with the discrete Fourier transform the z transform can be viewed as a mathematical operation that takes a set of points representing a time sequence and transforms them into a set of complex numbers. However a greater physical insight is obtained by approaching the z transform via the Laplace transform of a sampled signal.

7.5.1 Sampling and the z transform

Sampling is the process by which a continuous signal is converted into a discrete signal. Often this occurs naturally in a process, e.g., samples of blood are taken from a patient every fifteen minutes in a test to detect change in blood sugar levels. It is also performed deliberately, say in the conversion of a continuous signal into a discrete signal so that it can be represented in a digital computer (digitisation also occurs in amplitude level). In order to develop a mathematical model to represent the sampling action consider the simple switch shown in Figure 7.23. The switch closes periodically giving a sampling interval T and it stays closed for a time Δt which is assumed much smaller than T.

The result of applying a continuous signal $x(t)$ to such a switch would be to produce a series of narrow pulses. The width of the pulses would all be Δt but their

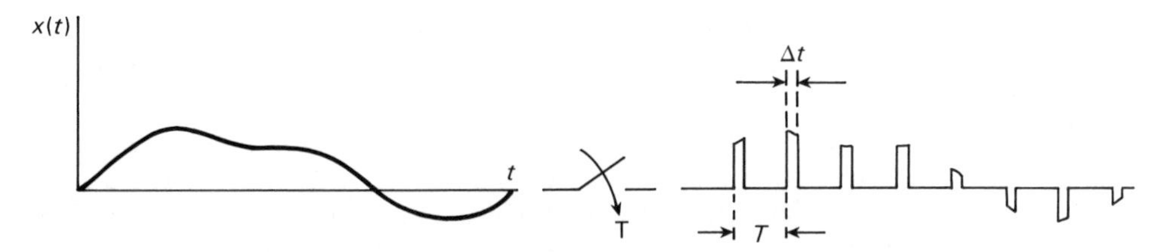

Figure 7.23 Illustration of the sampling action.

amplitude would vary from pulse to pulse and also within the duration of the individual pulses. A mathematical description of such a pulse train would be complex, a more useful description is to consider each of the short pulses as impulses having strength equal to $x(nT)\Delta t$ (the amplitude has been taken as a constant value $x(nT)$). Such an idealised sampling process is known as impulse sampling and denoting the resulting signal as $x^*(t)$ (not to be confused with complex conjugate), then

$$x^*(t) = \sum_{n=-\infty}^{+\infty} x(nT)\delta(t - nT)\Delta t$$

The Laplace transform $X^*(s)$ of this signal is

$$X^*(s) = \int_0^{\infty} \sum_{n=-\infty}^{+\infty} x(n)\delta(t - nT)\Delta t e^{-snT}\, dt$$

which by the sifting property of the impulse becomes

$$X^*(s) = \Delta t \sum_{n=0}^{\infty} x(n)e^{-snT} \tag{7.5.1}$$

Equation (7.5.1) can be simplified by making the substitution $z = e^{sT}$, or $z^{-1} = e^{-sT}$ giving

$$X^*(s) = \Delta t \sum_{n=0}^{\infty} x(n)z^{-n}$$

The portion of the expression

$$X(z) = \sum_{n=0}^{\infty} x(n)z^{-n} \tag{7.5.2}$$

is defined as the z transform of the sequence $x(n)$. (Strictly it is the one-sided z transform as it is possible to define the lower limit of the summation as $-\infty$ and produce a two-sided z transform.) The factor Δt (the width of the sampling pulse) has not been included in the definition of the transform. This is because of the following:

1. It can be regarded as a scaling factor and can be easily incorporated into the system following the sampler.

2. In practical systems a 'hold circuit' is often incorporated. This holds the signal at its sample value until the next sampling instant and this action does not depend upon Δt.
3. In many signal processing applications it is required to digitally process one set of stored points into another set having more desirable characteristics. There is no meaning to sampling pulse width in this application and the z transform as defined by eqn (7.5.2) forms a useful means of representing and processing the points.

Repeating the definition of the z transform

$$X(z) = \sum_{n=0}^{\infty} x(n) z^{-n}$$

The following notation will be used

$$X(z) \leftrightarrow x(n)$$
$$X(z) = Z\{x(n)\}$$
$$x(n) = Z^{-1}\{X(z)\}$$

7.5.2 Evaluation of the transform

Repeating the definition of the transform, given by eqn (7.5.2)

$$X(z) = \sum_{n=0}^{\infty} x(n) z^{-n}$$

it is evident that this is an infinite series

$$X(z) = x(0) + x(1)z^{-1} + x(2)z^{-2} + \ldots\ldots$$

If the signal $x(n)$ exists only for a finite number of points then the form of $X(z)$ is easily obtained.

z transform of the unit sample

For the unit sample

$$\begin{aligned} x(n) &= 1 \qquad n = 0 \\ &= 0 \qquad \text{otherwise} \end{aligned}$$

Hence, only the first term exists in eqn (7.5.2) and

$$X(z) = 1$$

z transform of the unit step

For the unit step

$$\begin{aligned} x(n) &= 1 \qquad n \geqslant 0 \\ &= 0 \qquad \text{otherwise} \end{aligned}$$

There are now an infinite number of terms in the series for $X(z)$.

$$X(z) = 1 + z^{-1} + z^{-2} + \ldots\ldots \qquad (7.5.3)$$

In order to express $X(z)$ in closed form the series of eqn (7.5.3) is recognised as a geometrical progression with the first term unity and ratio z^{-1}. The sum to N terms is given by

$$S_N = \frac{1 - z^{-N}}{1 - z^{-1}}$$

hence

$$X(z) = \lim_{N \to \infty} \frac{1 - z^{-N}}{1 - z^{-1}}$$

Whether or not this limit exists depends on the behaviour of z^{-N} as $N \to \infty$ and this in turn depends on the value of z. There is a range of z values, i.e, a region in the z plane, where

$$\lim_{N \to \infty} z^{-N} = 0$$

In this region the z transform exists and is given by

$$X(z) = \frac{1}{1 - z^{-1}}$$

$$= \frac{z}{z - 1}$$

Each z transform (as with the Laplace transform) is associated with a region of the complex plane where the transform exists. For greater detail the student is referred to more advanced texts on this subject.

A whole range of transforms can be obtained by similar calculation (and with the aid of some useful theorems). However, as with the Laplace transform, there is not usually the need to obtain the transform from first principles. Extensive transform tables exist and Table 7.3 lists the transforms for more common functions.

7.5.3 The shifting theorem of the z transform

A range of theorems exists concerning properties of the z transform, these in many cases are similar to the corresponding theorems of the Laplace transform. Such theorems are the ones concerned with linearity, convolution, scaling and other properties. Only one property will be derived here as it proves very useful when

Table 7.3 A short table of z transforms

Signal description	$x(n)$	$X(z)$
Unit sample	$\delta(n)$	1
Discrete unit step	$u(n)$	$\frac{z}{z-1}$
Discrete unit ramp	n	$\frac{z}{(z-1)^2}$
Discrete power series	a^n	$\frac{z}{z-a}$
Discrete exponential	e^{anT}	$\frac{z}{z-e^{aT}}$
Discrete sine wave	$\sin n\omega T$	$\frac{z \sin \omega T}{z^2 - 2z \cos \omega T + 1}$
Discrete cosine wave	$\cos n\omega T$	$\frac{z(z - \cos \omega T)}{z^2 - 2z \cos \omega T + 1}$

$X(z) = \sum_{n=0}^{\infty} x(n)\, z^{-n}$; all functions assumed zero for $n < 0$

applying the z transform to system analysis, this is the shift property of the z transform. Consider a time sequence $x(n)$ that has a z transform $X(z)$, this sequence is shown in Figure 7.24(a). The sequence $x(n)$ has been shown with values prior to $n = 0$, these will not affect the one-sided z transform as the limits of the summation are zero and infinity. Consider now the sequence delayed by k sample points, to form the sequence $x(n-k)$ shown in Figure 7.24(b) for $k = 2$. The z transform of the delayed sequence is

$$Z\{x(n-k)\} = \sum_{n=0}^{\infty} x(n-k)z^{-n} \tag{7.5.4}$$

Making the substitution $m = n - k$ into eqn (7.5.4)

$$\begin{aligned} Z\{x(n-k)\} &= \sum_{m=-k}^{\infty} x(m)z^{-(m+k)} \\ &= z^{-k} \sum_{m=-k}^{-1} x(m)z^{-m} + z^{-k} \sum_{m=0}^{\infty} x(m)z^{-m} \\ &= \text{initial condition} + z^{-k} X(z) \end{aligned}$$

where the initial condition contribution can be written

$$\text{initial condition} = x(-k) + z^{-1}x(-k+1) + \ldots\ldots\, z^{-k+1}x(-1)$$

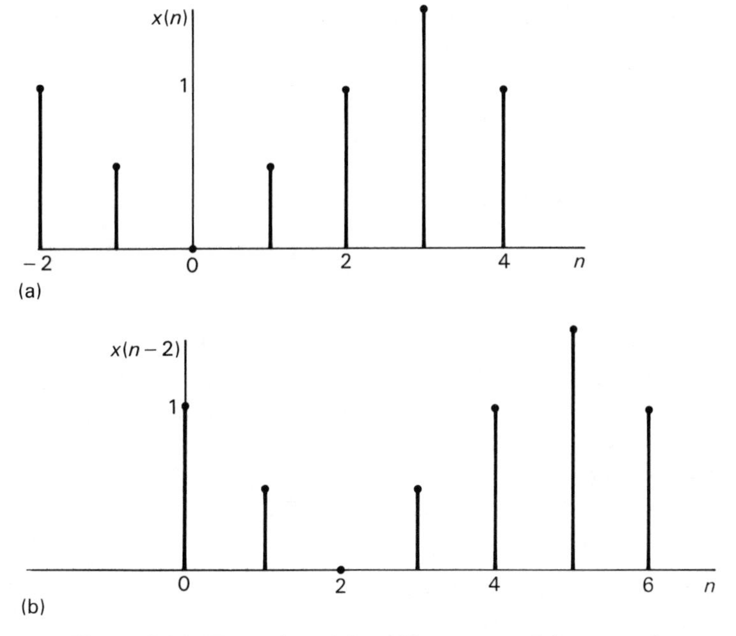

Figure 7.24 Illustration of the shift property of the z transform.

In the sequence shown in Figure 7.24, $x(-1)=1/2$, $x(-2)=1$, $k=2$ and the initial condition term becomes

$$1+\tfrac{1}{2}z^{-1}$$

If the signal $x(n)$ has the property $x(n)=0$ for $n<0$ then

$$Z\{x(n-k)\}=z^{-k}X(z)$$

This property of the z transform will be used in Section 7.5.5 for the solution of difference equations.

7.5.4 Inversion of the z transform

As with the Laplace transform the process of obtaining the time sequence $x(n)$ corresponding to a z transform $X(z)$ is known as inversion. The method used in the Laplace transform consisted of expressing the function via a partial fraction expansion as a sum of standard forms. Each term could then be inverted directly by reference to a table of transforms. This method can be used with a small modification to the inversion of the z transform. However alternative forms of inversion are also available with the z transform that have no parallel with methods used for inversion of the Laplace transform.

Inversion by partial fraction expansion

Consider the inversion of the z transform

$$X(z) = \frac{z}{(z-1)(z-0.5)}$$

This can be easily expressed by means of partial fractions as

$$X(z) = \frac{2}{z-1} - \frac{1}{z-0.5}$$

Although this is perfectly correct the two terms on the right-hand side do not appear in Table 7.3 as standard forms. They could be converted to standard form but this is not necessary if the following procedure is followed. Instead of expressing $X(z)$ in terms of partial fractions an expression for $X(z)/z$ is obtained

$$\frac{X(z)}{z} = \frac{1}{(z-1)(z-0.5)} = \frac{2}{z-1} - \frac{2}{z-0.5} \tag{7.5.5}$$

Then multiplying through eqn (7.5.5) by z gives

$$X(z) = \frac{2z}{z-1} - \frac{2z}{z-0.5} \tag{7.5.6}$$

Both terms on the right-hand side of eqn (7.5.6) are standard forms and the time sequence can be obtained as

$$x(n) = 2u(n) - 2(0.5)^n$$

giving the sequence

$$0, 1, 1.5, 1.75, 1.875, \ldots\ldots$$

Expansion as a power series

The definition of the z transform given by eqn (7.5.2) is

$$X(z) = \sum_{n=0}^{\infty} x(n)z^{-n} = x(0) + x(1)z^{-1} + x(2)z^{-2} + \ldots\ldots$$

This is a power series whose coefficients are the terms in the series for $x(n)$. If the z transform could be written out as such a series the values $x(n)$ could be obtained directly from the coefficients. The power series can be obtained by direct division of the numerator and denominator polynomials. Taking, as an example, the transform already inverted by the partial fraction expansion

$$X(z) = \frac{1}{z^2 - 1.5z + 0.5}$$

Arranging as a long division produces

$$
\begin{array}{rllll}
 & z^{-1} + 1.5z^{-2} + 1.75z^{-3} & & & \\
\hline
z^2 - 1.5z + 0.5) & z & & & \\
 & z & -1.5z^0 & +0.5z^{-1} & \\
\hline
 & & 1.5z^0 & -0.5z^{-1} & \\
 & & 1.5z^0 & -2.25z^{-1} & +0.75z^{-2} \\
\hline
 & & & 1.75z^{-1} & -0.75z^{-2} \\
 & & & 1.75z^{-1} & -2.625z^{-2} + 0.875z^{-2} \\
\hline
 & & & & 1.875z^{-2} - \ldots\ldots
\end{array}
$$

hence

$$\frac{z}{z^2 - 1.5z + 0.5} = 0z^0 + 1.0z^{-1} + 1.5z^{-2} + 1.75^{-3} + \ldots\ldots$$

and the coefficients, giving $x(n)$ are

$$x(n) = 0, 1, 1.5, 1.75, 1.875 \ldots\ldots$$

A third method of inverting the z transform is via the numerical solution of the difference equation it represents. This method will be considered as part of a more general framework in the next section.

7.5.5 Use of the z transform in the solution of difference equations

As introduced in Section 3.8.2 a linear discrete system can be described in general by the difference equation.

$$a_0 y(n) + a_1 y(n-1) + \ldots\ldots a_k y(n-k) = b_0 x(n) + b_1 x(n-1) + \ldots\ldots b_l x(n-l) \tag{7.5.7}$$

where $x(n)$ and $y(n)$ represent the input and output signals. In Section 3.9.1 it was shown that an algorithmic method could be used to obtain a numerical solution to such an equation. It will now be shown how the z transform can be used to obtain a closed form of solution.

Taking the z transform of all the terms in eqn (7.5.7) and using the delay theorem with the assumption that all the initial conditions are zero gives

$$a_0 Y(z) + a_1 z^{-1} Y(z) + \ldots\ldots a_k z^{-k} Y(z) = b_0 X(z) + b_1 z^{-1} X(z) + \ldots\ldots b_1 z^{-1} X(z)$$

This can be rearranged to give

$$\frac{Y(z)}{X(z)} = \frac{b_0 + b_1 z^{-1} + \ldots\ldots b_l z^{-l}}{a_0 + a_1 z^{-1} + \ldots\ldots a_k b^{-k}} \tag{7.5.8}$$

This ratio $Y(z)/X(z)$ is (compare with the Laplace transform for continuous systems) the transfer function of the discrete system. As with its counterpart for continuous systems the transfer function does not depend upon any specific input signal and is defined with all initial conditions as zero.
Writing

$$\frac{Y(z)}{X(z)} = G(z)$$

then

$$Y(z) = G(z)X(z)$$

Knowing the z transform of the input signal $X(z)$ the output time series can be obtained by inversion of $Y(z)$.

As the z transform of the unit sample is unity the transfer function can also be interpreted as the z transform of the system response to a unit pulse input. This property gives another method of inverting a z transform by considering the transform as the transfer function of a system with unit pulse input. The transfer function can be interpreted as a difference equation which can be solved numerically to obtain the required time series. As an illustration consider the z transform of a unit ramp; then

$$\begin{aligned} G(z) &= \frac{Tz}{(z-1)^2} \\ &= \frac{Tz^{-1}}{1 - 2z^{-1} + z^{-2}} \end{aligned}$$

This represents the difference equation

$$y(n) = Tx(n-1) + 2y(n-1) + y(n-2)$$

Taking the input as the unit sample

$$\begin{aligned} x(n) &= 1 \qquad n = 0 \\ &= 0 \qquad \text{otherwise} \end{aligned}$$

With zero initial conditions, $y(-1) = 0$ and $y(-2) = 0$ then numerical solution of this difference equation gives

$$y(n) = 0, T, 2T, 3T, 4T, \ldots\ldots$$

In order to illustrate the use of the z transform to obtain a closed form of solution an example will be given using difference equations introduced in Chapter 3.

EXAMPLE 7.5.1

In Chapter 3 the response of a continuous system described by the equation

$$T\frac{dv_o}{dt} + v_o = v_i$$

was approximated by a difference equation. Two approximations to differentiation were used leading to the difference equations

$$\text{(a)}\quad v_o(n) - \frac{T_1}{T+T_1}v_o(n-1) = \frac{T}{T+T_1}v_i(n) \tag{7.5.9}$$

$$\text{(b)}\quad v_o(n) - \left[\frac{2T_1-T}{2T_1+T}\right]v_o(n-1) = \frac{T}{2T_1+T}v_i(n) + \frac{T}{2T_1+T}v_i(n-1) \tag{7.5.10}$$

Obtain the solution to both these equations when the input is a 10 V step applied at $n = 0$ and the initial conditions are zero. Take the specific case where $T_1 = 1.0$, $T = 0.1$.

SOLUTION

(a) Here

$$\frac{T}{T+T_1} = 0.090909; \qquad \frac{T_1}{T+T_1} = 0.90909$$

and taking the z transform of eqn (7.5.9)

$$V_o(z) - 0.90909z^{-1}V_o(z) = 0.090909V_i(z)$$

As v_i consists of a 10 V step, $V_i(z) = 10z/(z-1)$ giving

$$\begin{aligned} V_o(z) &= \frac{0.090909}{1-0.90909z^{-1}}\frac{z}{z-1} \\ &= \frac{0.090909z^2}{(z-0.90909)(z-1)} \end{aligned}$$

Expressing $V_o(z)/z$ as partial fractions

$$\frac{V_o(z)}{z} = 0.090909\left[\frac{1}{(1-0.90909)}\frac{1}{(z-1)} - \frac{0.90909}{(1-0.90909)}\frac{1}{(z+0.90909)}\right]$$

$$V_o(z) = 10\left[\frac{z}{z-1} - 0.90909\frac{z}{z-0.90909}\right]$$

Inverting gives

$$v_o(n) = 10[1 - 0.90909(0.90909)^n]$$

(b)

$$\frac{2T_1 - T}{2T_1 + T} = 0.9047, \qquad \frac{T}{2T_1 + T} = 0.0476$$

Taking the z transforms of eqn (7.5.10) gives

$$V_o(z) - 0.9047z^{-1}V_o(z) = 0.0476V_i(z) + 0.0476z^{-1}V_i(z)$$

Again $V_i(z) = 10z/(z-1)$ giving

$$V_o(z) = \frac{0.476(z+1)z}{(z-0.9047)(z-1)}$$

$$= 0.476\left[\frac{21z}{z-1} - \frac{20}{z-0.9047}\right]$$

$$v_o(n) = 10[1 - 0.9524(0.9048)^n]$$

These expressions for $v_o(n)$ can be used to check the results in Tables 3.2 and 3.3 which were obtained numerically.

7.6 Pole-zero representation for discrete systems

In Section 7.4 it was shown how the transfer function of a continuous system could be interpreted in terms of the positions of its poles and zeros in the complex s plane. Many of the concepts introduced in this section apply equally well to a discrete system if the complex z plane is used instead of the s plane. However the interpretation of the system response, as given in Section 7.4.2 is very different in the z plane.

7.6.1 Interpretation of pole-zero diagram for discrete systems

The concept of the complex plane and poles and zeros is mathematical and much of the material of Section 7.4 will apply when the system is discrete and the plane under consideration is the z plane. The poles and zeros applicable are now those corresponding to the transfer function arising from the z transform.

The interpretation of the time response associated with a given pole-zero pattern will, however, be completely different when it is the z plane under consideration.

In factorised form the z transform can be written

$$G(z) = \frac{K(z-Z_1)(z-Z_2)\ldots\ldots(z-Z_m)}{(z-P_1)(z-P_2)\ldots\ldots(z-P_n)}$$

Taking the partial fraction expansion of $G(z)/z$ and multiplying the result by z gives $G(z)$ in the form

$$G(z) = \frac{K_1 z}{(z - P_1)} + \frac{K_2 z}{(z - P_2)} + \ldots\ldots \frac{K_n z}{(z - P_n)}$$

and inversion gives a time response

$$g(nT) = K_1(P_1)^n + K_2(P_2)^n + \ldots\ldots K_n(P_n)^n$$

This response applies to simple poles only. It has the form of the sum of weighted modes, each mode having the form $(P)^n$. The weighting of each mode is the coefficient in the partial fraction expansion of $G(z)/z$. Writing $(P)^n$ as $P(P)^{n-1}$ it can be seen that the response at the nth instant can be obtained from that at the $(n-1)$th instant by multiplying by P.

Considering first the case of real poles, $P = x$, then the responses have the following properties.

***x* positive**

$x < 1$	$x^n \to 0$	as $n \to \infty$
$x = 1$	$x^n = 1$	for all n
$x > 1$	$x^n \to \infty$	as $n \to \infty$

***x* negative**

$\lvert x \rvert < 1$	$x^n \to 0$ as $n \to \infty$, x^n alternates in sign
$\lvert x \rvert = 1$	$x^n = +1$ (n even), $x^n = -1$ (n odd)
$\lvert x \rvert > 1$	$x^n \to \infty$ as $n \to \infty$, x^n alternates in sign

For P complex the poles must occur in conjugate pairs with complex coefficients. This gives a response

$$x(n) = K(P)^n + K^*(P^*)^n$$

Letting $P = |P| \underline{/\theta}$ and $K = |K| \underline{/\varphi}$ then

$$x(n) = |K||P|^n \underline{/(n\theta + \varphi)} + |K||P|^n \underline{/-(n\theta + \varphi)} = 2|K||P|^n \cos(n\theta + \varphi)$$

Although $\cos(n\theta + \varphi)$ alternates in sign it is bounded by the values ± 1. The convergence or divergence of the sequence $x(n)$ is determined by $|P|^n$. If $|P| < 1$ then the sequence converges and the system is stable. Hence the stability boundary in the z plane is the unit circle. The form of the modes for different pole positions is shown in Figure 7.25.

The weighting of the modes can be found from the pole-zero diagram in a similar manner to that used for continuous systems. However the coefficients are those in the partial fraction expansion of $G(z)/z$ – this is equivalent to adding an additional pole at the origin to the pole-zero diagram $G(z)$ when calculating the weightings.

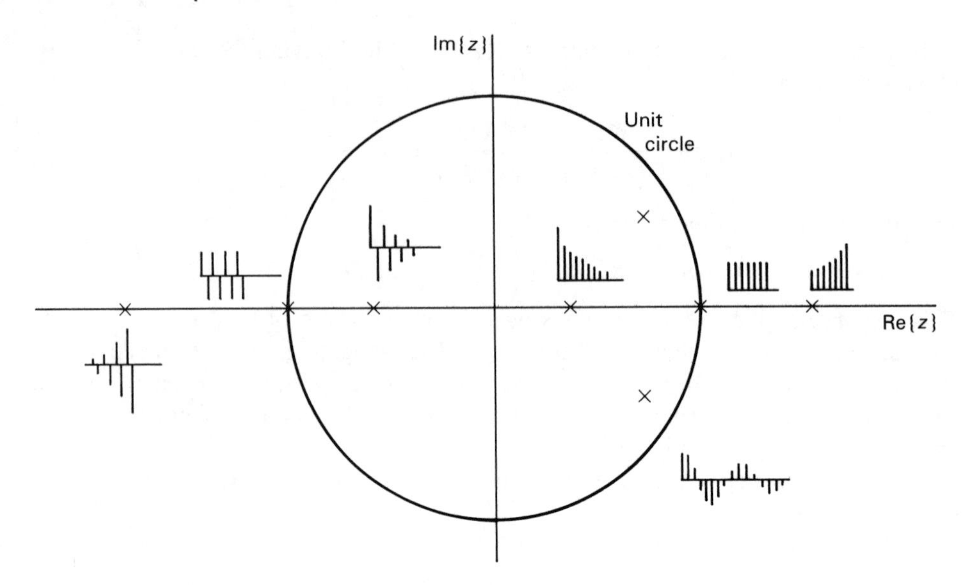

Figure 7.25 Form of response corresponding to pole positions in the z plane.

The frequency response of a discrete system can be obtained by substituting $z = e^{j\omega T}$ into the z transfer function. This is equivalent to evaluating $G(z)$ on the unit circle as shown in Figure 7.26.

From this interpretation of the frequency response function it is easy to observe the properties that were proved in Section 5.3.

The following example will illustrate the use of the complex plane to evaluate the response of a discrete system.

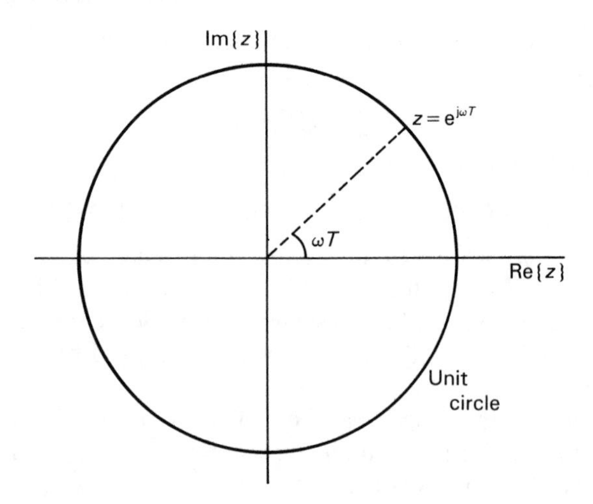

Figure 7.26 Frequency response of a discrete system obtained by evaluation around the unit circle in the z plane.

EXAMPLE 7.6.1

A discrete system has a z transfer function given by

$$G(z) = \frac{z}{(z+0.5)(z^2 - z + 0.61)}$$

(a) By use of the pole-zero diagram obtain an expression for the response of the system following a unit step applied at $n = 0$.
(b) Calculate graphically the magnitude and angle of the frequency response at a frequency equal to $1/12$ of the sampling frequency.

SOLUTION

The system transfer function $G(z)$ has the following:

A zero at $z = 0$.
A pole at $z = -0.5$.
A complex pole pair when $z^2 - z + 0.61 = 0$,
i.e., at $z = 0.5 \pm j0.6$.

(a) The z transform of a unit step is $z/(z-1)$ hence the output z transform $Y(z)$ is given by

$$Y(z) = \frac{z}{(z+0.5)(z^2 - z + 0.61)} \frac{z}{(z-1)}$$

The pole-zero diagram with relevant distances and angles is shown in Figure 7.27.

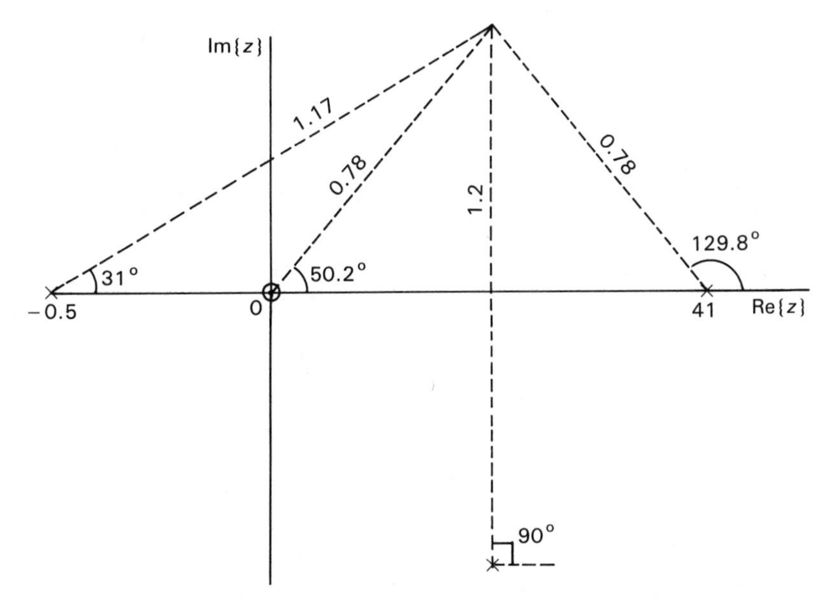

Figure 7.27 Pole-zero diagram for Example 7.6.1.

The modes are

$$(1)^n \text{ due to the pole at } z = +1$$
$$(-0.5)^n \text{ due to the pole at } z = -0.5$$
$$(0.78)^n\underline{/n50.2^\circ} \text{ and } (0.78)^n\underline{/-n50.2^\circ} \text{ due to the complex pole pair at}$$
$$z = 0.5 \pm j0.6$$

To obtain the weighting of the modes an additional pole is needed at $z = 0$. This has the effect of cancelling one of the zeros at the origin.

The weightings of the poles are as follows;

at $z = +1$,

$$\frac{1\underline{/0^\circ}}{1.5\underline{/0^\circ} \times 0.78\underline{/-50.2^\circ} \times 0.78\underline{/+50.2^\circ}} = 1.096\underline{/0^\circ}$$

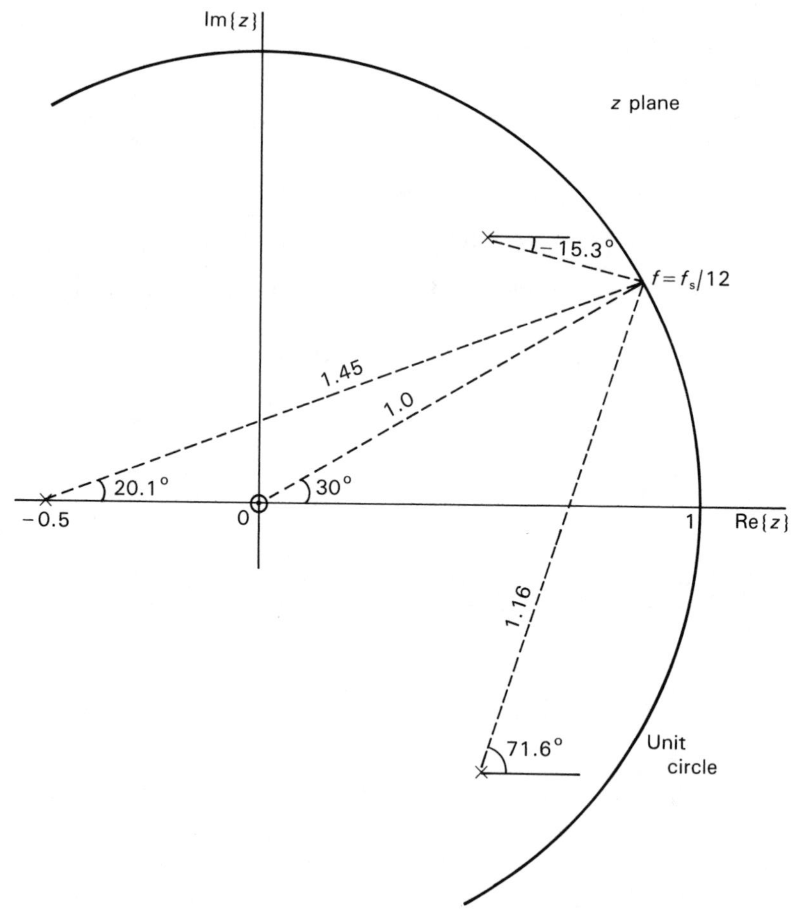

Figure 7.28 Evaluation of $G(z)$ at $z = 1\underline{/30^\circ}$.

at $z = -0.5$,

$$\frac{0.5\angle 180^\circ}{1.5\angle 180^\circ \times 1.17\angle -149^\circ \times 1.17\angle +149^\circ} = 0.244\angle 0^\circ$$

at $z = 0.5 + \text{j}0.6$,

$$\frac{0.78\angle 50.2^\circ}{0.78\angle 129.8^\circ \times 1.17\angle 31^\circ \times 1.2\angle 90^\circ} = 0.712\angle -159.4^\circ$$

at $z = 0.5 - \text{j}0.6$, $0.712\angle +159.4^\circ$ (conjugate of $0.712\angle -159.4^\circ$).
Hence the complete response is

$$y(n) = 1.096 + 0.244(-0.5)^n - 1.424(0.78)^n \cos(50.2n + 159.4^\circ)$$

The student should evaluate the response for values of n from 0 to 4 say and check the result against that obtained by one of the methods described earlier.

(b) When $f = f_s/12$, $\omega T = 2\pi/12$ which is an angle of 30°. Hence the point $z = \text{e}^{\text{j}\omega T}$ is $1\angle 30^\circ$ and is shown in Figure 7.28 with relevant magnitudes and angles indicated.

Frequency response at $f = f_s/12$ is

$$G(z)\big|_{z=\text{e}^{\text{j}\pi/6}} = \frac{1\angle 30^\circ}{1.45\angle 20.1^\circ \times 0.38\angle -15.3^\circ \times 1.16\angle 71.6^\circ}$$
$$= 1.56\angle -46.4^\circ$$

7.7 Summary

This chapter has introduced the Laplace transform by considering the inclusion of an exponential converging factor into the Fourier transform. This was done to extend the range of useful signals that possess transforms. Other benefits followed, the Laplace transform is simpler in form that the Fourier transform, its inversion is more easily accomplished and any initial conditions can be easily incorporated into the analysis.

The system transfer function, written in terms of the complex frequency s, forms a convenient method of characterising systems. This form of system description is useful if the transfer function is interpreted in terms of the position of its poles and zeros in the complex s plane. Distance and angle measurements on the plane enable the systems time and frequency response to be evaluated.

The z transform is the discrete equivalent of the Laplace transform. Like the Laplace transform it is very easy to use and it enables the time response of discrete systems to be easily evaluated. Again, interpreting the z transform via the position of its poles and zeros in the complex z plane enables system time and frequency responses to be evaluated graphically.

Problems

7.1 The following time signals are zero for $t < 0$. Determine their Laplace transforms.

(a) $e^{-t} + 2e^{-3t}$
(b) $t^2 + 3t + 2$
(c) $4 \sin 2t + 4 \cos 2t$
(d) $e^{-2t} \cos 4t$
(e) $\cos 2t + te^{-3t}$

7.2 Determine the Laplace transforms of the time signals shown in Figure 7.29.

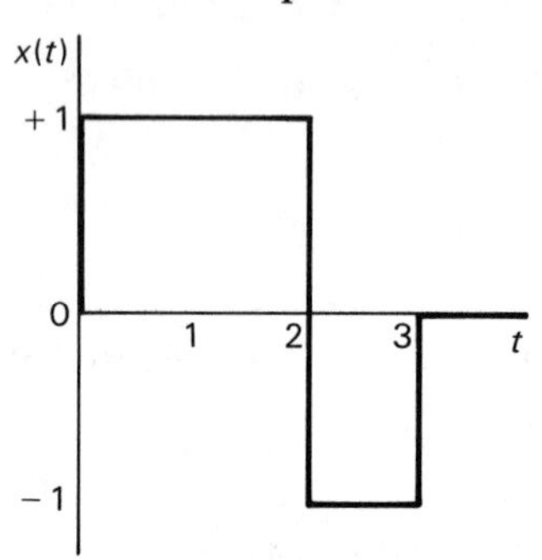

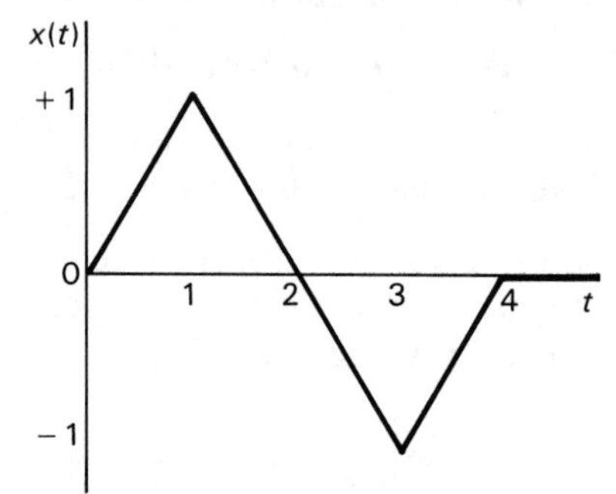

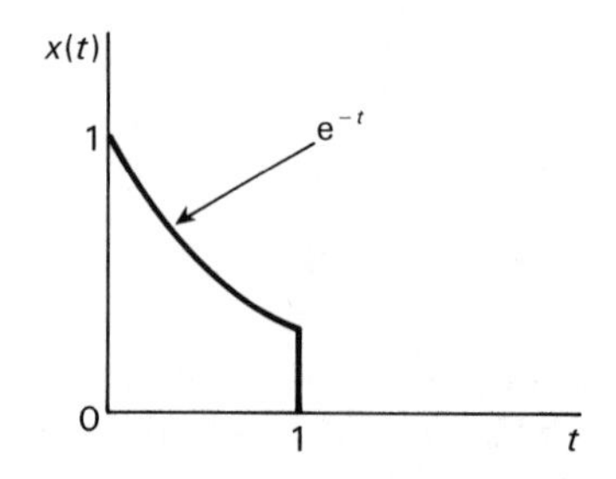

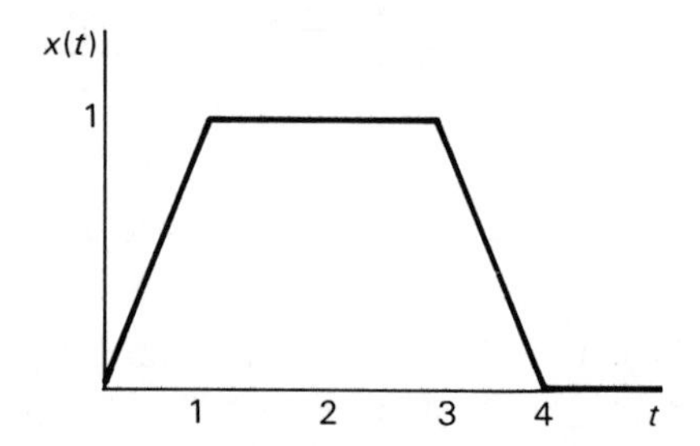

Figure 7.29

7.3 Show that the Laplace transform of a periodic function period T can be obtained by taking the Laplace transform over one period and dividing this by $(1 - e^{-sT})$.

Use this theorem to obtain the Laplace transforms of the periodic signals shown in Figure 7.30.

7.4 Determine the time functions corresponding to the following Laplace transforms

$$\frac{s}{(s+1)(s+2)}, \quad \frac{s^2+4s+7}{(s+1)^2(s+3)}$$

$$\frac{2(s+3)}{s^2+4}, \quad \frac{2(s+4)}{s^2+6s+13}$$

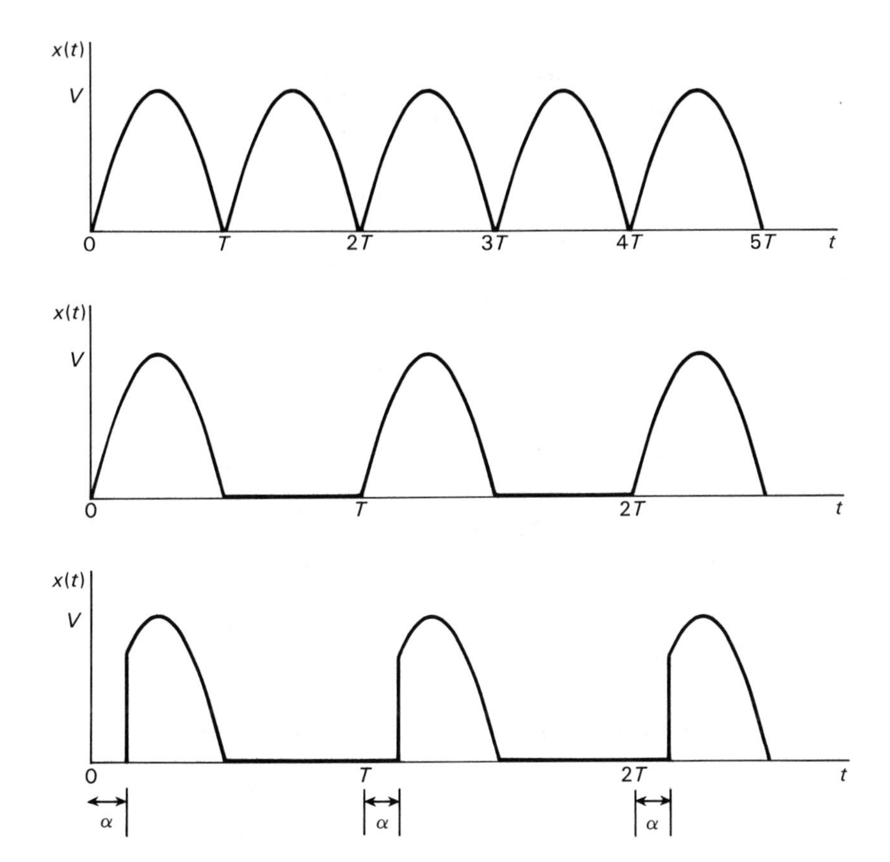

Figure 7.30

7.5 A system is described by the following differential equation

$$\frac{d^2y}{dt^2} + 7\frac{dy}{dt} + 12y = x$$

Determine the response of the system to: (a) a unit step; (b) a unit ramp, applied at $t = 0$. The initial conditions at $t = 0$ are

$$y = -2 \qquad \frac{dy}{dt} = 0$$

7.6 For the circuit shown in Figure 7.31 determine the variation in voltage $v_o(t)$ following the application of a 10 V step in input voltage $v_i(t)$ at $t = 0$. At this time the capacitor is charged to a voltage of 20 V (polarity as shown), and the current through the inductor is zero.

7.7 Determine expressions for the transfer function $V_o(s)/V_i(s)$ for the circuits shown in Figure 7.32.

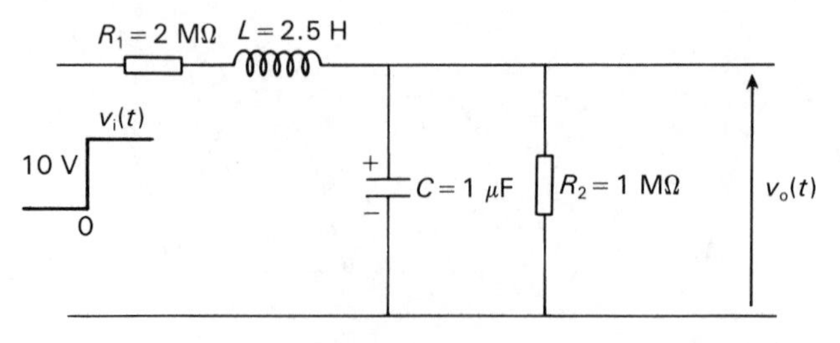

Figure 7.31

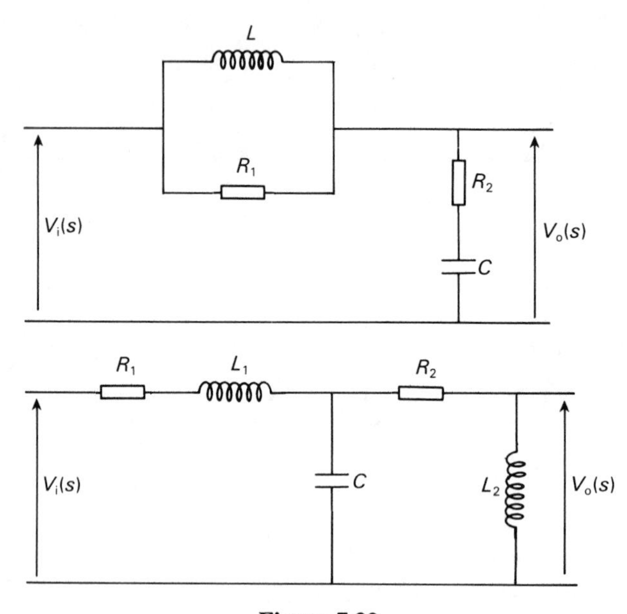

Figure 7.32

7.8 Figure 7.33 outlines a vibration isolator. This is a method of mounting a vibrating machine (e.g., a pump or a compressor) such that the effect of the forces reacting against the ground is minimised.

M represents the mass of the mounting and this is coupled to the ground via the spring, stiffness k_1. M_1, k_2 and B form an auxiliary damper. If the force from the machine is represented by $f(t)$ determine the transfer function relating this force to: (a) the force transmitted to the ground $f_g(t)$; (b) the displacement of the mount $x(t)$.

7.9 The circuit shown in Figure 7.34 is subject to an input signal

$$v_i(t) = 10 \sin \omega t \qquad t \geqslant 0$$
$$= 0 \qquad t < 0$$

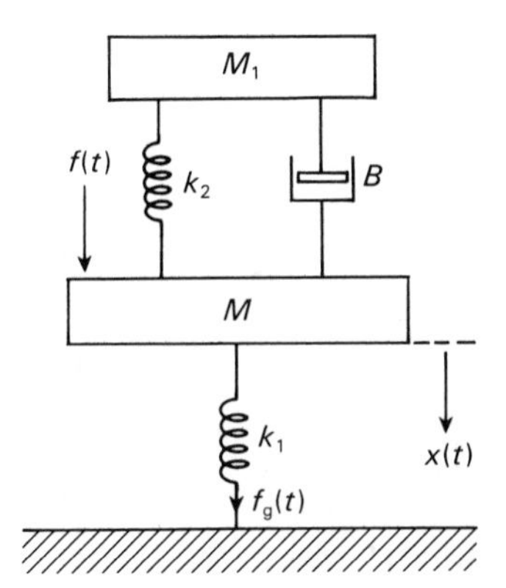

Figure 7.33

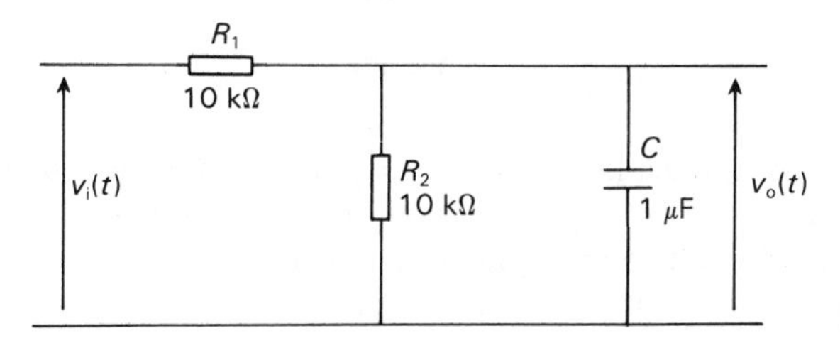

Figure 7.34

If $\omega = 200$ rad/s and the capacitor is initially uncharged determine an expression for the response $v_o(t)$. How does this result compare with that obtained by frequency response methods and account for the difference?

7.10 Figure 7.35(a) shows one stage of an a.c. coupled transistor amplifier and Figure 7.35(b) shows its low frequency equivalent circuit. Derive an expression for the response $v_o(t)$ when the input $v_i(t)$ is the pulse shown.

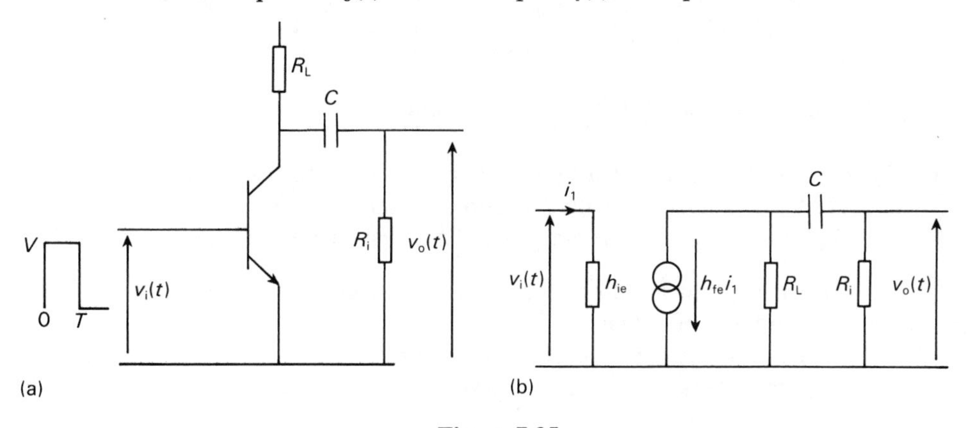

Figure 7.35

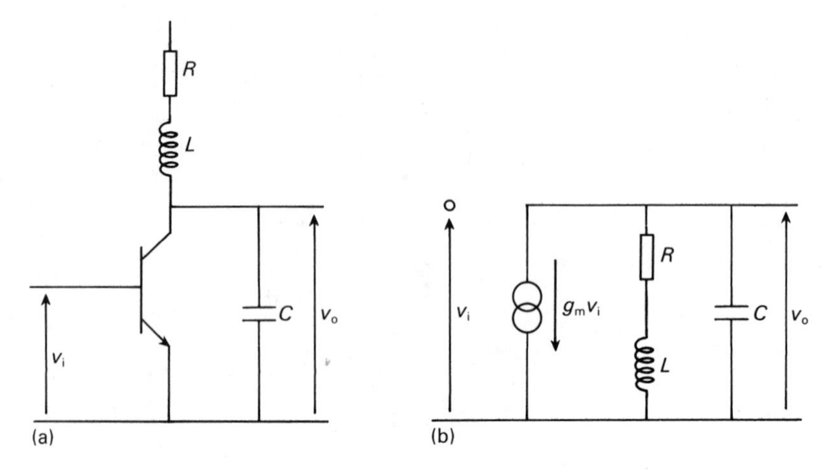

Figure 7.36

7.11 'Shunt peaking' is a method of improving the rise time in wide band amplifiers. An inductor is placed in series with the collector load resistor and this 'compensates' for the effect of shunt capacitance.

Figure 7.36(a) shows a simplified circuit of one stage of a video amplifier and Figure 7.36(b) is a small signal equivalent circuit. L represents the shunt peaking inductor and C the total stage shunt capacitance.

Derive the transfer function $V_o(s)/V_i(s)$ for the stage.

Given that $R = 2\ \text{k}\Omega$, $C = 50$ pF, $g_m = 20$ mA/V determine the response $v_o(t)$ to an input signal $v_i(t)$ consisting of a step of amplitude 50 mV, for the following values of L, $L = 10\ \mu$H, $L = 50\ \mu$H, $L = 200\ \mu$H.

7.12 A unit step is applied to a system at $t = 0$ and the resulting response is

$$y(t) = 1 - \frac{4}{17}\,e^{-5t} - \frac{e^{-1.2t}}{136}\,[104 \cos 1.6t + 178 \sin 1.6t]$$

Determine the transfer function of the system.

7.13 Figure 7.37 shows part of a hydraulic system to control elevator angle in an aircraft. A solenoid controls the position of a 'spool valve'.

The constants of the relevant parts of the system are as follows.

M = mass of solenoid plunger and valve 0.2 kg
k = stiffness of spring 720 N/m
B = dashpot damping constant 14.4 N/m s^{-1}
K_F = force per unit of solenoid current 1.8 N/A

Determine the transfer function relating valve position x and solenoid current i.

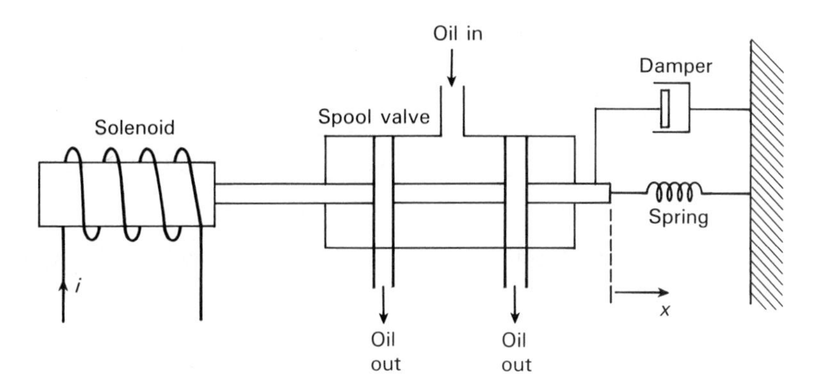

Figure 7.37

If a step of current of magnitude 1 A is applied, determine: (a) the percentage overshoot; (b) the time it takes to reach this overshoot.

7.14 Control systems are designed in many cases to make the output quantity follow the input quantity. Although this is not possible under transient conditions it is desirable that the output follows as closely as possible under steady state conditions (as $t \to \infty$). A measure of this ability is the error signal which is the difference between the input and the output signals.

A control system has an overall transfer function between output and input given by

$$\frac{Y(s)}{X(s)} = \frac{100}{s^2 + 14s + 100}$$

Without inverting the transform determine the steady state error in the system (as $t \to \infty$) for both unit step and unit ramp inputs.

7.15 A system as a transfer function, $H(s)$ is given by

$$H(s) = \frac{4(s + 2.5)}{s^2 + 2s + 10}$$

(a) Draw to scale the pole-zero diagram for the system.
(b) Using graphical methods on the pole-zero diagram determine: (i) the system impulse response; (ii) the system step response.

7.16 A feedback control system has an overall transfer function given by

$$\frac{Y(s)}{X(s)} = \frac{K}{s^2 + 4s + (3 + K)}$$

where K is the system gain.

Use the pole-zero diagram to calculate the following:

(a) The system step response when $K = 0.5$.
(b) The damping factor ζ and the undamped natural frequency ω_n when $K = 2$.
(c) The magnitude and angle of the system frequency response at $\omega = 2$ when $K = 5$.

7.17 Determine the z transforms of the signals shown in Figure 7.38. Express your answer as a ratio of polynomials in z.

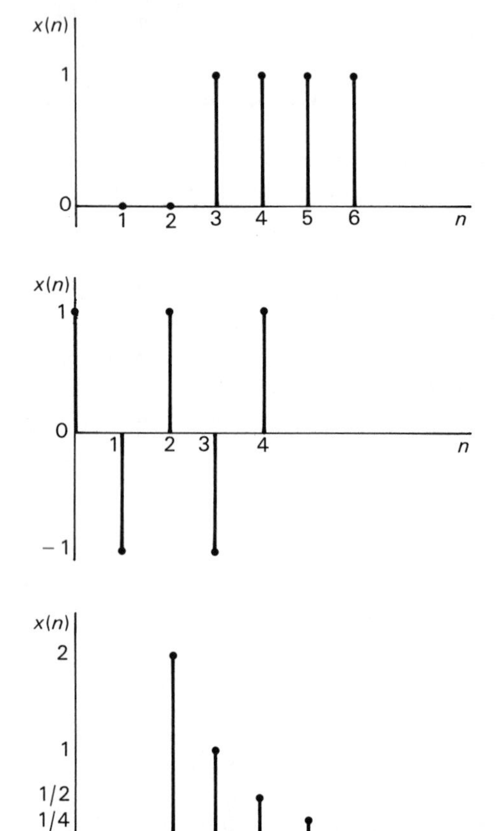

Figure 7.38

7.18 Determine the inverse z transform of the following functions. Do the inversion by the partial fraction expansion and check the first five values in the sequence by long division.

(a) $X(z) = \dfrac{z(5z + 0.5)}{z^2 - 0.25}$

(b) $X(z) = \dfrac{z(3z - 2.8)}{z^2 - 1.8z + 0.8}$

(c) $X(z) = \dfrac{z^3}{(z - 1)^2(z - 0.5)}$

(d) $X(z) = \dfrac{z(2z + 1)}{z^2 - 0.5z + 0.25}$

7.19 A first order digital filter is described by the following difference equation

$$y(n) = 2x(n) + 0.2y(n - 1)$$

Determine the unit sample response and the step response of the filter.

7.20 Fibonacci sequences find application in problems ranging from population growth in biology to optimum system design in control engineering. A Fibonacci sequence is a sequence of numbers such that each number is the sum of the previous two numbers. Different initial conditions will produce different sequences. Write down the difference equation describing the relationship between successive numbers in the sequence. Taking the first two numbers as 0 and 1 use z transforms to derive an expression for the nth number in the sequence.

7.21 A person deposits £10 000 in a long term savings account. The account pays an interest rate of 12% per annum but the capital cannot be withdrawn in the short term. To provide short term cash the investor takes half the annual interest and deposits this in a short term account paying an annual interest of 6%. The other half of the interest adds to the capital in the long term account.

(a) Determine an expression for the interest drawn from the long term account at the end of the nth year.
(b) Assuming no other deposits, and no withdrawals, determine an expression for the capital in the short term account at the end of the nth year.

7.22 A discrete system has a pulse transfer function given by

$$H(z) = \frac{(z - 1)}{z^2 - z + 0.5}$$

Mark the position of the poles and zeros of the function of the z plane. Using graphical methods obtain an expression for: (a) the unit sample response; (b) the response of the system to a discrete step input.

7.23 A discrete control system has an overall transfer function given by

$$H(z) = \frac{K}{z^2 - 0.5z + (K - 0.5)}$$

where K is the system gain. Use the pole-zero diagram to obtain the following:

(a) The system response to a discrete unit step when $K = 0.5$.
(b) The magnitude and angle of the system frequency response at a normalised frequency ωT of 1 rad when $K = 0.625$.
(c) The value of K that will just cause the system to become unstable.

MATLAB exercises

7.24 Use the command `invlaplace` in the Symbolic Math Toolbox to invert the Laplace transforms given in Problem 7.4.

7.25 This exercise uses the command `step` to obtain the step response of a system when initial conditions are present.

Use MATLAB to obtain the step response of the systems given in Problems 7.5 and 7.6 with the initial conditions as specified. Display the responses against time together with the theoretical responses.

Note After obtaining the relevant transfer functions, the step responses can be easily obtained by using the command `step`. However, initial conditions cannot be incorporated into this command and the response to these must be obtained separately. The contribution of the initial conditions, to the transform of the output response, can be interpreted as the impulse response of a system (the Laplace transform of an impulse is unity). Hence, this contribution can be obtained by use of the command `impulse`.

7.26 This exercise obtains the response of a system to a half sine wave input.

Repeat Problem 7.9 using a MATLAB program to obtain the required time responses. By plotting this response, compare it with the response obtained using frequency response methods.

Note The input half sine wave must be expressed as a vector and the command `lsim` used to obtain the required response.

7.27 This exercise investigates the response of a second order system to a step input.

Write a MATLAB program to determine the step response of the system given in Problem 7.13. Responses should be obtained for a range of values of B

and all plotted on the same graph. For each value of B the corresponding values of undamped natural frequency ω_n and damping factor ζ should be obtained.

Note The step response can easily be obtained by use of the command `step`. Although ω_n and ζ are easily calculated, they can also be obtained from the command `damp`.

7.28 This exercise obtains the poles and zeros of a transfer function and plots their positions on the complex plane.

Given the transfer function

$$G(s) = \frac{s^3 + s^2 - s + 15}{s^4 + 8s^3 + 29s^2 + 52s}$$

write a MATLAB program to determine the poles and zeros of this function and to plot their positions on the complex plane.

Note The poles and zeros of the transfer function can be obtained by using the command `roots` on the denominator and numerator polynomials. The pole-zero positions can be plotted by using the command `plot(Y)`, which will plot the imaginary part of `Y` against the real part of `Y`. By using the `hold on` command, each pole and zero can be plotted in turn to produce the complete pole-zero diagram.

7.29 This exercise determines the step response of a system by means of a partial fraction expansion.

Write a MATLAB program to determine the step response of the loudspeaker system analysed in Example 7.4.3. The method of solution to be used is that described in Section 7.4.3. The pole positions of the function describing the transform of the step response have to be located. These determine the associated modes of the time response. The modes are combined, with appropriate weightings, to give the step response.

Note For this example, the weightings of the modes (the coefficients in the partial fraction expansion) are the same as the residues of the function. Both the poles and the residues can be obtained directly from the command `[r,p,k]=residue (b,a)`.

Although this is the most direct method of obtaining the poles and their weightings, the student should also use the procedures of Example 7.4.3. The ability of MATLAB to handle complex numbers helps considerably in determining the weightings of the modes.

7.30 This exercise investigates methods of inverting the z transform.

Write MATLAB programs to invert the z transforms given in Problem 7.18. Three methods of inversion should be investigated.

(a) Obtain the unit sample responses of the systems described by the z transforms. This can be done by means of the command `dimpulse`.
(b) Perform long division of the polynomials forming the z transforms by means of the command `deconv`.
(c) By means of the partial fraction expansion, determine the time response by means of the system modes and their respective weightings.

Note The long division of the numerator by the denominator polynomials can be performed by the command `deconv`. This command will produce positive powers of z in the quotient plus a remainder. In order to give the required sequence of coefficients, the numerator must be multiplied by z^m before deconvolution; m determines the number of samples in the response.

The command `residue` can be used to determine the poles and the respective weightings to the modes. Remember: to obtain the response in standard form, the z function used is $X(z)/z$.

7.31 Write a MATLAB program to obtain (a) the unit sample response, (b) the unit step response of the system given in Problem 7.22. Use one, or all, of the methods of the previous exercise.

8

Filtering of Signals

8.1 Introduction

Filtering is the process that accepts the required components of a signal and rejects unwanted components. A filter is a system that produces the filtering process. Like any other system its action can be described in both the frequency and time domains. In this chapter the filter specification will be in terms of its frequency response, it will be required to pass certain bands of frequencies and reject others. Depending on these bands of frequencies the filter can be described as low-pass, high-pass, band-pass, band-stop. The ideal magnitude characteristic of these filter types are shown in Figure 8.1. The frequency range is divided into pass-band regions and stop-band regions and the frequencies that divide these regions are known as the cut-off frequencies.

Chapter 1 gave the example of a low-pass filter that could be used to remove high-frequency noise from a recording. An audio amplifier may also include a high-pass filter to remove 'rumble' and subsonic noise to prevent amplifier overload. The tuning system of a radio receiver is an example of a band-pass filter. The range of frequencies (carrier and sidebands) comprising the required stations are accepted, all other stations are rejected. A band-stop filter could be used to remove a single frequency such as mains hum from a signal.

The responses shown in Figure 8.1 are ideal responses and in practice only an approximation to these responses can be obtained. Obtaining the required approximation is the filter design problem. As well as a magnitude response the filters will also have a phase response. If it is required to control this response as well as the magnitude response this constraint will complicate the design problem.

Filters can be divided into analogue and digital filters depending on whether they operate on continuous or discrete signals. Analogue filters are further subdivided into passive and active filters.

Passive filters are usually constructed from inductors and capacitors (LC filters) or resistors and capacitors (RC filters). In this type of filter the input and output impedances are subject to certain constraints and this complicates the design problem. The consideration of passive filters (other than in very simple forms) are beyond the scope of this book.

Active filters contain an element that gives a power amplification. Modern active filters usually consist of resistance capacitor combinations connected in a feedback arrangement around an operational amplifier. The low output impedance of such a device means that simple filter sections can be easily cascaded to produce more complex filters.

Digital filters are normally implemented either on digital computer or, for simpler filters, on special purpose integrated circuit chips. Special purpose processors

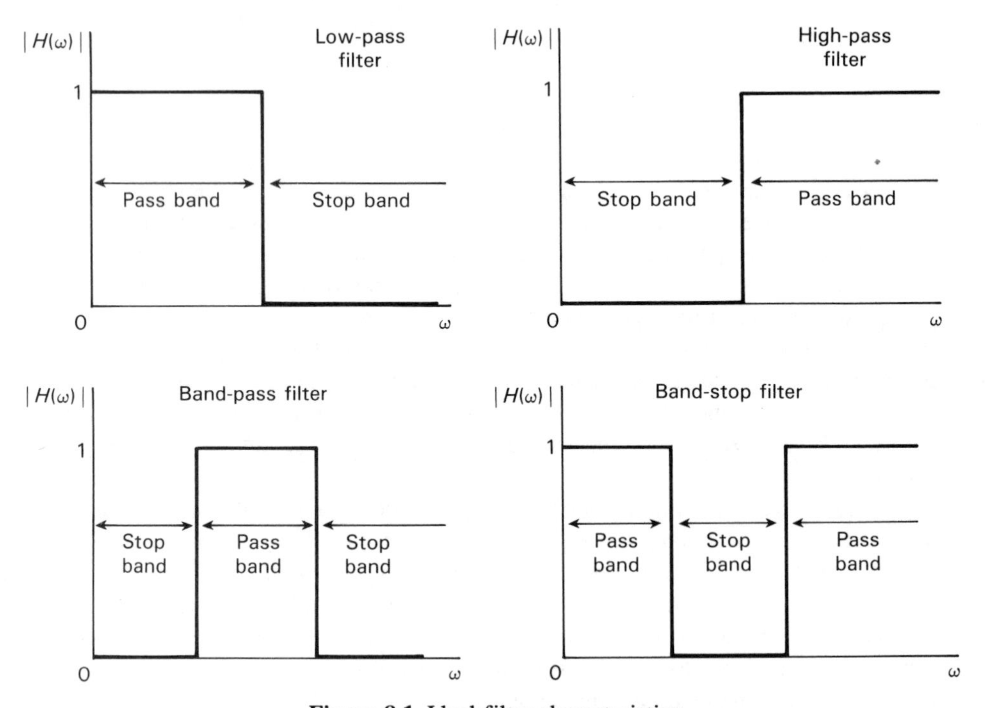

Figure 8.1 Ideal filter characteristics.

are available that can be incorporated into computers that are matched to real-time signal processing. It is beyond the scope of this book to discuss such methods of implementation. However the design of digital filters and some of the more general problems concerned with their implementation can be considered.

Many digital filter design methods start from an equivalent analogue response. It therefore seems logical to start with a consideration of analogue filters.

8.2 Ideal filters

In the introduction it has been stated that the responses shown for the various filters are ideal responses and can only be approximated in practice. It is instructive to consider why this is so, taking as an example the ideal low-pass filter. Remembering that a filter's frequency response and its impulse response form a Fourier transform pair the impulse response associated with a low-pass filter characteristic is shown in Figure 8.2.

This response is the response to a unit impulse applied at $t=0$. However the response to this impulse starts before $t=0$, it represents a non-causal system and cannot be implemented practically.

Suppose however a system could be constructed such that the impulse response was as shown in Figure 8.3, it is the impulse response of the ideal filter delayed by time τ.

Using the delay theorem of the Fourier transform such an impulse response would correspond to a frequency response

$$\begin{aligned} H(\mathrm{j}\omega) &= \mathrm{e}^{-\mathrm{j}\omega\tau} \qquad -\omega_c < \omega < +\omega_c \\ &= 0 \qquad \text{otherwise} \end{aligned}$$

The filter still has the ideal magnitude response but there is a linear phase shift in the pass band.

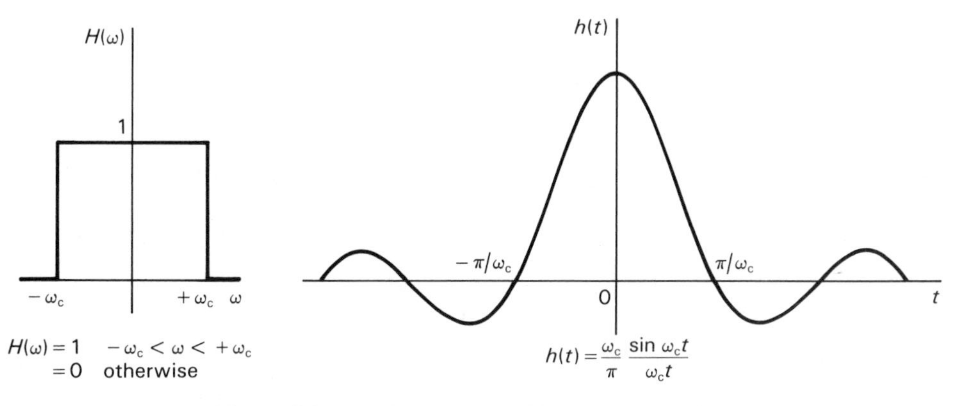

Figure 8.2 Impulse response of ideal low-pass filter.

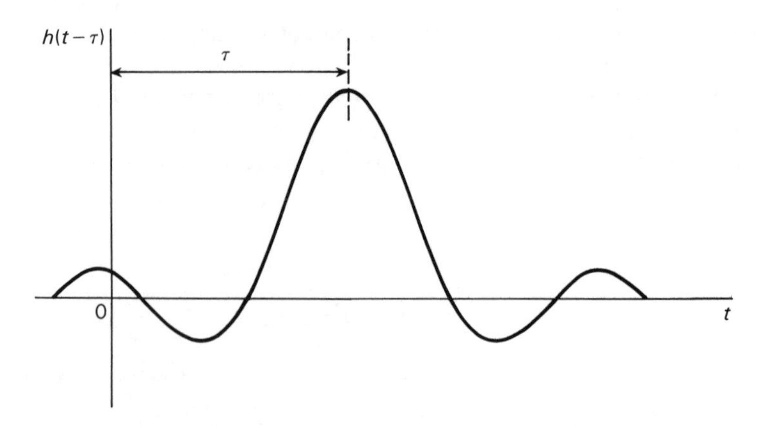

Figure 8.3 Time shift on filter impulse response.

The impulse response shown in Figure 8.3 however is still non-causal, it exists before $t = 0$. However by making τ larger (accepting more phase shift in the filter) it would intuitively appear that the effect of taking the part of response prior to $t = 0$ as zero would not be too serious. Although it cannot be proved here, the effect of truncating the impulse response in this way is to give a frequency response as shown in Figure 8.4. This differs from the ideal response in that the following occurs:

1. There are ripples in the response in both pass and stop band.
2. There is only a gradual transition from the pass to the stop band.

These factors will be discussed in more detail in the following sections.

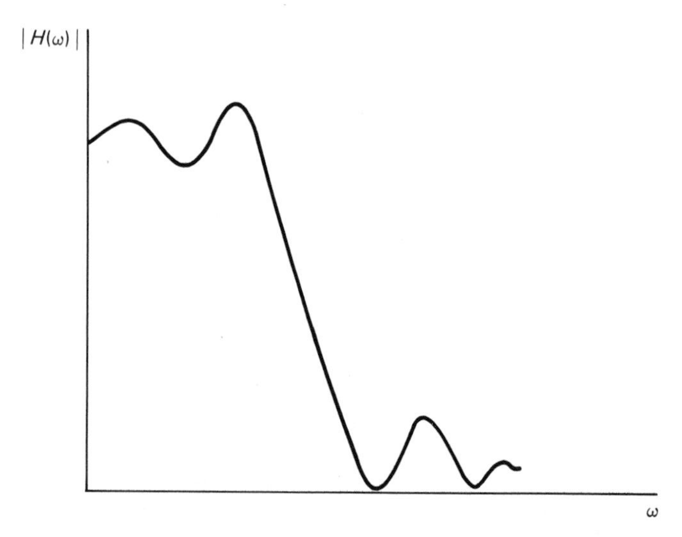

Figure 8.4 Frequency response of a practical low-pass filter.

8.3 Analogue filters

As stated in the introduction a number of filter types exist depending on the positions of the pass and stop bands. However high-pass, band-pass and band-stop filters can be obtained by suitable frequency transformations starting from a low-pass filter design. Hence it is only low-pass filters that are considered here.

In Chapter 5 the frequency responses of first and second order systems were considered and it was shown that these systems had responses that were low-pass.

Suppose it is required to approximate an ideal low-pass filter with cut-off frequency ω_c by a first order system having a frequency response function

$$H(j\omega) = \frac{1}{1 + j\omega/\omega_c}$$

The magnitude response of this function is shown in Figure 8.5.

As can be seen this response deviates markedly from the ideal response. Over the pass band the response falls and is 3 dB down at the cut-off frequency ω_c. The ideal filter would give infinite attenuation in the stop band. However the first order filter gives an attenuation characteristic that falls at 20 dB/decade giving no clearly defined stop band. In practice the stop band has to be defined as the region where the attenuation is greater than some specified figure. Suppose this figure was defined as −40 dB for the filter in question, this would define three regions of the frequency range. A pass band where the attenuation is less than 3 dB, a stop band where the

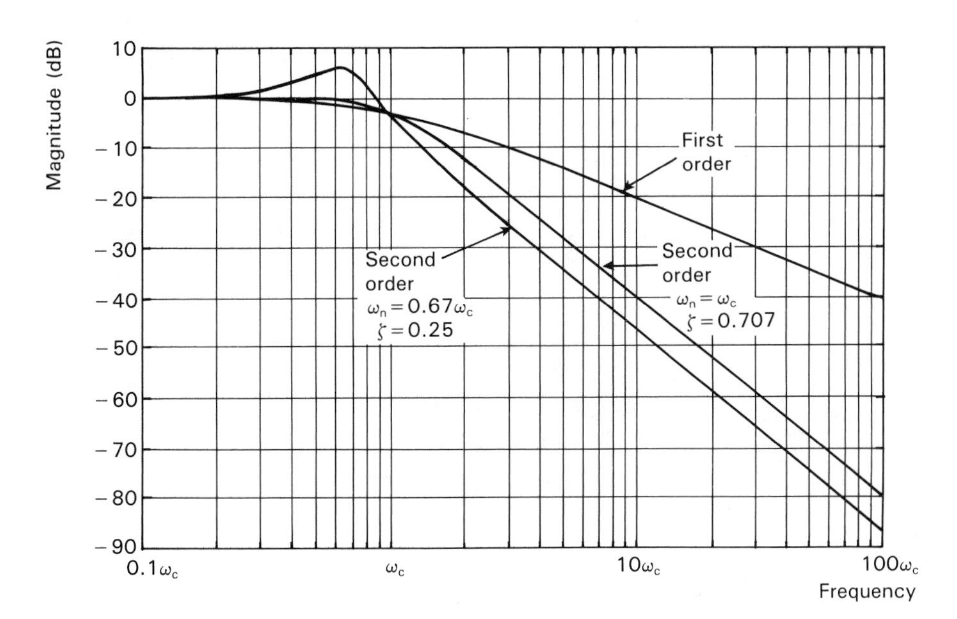

Figure 8.5 Response of first and second order long-pass filter.

attenuation is greater than 40 dB, and a transition region between the pass and the stop bands. For the filter in question the width of the transition region is $100\omega_c$.

Consider now a second order system

$$H(j\omega) = \frac{\omega_n^2}{(j\omega)^2 + 2\zeta\omega_n j\omega + \omega_n^2} \tag{8.3.1}$$

Two parameters can now be set to approximate the ideal response, ω_n and ζ. Suppose it is required to produce an attenuation of 3 dB at the cut-off frequency ω_c. Then for a given value of ζ this will determine ω_n. Considering two values of ζ, $\zeta = 0.707$, $(1/\sqrt{2})$ and $\zeta = 0.25$, then by using eqn (8.3.1) the corresponding values of ω_n are $\omega_n = \omega_c$ and $\omega_n = 0.673\omega_c$. The responses for these values of ω_n and ζ are also shown in Figure 8.5.

With $\zeta = 0.707$ a pass band is obtained that is flat within 3 dB up to the cut-off frequency. The attenuation then increases reaching a rate of 40 dB/decade. For 60 dB attenuation this gives a transition region or $32\omega_c$.

With $\zeta = 0.25$ the transition region is reduced to $22\omega_c$. However now the response in the pass band reaches a peak of 6 dB before falling to -3 dB at the cut-off frequency.

This discussion based on a first and second order system gives an indication of the results that can be obtained using higher order filters.

1. The higher the order of the filter the sharper the cut-off and the narrower the transition region.
2. By suitable choice of the coefficients in the frequency response function a narrower transition region can be obtained at the expense of peaks (ripple) occurring in the pass band.

In the next two sections two types of higher order filter will be discussed, the Butterworth filter, and the Chebyshev filter. The Butterworth response has no ripples in its pass band but it has a wider transition region than the same order Chebyshev response. However the Chebyshev filter obtains its narrow transition region at the expense of ripples in its pass or stop bands.

8.3.1 The Butterworth filter

The Buttersorth filter is most easily described by its magnitude response.

$$|H(j\omega)| = \frac{1}{\sqrt{1 + \left[\frac{\omega}{\omega_c}\right]^{2N}}} \tag{8.3.2}$$

where N is the order of the filter and ω_c is its cut-off frequency. This magnitude response is plotted in Figure 8.6.

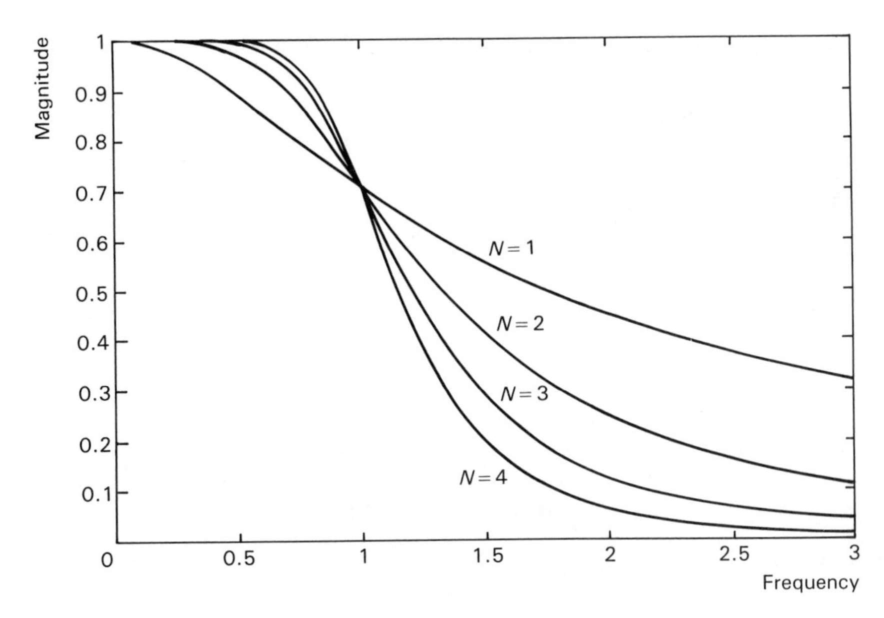

Figure 8.6 Magnitude response for Butterworth filters.

As can be seen, all the plots pass through the same point $\omega = \omega_c$ when the amplitude is 0.707 or -3 dB. The higher the order of the filter the sharper the transition from the pass band to the stop band.

Although eqn (8.3.2) describes the amplitude response of the filter it does not indicate how to construct a filter that will have this response. To do this the transfer function $H(s)$ is required and this can be obtained by consideration of the poles of $|H(s)|^2$.

$$|H(\mathrm{j}\omega)|^2 = \frac{1}{1 + \left[\dfrac{\omega}{\omega_c}\right]^{2N}}$$

$|H(s)|^2$ can be obtained by substituting $\omega = s/\mathrm{j}$ then

$$|H(s)|^2 = \frac{1}{1 + \left[\dfrac{s}{\mathrm{j}\omega_c}\right]^{2N}}$$

This has poles when

$$\left[\frac{s}{\mathrm{j}\omega_c}\right]^{2N} = -1$$

$$s = \mathrm{j}\omega_c(-1)^{1/2N}$$

(-1) can be expressed in exponential form, $-1 = e^{j\pi}$, then remembering there are $2N$ roots of $(-1)^{1/2N}$

$$(-1)^{1/2N} = e^{j[\pi/2N + k(2\pi/2N)]} \qquad k = 0, 1, \ldots\ldots (2N-1)$$

Writing j also in exponential form

$$j = e^{j\pi/2}$$

then

$$s = \omega_c \, e^{j\pi(N+1+2k)/2N} \tag{8.3.3}$$

Equation (8.3.3) gives the poles of $|H(s)|^2$, but to implement the filter the poles of $H(s)$ are required. However

$$|H(s)|^2 = H(s)H^*(s)$$

and on the imaginary axis $s = j\omega$ and $s^* = -j\omega = -s$, then

$$|H(s)|^2 = H(s)H(-s)$$

Hence all the poles in the left-hand half plane (corresponding to a stable system) are the required poles. These belong to the transfer function $H(s)$. It should be noted that there will never be poles on the imaginary axis as referring to eqn (8.3.3) the multiple of $\pi/2$ is $(N+1+2k)/N$ and this multiple can never be odd.

The required pole positions can be calculated directly from eqn (8.3.3), it should be noted that they all lie on the unit circle. However an easier method is to calculate the position of one pole ($k = 0$ say) and then note that all the poles are evenly spaced by an angle π/N.

From the pole positions the system transfer function can then be obtained, the procedure is illustrated in the following example.

EXAMPLE 8.3.1

In terms of the cut-off frequency ω_c, obtain the transfer function for a fourth order, low-pass, Butterworth filter.

SOLUTION

In this example $N = 4$. From eqn (8.3.3) the angle associated with a pole is

$$\theta = \frac{\pi}{2} \times \left[\frac{N+1+2k}{N}\right]$$

$$= \frac{5\pi}{8} + \frac{\pi}{4}k \qquad \text{for } N = 4$$

Hence the pole positions are

$$s_0 = \omega_c \underline{/112.5^\circ}, \qquad s_1 = \omega_c \underline{/\ 157.5^\circ}, \qquad s_2 = \omega_c \underline{/\ \ 202.5^\circ},$$

$$s_3 = \omega_c \underline{/247.5^\circ}, \qquad s_4 = \omega_c \underline{/-67.5^\circ}, \qquad s_4 = \omega_c \underline{/\ \ -22.5^\circ},$$

$$s_6 = \omega_c \underline{/\ 22.5^\circ}, \qquad s_7 = \omega_c \underline{/\ \ 67.5^\circ}.$$

These poles are shown in Figure 8.7.

The poles s_0 to s_3 are those associated with the required transfer function. To obtain this transfer function the pole positions must be converted to cartesian form.

$$s_0 = \omega_c(-0.3827 + \mathrm{j}0.9239) \qquad s_1 = \omega_c(-0.9239 + \mathrm{j}0.3827)$$

$$s_2 = \omega_c(-0.9239 - \mathrm{j}0.3827) \qquad s_3 = \omega_c(-0.3827 - \mathrm{j}0.9239)$$

The required transfer function takes the form

$$H(s) = \frac{K}{(s - s_0)(s - s_1)(s - s_2)(s - s_3)} \tag{8.3.4}$$

The constant K is included but the pole-zero diagram does not gives its value. In eqn (8.3.4) all the terms are of first order but the coefficients involved are complex, they could not be implemented practically. The poles s_0, s_3 and s_1, s_2 are complex

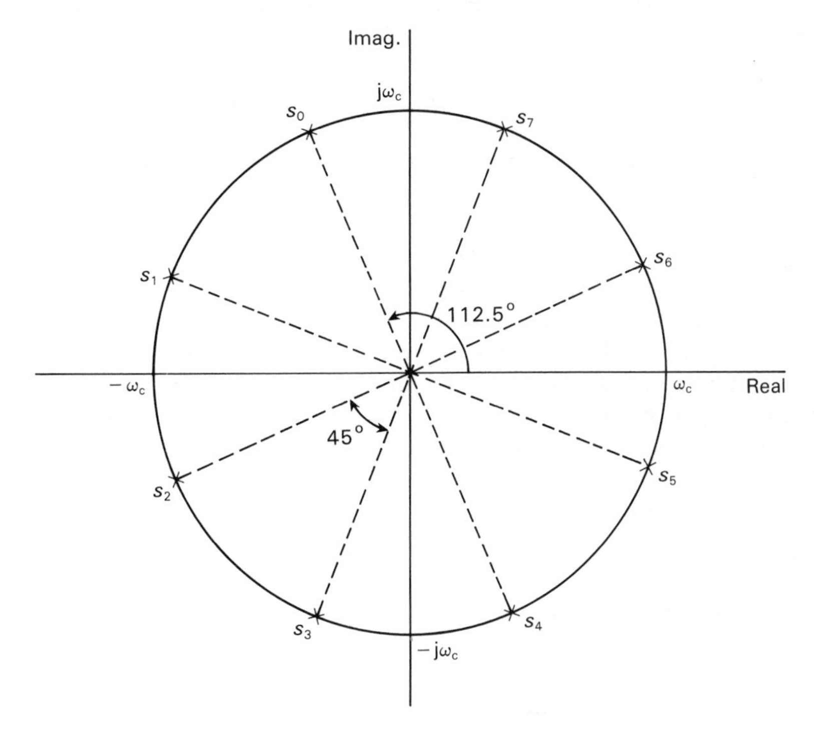

Figure 8.7 Pole positions for fourth order Butterworth filter.

conjugates, multiplying these together will give quadratic terms with real coefficients.

$$H(s) = \frac{K}{(s^2 + 0.7654\,\omega_c s + \omega_c^2)(s^2 + 1.8478\,\omega_c s + \omega_c^2)} \tag{8.3.5}$$

By multiplying together the quadratic terms the denominator can be expressed as a polynomial.

$$H(s) = \frac{K}{s^2 + 2.6131\,\omega_c s^3 + 3.4142\,\omega_c^2 s^2 + 2.6131\,\omega_c^3 s + \omega_c^4} \tag{8.3.6}$$

Before the filter can be implemented from either of these forms the constant K has to be evaluated. This can be done by noting that at $\omega = 0$ $(s = 0)$, $H(s) = 1$ hence $K = \omega_c^4$.

The form that is finally used, eqn (8.3.5) or eqn (8.3.6) depends on the method of implementation. For passive implementation the polynomial form, eqn (8.3.6) is more useful. For active implementation (see Section 8.4) the factorised form, eqn (8.3.5) is more useful.

It should be noted that the quadratic terms in eqn (8.3.5) can be obtained directly by noting the damping factor ζ and natural frequency ω_n associated with a complex pole pair. The factor can then be expressed using the standard form, $(s^2 + 2\zeta\omega_n + \omega_n^2)$.

If eqn (8.3.5) or eqn (8.3.6) is divided through by ω_c^4 then a new normalised variable can be written as s/ω_c. In this form extensive tables are available of the coefficients in both factorised and polynomial forms of the transfer function.

8.3.2 The Chebyshev filter

For a given order the Chebyshev filter has a higher rate of cut-off than the corresponding Butterworth filter. However instead of the gain falling monotonically there are ripples in the response. The form of the filter determines whether these ripples are in the pass or the stop band. Only the form having pass band ripple is considered here and its magnitude response is given by

$$|H(j\omega)| = \frac{1}{\sqrt{(1 + \varepsilon^2 T_n^2(\omega/\omega_c))}} \tag{8.3.7}$$

Here n is the order of the filter and T_n is the nth order Chebyshev polynomial. Table 8.1 lists these polynomials for orders 0 to 5 where $x = \omega/\omega_c$.

ε is a parameter that controls the amount of ripple in the pass band. The ripple has maximum value 1 and minimum value $1/\sqrt{(1 + \varepsilon^2)}$. Figure 8.8 shows the form of the response for $n = 4$ and $n = 5$.

Table 8.1 Chebyshev polynomials

Order n	*Chebyshev polynomial $T_n(x)$* $x = \omega/\omega_c$
0	1
1	x
2	$2x^2 - 1$
3	$4x^3 - 3x$
4	$8x^4 - 8x^2 + 1$
5	$16x^5 - 20x^3 + 5x$

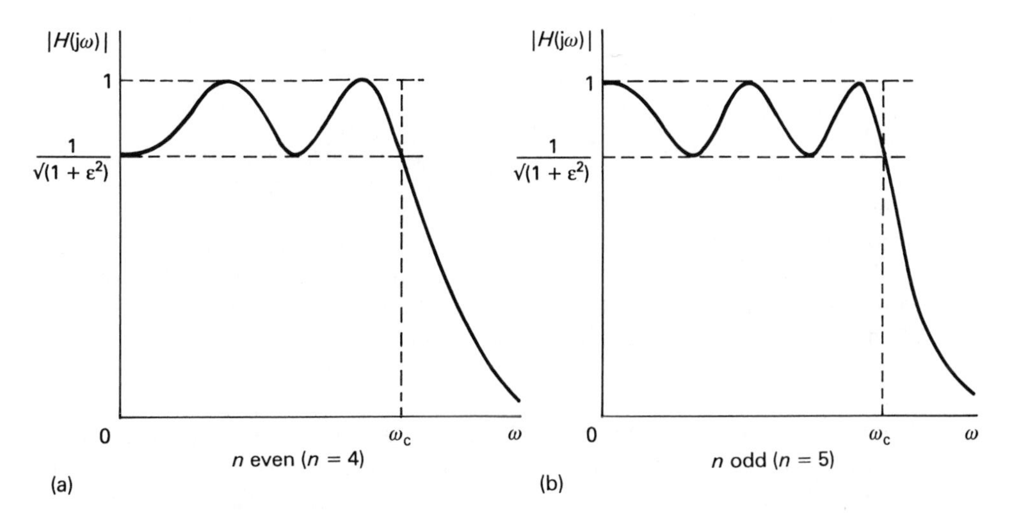

Figure 8.8 Response of Chebyshev filter for odd and even *n*.

In order to implement the filter physically a transfer function $H(s)$ is required. In theory this can be obtained in a manner similar to that used for the Butterworth filter. However the mathematics involved is much more complicated than that used for the Butterworth filter. This is because the pole positions involve the Chebyshev polynomial $T_n(\omega/\omega_c)$ and some knowledge of this function is required in the analysis. The pole positions are going to depend not only on the order of the filter but also on the parameter ε. It can be shown that the poles lie on an ellipse whose major and minor axes depend on the parameters of the filter.

Fortunately, from the point of view of filter design, there is a much easier way of obtaining the required transfer function. As with the Butterworth filter extensive tables exist of the coefficients for different orders of filters. However not only do these coefficients depend on the filter order but they also depend upon the parameter ε. However as ε governs the peak to peak ripple in the filter it is usual to quote the tables for different amounts of ripple. Table 8.2 is a very much shortened form giving the denominator of $H(s)$ in factorised form for filters up to order 5 and for peak–peak ripple of 0.5 dB and 1 dB.

Table 8.2 Factorised denominator polynomials for Chebyshev filters

n	$\varepsilon = 0.3493$ *Peak–peak ripple* = 0.5 dB
1	$s + 2.8628$
2	$s^2 + 1.4256s + 1.5162$
3	$(s + 0.6265)(s^2 + 0.6265s + 1.1424)$
4	$(s^2 + 0.3507s + 1.0635)(s^2 + 0.8467s + 0.3564)$
5	$(s + 0.3623)(s^2 + 0.2239s + 1.0358)(s^2 + 0.5862s + 0.4768)$
n	$\varepsilon = 0.5088$ *Peak–peak ripple* = 1.0 dB
1	$s + 1.9652$
2	$s^2 + 1.0978s + 1.1025$
3	$(s + 0.4942)(s^2 + 0.4941s + 1.9942)$
4	$(s^2 + 0.2791s + 0.9865)(s^2 + 0.6737s + 0.2794)$
5	$(s + 0.2895)(s^2 + 0.1789s + 0.9883)(s^2 + 0.4684s + 0.4293)$

The constant in the numerator of the transfer function will not be ω_c^n as in the case of the Butterworth filter. It can be chosen to obtain the correct zero frequency gain ($s = 0$) and its value will depend upon the order of the filter and ε.

The following example illustrates the design procedure.

EXAMPLE 8.3.2

Design a low-pass Chebyshev filter with 1 dB of ripple in the pass band. The cut-off frequency is to be 1 kHz and the attenuation must be at least 60 dB at five times this frequency.

SOLUTION

The magnitude of the response is given by

$$|H(j\omega)| = \frac{1}{\sqrt{(1 + \varepsilon^2 T_n^2(\omega/\omega_c))}}$$

Converting to decibels gives

$$\text{magnitude in dB} = -10 \log(1 + \varepsilon^2 T_n^2(\omega/\omega_c))$$

This must equal -60 when $\omega/\omega_c = 5$. Hence

$$10^6 = 1 + \varepsilon^2 T_n^2(5)$$

$$10^3/\varepsilon \approx T_n(5)$$

For 1 dB of ripple $\varepsilon = 0.5088$ giving a value of $T_n(5) = 1965.4$.

Evaluating $T_n(5)$ from Table 8.1 gives values $T_3(5) = 425$, $T_4(5) = 4801$. Hence $n = 4$ is the lowest order that will suffice.

From Table 8.2 the required transfer function is given by

$$H(s) = \frac{K}{(s^2 + 0.2791s + 0.9865)(s^2 + 0.6737s + 0.2794)}$$

This is in normalised form, to obtain the transfer function to be implemented the substitution s/ω_c must be used where $\omega_c = 2\pi \times 10^3$. This gives the transfer function as

$$H(s) = \frac{K \times 1.5585 \times 10^{15}}{(s^2 + 1.7536 \times 10^3 s + 38.9455 \times 10^6)(s^2 + 4.2330 \times 10^3 s + 11.0302 \times 10^6)}$$

The d.c. gain is obtained by substituting $s = 0$.

$$H(0) = \frac{K \times 1558.5}{38.9455 \times 11.0302} = 3.628K$$

If K is given a value of 0.2756 then at d.c. the gain will be unity (0 dB). However the ripples will then cause a rise of 1 dB (1.12 times) in the pass band, see Figure 8.8. Alternatively K can be given a value of 0.2456, the d.c. gain will then be 0.8912 (−1 dB) but will rise to unity (0 dB) in the pass band. Taking the latter value of K the magnitude response is plotted in Figure 8.9.

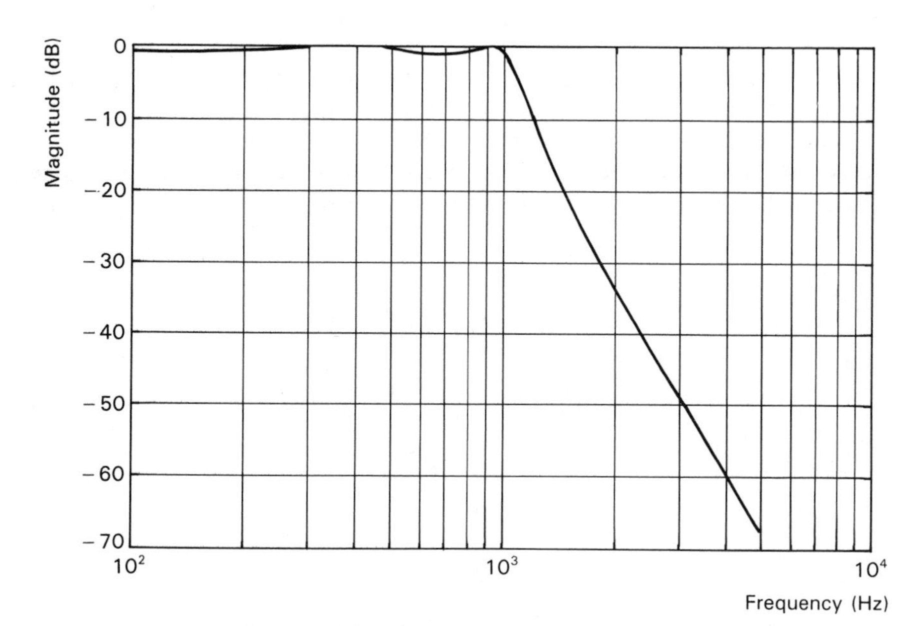

Figure 8.9 Frequency response for filter of Example 8.3.2.

8.4 Active implementation of filters

The Butterworth and Chebyshev transfer functions introduced in the last section can be implemented by active or passive circuits. The transfer functions have been expressed in a form that points to an implementation obtained by cascading first and second order functions. However in the passive implementation, loading between the stages will alter the coefficients of the factors. If an active implementation using operational amplifiers is chosen, then the low output impedance of these amplifiers will make the loading effects negligible. This is the approach that will be investigated in this section.

In factored form each factor represents a first or second order system. It has already been shown in Section 3.7.2 how operational amplifiers can be used to implement such systems for use in analogue computing. However these circuits are not very economical in their use of amplifiers and the first and the second order system can be constructed around a single operational amplifier.

Figure 8.10 shows the circuit that can be used to obtain a first order transfer function. Using nodal analysis at the amplifier input terminal and applying the virtual earth principle

$$\frac{V_i}{R_1} + \frac{V_o}{R_2} + V_o s C_2 = 0$$

$$\frac{V_o}{V_i} = -\frac{R_2}{R_1} \frac{1}{(1 + sC_2R_2)}$$

This is the transfer function of a first order system and by appropriate choice of component values the gain and time constant can be fixed.

To produce a second order factor requires a more complex circuit. Several amplifiers can be interconnected in a manner similar to that used for analogue computing in Section 3.7.2. Such circuits are available on a single chip and are

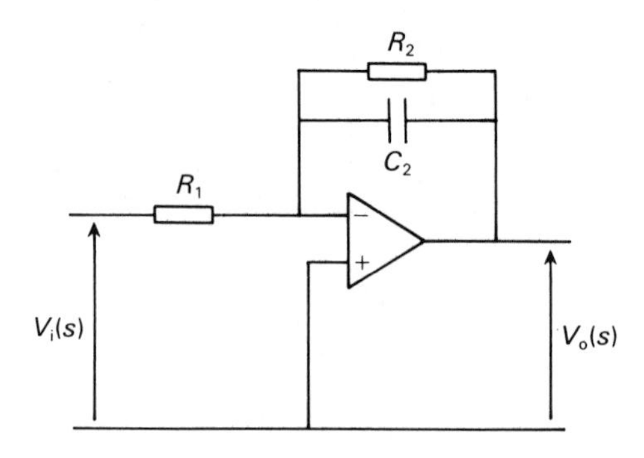

Figure 8.10 Operational amplifier used to produce a first order filter.

known as state variable filters. However a second order system can be built around a single operational amplifier as shown in Figure 8.11 (a).

Both the phase inverting and non-phase inverting inputs are used in this circuit. The virtual earth principle can still be applied but it is the voltage *between* the input terminals that is taken to be virtually zero. Both input terminals are regarded as having the same potential with respect to zero. This potential is V_o because of the unity feedback to the phase inverting input.

An alternative approach is to regard the unity feedback as producing a unity gain amplifier. However as the signal input is the non-phase inverting input the amplifier gain is +1. The circuit illustrating this method of approach is shown in Figure 8.11 (b).

Using Figure 8.11 (b) the transfer function $V_o(s)/V_i(s)$ can be obtained by nodal analysis at the points X and Y. At X

$$\frac{V_i - V}{R} + \frac{V_o - V_x}{R} + (V_o - V_x)sC_2 = 0 \qquad (8.4.1)$$

At Y

$$\frac{V_x - V_o}{R} + (0 - V_o)sC_1 = 0 \qquad (8.4.2)$$

From eqn (8.4.2)

$$V_x = (1 + sC_1R)V_o$$

Substituting this expression for V_x into eqn (8.4.1) leads to the required transfer function.

$$\frac{V_o(s)}{V_i(s)} = \frac{\dfrac{1}{R^2C_1C_2}}{s^2 + \dfrac{2s}{C_2R} + \dfrac{1}{R^2C_1C_2}}$$

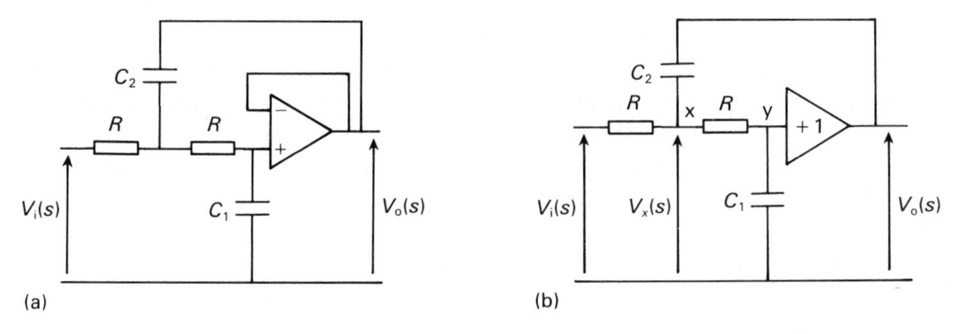

Figure 8.11 Operational amplifier used to produce a second order filter.

Comparison with the standard form gives

$$\omega_n = \frac{1}{R\sqrt{(C_1C_2)}}, \qquad \zeta = \sqrt{(C_1/C_2)}$$

The following example illustrates the use of this circuit in a multi-stage filter.

EXAMPLE 8.4.1

Determine the transfer function of a fifth order Butterworth filter with a cut-off frequency of 200 Hz. Show how the filter may be implemented using the circuits described.

SOLUTION

For a fifth order filter the poles are spaced at an angle of $360°/2 \times 5 = 36°$ around a circle radius ω_c ($\omega_c = 2\pi \times 200$). The position of one of the poles can be obtained by substituting $k = 0$ into eqn (8.3.3)

$$\begin{aligned} s_o &= 180 \times (2 \times 0 + 5 + 1)/2 \times 5 \\ &= 108° \end{aligned}$$

It then follows from the spacing of 36° that the angle of the poles in the left-hand half plane are

$$108°,\ 144°,\ 180°,\ 216°,\ 252°$$

these are shown in Figure 8.12.

The damping factors associated with the complex pole pairs are $\zeta = 0.3090$ and $\zeta = 0.8090$. This gives the required transfer function as

$$H(s) = \frac{\omega_n^5}{(s^2 + 0.6180\omega_n s + \omega_n^2)(s^2 + 1.6180\omega_n s + \omega_n^2)(s + \omega_n)}$$

where

$$\omega_n = 2\pi \times 200$$

To implement the filter two second order factors and a first order factor are required. Some choice is available regarding the method of choosing component values. However in a practical circuit, resistor values should not be so small as to cause undue loading nor so large as to cause problems with amplifier current off sets. Capacitors should not be so small that stray capacitance has an effect or so large that they are impractical.

The resistor value R will be chosen here as 1 kΩ.

Then as $\omega_n^2 = 1/R^2C_1C_2$ this gives

$$C_1C_2 = 0.6332 \times 10^{-12}$$

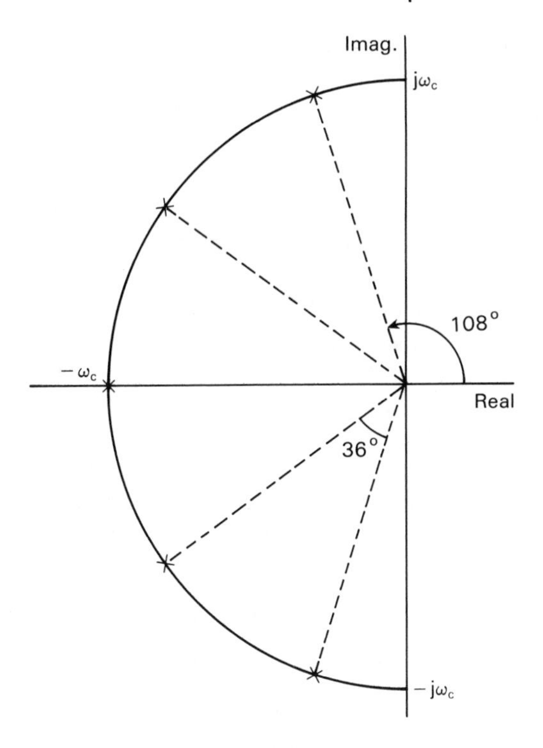

Figure 8.12 Pole positions for fifth order Butterworth filter.

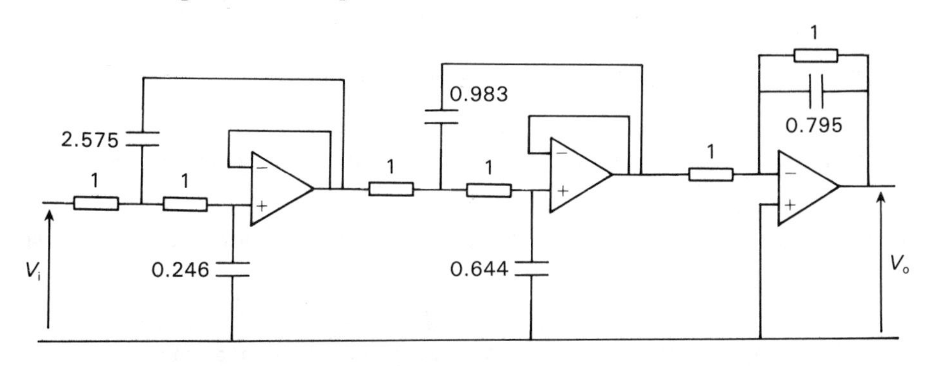

Figure 8.13 Circuit for fifth order Butterworth filter (values in kΩ and μF).

But $\zeta^2 = C_1/C_2$ hence for $\zeta = 0.3090$

$$\frac{C_1}{C_2} = 0.0955$$

giving $C_2 = 2.575\ \mu\text{F}$, $C_1 = 0.246\ \mu\text{F}$.

The corresponding values for the second order factor with $\zeta = 0.8090$ are $C_1 = 0.644\ \mu\text{F}$, $C_2 = 0.983\ \mu\text{F}$.

If a resistor of value 1 kΩ is also used in the first order section this will require a capacitor value of 0.795 μF to give the required time constant. The complete circuit is shown in Figure 8.13.

8.5 Digital filters

Paralleling the action of the analogue filter a discrete (digital) filter is a discrete system designed to reduce unwanted components in a (discrete) signal. Discrete systems have been discussed in both the time domain and the frequency domain in earlier chapters. In Section 3.8.2 the distinction was made between recursive or non-recursive discrete systems. Digital filters are described as recursive or non-recursive depending on the difference equation used to describe them.

An alternative description of these filters can be obtained by considering their response to a unit sample. A non-recursive filter can be described by a difference equation

$$y(n) = b_0 x(n) + b_1 x(n-1) + b_2 x(n-2) + \ldots\ldots b_k x(n-k)$$

Consider the response to the unit sample

$$\begin{aligned} x(n) &= 1 \qquad && n = 0 \\ &= 0 && \text{otherwise} \end{aligned}$$

This response can be obtained by considering the signals at time instants following $n = 0$, as shown in Table 8.3.

There are two points of interest concerning this response:

1. The response is of finite duration, it exists for k points.
2. The response consists of the coefficients of the difference equation describing the system.

Because of the finite duration of the response, non-recursive filters are also known as **F**inite **I**mpulse **R**esponse or FIR filters.

Table 8.3 Response of a non-recursive filter to a unit sample input

n	$x(n)$	$x(n-1)$	$x(n-2)$	......	$x(n-k)$	$y(n)$
0	1	0	0		0	b_0
1	0	1	0		0	b_1
2	0	0	1		0	b_2
⋮	⋮	⋮	⋮		⋮	⋮
k	0	0	0		1	b_k
$k+1$	0	0	0		0	0
$k+2$	0	0	0		0	0

Table 8.4 Response of a specific recursive filter to a unit sample input

n	$x(n)$	$x(n-2)$	$y(n-1)$	$y(n)$
0	1.0	0.0	0.00	1.00
1	0.0	0.0	1.00	1.00
2	0.0	1.0	1.00	0.75
3	0.0	1.0	0.75	0.50
4	0.0	0.75	0.50	0.3125
⋮	⋮	⋮	⋮	⋮

The second point of interest indicates a method of design for such filters – the impulse response gives the coefficients of the filter directly. This method of design will be investigated in the next section.

To obtain the general impulse response for the recursive filter is not so easy. A simple recursive system will be used by way of an example to illustrate this case. Consider a recursive filter described by the difference equation

$$y(n) = x(n) + y(n-1) - 0.25y(n-2)$$

Again the response can be obtained numerically by considering sampling instants starting at $n=0$. It is assumed that there are no initial conditions, when $n=0$, $y(n-1)=y(n-2)=0$. The steps are shown in Table 8.4.

This response will be infinite in duration, for this reason recursive filters are known as **I**nfinite **I**mpulse **R**esponse or IIR filters. There is no simple relationship between the impulse response and the filter coefficients in this type of filter.

Implementation of digital filters will be discussed more fully in Section 8.5.3. However from Chapter 3 it is known that the implementation of IIR filters requires a feedback structure and some of the paths can entail positive feedback. This makes the filter sensitive to parameter values and errors due to finite word length can have a drastic effect on filter performance.

FIR filters do not employ feedback and are much more stable structures. However for a given response FIR filters usually require more stages than IIR filters. One advantage of the FIR filter is that it can produce a linear phase characteristic – this is not possible with an IIR filter.

Methods of designing both types of filters will be investigated in the following sections.

8.5.1 Finite impulse response (FIR) filters

The last section introduced the concept of FIR filters. This type of filter is attractive from the design viewpoint because of the ease that one can obtain the filter coefficients from the sampled impulse response. The sampled impulse response can be obtained from the required frequency response via the inverse, discrete time Fourier transform. This transform, as introduced in Section 6.4.2, used the symbol θ to denote normalised frequency, $\theta = \omega T$, where T is the sampling time. However in most texts on digital filtering ω denotes normalised frequency and this convention will be used here.

Taking the low-pass filter with cut-off frequency ω_c (normalised) the impulse response $h(n)$ is given by

$$h(n) = \frac{1}{2\pi}\int_{-\omega_c}^{+\omega_c} e^{jn\omega}\,d\omega = \frac{\omega_c}{\pi}\,\frac{\sin n\omega_c}{n\omega_c}$$

This response is shown in Figure 8.14(a) for the case $\omega_c = \pi/4$.

As with the ideal analogue filter this response is non-causal and cannot be implemented physically. A causal response can be obtained by the following:

1. Truncating the impulse response to length $n = N$.
2. Time delaying the impulse response. In order to maintain the linear phase characteristic the symmetry of the response must be preserved and the required delay is $\alpha = (N-1)/2$. The case where $N = 13$ is shown in Figure 8.14(b), hence $\alpha = 6$.

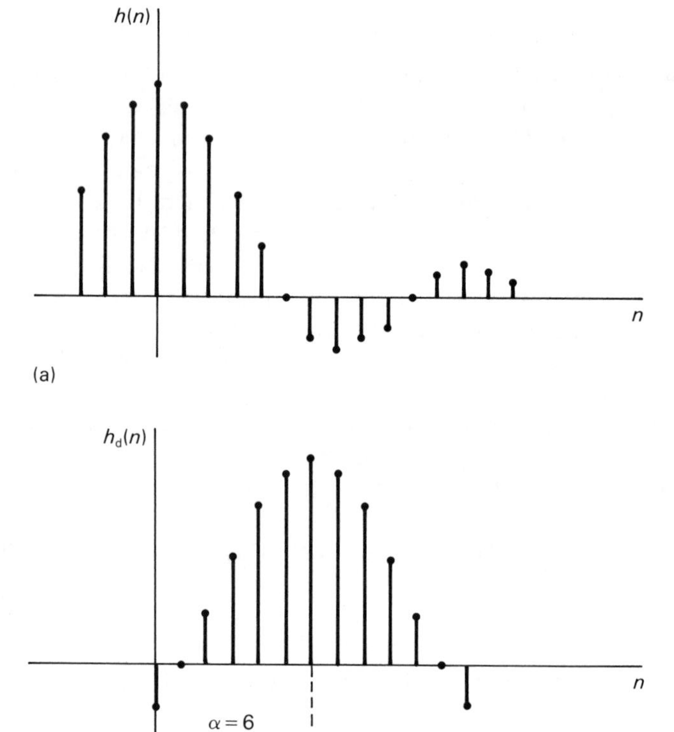

Figure 8.14 Unit sample response of ideal low-pass filter.

Hence the shifted truncated impulse response is given by

$$h_d(n) = \frac{\omega_c}{\pi}\frac{\sin(n-\alpha)\omega_c}{(n-\alpha)\omega_c} \qquad n = 0 \text{ to } N-1 \tag{8.5.1}$$
$$= 0 \qquad \text{otherwise}$$

where

$$\alpha = \frac{N-1}{2} \tag{8.5.2}$$

The case illustrated in Figure 8.14 has taken N to be odd, the delay α is then an integer number of samples. If N is even the delay is not an integer number of samples, however eqn (8.5.1) still gives the required coefficients.

Truncation of the impulse response now means that the ideal frequency response will not be obtained. As with the analogue filters the effect of the truncation is to give ripples in the pass and stop bands and a finite transition region.

The truncated response $h_d(n)$ can be considered to be obtained by multiplying the delayed impulse response by a 'window' function.

$$h_d(n) = h(n-\alpha)\mathrm{w}(n) \qquad \text{for all } n$$

where $\mathrm{w}(n)$ is the window

$$\mathrm{w}(n) = 1 \qquad 0 \leqslant n \leqslant N-1 \tag{8.5.3}$$
$$= 0 \qquad \text{otherwise}$$

Figure 8.15(a) shows the function w(n) and because of its shape, it is termed a rectangular or boxcar window. Because it is the abrupt truncation of the impulse response by the rectangular window that produces the ripples in the frequency response alternative windows can be used that 'taper' the response to zero at each end. Figure 8.15(b) and Figure 8.15(c) show two such windows, the Bartlett window tapers the response linearly, the Hanning uses a cosine taper. The formulae describing these windows are:

Bartlett

$$\mathrm{w}(n) = 2n/(N-1) \qquad 0 \leqslant n \leqslant (N-1)/2$$
$$= 2 - 2n/(N-1) \qquad (N-1)/2 \leqslant n \leqslant N-1 \tag{8.5.4}$$
$$= 0 \qquad \text{otherwise}$$

Hanning

$$\mathrm{w}(n) = 1 - \cos(2\pi n/(N-1))/2 \qquad 0 \leqslant n \leqslant N-1 \tag{8.5.5}$$
$$= 0 \qquad \text{otherwise}$$

Although these windows improve the attenuation in the stop band they do so at the expense of an increased transition region. Other windows are available that give even more attenuation but again at the expense of even wider transition regions.

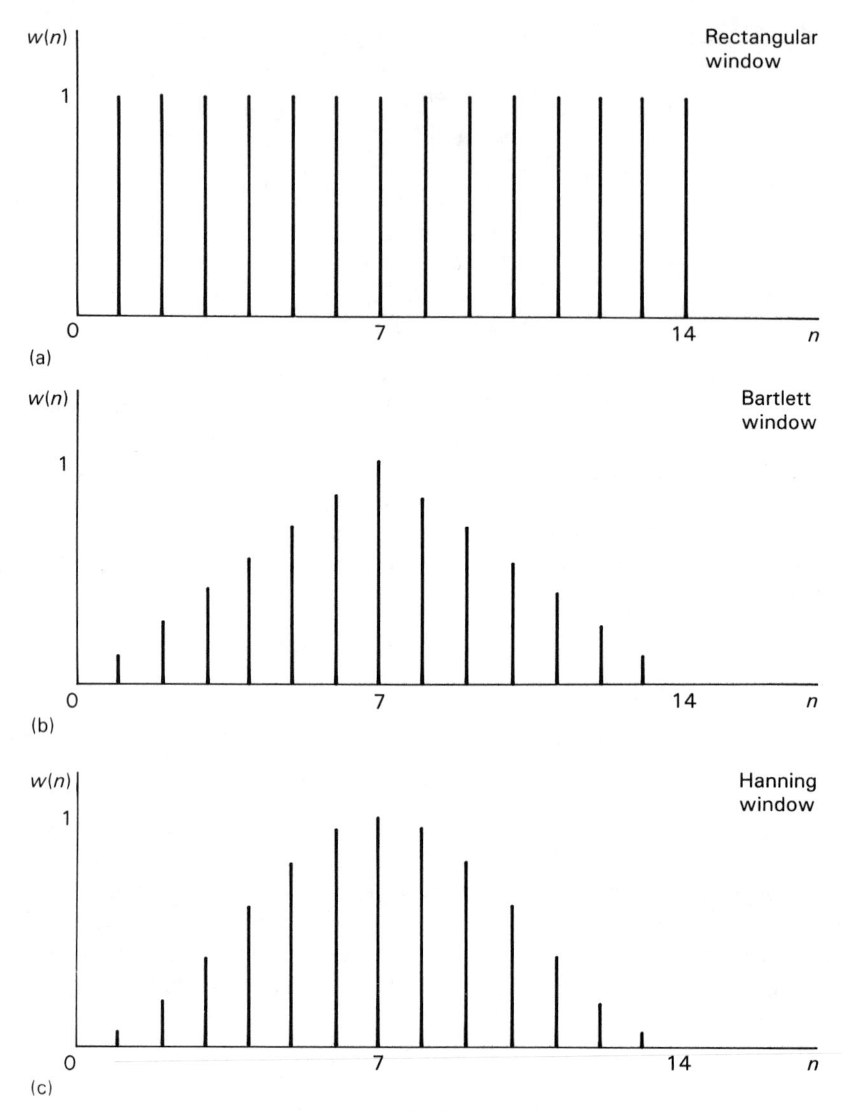

Figure 8.15 Window functions.

The following example illustrates the design of the FIR filter and the use of windows.

EXAMPLE 8.5.1

A low-pass digital filter is to be designed using a sampling frequency of 10 kHz and the cut-off frequency is to be 2 kHz. Determine the values of the coefficients for a fifteen stage filter: (a) using a rectangular window; (b) using a Hanning window. Compare the frequency response obtained for the filter using both these windows.

Table 8.5 Coefficient values for the FIR filter of Example 8.5.1

n	b_n *Rectangular*	$w(n)$ *Hanning*	b_n *Hanning*
0	0.0267	0.0000	0 0000
1	0.0505	0.0495	0.0025
2	0.0001	0.1883	0.0000
3	−0.0357	0.3888	−0.0294
4	−0.0624	0.6114	−0.0381
5	0.0935	0.8119	0.0759
6	0.3027	0.9506	0.2873
7	0.4000	1.0000	0.4000

SOLUTION

Using non-dimensional frequency the cut-off frequency $\omega_c = 0.4\pi$.

For a fifteen stage filter, $N = 15$, the delay α is given by eqn (8.5.2) $\alpha = (N-1)/2 = 7$. The points on the impulse response are given by eqn (8.5.1)

$$h_d(n) = \frac{\omega_c}{\pi}\frac{\sin(n-\alpha)\omega_c}{(n-\alpha)\omega_c} = \frac{\sin 0.4\pi(n-7)}{\pi(n-7)} \qquad n = 0 \text{ to } 14$$

With a rectangular window these are also the coefficients of the filter, $b_n = h_d(n)$, $n = 0$ to 14.

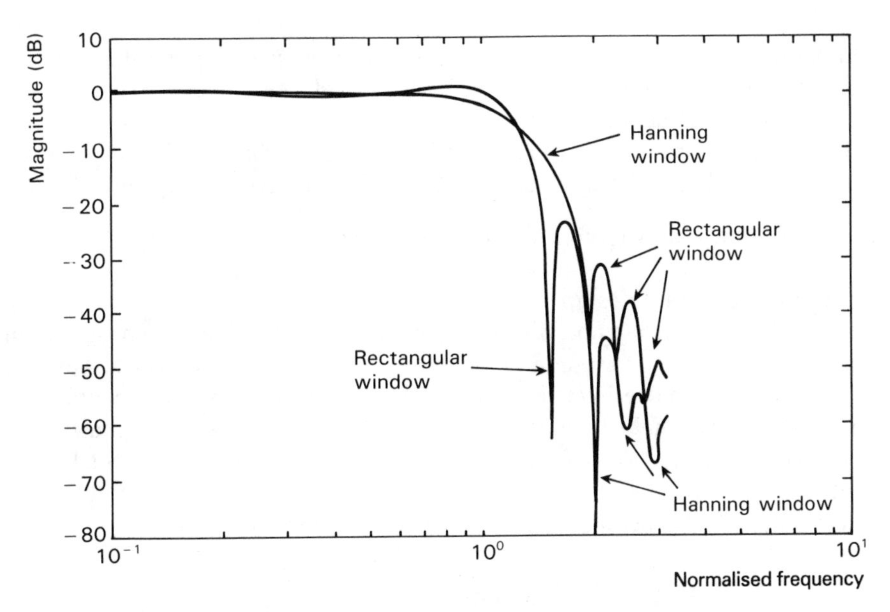

Figure 8.16 Frequency responses of FIR filter for Example 8.5.1.

When the Hanning window is used the values $h_d(n)$ have to be modified by the window function $w(n)$. This is given by eqn (8.5.5), which for $N = 15$ becomes

$$w(n) = [1 - \cos(\pi n/7)]/2 \qquad 0 \leqslant n \leqslant 14$$

Table 8.5 gives the coefficient values b_n for both rectangular and Hanning windows. The values have been given for n, 0 to 7 only, the remaining values are symmetric about $n = 7$, i.e., $b_8 = b_6$, $b_9 = b_5$ etc.

The filter transfer function is given by

$$H(z) = \sum_{n=0}^{14} b_n z^{-n}$$

The frequency response is obtained by substituting $z = e^{j\omega}$ into this equation (ω is normalized frequency). The magnitude response of the filter is shown in Figure 8.16. The use of the Hanning window has given greater attenuation in the stop band and has reduced the ripple in the pass band. These benefits have been obtained at the expense of a wider transition region.

8.5.2 Infinite impulse response (IIR) filters

The two methods to be described for the design of these filters take as their starting point the frequency response of an equivalent analogue filter. The design methods are as follows:

1. Impulse invariant method. This produces a recursive digital filter having an impulse response that is a sampled version of the impulse response of the equivalent analogue filter.
2. Bilinear transform method. This method makes a direct substitution into the transfer function of the equivalent analogue filter to give the required z form.

These methods will now be considered in greater detail.

Impulse invariant design

The FIR filter design method discussed in Section 8.5.1 was based on obtaining an impulse response corresponding to the desired frequency response. This is also the basis of the impulse invariant method of design. However the coefficients in the required filter cannot be obtained directly from the impulse response. Instead a recursive form is considered that produces the required sampled impulse response.

Consider a first order analogue filter having a transfer function

$$H(s) = \frac{1}{1 + sT_1} \tag{8.5.6}$$

By inverting this Laplace transform the corresponding impulse response is

$$h(t) = \frac{1}{T_1} e^{-t/T_1} \qquad t > 0$$

Sampling this response with sampling time T gives

$$h(n) = \frac{1}{T_1} e^{-nT/T_1} \qquad n > 0$$

Taking the z transform gives

$$H(z) = \frac{1}{T_1} \cdot \frac{z}{z - e^{-T/T_1}}$$

Would a digital filter having this z transform give the frequency response of the analogue filter given by eqn (8.5.6)? This question has been answered in Section 6.5, the frequency responses would correspond if the following conditions were satisfied.

1. $H(z)$ were scaled by the sampling time T.
2. The effects of aliasing were negligible.

It is the latter point that causes the problem and causes the departure of the response from the required analogue response. In order to avoid the effects of aliasing, the sampling frequency should be significantly higher than the filter cut-off frequency. Interpretation of 'significantly' can be quite difficult for higher order filters and some form of computer aided design technique is useful.

Considering now the more general case of an nth order filter. Partial fractions can be used to express the transfer function as a sum of the first order terms.

$$H(s) = \frac{k_1}{(s - P_1)} + \frac{k_2}{(s - P_2)} + \cdots\cdots \frac{k_n}{(s - P_n)}$$

P_1, P_2, P_n, are the poles of the transfer function $H(s)$. It is assumed that the poles are all simple poles giving this form of expansion. The poles however need not be real (if they are complex they will occur in conjugate pairs), in this case the associated coefficient in the partial fraction expansion will also be complex.

Now the design method follows that of the first order system

Transfer function $H(s)$	$\longrightarrow$	Impulse response $h(t)$	$\longrightarrow$	Sampled impulse response $h(nT)$	$\longrightarrow$	z transfer function $H(z)$

This gives

$$H(z) = \frac{k_1 z}{(z - e^{P_1 T})} + \frac{k_2 z}{(z - e^{P_2 T})} + \cdots\cdots \frac{k_n}{(z - e^{P_n T})} \tag{8.5.7}$$

The scaling factor T may or may not be included in this expression. If it is included it does not mean that the frequency response of $H(z)$ will coincide with that of $H(s)$ (even at $\omega = 0$) because of aliasing effects. In practice often the scaling is adjusted in the overall system gain, perhaps to make the magnitude of the frequency response correct at some significant frequency.

This method of design will be illustrated by the following example.

EXAMPLE 8.5.2

The following transfer function is for a third order Butterworth filter with a cut-off frequency of 100 rad/s.

$$H(s) = \frac{10^6}{(s + 100)\,(s^2 + 100s + 10^4)}$$

Use the impulse invariant design method to obtain the transfer function $H(z)$ of the equivalent digital filter using a sampling frequency $\omega_s = 400\,\text{rad/s}$.

SOLUTION

The poles of $H(s)$ are at

$$s = -100$$

$$s = \frac{-100 \pm \sqrt{(10^4 - 4 \times 10^4)}}{2}$$

$$= -501(1 \pm j\sqrt{3})$$

Hence

$$H(s) = \frac{A}{(s + 100)} + \frac{B}{(s + 50 + j50\sqrt{3})} + \frac{B^*}{(s + 50 - j50\sqrt{3})}$$

$$A = \frac{10^6}{(10^4 - 10^4 + 10^4)} = 100$$

$$B = \frac{10^6}{(-50 - j50\sqrt{3} + 100)\,(-50 - j50\sqrt{3} + 50 - j50\sqrt{3}}$$

$$= \frac{200}{(-3 - j\sqrt{3})}$$

$$B^* = \frac{200}{(-3 + j\sqrt{3})}$$

Giving

$$H(s) = \frac{100}{s + 100} - \frac{200}{(3 + j\sqrt{3})\,(s + 50 + j50\sqrt{3})} - \text{conjugate terms}$$

The equivalent z transform is given by eqn (8.5.7)

$$H(z) = 100\mathrm{T}\left[\frac{z}{z - \mathrm{e}^{-100T}} - \frac{2z}{(3 + \mathrm{j}\sqrt{3})\,(z - \mathrm{e}^{-50T}\mathrm{e}^{-\mathrm{j}50\sqrt{3}T})} - \text{conjugate}\right]$$

Sampling frequency = 400 rad/s = $2\pi/T$, $T = \pi/200$, $100T = 0.5\pi$, $50T = 0.25\pi$, $50\sqrt{3}T = 1.36$. Inserting these constants the expression for $H(z)$ becomes

$$H(z) = 0.5\pi\left[\frac{z}{z - 0.21} - \frac{2z}{(3 + \mathrm{j}\sqrt{3})\,(z - 0.455\mathrm{e}^{-\mathrm{j}1.36})} - \text{conjugate}\right]$$

But

$$\mathrm{e}^{-\mathrm{j}1.36} = \cos(-1.36) + \mathrm{j}\,\sin(-1.36) = 0.209 - \mathrm{j}0.978$$

giving

$$H(z) = 0.5\pi\left[\frac{z}{z - 0.21} - \frac{2z}{(3 + \mathrm{j}\sqrt{3})\,(z - 0.095 + \mathrm{j}0.445)} - \text{conjugate}\right]$$

The last two terms are in the form of a complex number plus its conjugate, these combine to give twice its real part. Hence the expression for $H(z)$ becomes

$$H(z) = \frac{1.57}{1 - 0.21z^{-1}} - \frac{1.57 - 0.55z^{-1}}{1 - 0.19z^{-1} + 0.21z^{-2}}$$

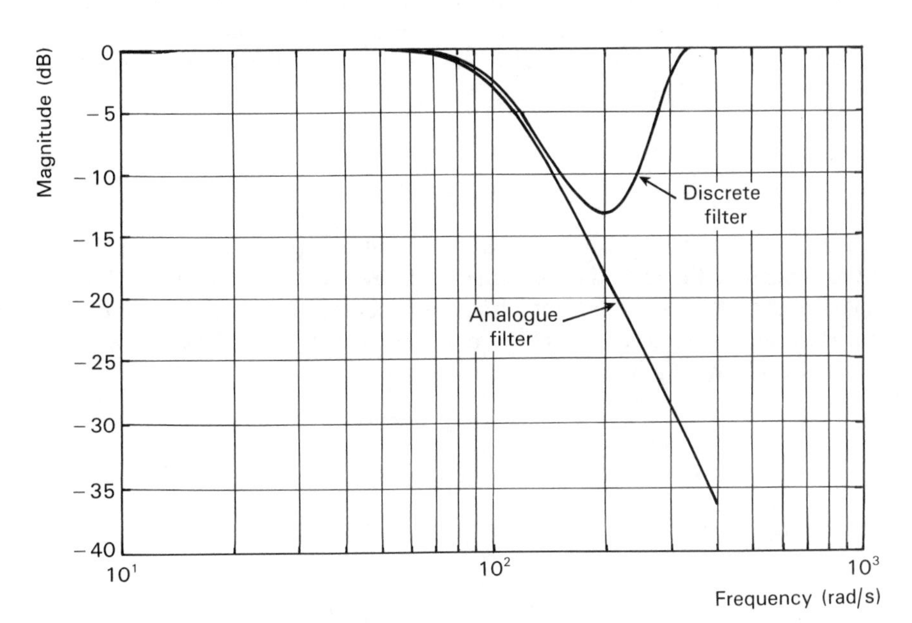

Figure 8.17 Frequency response of IIR filter for Example 8.5.2.

The frequency response of the filter, together with of the basic analogue filter is shown in Figure 8.17.

Design by the bilinear transform

In Section 3.9.2 the problem of digital simulation was considered. For a first order system it was shown that the differential equation representing the system can be approximated by a first order difference equation. This equivalent system can be considered to be an equivalent digital filter. Consider the operation of integration

$$y = \int_{t_0}^{t} x(t)\,\mathrm{d}t$$

By use of the mid-ordinate rule this operation can be replaced by a discrete equivalent. Making $t - t_0$ equal to the sampling time T then

$$y(n) = \frac{T(x(n) + x(n-1))}{2} + y(n-1)$$

Using the z transform, $Y(z)$ can be expressed

$$y(z) = \frac{T}{2}\,\frac{(1 + z^{-1})}{(1 - z^{-1})}\,X(z)$$

Figure 8.18 shows the analogue and discrete operations in transfer function form.

This correspondence has been obtained for one integrator. However a system can be completely realised with integrators as the only dynamic element in state variable form. By following this approach it can be shown that the equivalent discrete form can be obtained by making the substitution

$$s = \frac{2}{T}\,\frac{(1 + z^{-1})}{(1 - z^{-1})} \qquad (8.5.8)$$

in the transfer function. The transformation defined by eqn (8.5.8) is a special case of the more general mathematical transformation known as the bilateral transform.

Having obtained the equivalent z transfer function its frequency response can be obtained by substitution of $z = \mathrm{e}^{\mathrm{j}\omega T}$ (note ω is not normalised here). This is equivalent to substituting into $H(s)$,

$$s = \frac{2}{T}\,\frac{(1 - \mathrm{e}^{-\mathrm{j}\omega T})}{(1 + \mathrm{e}^{-\mathrm{j}\omega T})}$$

This expression for s can be simplified by writing it in the form

$$s = \frac{2}{T}\,\frac{(\mathrm{e}^{-\mathrm{j}\omega T/2}\,\mathrm{e}^{\mathrm{j}\omega T/2} - \mathrm{e}^{-\mathrm{j}\omega T/2}\,\mathrm{e}^{-\mathrm{j}\omega T/2})}{(\mathrm{e}^{-\mathrm{j}\omega T/2}\,\mathrm{e}^{\mathrm{j}\omega T/2} + \mathrm{e}^{-\mathrm{j}\omega T/2}\,\mathrm{e}^{-\mathrm{j}\omega T/2})}$$

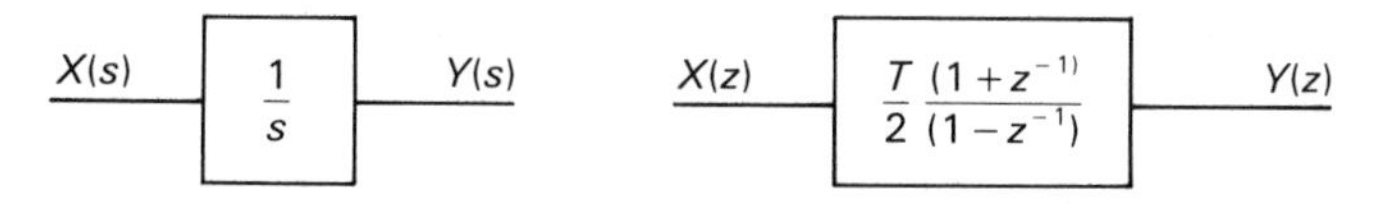

Figure 8.18 Analogue integration and its discrete equivalent.

Dividing through by $e^{-j\omega T/2}$ the expression becomes

$$s = j\frac{2}{T}\frac{\sin(\omega T/2)}{\cos(\omega T/2)} = \frac{j2}{T}\tan\frac{\omega T}{2}$$

This has no real part, when substituted in $H(s)$ it will produce a frequency response function $H(j\omega_d)$ where

$$\omega_d = \frac{2}{T}\tan\frac{\omega T}{2}$$

The operation corresponds to taking the frequency response of the analogue filter and then 'distorting' or 'warping' the frequency scale to produce the response of the equivalent digital filter.

This can be illustrated by means of a first order system having a frequency response function

$$H(j\omega) = \frac{1}{1 + j(\omega/\omega_c)}$$

and taking the sampling frequency ω_s to be three times the cut-off frequency ω_c

$$\omega_d = \frac{2}{T}\tan\frac{\omega T}{2} = \frac{3\omega_c}{\pi}\tan\frac{\omega\pi}{\omega_c}$$

Hence the digital filter has a frequency response given by

$$H(j\omega_d) = \frac{1}{1 + j(\omega_d/\omega_c)}$$

The magnitude of this response is shown in Figure 8.19 where it is obtained by warping the frequency scale of the response $|H(j\omega)|$.

Some correction can be made for the frequency warping by 'pre-warping' the cut-off frequency ω_c. The cut-off frequency ω_{cd} is obtained as

$$\omega_{cd} = \frac{2}{T}\tan\frac{\omega_c T}{2}$$

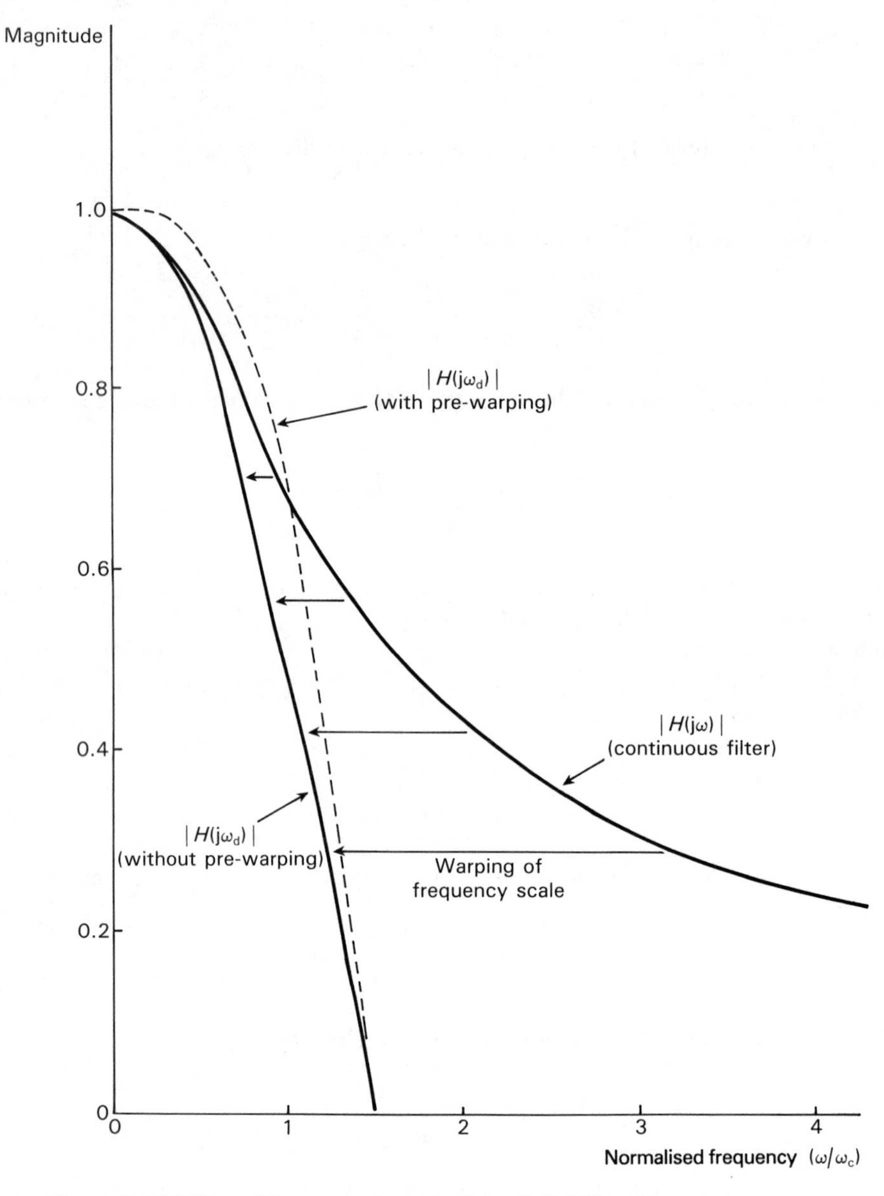

Figure 8.19 Effect of frequency 'warping' on digital filter frequency response.

and the digital filter now has a response

$$H(j\omega_d) = \frac{1}{1 + j(\omega_d/\omega_{cd})}$$

This is also plotted on the magnitude response of Figure 8.19.

In a more general filter the specification may include more than one cut-off or transition frequency and all these frequencies are subject to pre-warping.

The following example illustrates the use of the bilinear transform in the design of a third order Chebyshev filter.

EXAMPLE 8.5.3

A third order Chebyshev filter has a normalized transfer function.

$$H(s) = \frac{0.4913}{(s + 0.4942)\,(s^2 + 0.4941\,s + 0.9942)}$$

and its cut-off frequency is 200 rad/s. It is required to construct an equivalent digital filter, using the bilinear transform method, with a sampling frequency of 600 rad/s. Determine the required z transfer function: (a) without pre-warping of the cut-off frequency; (b) with pre-warping of the cut-off frequency. Compare the magnitude response of these two filters with that of the analogue filter.

SOLUTION

Although the algebra involved in the bilinear transform method is straightforward it can be rather tedious. Also care must be taken with the numerical work. Often small coefficients can appear as the difference between large coefficients. Sufficient decimal places must be maintained to prevent this causing undue error.

The transfer function must be 'un-normalized' by replacing s by s/ω_c where ω_c is the cut-off frequency. Then the bilinear transform

$$s = \frac{2}{T}\,\frac{(1 - z^{-1})}{(1 + z^{-1})}$$

is used.

Writing $T = 2\pi/\omega_s$, where ω_s is the sampling frequency used, gives

$$s = x\,\frac{(1 - z^{-1})}{(1 + z^{-1})}$$

where $x = \omega_s/\pi\omega_c$. Hence

$$H(z) = \frac{0.4913}{\left[\dfrac{x(1 - z^{-1})}{(1 + z^{-1})} + 0.4942\right]\left[\dfrac{x^2(1 - z^{-1})^2}{(1 + z^{-1})^2} + \dfrac{0.4941x(1 - z^{-1})}{(1 + z^{-1})} + 0.9942\right]}$$

Multiplying through by $(1 + z^{-1})^3$ and collecting terms gives

$$H(z) = \frac{0.4913(1 + z^{-1})^3}{\begin{array}{l}[(x + 0.4942) + (0.4942 - x)z^{-1}] \\ \times\, [(x^2 + 0.4941x + 0.9942) + (1.9884 - 2x^2)z^{-1} + (x^2 - 0.4941x + 0.9942)z^{-2}]\end{array}}$$

Without pre-warping $\omega_s/\omega_c = 3$ and $x = 0.9549$ giving

$$H(z) = \frac{0.1426(1+z^{-1})^3}{(1-0.3179z^{-1})(1+0.0692z^{-1}+0.6032z^{-2})} \tag{8.5.9}$$

With pre-warping the cut-off frequency changes to ω_{cd}

$$\omega_{cd} = \frac{2}{T}\tan\frac{\omega_c T}{2}$$

and substituting $\omega_c = \omega_s/3$ and $T = 2\pi/\omega_s$ gives

$$\omega_{cd} = \frac{\omega_s}{\pi}\tan\frac{\pi}{3}$$

$$x = 1/\tan\frac{\pi}{3} = 0.5773$$

Then

$$H(z) = \frac{0.2843(1+z^{-1})^3}{(1-0.0776z^{-1})(1+0.8196z^{-1}+0.6462z^{-2})} \tag{8.5.10}$$

The frequency responses corresponding to the transfer functions given by eqns (8.5.9) and (8.5.10) are plotted in Figure 8.20. The transfer functions given by these

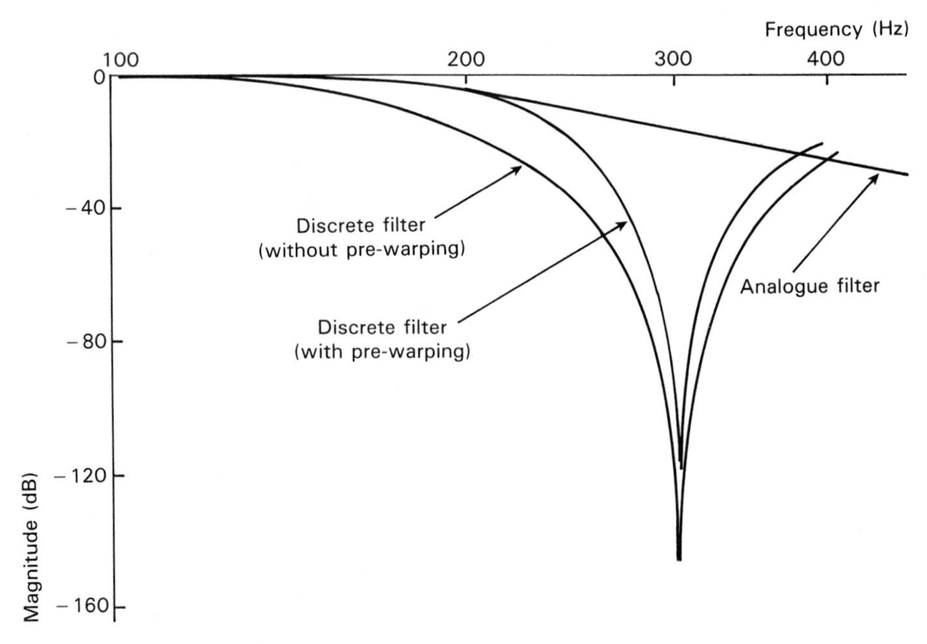

Figure 8.20 Frequency responses of filter in Example 8.5.3.

equations could be expressed as ratio of polynomials. The form chosen depends upon the method of realisation and this will be the subject of the next section.

8.5.3 Realisation of digital filters

The design methods discussed in the last section produced a description of the required filter in the form of a transfer function $H(z)$. This transfer function can be interpreted as a difference equation and implemented by using delay elements and weighted additions as described in Section 3.9.2. Take as an example a digital filter having a transfer function.

$$H(z) = \frac{Y(z)}{X(z)} = \frac{1 + z^{-1} + 0.5z^{-2}}{1 - 0.75z^{-1} + 0.125z^{-2}} \tag{8.5.11}$$

which gives a difference equation

$$y(n) = x(n) + x(n-1) + 0.5x(n-2) + 0.75y(n-1) - 0.125y(n-2)$$

This can be implemented as shown in Figure 8.21 when a delay element has been indicated by the operation z^{-1}.

Consider now the transfer function $H(z)$ written in terms of an intermediate variable $U(z)$

$$H(z) = \frac{U(z)}{X(z)}\,\frac{Y(z)}{U(z)}$$

where

$$\frac{U(z)}{X(z)} = \frac{1}{1 - 0.75z^{-1} + 0.125z^{-2}}$$

$$\frac{Y(z)}{U(z)} = 1 + z^{-1} + 0.5z^{-2}$$

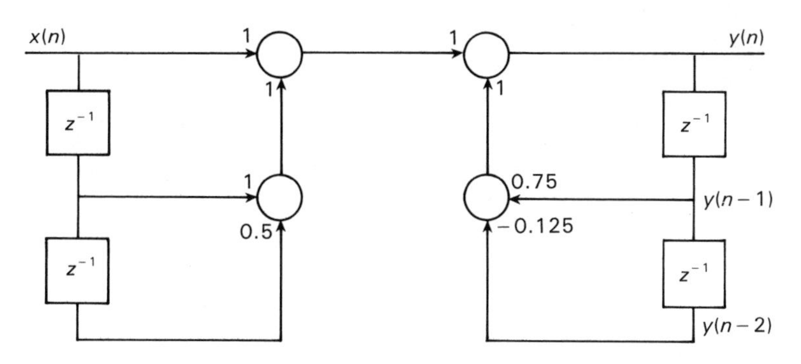

Figure 8.21 Implementation of digital filter.

These equations can be implemented by two cascaded systems as shown in Figure 8.22(a).

However there is no need to produce two sets of the delayed variables $u(n-1)$ and $u(n-2)$, the realisation can be re-drawn as shown in Figure 8.22(b).

It should be emphasised that the realisations shown in Figure 8.21 and Figure 8.22(b) both give the same transfer function, the $H(z)$ as defined by eqn (8.5.11). The practical implications for these realisations are however different. The realisation of Figure 8.21 requires four delay elements (memories) while that of Figure 8.22(b) requires only two. In the first realiation all the variables sum to give $y(n)$ and the order of this summation would not matter. However in the second realisation the intermediate variable $u(n)$ is required and the order of the summation is important.

The realisations of Figure 8.21 and Figure 8.22 are known as direct realisations. By manipulation of the transfer function $H(z)$ it is possible to produce an infinite number of realisations and all these are theoretically equivalent. However when a number of practical factors are taken into account these realisations can give very different practical results. Some of the more important of these factors are as follows.

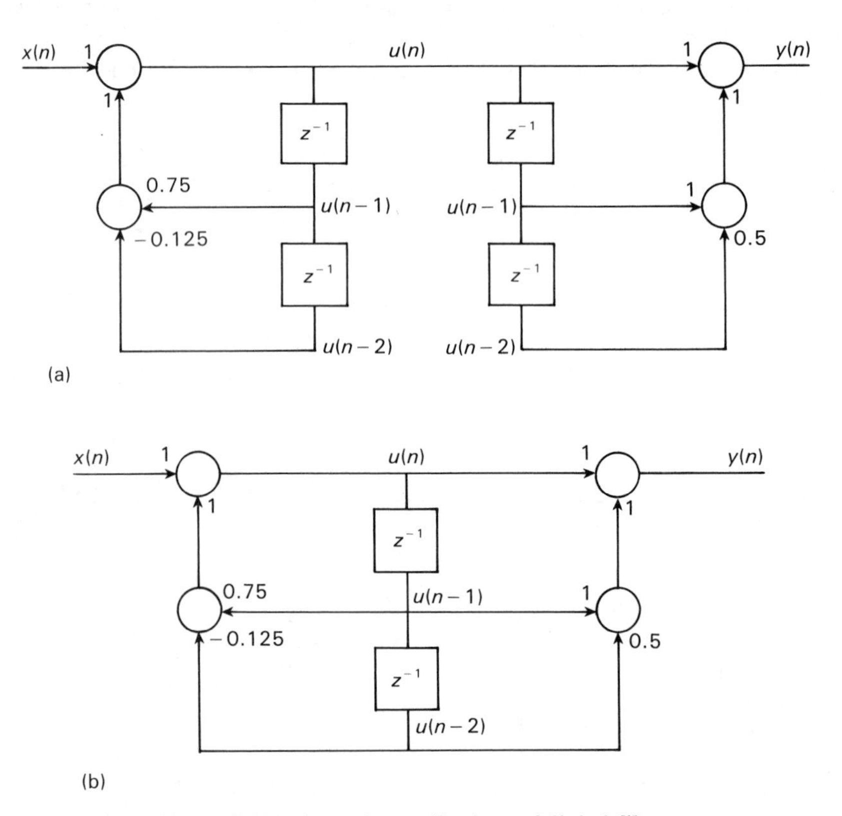

Figure 8.22 Alternative realisations of digital filters.

The majority of digital filters are going to be implemented by some form of digital computer. This means that the coefficients and variables will be represented with finite precision. It is important to be able to evaluate how this representation will affect the overall performance of the filter and to choose a realisation that is at least sensitive to its effects.

Memory is required not only to store coefficient values but also to store previous values of input and output variables. As shown in the last example different realisations can produce different memory requirements and this may or may not be a relevant factor depending on the physical system used for filter implementation.

The computing complexity required can be influenced by the realisation adopted. The number of multiplications required, the number of times a number is retrieved from memory, the sequence in which operations have to be performed, all these factors are influenced by the realisation chosen.

The factors discussed in the previous paragraphs interact with the hardware used for the implementation. Is the filter to be implemented on a main frame, a micro-computer, a custom built chip? Are special processors available, can parallel processing be used? A full discussion of possible realisations with regard to all the factors outlined is well beyond the scope of this book. The two methods of realisation that have been described are known as direct realisations. It can be shown that as the number of poles and zeros in the transfer function $H(z)$ increases then these realisations are very sensitive to errors due to finite word length. Hence such realisations are usually restricted to functions having at the most two poles and two zeros, biquadratic functions. Two realisations will be considered that show how high order transfer functions can be constructed using the biquadratic factors as unit building blocks.

Cascade realisations

In a cascade realisation the required transfer function $H(z)$ is represented as a product of factors as shown in Figure 8.23

$$H(z) = H_1(z)H_2(z) \ldots\ldots H_k(z)$$

There are usually many ways in which the numerator and denominator can be factorised to produce this form. However in order to reduce the sensitivity to finite precision effects, the number of poles and zeros in each section should be kept to a minimum. The most usual form is the biquadratic factor

$$H_i(z) = \frac{b_{0i} + b_{1i}z^{-1} + b_{2i}z^{-2}}{1 + a_{1i}z^{-1} + a_{2i}z^{-2}}$$

Theoretically the first order factors in the numerator and denominator of $H(z)$ can be arbitrarily grouped in pairs to form the biquadratic factors. However as the coefficients in the biquadratic factors must be real complex conjugate factors must be taken as a pair. For real first order factors there is some choice in the pairings, although theoretically identical, different pairings can produce different practical

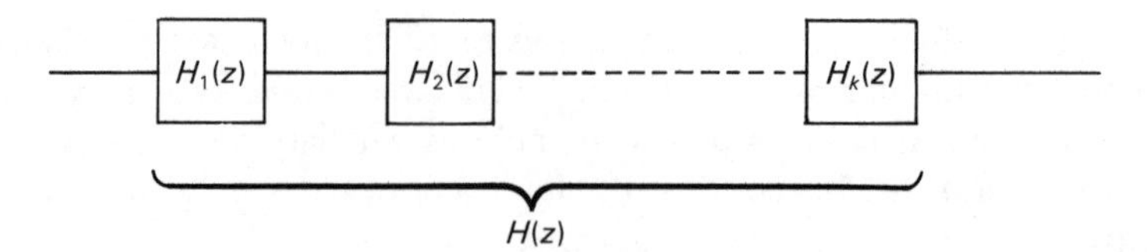

Figure 8.23 Cascade realisations of transfer function.

results. If the order of numerator or denominator is odd there will be a first order factor unpaired. This can either be realised as a first order factor or by setting the coefficient b_{2i} or a_{2i} to zero in the biquadratic factor.

Parallel realisation

As its name implies this realisation is formed from a number of parallel paths as shown in Figure 8.24.

The transfer function of each path can be obtained by expressing the function $H(z)$ as a partial fraction expansion. To avoid complex coefficients the expansion takes the form

$$H(z) = H_1(z) + H_2(z) + \ldots\ldots H_k(z)$$

where the ith term takes the form of a linear numerator and a quadratic denominator.

$$H_i(z) = \frac{b_{0i} + b_{1i}z^{-1}}{1 + a_{1i}z^{-1} + a_{2i}z^{-2}}$$

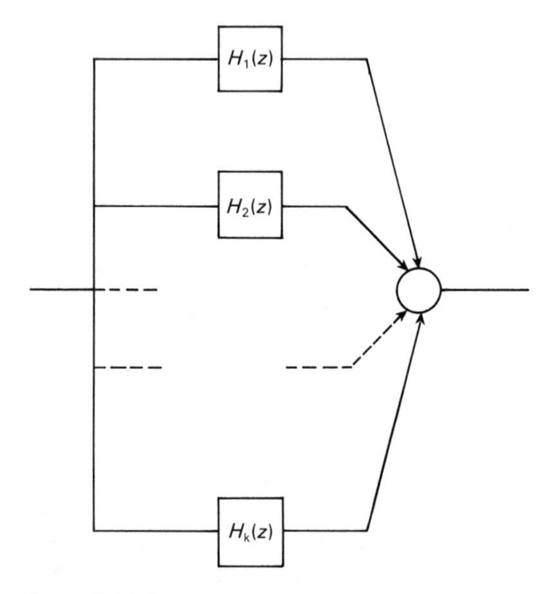

Figure 8.24 Parallel realisation of transfer function.

As with the cascade form the groupings of the poles to form the denominator is to some extent arbitrary although poles that are complex conjugates must be grouped to give real coefficients. By making some of the coefficients zero a linear factor or a constant term can be produced.

As stated earlier both the cascade and parallel structures give reduced sensitivity to finite word length effects when compared with direct structures. Whether the cascade or the parallel structures are superior in this respect depends upon the individual filter under consideration. However the parallel form has the advantage that there is more freedom regarding the order in which the computations are performed and this form is more suited to parallel processing.

The following example takes the transfer function developed in Example 8.5.2 and develops both a cascade and a parallel realisation.

EXAMPLE 8.5.4

In Example 8.5.2 the following transfer function was obtained for a digital filter

$$H(z) = \frac{1.57}{1 - 0.21z^{-1}} - \frac{1.57 - 0.55z^{-1}}{1 - 0.19z^{-1} + 0.21z^{-2}}$$

Show how this transfer function can be realised in both parallel and cascade forms.

SOLUTION

As given, the transfer function is in a convenient form for parallel realisation. This realisation is shown in Figure 8.25.

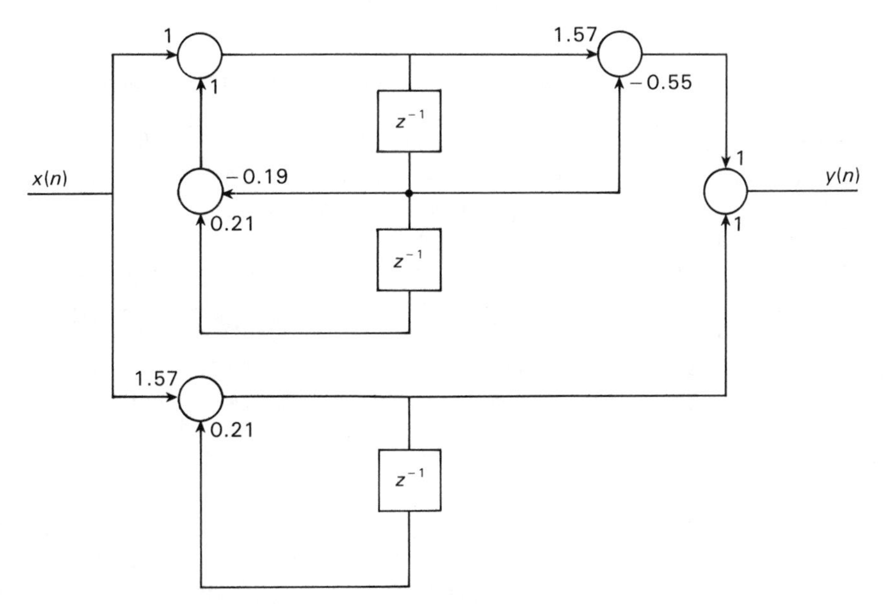

Figure 8.25 Parallel realisation for Example 8.5.4.

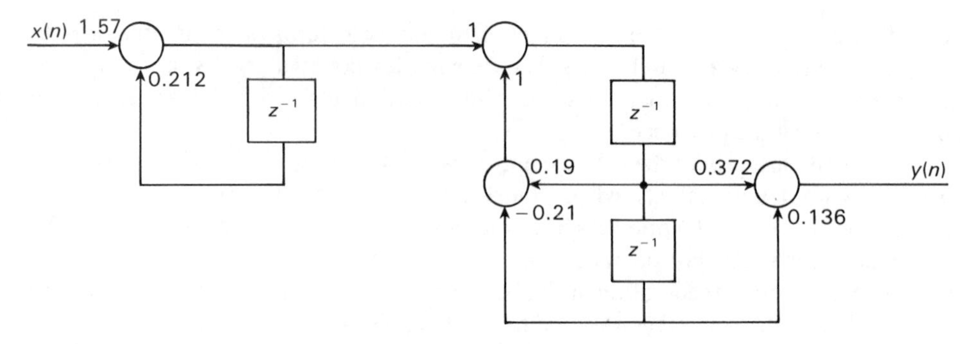

Figure 8.26 Cascade realisation for Example 8.5.4.

For cascade realisation the terms must be combined giving

$$H(z) = \frac{1.57(0.372z^{-1} + 0.136z^{-2})}{(1 - 0.21z^{-1})(1 - 0.19z^{-1} + 0.21z^{-2})}$$

Some choice is available regarding the grouping of the numerator and denominator terms. A possible cascade realisation is shown in Figure 8.26.

8.6 Summary

This chapter has considered the filtering of signals from the frequency domain viewpoint. In the frequency domain an ideal filter gives infinite attenuation over certain bands of frequencies and zero attenuation over others. Ideal filters cannot be realised practically and transition regions must be accepted that have finite, non-zero attenuation. It is the form of compromise chosen between the ideal and practical responses that leads to a wide range of design methods.

Analogue filters operate on analogue signals and two design methods have been investigated for these filters. Considering only low-pass filters, a Butterworth response gives no ripples in the pass band but a relatively wide transition region. For the same order filter the Chebyshev response gives a narrower transition region but at the expense of ripples in the pass band. Methods of design, and the implementation of the filters by active circuits have been considered.

Digital filters operate on discrete signals. These filters can be divided into finite impulse response (FIR) and infinite impulse response (IIR) filters. Design methods have been considered for both these types of filter. Practical problems of implementing the filters have been discussed and different realisations derived.

Problems

8.1 It is required to construct a low-pass Butterworth filter having a cut-off frequency of 1 kHz. Determine the pole positions and hence the transfer functions for both a third and fourth order filter. Express the transfer function in both polynomial and quadratic factored forms.

Compare the attenuation of the two filters (in dB) at a frequency equal to twice the cut-off frequency.

8.2 A low-pass Butterworth filter is to have a cut-off frequency of 20 kHz. Its attenuation at three times the cut-off frequency is to be at least 25 dB. Obtain the transfer function of the lowest order filter that will meet this specification.

8.3 The circuit shown in Figure 8.27 is the equivalent circuit of the coupling between stages of a transistor amplifier. The inductor L provides 'series peaking' frequency compensation. If $C = 30$ pF and $L = 120$ μH determine the value of R such that the stage gives a second order Butterworth response.

If two such stages are cascaded, each stage having the L and C values above, determine the values of R in each stage to give a fourth order Butterworth response.

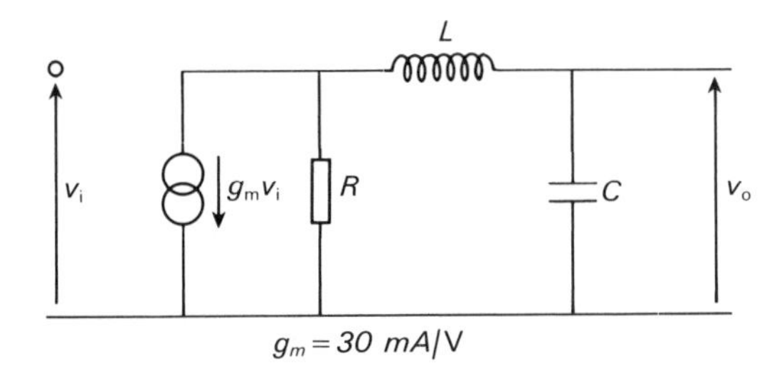

Figure 8.27

8.4 High-pass filters can be derived from low-pass designs by using a transformation of s. Specifically if the normalised low-pass frequency response function is used then replacing s by ω_c/s will produce a high-pass filter with cut-off frequency ω_c. Derive the transfer function for a third order, high-pass Butterworth filter with a cut-off frequency of 200 Hz. Evaluate the magnitude of this function at frequencies equal to twice the cut-off frequency and half the cut-off frequency.

8.5 Design a low-pass Chebyshev filter of order three having a cut-off frequency of 10 kHz and 1 dB of ripple in the pass band. What is the attenuation at a frequency equal to five times the cut-off frequency?

8.6 A low-pass Chebyshev filter is to have a cut-off frequency of 50 Hz and a pass band ripple of 0.5 dB. The filter is to be the lowest order that will give an attenuation of at least 20 dB at twice the cut-off frequency. Obtain the design for such a filter and obtain the attenuation at twice the cut-off frequency.

8.7 By cascading the circuits shown in Figures 8.10 and 8.11(a) produce an active implementation of a third order low-pass Chebyshev filter. The ripple in the pass band is to be 0.5 dB and the cut-off frequency is to be 1 kHz. Use the resistor values of 1 kΩ throughout.

8.8 The circuit shown in Figure 8.28 is an alternative method of implementing a second order term in active filters. Show that this circuit produces a second order term with

$$\omega_n = \frac{1}{RC} \qquad \zeta = \frac{(3-K)}{2},$$

$$\text{Zero frequency gain} = K, \qquad K = 1 + \frac{R_1}{R_2}$$

Using resistor values of 5 kΩ determine the values of C and K to produce a second order Butterworth response with a cut-off frequency of 1.5 kHz. (The gain can be made unity at $\omega = 0$ by use of a simple attenuator.)

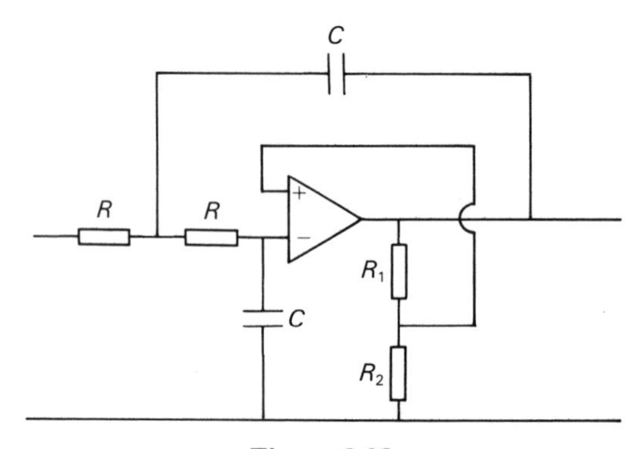

Figure 8.28

8.9 A second order high-pass transfer function can be obtained by the circuit shown in Figure 8.29. Obtain values of ω_n and ζ in terms of the circuit components.

Given $C = 0.1$ μF obtain the values of R to obtain a second order, high-pass Butterworth response with a cut-off frequency of 500 Hz. (Use the transformation given in Problem 8.4.)

8.10 A low-pass FIR digital filter is to have a cut-off frequency of 100 Hz. The sampling frequency is to be 1 kHz. Obtain the coefficients of the filter: (a) using a rectangular window; (b) using a Bartlett window.

Obtain the attenuation of the filter at 300 Hz for both these windows.

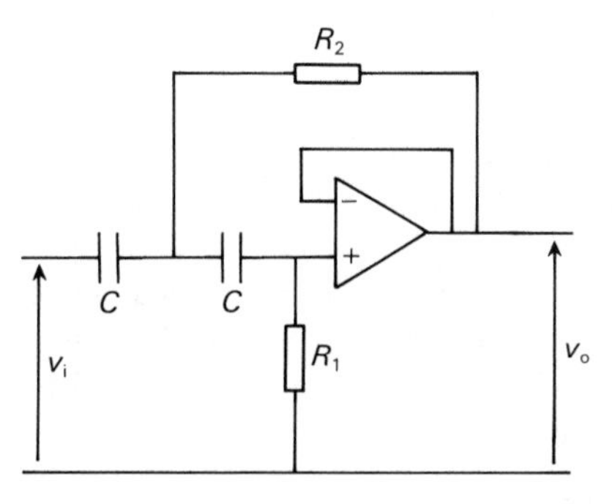

Figure 8.29

8.11 The frequency response of an ideal differentiator is shown in Figure 8.30 (normalised frequency ω). Design a 5 stage and a 9 stage FIR filter to give this response. Calculate the magnitude and phase of the response at $\omega = \pi/2$ and compare this with the required response.

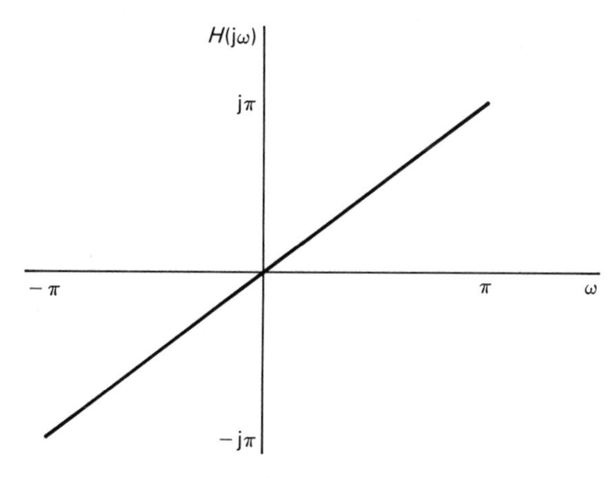

Figure 8.30

8.12 It is required to construct a digital filter whose frequency response corresponds to that of an analogue integrator

$$H(s) = \frac{1}{s}$$

The sampling frequency is to be 1 kHz. Obtain the transfer function $H(z)$, of the equivalent digital filter by: (a) impulse invariant design; (b) the bilinear transformation. Calculate the magnitude and phase of the two digital filters at a frequency of 200 Hz. Compare these with the corresponding figures for the analogue filter.

8.13 A second order low-pass Chebyshev filter has a cut-off frequency of 100 Hz and the ripple in the pass band is to be 0.5 dB. Obtain the transfer function of the equivalent digital filter using the impulse invariant method of design. Use a sampling frequency of 1 kHz.

8.14 A third order low-pass Butterworth filter has a cut-off frequency of 20 Hz. The equivalent digital filter is required where the sampling frequency is 100 Hz. Determine the equivalent filter by means of the bilinear transform, (a) without pre-warping the frequency (b) with pre-warping.

8.15 A second-order Butterworth filter (cut-off frequency 1 rad/s) has a transfer function

$$H(s) = \frac{1}{s^2 + \sqrt{2}s + 1}$$

If the sampling frequency is 1 Hz determine the corresponding z forms by: (a) impulse invariant design; (b) use of the bilinear transform.

Obtain the magnitude of the frequency response of the analogue filter and the two digital filters at a frequency of 2 rad/s.

8.16 A digital filter has a transfer function given by

$$H(z) = \frac{z^2 + z}{z^3 - 0.7z^2 + 0.15z - 0.025}$$

Given that $(z - 0.5)$ is a root of the denominator obtain the following realisations of the filter:

(a) A direct realisation using as few delay elements as possible.
(b) A cascade realisation.
(c) A parallel realisation.

MATLAB exercises

8.17 This exercise examines the frequency responses of different orders of Butterworth low-pass filters.

Using eqn (8.3.2), write a MATLAB program to plot the magnitude of the frequency response of a low-pass Butterworth filter with cut-off frequency 1 rad/s. Obtain plots for filter orders 1–5 and superimpose these onto a single display using a frequency range 0–5 rad/s.

Check these responses against those obtained using the MATLAB command `butter(n,Wn,'s')`.

8.18 This exercise investigates the pole positions for a low-pass Butterworth filter.

The poles of an Nth order Butterworth filter are the poles of the function

$$|H(s)|^2 = \frac{1}{1 + (s/\mathrm{j}\omega_c)^{2N}}$$

that lie in the left half plane.

Write a MATLAB program that will determine the pole positions of an Nth order filter, with $\omega_c = 1$, and display these poles on a plot representing the left half complex plane. Add to this display a plot representing a unit semi-circle and confirm that all the poles lie on this contour.

Use the MATLAB command `[z,p,k]=butter(N,Wn,'s')` to obtain the required pole positions and confirm that these agree with the positions already calculated.

Note The poles of $|H(s)|^2$ can be determined by using the command `r=roots(p)` where `p` is the vector of the coefficients of the denominator polynomial. The plot of the pole positions can be obtained by plotting the imaginary part of the pole against its real part. The unit circle can be added by use of the equation

$$x^2 + y^2 = 1$$

8.19 This exercise examines the responses of Chebyshev low-pass filters.

Using eqn (8.3.7), write a MATLAB program to display magnitude responses of Chebyshev filters with a cut-off frequency of 1 rad/s. For a ripple factor $\varepsilon = 0.1$, plot responses, all on the same display, for filter orders 1 to 5 over a frequency range 0 to 5 rad/s.

Repeat for a third order filter using values of ε of 0.5, 1.0, 1.5 and 2.0.

Plot the response of a third order filter, $N = 3$, with $\varepsilon = 0.5$. Superimpose on this the response obtained by using the MATLAB call `cheby1 (N,Rp,1,'s')` where `Rp` is the peak–peak ripple expressed in dB.

Note A problem of writing the required program is that of calculating the values of the Chebyshev polynomials. Tables of the polynomials are available (as in Table 8.1) and these could be stored within the program. However, this process would become laborious if high polynomial orders are required. An alternative is to use the recurrence relationship between polynomials

$$T_{(n+1)}(x) = 2xT_n(x) - T_{n-1}(x)$$

with

$$T_0(x) = 1 \quad \text{and} \quad T_1(x) = x$$

This relationship can be easily programmed into MATLAB treating the right hand side as a product of the vectors $[2x - 1]$ and $[T_n \ T_{n-1}]^{\mathrm{T}}$.

8.20 This exercise plots, on the complex plane, the positions of the poles of a low-pass Chebyshev filter.

The procedure is similar to Exercise 8.18 which was based on a Butterworth filter. However, it is not easy to obtain the required pole positions for a Chebyshev filter and the command `[z,p,k]=cheby1(n,Rp,Wn,'s')` is used. The poles do not lie on a circle but on an ellipse whose equation is

$$\frac{x^2}{a^2} + \frac{y^2}{b^2} = 1$$

where

$$a = \sinh\left[\frac{1}{N}\sinh^{-1}\left(\frac{1}{\varepsilon}\right)\right] \qquad b = \cosh\left[\frac{1}{N}\sinh^{-1}\left(\frac{1}{\varepsilon}\right)\right]$$

Plot the portion of the ellipse that lies in the left half plane and verify that the filter poles lie on this contour for arbitrary N and ε.

8.21 This exercise investigates the frequency response of an FIR low-pass digital filter.

The impulse response, $h(n)$, of such a filter is given by

$$h(n) = \frac{\omega_c}{\pi}\,\frac{\sin(n\omega_c)}{n\omega_c}$$

where ω_c is the normalised cut-off frequency.

(a) Plot the impulse response and frequency response (gain and phase) of a low-pass FIR filter with $\omega_c = 0.1$, when n covers a large range, -1000 to $+1000$ say.

(b) Investigate the effect of truncating the impulse response by restricting n to the range -5 to $+5$.

(c) Investigate the effect of truncating the impulse response but not keeping its symmetry by letting n take the range -3 to $+5$.

(d) Returning to the symmetrical impulse response, investigate the effect of using Hanning and triangular windows.

Note The impulse responses obtained exist for negative sample values and, as such, are unrealisable. However, as the index of an element in the vector representing the impulse response cannot be negative, this vector represents a delayed impulse response which is realisable.

The frequency response of the digital filter can be obtained by using the command `freqz`. It should be noted that this command uses a frequency scale that is normalised to half the sampling frequency. The command `freqz` is not very flexible in its portrayal of the response. In this exercise it is useful to compare responses by plotting several on one graph. In later problems it will be a requirement to plot responses from both analogue and digital filters together for comparison purposes. For this reason, the program written for Exercise 5.25 is best used here.

8.22 This exercise investigates the impulse invariant method of digital filter design.

By use of the command `[b,a]=butter(n,Wn,'s')`, determine the transfer function of a fifth order, low-pass Butterworth filter having a cut-off frequency of 100 rad/s. Write a MATLAB program using the impulse invariant design method to produce the pulse transfer function of the equivalent digital filter using a sampling frequency of 100 Hz.

Check the coefficients obtained for the digital filter by use of the command `impinvar` and plot the magnitude responses of both analogue and digital filters on the same graph.

Obtain the impulse response of both the analogue and digital filters and plot them together for comparison purposes.

Note The impulse invariant design method is outlined under the command `impinvar` in Version 4 of the user's guide to the Student Edition of MATLAB. The command `residue` can be used to express the analogue filter's transfer function as a partial fraction expansion. The s domain poles are then replaced by the z domain poles and the new transfer function obtained using `residuez`.

8.23 This exercise compares the frequency responses of FIR and IIR filters.

Using the impulse invariant design method, obtain the coefficients of a digital filter based on a fifth order Butterworth filter having a cut-off frequency of 1 Hz and sampling frequency 10 Hz.

Obtain the coefficients of an 11 stage FIR filter with cut-off and sampling frequencies as above.

Compare the frequency responses (both magnitude and phase) of both these filters and examine the effect of using different windows in the FIR filter design.

8.24 This exercise illustrates the bilinear design method for digital filters.

Using the command `[b,a]=butter(n,Wn,'s')`, obtain the coefficients in the transfer function of a fifth order, low-pass, Butterworth filter with a cut-off frequency of 100 rad/s.

Write a MATLAB program to use the bilinear transform method to obtain the coefficients in a digital filter based upon the Butterworth design. Use a sampling frequency of 2000 rad/s. Verify these coefficients by use of the command `bilinear`. Plot the magnitude responses of both the analogue and digital filters on the same graph.

Note The outline of a general method of design is given under the command `bilinear` in Version 4 of the user's guide to the Student Edition of

MATLAB. The method transforms the pole positions of the continuous transfer function into the pole positions of the discrete transfer function by use of the bilinear transform. The pole positions can be obtained from the command `roots` and, after suitable transformations, the polynomial in z can be obtained by using the command `poly`. Suitable scaling is then required and zeros added at $z = 1$.

8.25 This exercise illustrates the 'warping' of the frequency axis in the frequency response of digital filters designed by the bilinear transformation method.

(a) With the aid of the command `cheby1(n,Rp,Wn,'s')`, plot the magnitude response of a fifth order Chebyshev low pass filter with 2 dB of ripple and a cut-off frequency of 1 rad/s. With the aid of the command `bilinear(num,den,Fs)`, obtain the frequency response of the equivalent digital filter using a sampling frequency of $3/2\pi$ Hz. Plot the magnitude of this response on the same display as that of the analogue filter. Note, by inspection, how the response of the digital filter can be obtained from that of the analogue filter by a non-linear shift in the frequency axis.

(b) By suitable transformation of the frequency axis used for the analogue filter response, show that the response of the digital filter is obtained.

(c) Use the command `bilinear(num,den,Fs,Fp)` to 'pre-warp' the response of the digital filter. Choose a range of values of `Fp` and show that the responses of the analogue and digital filters are equal at this frequency.

9

Random Signals

9.1 Introduction

The concept of a random signal was introduced very briefly in Chapter 1, Example 1.1. The signal concerned was generated by the particle size of the material forming the disc in a vinyl recording. It was typified by the fact that it was not predictable and would differ in different pressings of the disc. This contrasts with a deterministic signal which can be described either graphically or by a mathematical formula, and hence where its variation is precisely known.

Expanding on this idea many signals are generated by processes that are extremely complex, are not completely known and are not repeatable. For example, although the basic dynamics associated with the tossing of a coin are well known it would be extremely difficult to analyse the process in order to predict the outcome head or tail. It would also be an impossible practical task to repeat the operation exactly in order to obtain the same result. Signals generated by such processes are random signals. Examples of random signals are road noise (vibration) caused by the particles forming the material of the road surface, background noise in buildings, wind gusts affecting the aerodynamics of an aircraft, noise in electronic equipment due to electronic motion.

Because of the inherent difficulties in describing these random signals it would make life easier if they could be ignored in the design of systems and the design concentrated on methods using deterministic signals only. Unfortunately this is not the case. For a designer of motor vehicle suspension systems the nature of the random road surface is very important. When designing sound insulation for offices some method of describing the random interfering sound is required. Hence methods of system analysis are required where the signals applied to the system are random.

A characteristic of a random signal is that it cannot be represented by a mathematical equation expressing how its amplitude depends upon time. In this chapter alternative methods of description are considered. It would be convenient if these methods followed closely those used for deterministic signals.

At first sight this does not seem possible; if the value of the input signal, at a specific time instant, is unknown, how can one make any sort of analysis to obtain the system response? However, although the precise value of a random signal may be unknown, one can often make statements regarding its probable value. For example, considering the displacement of a wheel on a motor vehicle due to road irregularities, one might say that a displacement of 5 mm is more probable than a displacement of 1 mm and is even more probable than a displacement of 2 cm. Hence, a description of a random signal based on the probability of its lying within certain ranges may be possible. The early part of this chapter considers this approach and introduces some of the required aspects of probability theory.

A problem with the probabilistic description of a signal is that it becomes difficult to derive the system response to a signal described in this manner. System analysis becomes easier if non-probabilistic measures can be used for the signals. Examples of such measures are the mean value and the mean square value of a signal. These are examples of the 'moment' of a random signal and they play an important part in random signal analysis.

As with deterministic signals, it is often the relationship between two signals that is of importance. The concepts of probability and moments can be extended to concepts of joint probability and joint moments. It is a joint moment that gives rise to the important concept of correlation functions. Consider again the random displacement of a car wheel due to road surface irregularities and compare two signals, the displacement of a front wheel and the displacement of the corresponding rear wheel. Although these are both random signals, there is a connection between them: one is a delayed version of the other. A measure of this 'connectedness' is given by the correlation function.

One of the most useful forms of signal description and system analysis considered in earlier chapters was that of the frequency response. There are basic problems defining the frequency content of a random signal and they will be considered in this chapter. However, frequency response methods are one of the most successful means of analysing random signals and the following chapter (spectral analysis) is devoted entirely to the topic.

9.2 *Description of random signals*

9.2.1 *Some properties of random signals*

In order to illustrate some of the properties of random signals a very simple random signal is considered. Although it is rather an artificial signal the generating mechanism is easily understood making it very suitable for introductory purposes.

Consider a signal $x(t)$ generated as follows. Every 10 seconds a coin is tossed, if the result is heads a voltage is set at $+1$ V for the following 10 seconds. If the result is tails the voltage is set at -1 V. How would the signal $x(t)$ vary with time? This question cannot be answered; it is a random signal and there is no mathematical formula for $x(t)$. However, one could specify a number of possible forms the signal might take. Figure 9.1 shows three possible signals, $x_1(t)$, $x_2(t)$, $x_3(t)$.

The signals $x_1(t)$, $x_2(t)$, $x_3(t)$, etc., are known as *realisations* of the signal $x(t)$. The complete set of realisations (infinite in this case) is known as an *ensemble.*

At a given time, say t_1, a 'collection' of ensemble values could be obtained, $x_1(t_1)$, $x_2(t_1)$, $x_3(t_1)$, etc.; these are $+1$, $+1$, -1 in the figure. Describing this collection is now similar to a problem in statistics, e.g. describing a set of shoe sizes taken from a sample of the population. One could calculate an average or mean value; for a very large number of realisations there would be as many $+1$ values as -1 values and the average would be zero.

The averages calculated have been taken over the realisations in the ensemble. They are known as *ensemble averages*; the time t_1 is arbitrary, and the ensemble averages do not change with time. A signal whose ensemble averages do not change with time is known as a *stationary* signal.

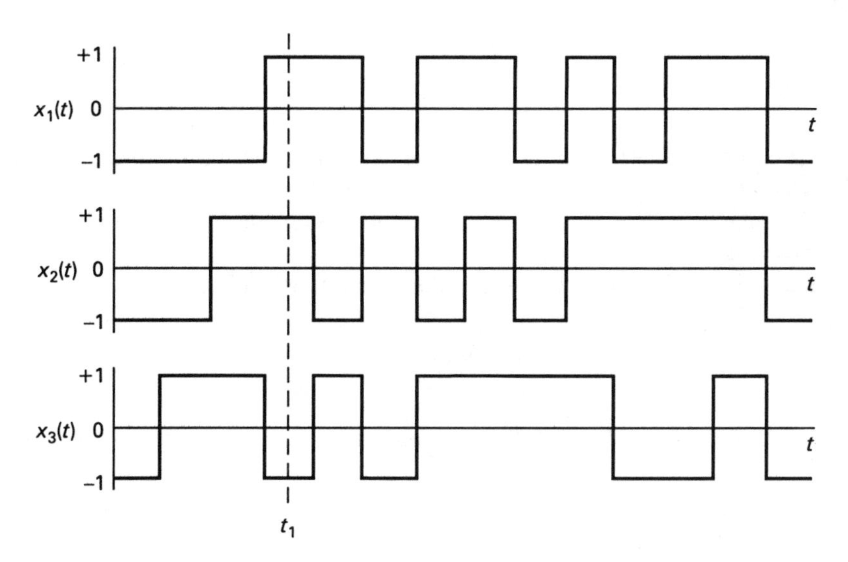

Figure 9.1 Some possible realisations of a random signal.

For the signal considered the random mechanism determining the signal value for different realisations, at a given time, is identical to that giving the values at different times over one realisation. If the mean and mean square values were calculated over one realisation, treating it as a deterministic signal, it would yield identical results to those obtained by ensemble averaging. A signal that has this property is termed an *ergodic* signal. For a signal to be ergodic it must be stationary.

All the signals considered in this and the following chapter will be ergodic and stationary.

9.2.2 Probability and random variables

In the previous section it was emphasised that the variation of a random signal with time cannot be described by a mathematical formula. This does not imply that nothing is known about the amplitude variation of such a signal. Considering the signal generated by the coin tossing mechanism, its value at any time instant can only be +1 V or −1 V. The specific value of the voltage cannot be predicted but the probabilities of its taking the alternative values are known. A variable whose value depends on the outcome of events governed by probability is a random variable.

The notion of probability is related to the outcome of some experiment or trial (e.g. the tossing of a coin). If a trial is repeated N times and an event X occurs N_x times then the relative frequency of this event is given by the ratio N_x/N. The probability of the event X, written $P(X)$, can be defined as the relative frequency of the event as $N \longrightarrow \infty$:

$$P(X) = \lim_{N \longrightarrow \infty} \frac{N_x}{N}$$

This definition obviously implies that repeated trials must be performed. This is a useful definition to use to give an estimate of the probability of an event for measured data. However, it cannot be used to predict the probability of an event from knowledge of the nature of the trial.

An alternative definition (known as the classical definition) is

$$P(X) = \frac{N_x}{N}$$

N is the number of *equally probable* outcomes of the trial N_x number of outcomes of event X.

EXAMPLE 9.2.1

What is the probability of obtaining an even number on one throw of a six sided die?

SOLUTION

The 'relative frequency' definition would entail repeating the experiment a large number of times. A possible outcome might be:

10 trials, even numbers 7 times, $P(\text{even}) = 0.7$

100 trials, even numbers 48 times, $P(\text{even}) = 0.48$

1000 trials, even numbers 493 times, $P(\text{even}) = 0.493$

Using the classical definition one reasons that there are six equally probable (assuming a fair die) outcome of a single throw. Three of these represent the event of interest, an even number. Hence

$$P(\text{even}) = \tfrac{3}{6} = 0.5$$

9.2.3 *Independent events and conditional probability*

Consider a repeated trials approach to the last problem and suppose in N trials the number of twos, fours and sixes thrown were N_2, N_4 and N_6 respectively. Then

$$\begin{aligned} P(\text{even}) &= P(\text{'two' or 'four' or 'six'}) \\ &= \lim_{N\to\infty}\left(\frac{N_2 + N_4 + N_6}{N}\right) \\ &= \lim_{N\to\infty}\left(\frac{N_2}{N} + \frac{N_4}{N} + \frac{N_6}{N}\right) \\ &= P(\text{two}) + P(\text{four}) + P(\text{six}) \end{aligned}$$

This result can be generalised providing the events are *mutually exclusive*. In the example the event 'throwing a two' and the event 'throwing a four' cannot both occur in a single trial – they are mutually exclusive (the events 'an even throw' and 'a number greater than three' are not mutually exclusive).

More generally, two events are mutually exclusive if the occurrence of one event precludes the occurrence of the other. For two mutually exclusive events A and B,

$$P(\text{A or B}) = P(\text{A}) + P(\text{B})$$

Often the probability of an event associated with a repeated trial is required, e.g. what is the probability of obtaining two successive throws of heads in a coin tossing experiment? All the equally probable outcomes of the experiment are {head head} {head tail} {tail head} {tail tail}. Of these, there is only one required outcome {head head}; hence the probability of this outcome is 1/4.

This approach can be generalised. Suppose there are N_1 possible outcomes of the first trial in which there are N_x occurrences of the event X and there are N_2 possible outcomes of the second trial, in which there are N_y occurrences of the event

Y. In an experiment consisting of the combined trials, there are N_1N_2 possible outcomes in which there are N_xN_y occurrences of the events X and Y. Then

$$P(X \text{ and } Y) = \frac{N_xN_y}{N_1N_2} = \frac{N_x}{N_1} \times \frac{N_y}{N_2}$$

However, N_x/N_1 is the probability of the event X in the first trial and N_y is the probability of the event Y in the second trial. Hence

$$P(X \text{ and } Y) = P(X)P(Y)$$

The probability of the event X and Y is the probability of the event X multiplied by the probability of event Y. This is an important result, but it is also subject to an important proviso: the events must be independent. Independence means that the outcome of one trial cannot influence the outcome of the second trial. This is true for the coin tossing experiment where the result of one throw of the coin cannot affect the following throw and

$$P(\text{'head' and 'head'}) = \tfrac{1}{2} \times \tfrac{1}{2} = \tfrac{1}{4}$$

However, consider the following experiment. A bag contains two red and two black balls. Two balls are drawn, at random, in succession (the first ball is not replaced before the second is drawn). What is the probability that two red balls are drawn?

The probability of getting a red ball on the first draw is $1/2$ (if the ball was replaced this would also be the probability of the second ball being red). On the second draw this changes, because there are now two black balls and one red ball in the bag and the probability of picking red is $1/3$. Because this probability is conditional upon the outcome of the first trial, it is termed a *conditional probability*. The overall probability of picking two red balls becomes $1/6$ $(1/2 \times 1/3)$. This can be confirmed by following reasoning similar to that of the independent trials.

In general

$$P(X \text{ and } Y) = P(X)P(Y \mid X)$$

where $P(Y \mid X)$ is the probability of the event Y conditional upon the outcome of event X. If the trials are independent then $P(Y \mid X) = P(Y)$ and

$$P(X \text{ and } Y) = P(X)P(Y)$$

9.2.4 The probability density function

Returning to the random variable represented by the voltage generated by the coin tossing mechanism, because the voltage can only take on discrete values $(+1$ or $-1)$ it is known as a discrete random variable. Associated with this random variable is a discrete probability density function showing the probability of the random variable taking any value within its range. The probability density function for the signal discussed is shown in Figure 9.2.

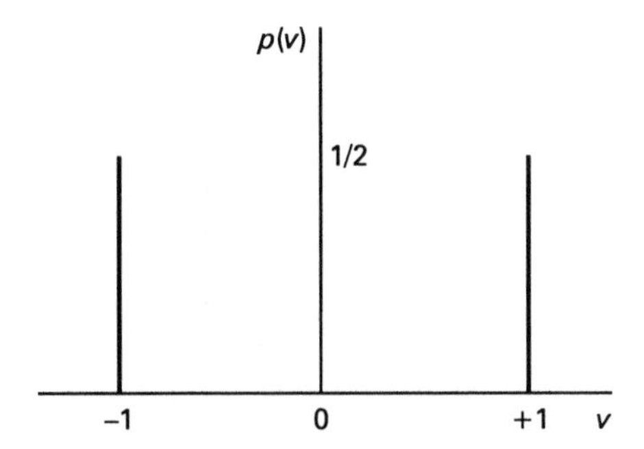

Figure 9.2 Probability density function of signal shown in Figure 9.1.

It should be noted that because of the mechanism generating the voltage v the values $+1$ and -1 are mutually exclusive:

$$P(-1) + P(+1) = 1$$

a probability of 1 representing a certain event.

This can be extended to the more general case so that for any discrete probability density function

$$\sum_i p(x_i) = 1$$

where the summation extends over all the values of the random variables x that have non-zero probability.

EXAMPLE 9.2.2

Two dice are thrown. Obtain the probability density function for the random variable representing the sum of the two numbers shown.

SOLUTION

The six possible outcomes of the first die can be combined with the six possible outcomes of the second die to produce 36 possibilities. These are all equally likely to occur with the probability $1/36$.

Table 9.1 Sum of die throws for example 9.2.2

		Throw of die 1					
		1	2	3	4	5	6
Throw of die 2	1	2	3	4	5	6	7
	2	3	4	5	6	7	8
	3	4	5	6	7	8	9
	4	5	6	7	8	9	10
	5	6	7	8	9	10	11
	6	7	8	9	10	11	12

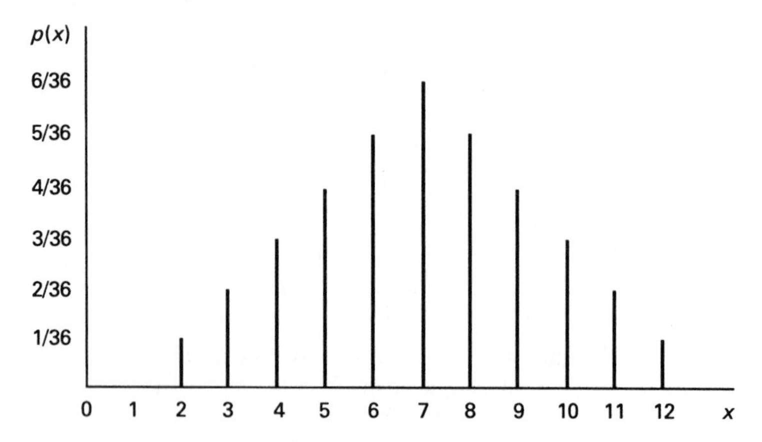

Figure 9.3 Probability density function for Example 9.2.2.

The sum of the throws can be conveniently represented by the Table 9.1 where the entries correspond to the sum of the throws.

Each entry has a probability of $1/36$ and, because they are mutually exclusive, the number of entries of a given total multiplied by $1/36$ gives the probability of that total. This gives the probability density function shown in Figure 9.3.

It can be seen that

$$\sum_i p(x = i) = 1$$

If the random variable concerned is continuous the notion of probability density function as defined for the discrete case cannot be applied. The variable ranges over an infinite number of values and the probability of it taking any one value is always zero. However, one could now consider the probability of the variable being within a certain range. This concept can be illustrated by a relatively simple example.

Suppose an electrical component manufacturer is producing resistors with a nominal value of 1 kΩ. The manufacturer wishes to check the actual values of a large batch of these resistors and measures the resistors to four figure accuracy. Typical measurements could be 982.3 Ω, 1001.8 Ω, 992.7 Ω, etc. At first sight it may appear that this could be plotted as a discrete distribution with intervals set by the accuracy of the measurements. However, unless a very large sample batch of resistors were available, this would give a density function whose amplitude would give little indication of the way the resistor values were spread around the nominal value. However, if the relative frequency of the resistor values lying within a certain range were plotted, then a meaningful result could be obtained. Suppose that for 1000 resistors the results shown in the Table 9.2 were obtained.

If the number of resistors in the sample were increased the relative frequency of a resistor being in a given range would approach the probability of this event and it is

Table 9.2 Relative frequency of resistor values

Range (Ω)	*No. of resistors*	*Relative frequency (probability)*	*Probability/range (probability/density)*
881–900	18	0.018	0.0009
901–920	32	0.032	0.0016
921–940	63	0.063	0.0032
941–960	95	0.095	0.0048
961–980	134	0.134	0.0067
981–1000	158	0.158	0.0079
1001–1020	165	0.165	0.0083
1021–1040	130	0.130	0.0065
1041–1060	97	0.097	0.0049
1061–1080	60	0.060	0.0030
1081–1100	30	0.030	0.0015
1101–1120	18	0.018	0.0006

useful to regard the relative frequency figures as probabilities. This information can be plotted in the form of a histogram where a rectangle is used to indicate the probability of a certain range; however, the probability will be indicated by the *area* of the rectangle, not its amplitude. This is shown in Figure 9.4.

What does the amplitude of these rectangles represent? It is equal to area/base width and has the dimension of probability/Ω range; it is a *probability density*.

The range chosen to set the base of the rectangles is arbitrary and could be subdivided. If the range 1001–1020 were subdivided into ranges 1001–1010 and 1011–1020 one would expect the number of resistors to divide approximately equally, e.g. 81 and 84 say. The probability associated with the corresponding rectangles approximately halves but the probability density would not show much change. One could envisage the situation where a very large number of resistors were used and the intervals used were further subdivided. In the limit as the interval tends to zero the histogram tends to a smooth curve, this curve being the continuous probability density function.

More generally for a random variable x there is an associated probability density function $p(x)$ such that the probability of x lying in the interval x to $x+\mathrm{d}x$ is $p(x)\,\mathrm{d}x$. This is shown in Figure 9.5.

The probability of the variable lying between values a and b is given by

$$P(a < x < b) = \int_a^b p(x)\,\mathrm{d}x$$

The total area represents the probability that the value of the variable x lies between $-\infty$ and $+\infty$. This must be 1.

$$P(-\infty \leqslant x \leqslant +\infty) = \int_{-\infty}^{+\infty} p(x)\,\mathrm{d}x = 1$$

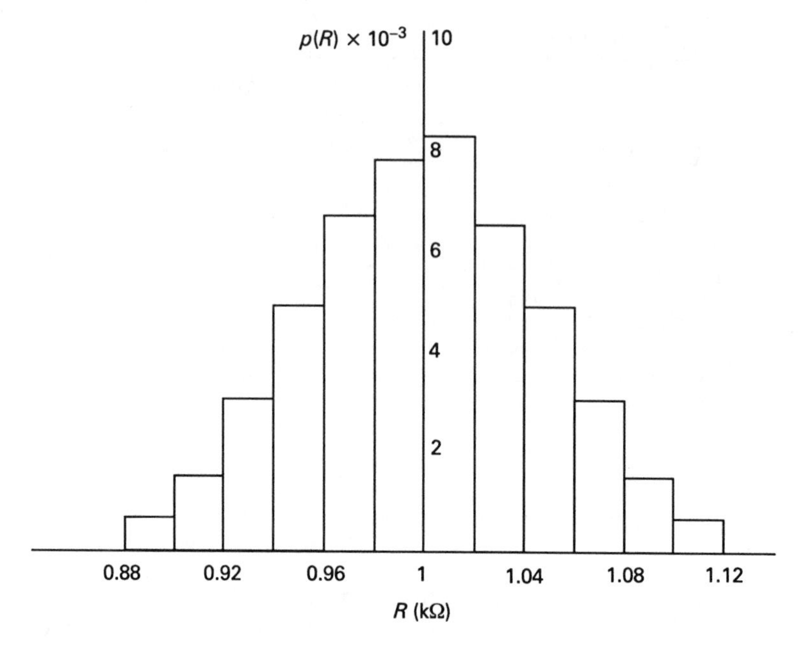

Figure 9.4 Histogram for measured resistor values.

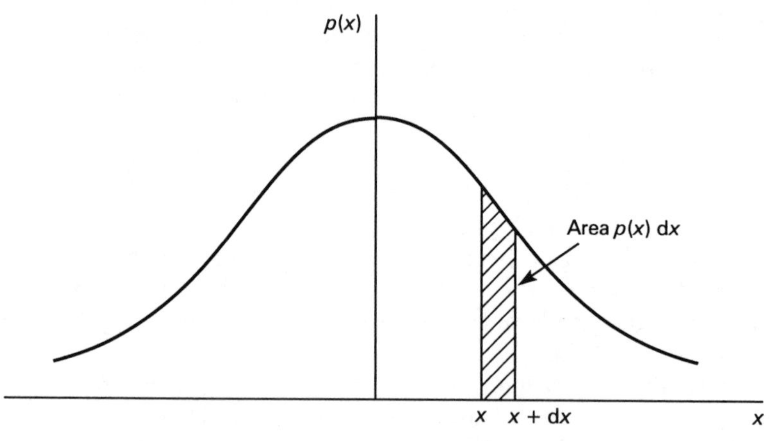

Figure 9.5 Continuous probability density function.

Examples of continuous probability density functions

1. The uniform distribution.
   This occurs when there is any equal probability of a variable occurring in a given range. Outside of the range the probability is zero. This is shown in Figure 9.6.
   Note that the value of $p(x)$ over its non-zero range is fixed by the fact that the area under the curve must be unity.

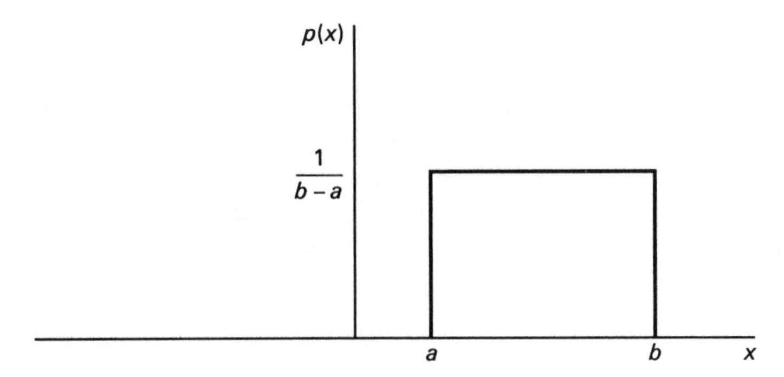

Figure 9.6 Uniform probability density function.

$$\int_{-\infty}^{+\infty} p(x) = \frac{1}{b-a}$$

2. The normal or Gaussian distribution.
   The probability density of this distribution is given by

$$p(x) = \frac{1}{\sqrt{(2\pi)}\sigma} e^{-(x-\mu)^2/2\sigma^2}$$

The general shape of the curve is of the form e^{-x^2} (an even function). The constants μ and σ give the position of the axis of symmetry and a measure of the spread of the curve. The significance of these two parameters will be discussed further in Section 9.2.6. The curve is shown in Figure 9.7 for three values of σ.

The total area under the curve should be unity and the multiplying factor $1/\sqrt{(2\pi)}\sigma$ ensures this. In general the integral giving the area under part of the

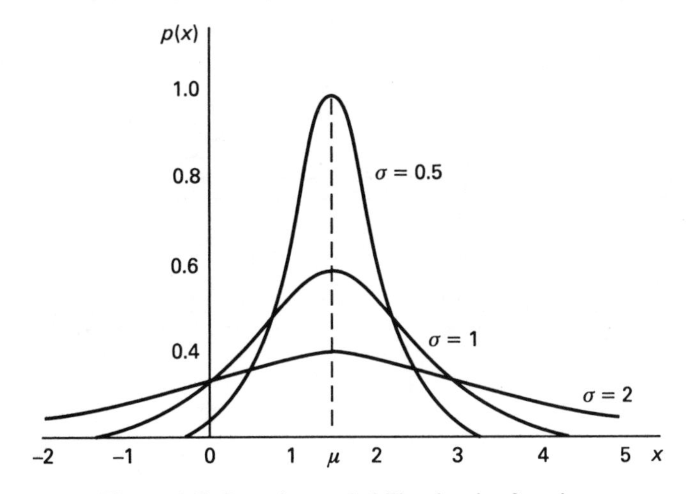

Figure 9.7 Gaussian probability density function.

curve cannot be expressed in closed form; tables are available giving this area in many books on statistics.

The Gaussian distribution describes many random physical processes. The reason for this can be explained by the central limit theorem. This states that a probability density function of a random variable that is a linear combination of other independent random variables tends to become Gaussian as the number of variables increases. This does not depend upon the component random variables being normally distributed.

9.2.5 Probability density functions from time records

In Section 9.2.1 the concept of ergodicity was introduced. If this can be applied to the probability density function then it should be possible to obtain this from a single time record. For the signal to be ergodic it must be stationary.

Consider a random signal $x(t)$ as shown in Figure 9.8. Over a given time T consider the amount of time the signal spends between levels x_1 and $x_1 + \Delta x_1$. These are shown in Figure 9.8 as t_1, t_2, t_3, t_4, etc. If this total time is Δt

$$\Delta t = t_1 + t_2 + t_3 + t_4 + \cdots$$

As $T \rightarrow \infty$ the probability that x lies within the range x_1 to $x_1 + \Delta x_1$ becomes

$$P(x_1 < x \leqslant x_1 + \Delta x_1) = \lim_{T\rightarrow\infty} \frac{\Delta t}{T}$$

The result obtained will depend on the interval Δx. As $\Delta x \rightarrow 0$ the time spent in this interval would become zero as would the associated probability.

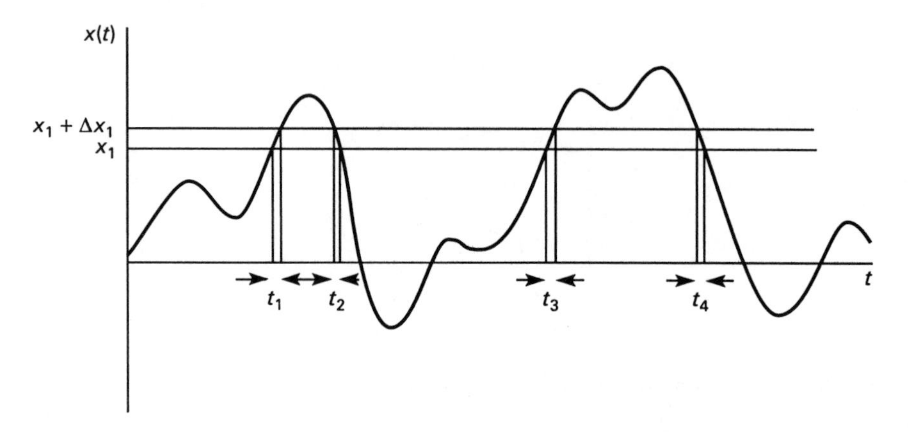

Figure 9.8 Probability density function from time records.

By introducing the concept of probability density and assuming this is constant throughout the interval Δx then

$$p(x_1)\Delta x_1 = \lim_{T\to\infty} \frac{\Delta t}{T} \tag{9.2.1}$$

This equation forms the basis of an instrument for the practical measurement of the probability density function of a signal. For a sufficiently long record the time Δt would be measured for a given level x_1. Hence, $p(x_1)$ could be estimated. This procedure would have to be repeated over the range of x.

Using eqn (9.2.1) enables the probability density function of a signal to be obtained if an analytical description of a signal is available. At first sight this statement contradicts the concept of a random process where no analytical description of the process is possible. However a deterministic signal can be regarded as one realisation of an ensemble of such signals with random time shifts.

The following example will illustrate this point.

EXAMPLE 9.2.3

A single sine wave can be regarded as one realisation of a random process.

$$x(t) = A\sin(\omega t + \varphi)$$

where φ is a random variable uniformly distributed in the range $-\pi$ to $+\pi$.

(a) Determine the probability density function of this signal.
(b) Determine the probability that the signal is in the range (i) 0 to $0.1A$, (ii) $0.9A$ to A.

SOLUTION

(a) One realisation of the signal is shown in Figure 9.9. The phase φ is constant for one realisation.

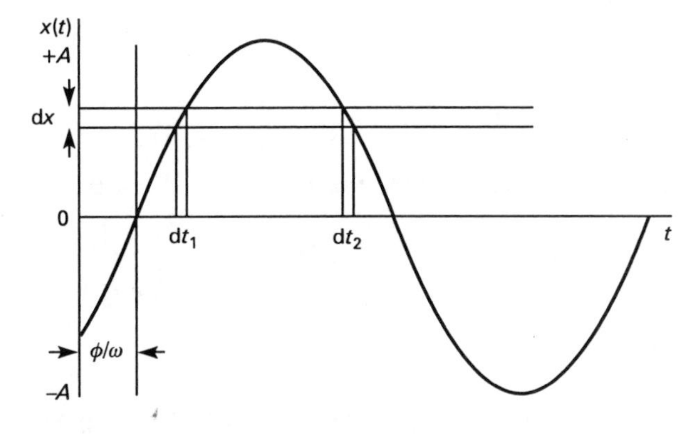

Figure 9.9 Sine wave as a realisation of a random signal.

The time $\mathrm{d}t$ spent in interval $\mathrm{d}x$ is $(\mathrm{d}t_1 + \mathrm{d}t_2)$. This time can be found by differentiating the expression for $x(t)$:

$$x(t) = A \sin(\omega t + \varphi)$$

$$\frac{\mathrm{d}x}{\mathrm{d}t} = A\omega \cos(\omega t + \varphi)$$

$$\mathrm{d}t = \frac{2\,\mathrm{d}x}{A\omega \cos(\omega t + \varphi)}$$

The probability of the signal being in this interval during one complete cycle is

$$\frac{2\mathrm{d}x}{TA\omega \cos(\omega t + \varphi)}$$

and this equals $p(x)\,\mathrm{d}x$. Hence

$$p(x) = \frac{2}{TA\omega \cos(\omega t + \varphi)}$$

Unfortunately $p(x)$ is given as a function of t and not x. However,

$$\cos(\omega t + \varphi) = \sqrt{(1 - \sin^2(\omega t + \varphi))}$$
$$= \sqrt{(1 - (x/A)^2)}$$

Also $\omega = 2\pi/T$ giving

$$p(x) = \frac{1}{A\pi\sqrt{(1 - (x/A)^2)}} \qquad x \leqslant |A| \qquad (9.2.2)$$
$$= 0 \qquad \text{otherwise}$$

$p(x)$ is plotted against x in Figure 9.10.

(b) The probability of x lying between the limits a and b is given by

$$\int_a^b p(x)\,\mathrm{d}x = \int_a^b \frac{\mathrm{d}x}{A\pi\sqrt{(1 - (x/A)^2)}}$$

This integral can be evaluated by the substitution $x = A \sin \theta$ giving

$$\int_a^b p(x)\mathrm{d}x = \frac{1}{\pi}\left[\sin^{-1}\left(\frac{a}{A}\right) - \sin^{-1}\left(\frac{b}{A}\right)\right]$$

(i) For the range 0 to $0.1A$, $a = 0.1A$, $b = 0$

$$P(0 \leqslant x \leqslant 0.1A) = \frac{1}{\pi}[\sin^{-1}(0.1)] = 0.032$$

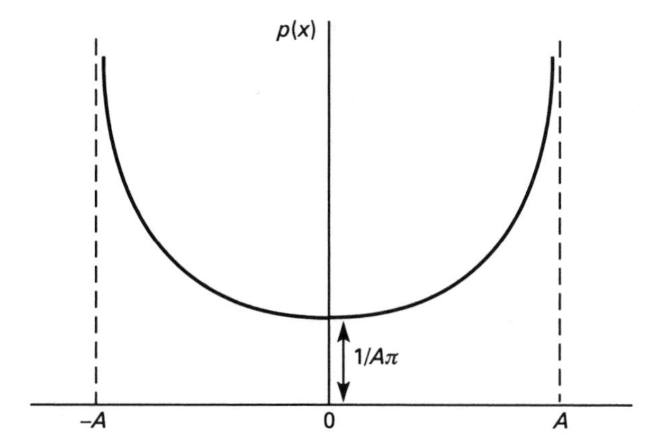

Figure 9.10 Probability density function for a sine wave.

(ii) For the range $0.9A$ to A, $a = A$, $b = 0.9A$

$$P(0.9A \leq x \leq A) = [\sin^{-1}(1) - \sin^{-1}(0.9)] = 0.144$$

As expected the probability of $x(t)$ being in a given interval is much greater at the peak of the curve than near $x = 0$. The sine function has its greatest rate of change at $x = 0$ and spends less time in the given interval there than when around the peak value.

It should be noted that the probability lying between $-A$ and $+A$ is given by

$$\int_{-A}^{+A} p(x)\,\mathrm{d}x = \frac{1}{\pi}[\sin^{-1}(1) - \sin^{-1}(-1)]$$
$$= 1$$

This is expected: the area under any probability density function is unity.

9.2.6 *Mean and mean square values, variance*

The concept of mean and mean square value as applied to deterministic signals has already been covered in Chapter 2. These are important properties of a signal and it would be useful if these measures could be applied to random signals. The definition of ergodicity implies that these averages can be derived from a single signal realisation, provided a mathematical description of the realisation is available. If this is not the case then the averages must be obtained from the signal's probability density function. This process will be described first for a process with a discrete probability density function and then for the continuous case.

Consider the binary signal described in Section 9.2.1, the value of the signal at any instant being dependent upon the throw of a coin. However, to make the signal more interesting two changes will be made to the generation process.

1. Let us assume the coin is not 'fair' and the probability of it landing heads is twice that of it landing tails.
2. If the coin lands heads the signal will be held at the +5 V level for the next second, and if tails it will be held at −2 V.

Hence the probabilities of the levels +5 V and −2 V can be written

$$P(5) = 2/3 \qquad P(-2) = 1/3$$

One realisation of the signal could be as shown in Figure 9.11.

The mean value of this signal is given by

$$\mu = \lim_{T \to \infty} \frac{1}{T} \int_0^T v(t)\, \mathrm{d}t$$

$$= \lim_{T \to \infty} \frac{1}{T} (5T_1 - 2T_2)$$

where T_1 is the time where $v(t)$ remains at 5 V and T_2 is the time it remains at −2 V.

However, as $T \to \infty$

$$T_1 \to P(+5)T \qquad T_2 \to P(-2)T$$

$$\mu = \frac{1}{T} [5 \times P(+5) \times T - 2 \times P(-2) \times T]$$

$$= 5 \times \left(\frac{2}{3}\right) - 2 \times \left(\frac{1}{3}\right)$$

$$= \frac{8}{3}$$

The method used for this example can be extended to cover the more general case where the sample signal from an ergodic random process can take N discrete values with the value x_i occurring with probability $p(x_i)$. Over a long time interval T

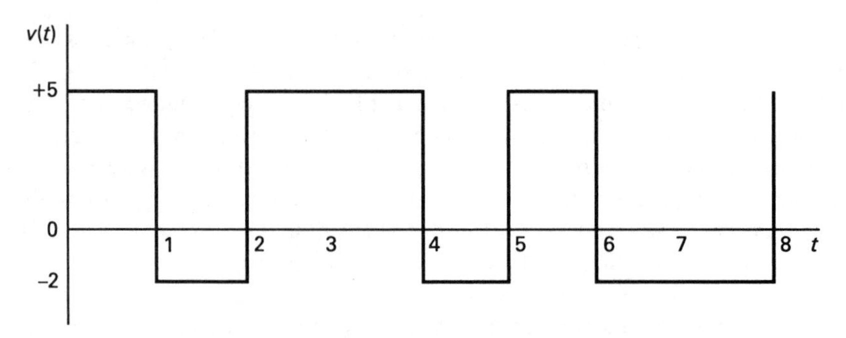

Figure 9.11 One realisation of a binary random signal.

the time spent at value x_i is $p(x_i)T$ and this gives a contribution to the integral defining the mean value of $x_i p(x_i)T$. Hence

$$\mu = \lim_{T\to\infty} \frac{1}{T}\int_0^T x(t)\,\mathrm{d}t$$
$$= \lim_{T\to\infty} \frac{1}{T}\sum_{i=1}^{N} x_i p(x_i)T$$
$$= \sum_{i=1}^{N} x_i p(x_i) \tag{9.2.3}$$

The derivation for a continuous random process follows using a similar argument. The probability that the signal lies in the interval x to $x+\mathrm{d}x$ is $p(x)\,\mathrm{d}x$ where $p(x)$ is the continuous probability density function. Over a long time period T the time $\mathrm{d}t$ spent in this interval is $p(x)\,\mathrm{d}x\,T$ and this gives a contribution to the integral of $xp(x)\,\mathrm{d}x\,T$

$$\mu = \lim_{T\to\infty} \frac{1}{T}\int_0^T x(t)\,\mathrm{d}t$$
$$= \lim_{T\to\infty} \frac{1}{T}\int_{-\infty}^{+\infty} xp(x)\,\mathrm{d}x\,T$$
$$= \int_{-\infty}^{+\infty} xp(x)\,\mathrm{d}x \tag{9.2.4}$$

The mean square value of a signal gives a measure of the power in the signal. For a random ergodic process it can be obtained from a single realisation $x(t)$ as

$$\overline{x^2} = \lim_{T\to\infty} \frac{1}{T}\int_0^T x^2(t)\,\mathrm{d}t$$

The mean square value can be obtained from the probability density function of $x(t)$ following the methods described previously to obtain the mean value. The contribution to the time integral becomes

$$x_i^2 p(x_i)T$$

for the discrete distribution and $x^2 p(x)\,\mathrm{d}x\,T$ for the continuous distribution. This leads to the formulae

$$\overline{x^2} = \frac{1}{N}\sum_{i=1}^{N} x_i^2 p(x_i) \tag{9.2.5}$$

$$\overline{x^2} = \int_{-\infty}^{+\infty} x^2 p(x)\,\mathrm{d}x \tag{9.2.6}$$

for discrete and continuous distributions respectively.

Another useful measure of a random signal is the power contained in the varying component, i.e. the total power minus the power in the d.c. component. This is a measure of the spread of a signal around its mean level. It is known as the *variance* of the signal and it is denoted by σ^2:

$$\sigma^2 = \overline{x^2} - \mu^2$$

The variance can also be written as

$$\sigma^2 = \lim_{T\to\infty} \frac{1}{T}\int_0^T [x(t) - \mu]^2 \mathrm{d}t$$

Expanding

$$\begin{aligned}\sigma^2 &= \lim_{T\to\infty} \frac{1}{T}\int_0^T x^2(t)\mathrm{d}t - 2\mu \lim_{T\to\infty} \frac{1}{T}\int_0^T x(t)\mathrm{d}t + \lim_{T\to\infty} \frac{1}{T}\int_0^T \mu^2 \,\mathrm{d}t \\ &= \overline{x^2} - 2\mu^2 + \mu^2 \\ &= \overline{x^2} - \mu^2\end{aligned}$$

Following the methods used for mean and mean square values, the variance can be expressed in terms of the probability density function as

$$\sigma^2 = \int_{-\infty}^{+\infty} (x - \mu)^2 \, p(x)\,\mathrm{d}x \qquad (9.2.7)$$

The probability density function used in this section has described the amplitude distribution of a random signal. Because the process has been assumed stationary, the probability density function and the associated measures of mean, mean square and variance do not depend upon time. The probability density function is of course more general than just applying to an amplitude distribution: it can apply to any random variable. In this context the mean and variance give information on the general shape of the probability density function.

The mean value gives a 'centre of gravity' of the function and the variance gives the 'spread' around this centre of gravity. Figure 9.12 shows the effect of changing these parameters for a given probability density function. Figure 9.12(a) shows the effect of changing the mean value with the variance constant; Figure 9.12(b) shows the effect of changing the variance with the mean value constant.

The formulae derived for mean, mean square and variance are rather cumbersome involving integrals (summations) and limits. In later sections other averaging involving more than one random variable will be introduced. It is useful to have a shorthand notation to represent the averaging process and this can be done by means of the *expectation operator*.

If x is a random variable and $f(x)$ is a function of this variable, then the expectation of $f(x)$, written $E\{f(x)\}$ is defined as

$$E\{f(x)\} = \int_{-\infty}^{+\infty} f(x)p(x)\,\mathrm{d}x$$

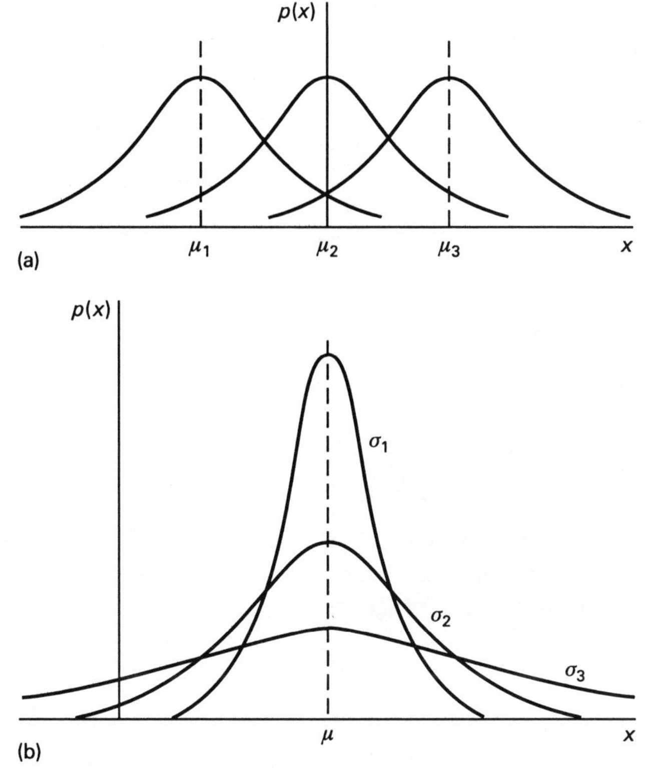

Figure 9.12 The variation of mean and variance of a probability density function.

Hence, if $f(x) = x$, then

$$E\{x\} = \int_{-\infty}^{+\infty} xp(x)\, \mathrm{d}x = \mu$$

Similarly

$$E\{x^2\} = \overline{x^2}$$

$$E\{(x-\mu)^2\} = \sigma^2$$

The mean, mean square and variance are examples of the *moments* of a random variable. In general, the kth moment is defined as $E\{x^k\}$ and the kth central moment as $E\{(x-\mu)^k\}$. Hence, the mean value is the first moment, the mean square value the second moment and the variance the second central moment.

The following example illustrates the method of calculation of mean, mean square and variance.

EXAMPLE 9.2.4

(a) Two random variables x_1 and x_2 have the following uniform probability density functions:

$$p(x_1) = \text{constant} \qquad -2 \leqslant x \leqslant +2$$
$$= 0 \qquad \text{otherwise}$$
$$p(x_2) = \text{constant} \qquad -1 \leqslant x \leqslant +3$$
$$= 0 \qquad \text{otherwise}$$

For both these variables calculate their mean, mean square and variance.

(b) In Example 9.2.3 a sine wave was treated as one realisation of a random process. Its probability density function was obtained as

$$p(x) = \frac{1}{A\pi\sqrt{(1-(x/A)^2)}} \qquad x \leqslant |A|$$

where A is the peak amplitude of the sine wave. From the probability density function, obtain the mean and variance of this signal and compare with the results obtained from time averages.

SOLUTION

(a) Remembering that the area underneath the probability density function must be unity, then

$$p(x_1) = 0.25 \qquad -2 \leqslant x \leqslant +2$$
$$= 0 \qquad \text{otherwise}$$
$$p(x_2) = 0.25 \qquad -1 \leqslant x \leqslant +3$$
$$= 0 \qquad \text{otherwise}$$

To calculate the mean value

$$\mu = \int_{-\infty}^{+\infty} xp(x)\,\mathrm{d}x$$

For variable x_1

$$\mu = \int_{-2}^{+2} 0.25x_1\,\mathrm{d}x_1 = 0$$

A similar calculation for variable x_2 gives $\mu = 1$. These results could have been obtained by inspection from the symmetry of the probability density functions.

To calculate the mean square value

$$\overline{x^2} = \int_{-\infty}^{+\infty} x^2 p(x)\,\mathrm{d}x$$

For variable x_1

$$\overline{x^2} = \int_{-2}^{+2} 0.25 x_1^2 \, dx_1 = 4/3$$

A similar calculation for variable x_2 gives $x^2 = 7/3$.

To calculate the variance

$$\sigma^2 = \int_{-\infty}^{+\infty} (x - \mu)^2 p(x) \, dx$$

For variable x_1, $\mu = 0$ and the variance equals the mean square value, $\sigma^2 = 4/3$. For variable x_2

$$\begin{aligned}
\sigma^2 &= \int_{-1}^{+3} 0.25 (x_2 - 1)^2 \, dx_2 \\
&= 0.25 \left(\int_{-1}^{+3} x_2^2 \, dx_2 - 2 \int_{-1}^{+3} x_2 \, dx_2 + \int_{-1}^{+3} 1 \, dx_2 \right) \\
&= 0.25 (28/3 - 8 + 4) \\
&= 4/3
\end{aligned}$$

It should be noted that the variances of the density functions for x_1 and x_2 are equal. This result is to be expected: the shapes of $p(x_1)$ and $p(x_2)$ are identical; they only differ by a shift in their mean values.

The last result could have been obtained more easily by using the relationship

$$\sigma^2 = \overline{x^2} - \mu^2$$

(b) Given

$$p(x) = \frac{1}{A\pi \sqrt{(1 - (x/A)^2)}}$$

then

$$\begin{aligned}
\mu &= \int_{-\infty}^{+\infty} x p(x) \, dx \\
&= \int_{-A}^{+A} \frac{x}{A\pi \sqrt{(1 - (x/A)^2)}} \, dx
\end{aligned}$$

Because the integrand is an odd function, integration between the limits $-A$ and $+A$ gives zero mean. Because of this zero mean, the mean square value and the variance are equal. Then

$$\begin{aligned}
\sigma^2 &= \int_{-\infty}^{+\infty} (x - \mu)^2 p(x) \, dx \\
&= \int_{-A}^{+A} \frac{x^2}{A\pi \sqrt{(1 - (x/A)^2)}} \, dx
\end{aligned}$$

This integral can be obtained from tables of standard integrals or by making the substitution $x = A\sin\theta$, $dx = A\cos\theta\, d\theta$:

$$\sigma^2 = \int_{-\pi/2}^{+\pi/2} \frac{A^2\sin^2\theta}{A\pi\cos\theta} A\cos\theta\, d\theta$$

$$= \frac{A^2}{\pi}\int_{-\pi/2}^{+\pi/2} \tfrac{1}{2}(1-\cos 2\theta)\, d\theta$$

$$= \frac{A^2}{\pi}\left\{[\theta]_{-\pi/2}^{+\pi/2} - \tfrac{1}{2}[\sin 2\theta]_{-\pi/2}^{+\pi/2}\right\}$$

$$= \frac{A^2}{2}$$

These results can be confirmed by calculating the averages for the time signal $x(t) = A\sin\omega t$. As the signal is periodic, the averages can be taken over one period T. It should be noted that $\omega T = 2\pi$.

$$\mu = \frac{1}{T}\int_0^T A\sin\omega t\, dt$$

$$= \frac{A}{T}\left[-\frac{1}{\omega}\cos\omega t\right]_0^T$$

$$= 0$$

For zero mean

$$\sigma^2 = \frac{1}{T}\int_0^T A^2\sin^2\omega t\, dt$$

$$= \frac{A^2}{T}\int_0^T \tfrac{1}{2}(1-\cos 2\omega t)\, dt$$

$$= \frac{A^2}{2T}\left\{[t]_0^T - \frac{1}{2\omega}[\sin 2\omega t]_0^T\right\}$$

$$= \frac{A^2}{2}$$

9.3 *Joint random variables*

The random processes considered so far in this chapter have been described in terms of their probability density functions and the associated parameters of mean, mean square and variance. Because the processes are stationary, these parameters do not vary with time. What has not been considered, so far, is whether the values taken by the process at any two time instants are related; are the random variables describing the process at the two instants independent?

In the examples where the processes have been generated by a coin tossing procedure then the random variables associated with the outcome of the procedure are independent. However, suppose this process is the input to a system that is not memoryless; then the output of the system depends upon the weighted integral of past inputs. Within the time scale covered by the system's impulse response, the values of the output process at any two time instants are not independent.

Consider a simple example to illustrate this. Again take a process generated by a coin tossing mechanism, but this time consider a discrete process. In each sample an outcome 'heads' will give a value +1 V and 'tails' will give −1 V. A possible realisation $x(n)$ is shown in Figure 9.13(a)

Suppose the process $x(n)$ is fed into a discrete system whose output $y(n)$ is given by

$$y(n) = \frac{x(n) + x(n-1)}{2}$$

The output $y(n)$ is shown in Figure 9.13(b) and at any instant is the average of the present and previous input samples. It should be noted that $y(n)$ is still a random process and the signal shown in Figure 9.13(b) is one realisation of this process.

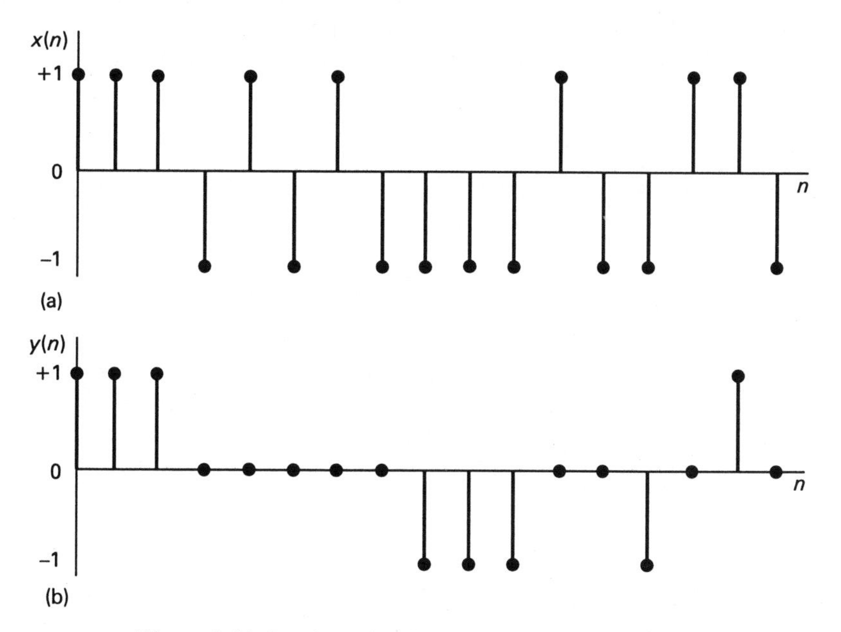

Figure 9.13 Random signals at system input and output.

However, there is an important difference between the signals $x(n)$ and $y(n)$. For the signal $x(n)$ the value taken at any instant does not depend upon previous values of the signal. This is not true for $y(n)$; for instance, in this signal a $+1$ V value never follows a -1 V value and vice versa. In order to put a measure on this 'connectedness' the concept of joint probability will be considered.

9.3.1 Joint probability density function

The concept of joint probability has been introduced earlier in this chapter. This concept will now be used to investigate the relationship between successive samples in the discrete signals $x(n)$ and $y(n)$ introduced in the previous section.

The values of the sample instants for the process $x(n)$ can take two values, $+1$, -1, each with a probability of $1/2$. Because the random variables at each instant are independent, then the joint probabilities for all combinations of successive values take an equal value of $1/2 \times 1/2 = 1/4$. The joint probabilities are illustrated graphically in Figure 9.14.

The function that describes $p(x(n), x(n+1))$ for all possible values of $x(n)$ and $x(n+1)$ (graphical in this case) is the joint probability density function. It has similarities to the probability density function for a single variable; in particular the total area under the function is one.

The joint probability density function for successive samples of the signal $y(n)$ requires more care to evaluate. Table 9.3 shows how values $y(n-1)$, $y(n)$ are generated from all possible combinations $x(n-2)$, $x(n-1)$, $x(n)$.

From this table it is seen that the probabilities that $y(n)$ should take values -1, 0, $+1$ are $1/4$, $1/2$, $1/4$ respectively. The probabilities of $y(n)$ following a value $y(n-1)$ are shown by the diagram of Figure 9.15(a).

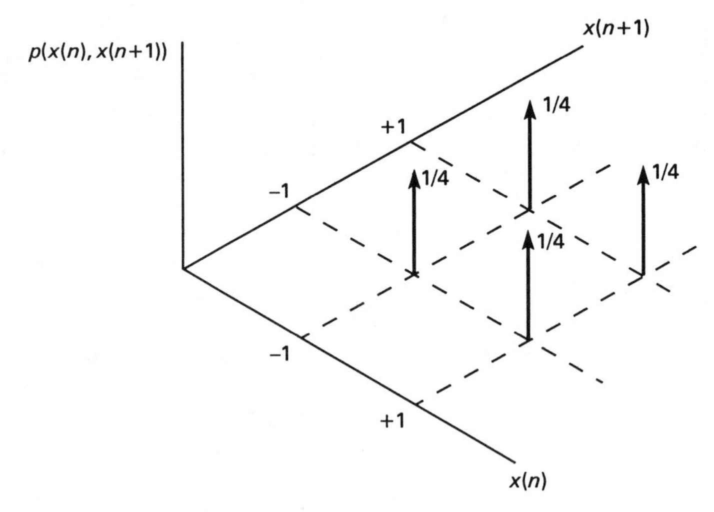

Figure 9.14 A joint probability density function.

Table 9.3 Generation of $y(n-1)$ and $y(n)$

$x(n-2)$	$x(n-1)$	$x(n)$	$y(n-1)$	$y(n)$
−1	−1	−1	−1	−1
−1	−1	+1	−1	0
−1	+1	−1	0	0
−1	+1	+1	0	+1
+1	−1	−1	0	−1
+1	−1	+1	0	0
+1	+1	−1	+1	0
+1	+1	+1	+1	+1

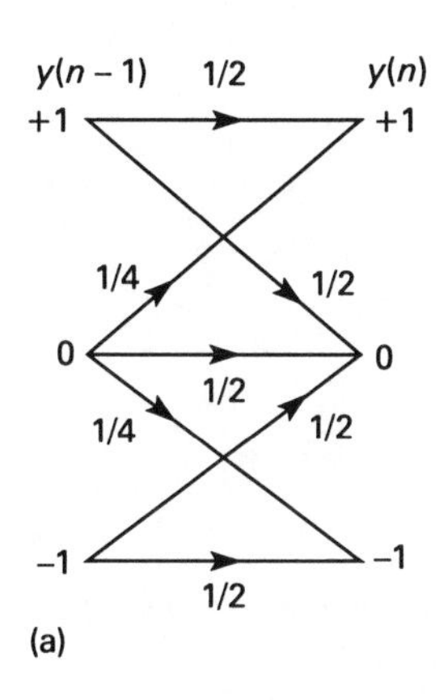

$y(n-1)$ \ $y(n)$	−1	0	+1
−1	1/8	1/8	0
0	1/8	1/4	1/8
+1	0	1/8	1/8

(b)

Figure 9.15 Joint probabilities for random variables $y(n-1)$, $y(n)$.

The joint probability of $(y(n), y(n-1))$ is shown in the table of Figure 9.15(b). It should be noted that if the function were plotted the total area under the function would be one. Also it should be noted that $p(y(n), y(n-1)) \neq p(y(n)) \times p(y(n-1))$.

For continuous random variables the definition of joint probability density has to be modified as in the case of a single random variable:

$$p(x, y)\,\mathrm{d}x\,\mathrm{d}y = P(x < X \leqslant x + \mathrm{d}x,\ y < Y \leqslant y + \mathrm{d}y)$$

The joint probability density function contains information regarding the dependence or non-dependence between two random variables. It could be applied to the random variables at any two time instants in a process. However, it is not a very convenient measure and a more useful measure can be obtained by extending the concept of moments to two random variables.

9.3.2 Moments of two random variables

The concept of expected value applied to a function of a random variable was introduced in Section 9.2.5. The concept can be extended to functions of two random variables

$$E\{f(x_1y)\} = \int_{-\infty}^{+\infty}\int_{-\infty}^{+\infty} f(x, y)\, p(x, y)\, \mathrm{d}x\,\mathrm{d}y$$

for continuous variables and

$$E\{f(x_1y)\} = \sum_j \sum_k f(x_j, y_k)\, p(x_j, y_k)$$

for discrete variables.

The joint moments of two random variables can be defined in terms of their expected values. The most important are the second moments

$$\begin{aligned} m_{11} &= E\{x, y\} \\ &= \int_{-\infty}^{+\infty}\int_{-\infty}^{+\infty} xy\, p(x, y)\, \mathrm{d}x\, \mathrm{d}y \end{aligned}$$

and the second central moment

$$\begin{aligned} \mu_{11} &= E\{(x-\mu_x)(y-\mu_y)\} \\ &= \int_{-\infty}^{+\infty}\int_{-\infty}^{+\infty} (x-\mu_x)(y-\mu_y)\, p(x, y)\, \mathrm{d}x\, \mathrm{d}y \end{aligned}$$

The second moment, m_{11}, is known as the *correlation* between x and y and is written R_{xy}. The second central moment, μ_{11} is known as the *covariance* between x and y and is written σ_{xy}^2.

From R_{xy} the definition of uncorrelated random variables is obtained. Two random variables x and y are uncorrelated if

$$R_{xy} = E\{xy\} = E\{x\}E\{y\}$$

If the variables are independent, $p(x, y) = p(x)p(y)$, then

$$\begin{aligned} R_{xy} &= \int_{-\infty}^{+\infty}\int_{-\infty}^{+\infty} xy\, p(x, y)\, \mathrm{d}x\,\mathrm{d}y \\ &= \int_{-\infty}^{+\infty} xp(x)\, \mathrm{d}x \int_{-\infty}^{+\infty} yp(y)\, \mathrm{d}y \\ &= E\{x\}\, E\{y\} \end{aligned}$$

proving that if the variables are independent they are also uncorrelated. However, the converse is not true. Uncorrelatedness is a weaker condition than independence; it is an integral property of $p(x, y)$, whereas independence must be true for every x and y.

EXAMPLE 9.3.1

Determine the correlation between successive values for the processes $x(n)$ and $y(n)$ defined at the start of Section 9.3.

SOLUTION

For the process $x(n)$, $p(x(n),\ x(n+1))$ is 1/4 for all values of $x(n)$ and $x(n+1)$. Hence

$$\begin{aligned} R_{x(n)x(n+1)} &= (-1)\times(-1)\times\tfrac{1}{4} + (-1)\times(+1)\times\tfrac{1}{4} + (+1)\times(-1)\times\tfrac{1}{4} + (+1)\times(+1)\times\tfrac{1}{4} \\ &= 0 \end{aligned}$$

As the expected values of both random variables are zero, the variables are uncorrelated. This result could have been deduced directly from the property that the random variables $x(n)$, $x(n+1)$ are independent.

For the process $y(n)$ the joint probability density is shown in Figure 9.15. Hence

$$\begin{aligned} R_{y(n)y(n+1)} &= (-1)\times(+1)\times\tfrac{1}{8} + (-1)\times(0)\times\tfrac{1}{8} + (-1)\times(+1)\times 0 \\ &\quad + (0)\times(-1)\times\tfrac{1}{8} + (0)\times(0)\times\tfrac{1}{4} + (0)\times(+1)\times\tfrac{1}{8} \\ &\quad + (+1)\times(-1)\times 0 + (+1)\times(0)\times\tfrac{1}{8} + (+1)\times(+1)\times\tfrac{1}{8} \\ &= \tfrac{1}{8} + \tfrac{1}{8} = \tfrac{1}{4} \end{aligned}$$

The expected values of both $x(n)$ and $y(n)$ are zero, and hence as

$$E\{y(n),\ y(n+1)\} \neq E\{y(n)\}E\{y(n+1)\}$$

the variables are correlated.

9.3.3 Autocorrelation and cross-correlation functions

In many practical situations where signals appear random there is in fact a pattern hidden among the randomness. Part of the random signal matches another portion of the signal or a portion of a different random signal. A measure of this matching within a signal is given by the autocorrelation function of the signal. Matching between different signals is given by the cross-correlation function. Some examples of the use of correlation are:

1. Analysis of surface finish. If, after a machining operation, the surface of a material was examined under a microscope, it would appear to be composed of random irregularities. However, there can be hidden patterns, not directly visible, that give clues to imminent machine tool failure or wear in the bearings of the machine. Such patterns can be detected and classified by use of the autocorrelation function of the signal describing the surface irregularity.
2. Detection of radar echoes. When a radar system detects a target, the reflected signal may be small and hidden among high levels of random background noise. By shaping the transmitted pulse, and then correlating the reflected signal against this known shape, the required signal can be extracted from the background noise.
3. Alignment of scan images for medical diagnosis. Detection of brain dysfunction can be made from brain scans by comparing the images from the left and right

hand halves of the brain. However, in order to interpret these images correctly, it is important that they are correctly aligned around the axis of symmetry. A new image is formed by reflecting half the original image around an axis. By correlating this image with the original image, the correct axis of symmetry can be found.

In the previous example the correlation between successive values in a discrete random process was obtained. This correlation could be obtained, not only between successive instants in the process, but between values at any two sampling instants. The result would be a function of the two sampling instants chosen, n_1, n_2. This function is the autocorrelation function of the process and for the process $x(n)$ is written $R_{xx}(n_1, n_2)$. If the process is stationary a shift of the time origin will not affect R_{xx}. The autocorrelation function depends only on the difference $k = (n_1 - n_2)$:

$$R_{xx}(k) = \lim_{N \to \infty} \frac{1}{2N} \sum_{n=-N}^{+N} x(n-k)\, x(n) \tag{9.3.1}$$

The corresponding expression for a continuous process is

$$R_{xx}(\tau) = \lim_{T \to \infty} \frac{1}{2T} \int_{-T}^{+T} x(t-\tau)\, x(t)\, \mathrm{d}t \tag{9.3.2}$$

The subscript xx indicates that the correlation is between time instants in the *same* process. Often, what is of interest is the correlation between time instants in different processes. This leads to the concept of a cross-correlation function. The cross-correlation function $R_{xy}(k)$ between two processes $x(n)$ and $y(n)$ is given by

$$R_{xy}(k) = \lim_{N \to \infty} \frac{1}{2N} \sum_{n=-N}^{+N} x(n-k) y(n) \tag{9.3.3}$$

For continuous processes $x(t)$ and $y(t)$ it is given by

$$R_{xy}(\tau) = \lim_{T \to \infty} \frac{1}{2T} \int_{-T}^{+T} x(t-\tau) y(t)\, \mathrm{d}t \tag{9.3.4}$$

EXAMPLE 9.3.2

(a) Determine the autocorrelation function of the random signal generated from the coin tossing experiment of Section 9.2.1. Take the levels as $\pm V$ and the time between the switching of levels as T.
(b) Determine the autocorrelation function of the random process given by

$$x(t) = X \sin(\omega_0 t + \varphi)$$

where φ is a random variable, uniformly distributed in the range $-\pi$ to $+\pi$.

(c) Determine the cross-correlation function between the process $x(t)$ given in part (b) and the process

$$y(t) = Y \sin(\omega_o t + \varphi + \theta)$$

where φ is as in part (b).

SOLUTION

(a) The process is stationary, and hence its autocorrelation function is given by

$$R_{xx}(\tau) = E\{x(t-\tau)x(t)\}$$

The process is also ergodic and its autocorrelation function can be obtained by time averaging or ensemble averaging. The former method is the more straightforward, but the latter will be used here as it illustrates better the general principles involved.

One realisation, $x(t)$, of the process is shown in Figure 9.16, together with the delayed signal $x(t-\tau)$.

At an arbitrary time t_1 the expected value of the product $x(t_1-\tau)x(t)$ is required, the expectation being taken over an ensemble of all possible products. For convenience let $x_1 = x(t_1)$ and $x_2 = x(t_1-\tau)$; then

$$R_{xx}(\tau) = E\{x(t-\tau)x(t)\} = \int_{-\infty}^{+\infty}\int_{-\infty}^{+\infty} x_1 x_2 p(x_1, x_2)\,\mathrm{d}x_1\,\mathrm{d}x_2$$

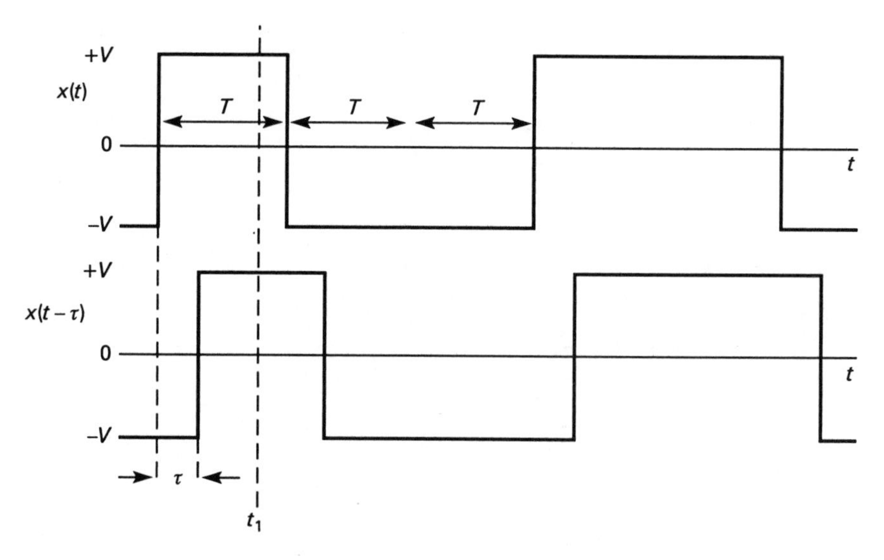

Figure 9.16 Random process for Example 9.3.2.

As x_1 and x_2 can only take values $\pm V$ this expectation becomes a summation:

$$\begin{aligned} R_{xx}(\tau) &= (+V)(+V)P(x_1 = +V,\, x_2 = +V) \\ &\quad + (+V)(-V)P(x_1 = +V,\, x_2 = -V) \\ &= (-V)(+V)P(x_1 = -V,\, x_2 = +V) \\ &\quad + (-V)(-V)P(x_1 = -V,\, x_2 = -V) \end{aligned}$$

This expression can be simplified as

$$\begin{aligned} R_{xx}(\tau) &= (+V^2)P(x_1 \text{ and } x_2 \text{ have the same sign}) \\ &\quad + (-V^2)P(x_1 \text{ and } x_2 \text{ have differing signs}) \end{aligned}$$

For the calculation of the joint probability, three distinct cases are considered: (i) $\tau = 0$, (ii) $|\tau| \leqslant T$, (iii) $|\tau| > T$.

For $\tau = 0$ the probability that x_1 and x_2 have the same sign is one. (The probability of them having differing signs is then zero.) Hence

$$R_{xx}(0) = V^2$$

For $|\tau| < T$ Figure 9.17 illustrates the cases that give the same sign to x_1 and x_2 (τ has been taken as positive here).

In Figure 9.17(a) the shift τ is not sufficient to cause a switching transition to reach t_1. The probability of this occurring over the ensemble is $(T - \tau)/T$. It does not matter whether or not there is a level change at $t = nT$.

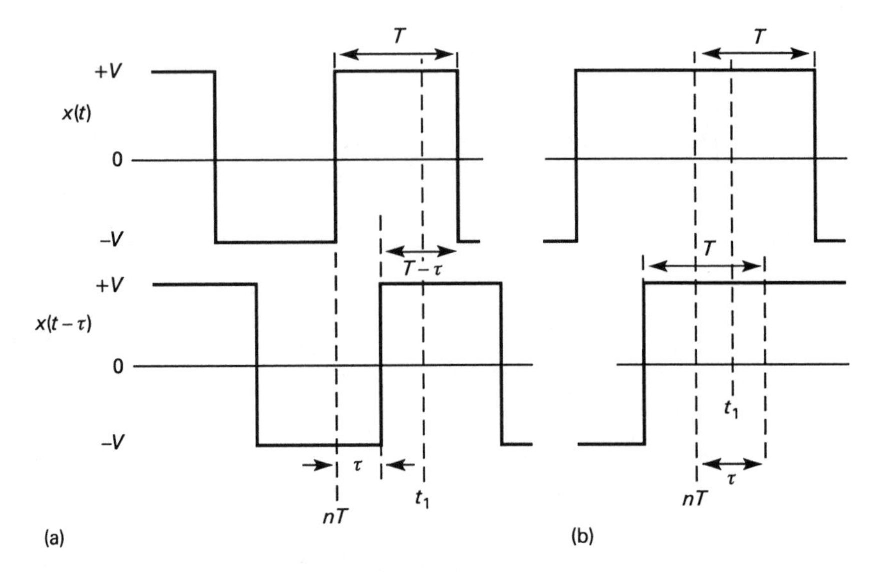

Figure 9.17 Sign changes between signal and its delayed version.

In Figure 9.17(b) the shift τ is such as to cause a switching transition to reach t_1. The probability of this occurring is τ/T. However, at the switching transition the level does not change and the probability of this occurring is 1/2.

$$P(x_1 \text{ and } x_2 \text{ same sign}) = \frac{T-\tau}{T} + \frac{1}{2}\frac{\tau}{T}$$
$$= 1 - \frac{\tau}{2T}$$

Hence

$$P(x_1 \text{ and } x_2 \text{ differing signs}) = 1 - \left(1 - \frac{\tau}{2T}\right)$$
$$= \frac{\tau}{2T}$$

$$R_{xx}(\tau) = (+V^2)\left(1 - \frac{\tau}{2T}\right) + (-V^2)\frac{\tau}{2T}$$
$$= V^2\left(1 - \frac{\tau}{T}\right)$$

The analysis has assumed that τ is positive; similar analysis shows that the result is the same for τ negative. The two results can be expressed as

$$R_{xx}(\tau) = V^2\left(1 - \frac{|\tau|}{T}\right)$$

For the case where $|\tau| > T$ the values of x_1 and x_2 are due to independent throws of a die. Hence the probability of x_1 and x_2 having the same sign is 1/2 and the probability of them having differing signs is also 1/2. Hence

$$R_{xx}(\tau) = (+V^2)\tfrac{1}{2} + (-V^2)\tfrac{1}{2}$$
$$= 0$$

A plot of the autocorrelation function is given in Figure 9.18.

(b) The process is both stationary and ergodic. The autocorrelation function can be calculated from consideration of one realisation with φ taken as constant.

Then

$$x(t) = X\sin(\omega_0 t + \varphi)$$
$$x(t-\tau) = X\sin(\omega_0[t-\tau] + \varphi)$$
$$R_{xx}(\tau) = \lim_{T\to\infty}\frac{1}{2T}\int_{-T}^{+T} X\sin(\omega_0 t + \varphi)X\sin(\omega_0([t-\tau] + \varphi)\,dt$$

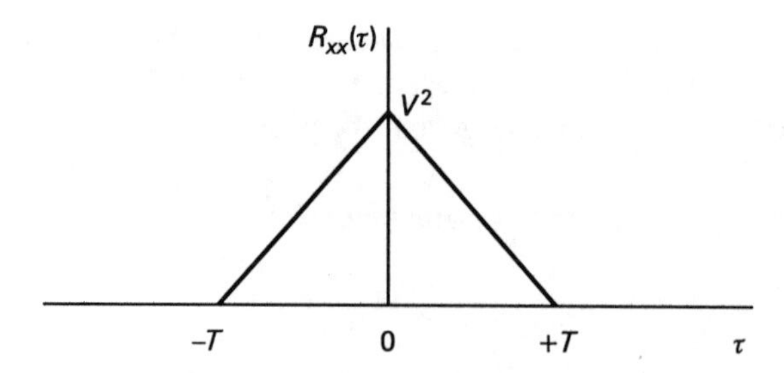

Figure 9.18 Autocorrelation function for Example 9.3.2(a).

Using the trigonometric identity

$$\sin A \sin B = \tfrac{1}{2}\,[\cos(A-B) - \cos(A+B)]$$

$$R_{xx}(\tau) = \lim_{T\to\infty} \frac{X^2}{2T}\int_{-T}^{+T} [\cos\omega_0\tau - \cos(2\omega_0 t - \omega_0\tau + 2\varphi)]\,\mathrm{d}t$$

$$= \frac{X^2}{2}\cos\omega_0\tau - \lim_{T\to\infty}\frac{X^2}{4\omega_0 T}[\sin(2\omega_0 t - \omega_0\tau + 2\varphi]_{-T}^{+T}$$

As $T\to\infty$ the numerator of the second term is bounded, but the denominator approaches ∞. Hence the second term approaches zero as $T\to\infty$ and

$$R_{xx}(\tau) = \frac{X^2}{2}\cos\omega_0\tau$$

$R_{xx}(\tau)$ is plotted in Figure 9.19(a).

To obtain $R_{xx}(-\tau)$ the sign of τ is changed throughout. The second term in the expression for $R_{xx}(\tau)$ still goes to zero. Hence

$$R_{xx}(-\tau) = \frac{X^2}{2}\cos(-\omega_0\tau)$$

$$= \frac{X^2}{2}\cos\omega_0\tau$$

or

$$R_{xx}(-\tau) = R_{xx}(\tau)$$

(c) Here

$$R_{xy}(\tau) = \lim_{T\to\infty}\frac{1}{2T}\int_{-T}^{+T} X\sin(\omega_0 t + \varphi)Y\sin(\omega_0[t-\tau] + \varphi + \theta)\,\mathrm{d}t$$

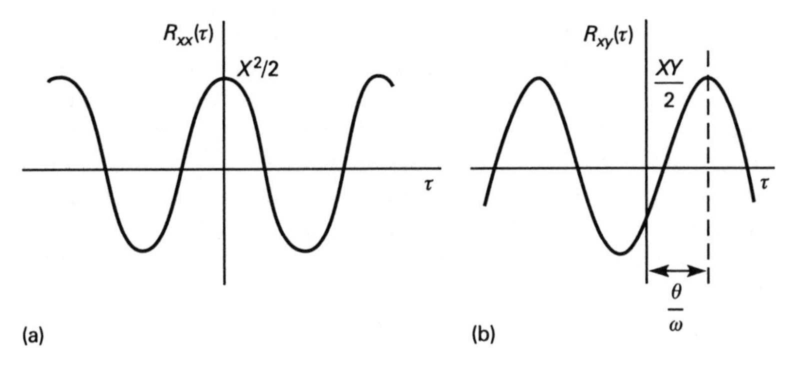

Figure 9.19 Autocorrelation and cross-correlation functions for Examples 9.3.2(b) and 9.3.2(c).

Using the trigonometrical identity of part (b)

$$R_{xy}(\tau) = \lim_{T\to\infty} \frac{1}{2T}\frac{XY}{2}\int_{-T}^{+T} X\cos(\omega_0\tau - \theta)\,dt$$

$$- \lim_{T\to\infty} \frac{1}{2T}\frac{XY}{2}\int_{-T}^{+T} \cos(2\omega_0 t - \omega_0\tau + 2\varphi + \theta)\,dt$$

As $T\to\infty$ the second term approaches zero and the first term gives

$$R_{xy}(\tau) = \frac{XY}{2}\cos(\omega_0\tau - \theta)$$

$R_{xy}(\tau)$ is plotted in Figure 9.19(b). Now

$$R_{yx}(\tau) = \lim_{T\to\infty} \frac{1}{2T}\int_{-T}^{+T} X\sin(\omega_0[t-\tau]Y\sin(\omega_0 t + \varphi + \theta)\,dt$$

Following the previous procedure

$$R_{yx}(\tau) = \frac{XY}{2}\cos(\omega_0[-\tau] - \theta)$$
$$= R_{xy}(-\tau)$$

9.3.4 Properties of the correlation functions

These properties will be given without formal proof. However, where appropriate, they will be illustrated by reference to part (b) of the previous example.

1. $R_{xx}(0)$ is the mean squared value of the signal. This follows by putting $\tau = 0$ into eqn (9.3.2). In the previous example $R_{xx}(0) = X^2/2$, which is the mean squared value of a sinusoid amplitude X.

2. $R_{xx}(-\tau) = R_{xx}(\tau)$. The autocorrelation function is an even function.

   This property has been demonstrated for a specific function in the previous example. In general, because the process is stationary, the autocorrelation function only depends upon the time difference $|\tau|$. It does not matter if τ represents a shift forward or a shift backward in time.
3. If the signal contains a periodic component its autocorrelation will also contain a periodic component having the same frequency as the signal component.

   This property is demonstrated in the single sinusoid of the previous example. It should be noted that, although the frequency of the original component is contained in the autocorrelation function, all phase information has been lost.
4. If the autocorrelation function is not periodic, its maximum value occurs at $\tau = 0$. As $\tau \to \infty$, $R_{xx}(\tau \to 0)$ provided the process has zero mean value.

   When $\tau = 0$ the process is correlating with itself and intuitively one would expect the correlation to be greatest for this value of τ. As $\tau \to \infty$ one would expect the random components in the process to have little dependence upon each other; hence, as $\tau \to \infty$, $R_{xx}(\tau) \to 0$. However, if the mean value of the process is not zero, one would expect this to correlate with itself no matter what the value of τ.

Properties of the cross-correlation function

The cross-correlation function does not have as many general properties as the autocorrelation function. As can be seen from Example 9.3.2(c), it is not in general an even function and its maximum does not occur at $\tau = 0$. One general property that it does have is that $R_{xy}(-\tau) = R_{yx}(\tau)$.

Example 9.3.2(c) also illustrates that if the processes $x(t)$ and $y(t)$ have periodic components with the same frequency then a component of this frequency will occur in $R_{xy}(\tau)$. Unlike the autocorrelation function, the phase relationship between the components is available from the cross-correlation function.

Electrical interpretation of correlation functions

A useful interpretation of auto- and cross-correlation functions can be obtained by considering the signals involved as current and voltage signals applied to an electrical circuit as shown in Figure 9.20.

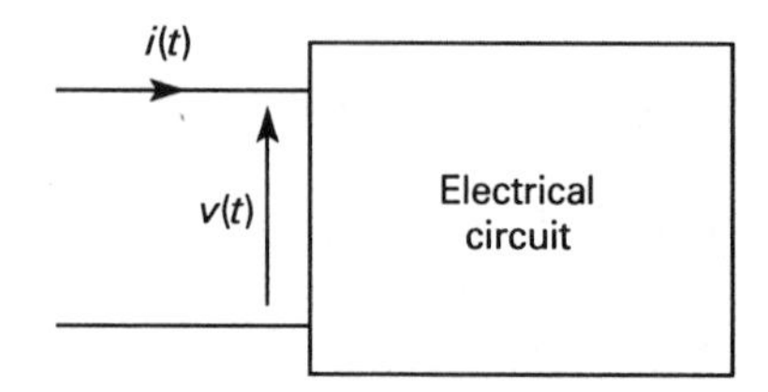

Figure 9.20 Electrical circuit for interpretation of correlation functions.

To interpret the autocorrelation function, $x(t)$ is taken as the voltage $v(t)$ and $i(t)$ is a delayed version of this signal, $i(t) = x(t-\tau)$. The autocorrelation function then represents the average power dissipated in this circuit for a given time delay τ.

To interpret the cross-correlation function $v(t) = x(t)$ and the current is a separate signal $y(t)$. The cross-correlation function then represents the average power dissipated for a given time delay τ between the signals.

If the signals concerned are single frequency sinusoids, the interpretation of these results corresponds to that derived from a.c. circuit theory. For $\tau = 0$ the circuit is purely resistive and dissipates power. For τ corresponding to phase shifts of $\pm 90°$ the circuit is purely reactive and dissipates no power. For τ corresponding to 180° the power is negative and the circuit acts as a power source. These results correspond to the variation in the auto- and cross-correlation functions as shown in Figure 9.19.

9.3.5 Autocorrelation function for the sum of two signals

It is often useful to be able to express the autocorrelation function of the sum of two signals in terms of the autocorrelation functions of the individual signals.

Suppose

$$z(t) = x(t) + y(t)$$

Then

$$R_{zz}(\tau) = \lim_{T\to\infty} \frac{1}{2T}\int_{-T}^{+T} z(t)z(t-\tau)\,\mathrm{d}t$$

$$\begin{aligned} z(t)z(t-\tau) &= [x(t)+y(t)][x(t-\tau)+y(t-\tau)] \\ &= x(t)x(t-\tau)+y(t)y(t-\tau) \\ &\quad + x(t)y(t-\tau)+y(t)x(t-\tau) \end{aligned}$$

Substituting into the expression for $R_{zz}(\tau)$ gives

$$R_{zz}(\tau) = R_{xx}(\tau) + R_{yy}(\tau) + R_{xy}(\tau) + R_{yx}(\tau)$$

In general the autocorrelation of the sum of two signals depends upon the cross-correlation between the signals. If $x(t)$ and $y(t)$ are uncorrelated, $R_{xy}(\tau) = R_{yx}(\tau) = 0$ and

$$R_{zz}(\tau) = R_{xx}(\tau) + R_{yy}(\tau)$$

This relationship is useful in practice for the detection of a sinusoidal signal $x(t)$ in the presence of noise. A noise signal $n(t)$ is a random signal and it has the property that $R_{nn}(\tau)\to 0$ as $\tau\to\infty$. The autocorrelation function of the signal $x(t)$, however, is periodic. Hence, if the noisy signal $z(t)$ is given by

$$z(t) = x(t) + n(t)$$

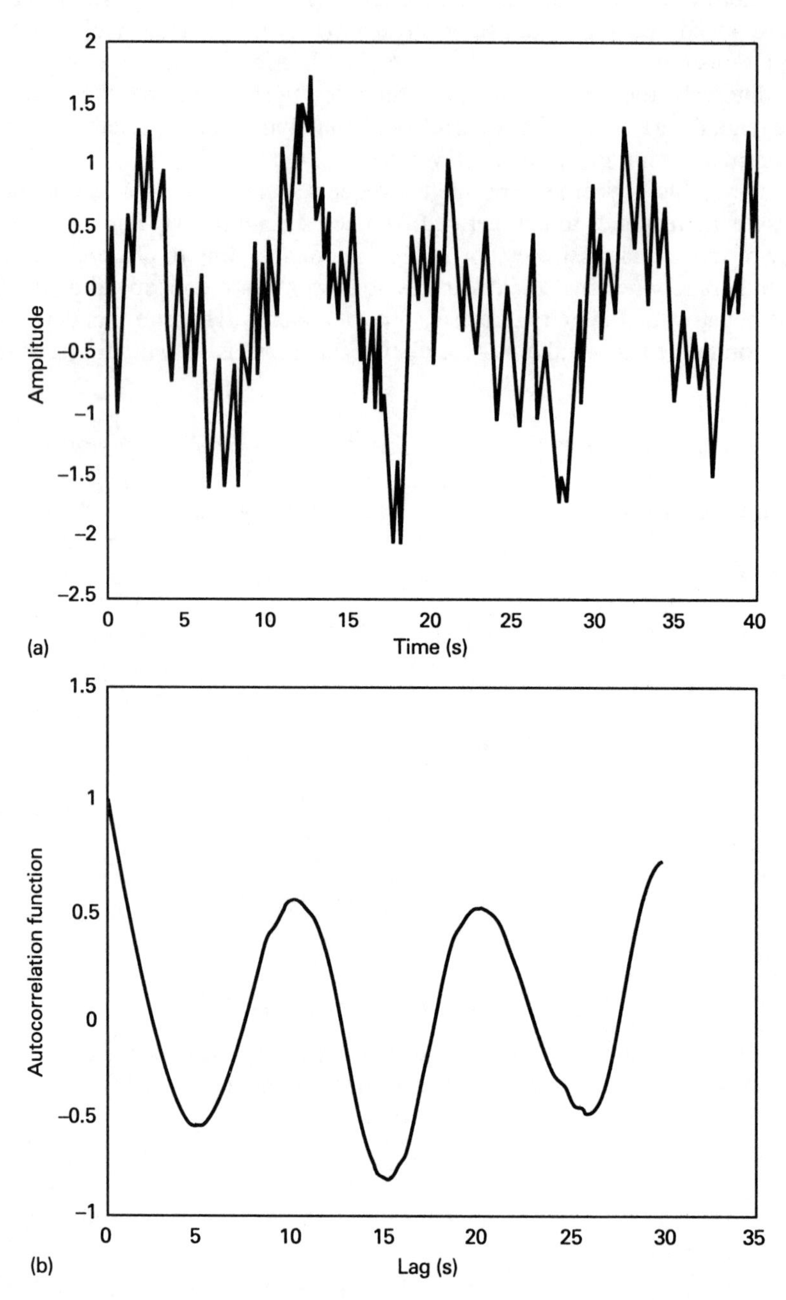

Figure 9.21 Autocorrelation function of a sinusoidal signal plus noise.

and $x(t)$ and $n(t)$ are uncorrelated, then

$$R_{zz}(\tau) = R_{xx}(\tau) + R_{nn}(\tau)$$

For large enough τ, $R_{nn}(\tau)$ will be negligible and the periodic component $R_{xx}(\tau)$ is evident.

Figure 9.21 illustrates this using computer generated data. Figure 9.21(a) is a sinusoidal signal plus a noise signal; the sinusoid is not detectable by eye. Figure 9.21(b) shows the autocorrelation function of this signal; for large τ the sinusoid is clearly visible.

9.4 Frequency domain description

As seen in earlier chapters, the use of the frequency domain via Fourier series and Fourier transforms is a very powerful method for describing deterministic signals. It would be useful if this form of description could be extended to include random signals. However, there are difficulties in applying the Fourier transform to a time signal that is random.

Consider two random signals $x(t)$, $y(t)$. Three realisations of each of these signals are shown in Figure 9.22.

From the visual appearance of these signals it appears that signal $y(t)$ has a higher frequency content in its spectrum than signal $x(t)$. However, two problems present themselves when trying to obtain the frequency content of these signals via the Fourier transform:

1. The integral defining the Fourier transform will not converge for a random signal.

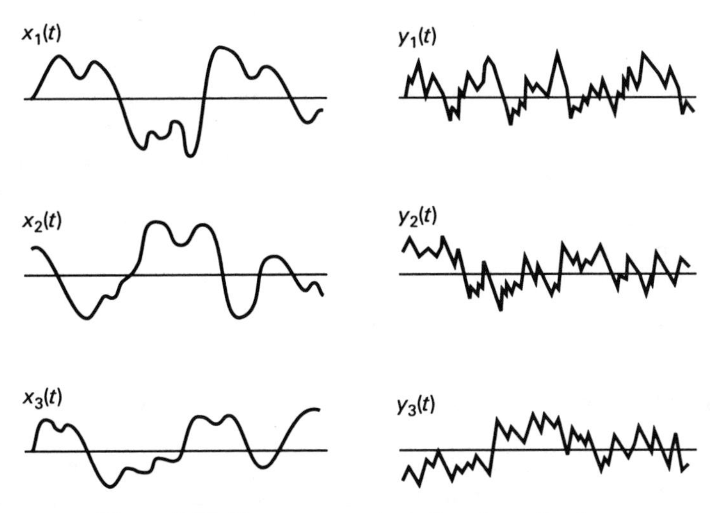

Figure 9.22 Frequency content of random signals.

2. Depending upon the realisation used, a different spectrum would be obtained from the transforms.

These points will be considered in more detail in the following section.

9.4.1 Power spectral density

Considering first the problem of convergence, in Section 6.2.3 the problem of convergence when determining the Fourier transforms of deterministic signals was mentioned. The concepts of power and energy signals were introduced and for one realisation of a random process the power in the signal can be defined as

$$P = \lim_{T\to\infty} \frac{1}{2T} \int_{-T}^{+T} x^2(t)\, \mathrm{d}t$$

The signal $x(t)$ over a finite time range $-T$ to $+T$ has a Fourier transform and Parseval's theorem can be applied, so

$$\int_{-T}^{+T} x_T^2(t)\, \mathrm{d}t = \frac{1}{2\pi} \int_{-\infty}^{+\infty} |X_T(\mathrm{j}\omega)|^2 \mathrm{d}\omega$$

where $X_T(\mathrm{j}\omega)$ is the Fourier transform of the signal $x_T(t)$ defined as

$$\begin{aligned} x_T(t) &= x(t) \qquad && -T \leqslant t \leqslant +T \\ &= 0 && \text{otherwise} \end{aligned}$$

The total power in the signal is now obtained by letting $T \to \infty$:

$$P = \lim_{T\to\infty} \int_{-T}^{+T} \frac{|X_T(\mathrm{j}\omega)|^2}{2T} \mathrm{d}\omega$$

The quantity $|X_T(\mathrm{j}\omega)|^2/2T$ when integrated with respect to frequency gives the total power in the signal. $\{X_T(\mathrm{j}\omega)|^2/2T$ when $T \to \infty$ represents a power spectral density giving a measure of how the power in the signal is distributed in frequency. The units of $\{X_T(\mathrm{j}\omega)|^2/2T$ are (amplitude/frequency)^2/time, which equals (amplitude)^2/frequency, which equals power/frequency. Hence, denoting the power spectral density as $S_{xx}(\omega)$, it can be defined as

$$S_{xx}(\omega) = \lim_{T\to\infty} \frac{|X_T(\mathrm{j}\omega)|^2}{T} \qquad (9.4.1)$$

with the proviso that some form of averaging is required to obtain a meaningful result.

The problem still remains that the power spectral density obtained is that from one realisation only. In order to make it representative of the process, some form of ensemble average is required. It might be argued that, because the process is ergodic, taking the limit $T \to \infty$ would produce the required result. However, there is a real problem in defining what is meant by a limit when considering a random process. Detailed consideration of this problem is beyond the scope of this book, but it will be examined further in the next chapter when the problem of measurement of power spectra is considered.

The problem of the limit in the previous paragraph can be avoided by adopting a different approach to the power spectrum of a random process. In Section 9.3.3 it was shown that the autocorrelation function of a random signal containing a periodic component is also periodic. This periodicity can be interpreted in the frequency domain by taking the Fourier transform of the autocorrelation function. Hence, if $R_{xx}(\tau)$ is the autocorrelation function of a random signal $x(t)$, then an alternative definition of power spectral density is

$$S_{xx}(\omega) = \int_{-\infty}^{+\infty} R_{xx}(\tau)\, e^{-j\omega\tau}\, dx \qquad (9.4.2)$$

The inverse transform gives

$$R_{xx}(\tau) = \frac{1}{2\pi} \int_{-\infty}^{+\infty} S_{xx}(\omega)\, e^{-j\omega\tau}\, d\omega$$

For $\tau = 0$

$$R_{xx}(0) = \frac{1}{2\pi} \int_{-\infty}^{+\infty} S_{xx}(\omega)\, d\omega \qquad (9.4.3)$$

But $R_{xx}(0)$ is the average power in the signal and hence integrating $S_{xx}(\omega)$ over all frequencies gives the signal power; $S_{xx}(\omega)$ is a power spectral density. The units of $R_{xx}(\tau)$ are (amplitude)^2 or power, and taking its Fourier transform gives the units of $S_{xx}(\omega)$ as power/frequency.

If, in eqn (9.4.3), the integral is performed with respect to frequency in Hz, f, rather than angular frequency, ω, then $\omega = 2\pi f$, $d\omega = 2\pi\, df$ and

$$R_{xx}(0) = \int_{-\infty}^{+\infty} S_{xx}(f)\, df$$

We have considered two approaches to obtain a definition of power spectral density for a random process:

1. By consideration of the Fourier transform of a random process $x(t)$,

$$S_{xx}(\omega) = \lim_{T \to \infty} \frac{1}{2T} |X_T(j\omega)|^2$$

where $X_T(\omega)$ is the Fourier transform of $x_T(t)$. Some form of averaging is required to make this definition meaningful.

2. By consideration of the Fourier transform of the autocorrelation function of the process

$$S_{xx}(\omega) = \int_{-\infty}^{+\infty} R_{xx}(\tau)\mathrm{e}^{-\mathrm{j}\omega\tau}\,\mathrm{d}\tau$$

It should be noted that $S_{xx}(\omega)$ is a real function. This is apparent from definition 1 as $|X_T(\mathrm{j}\omega)|^2$ is real. In definition 2, $R_{xx}(\tau)$ is an even function; hence its Fourier transform is real.

No formal proof will be given to show that these two definitions give rise to the same spectral density. The following example will illustrate the two approaches for a specific process.

EXAMPLE 9.4.1

Taking the process of Example 9.3.2(b), determine its power spectral density by the two methods given in the previous paragraph.

SOLUTION

1. Power spectral density via the autocorrelation function.
   In the solution to Example 9.3.2(b) the autocorrelation function of the process was obtained as

$$R_{xx}(\tau) = \frac{X^2}{2}\cos\omega_0\tau$$

The power spectral density is the Fourier transform of this function and, using Table 6.1, gives

$$S_{xx}(\omega) = \frac{X^2\pi}{2}[\delta(\omega - \omega_0) + \delta(\omega + \omega_0)]$$

Checking the total power in the spectrum

$$R_{xx}(0) = \frac{1}{2\pi}\int_{-\infty}^{+\infty}\frac{X^2\pi}{2}[\delta(\omega - \omega_0) + \delta(\omega + \omega_0)]\,\mathrm{d}\omega$$

and recalling that the area of the impulse function is the strength of the impulse,

$$R_{xx}(0) = \frac{X^2}{2}$$

2. Power spectral density via the direct transforms of the process.
   One realisation of the process for fixed φ is

$$x(t) = X\sin(\omega_o t + \varphi)$$

The power spectral density is

$$S_{xx}(\omega) = \lim_{T\to\infty} \frac{1}{2T} |X_T(j\omega)|^2$$

where $X_T(j\omega)$ is the Fourier transform of $x(t)$ over the interval $-T$ to $+T$. Then

$$|X_T(j\omega)|^2 = X_T^*(j\omega)X_T(j\omega)$$

and $X_T^*(j\omega)$ is the Fourier transform of $x(-t)$; hence there is no problem in obtaining the required Fourier transforms. However, the transforms contain impulse functions and the concepts involved in multiplying these functions cause some difficulty.

An alternative approach is to use the convolution property of the Fourier transform (Table 6.2)

$$X_T^*(j\omega)X_T(j\omega) = F\{x_T(-t) * x_T(t)\}$$

Remembering that $x_T(t)$ is zero outside of the range $-T$ to T, the convolution can be written as

$$\begin{aligned} x_T(-t) * x_T(t) &= \int_{-T}^{+T} X\sin(\omega_0[-\tau] + \varphi)X\sin(\omega_0[t-\tau] + \varphi)\,d\tau \\ &= \frac{X^2}{2}\int_{-T}^{+T} [\cos(-\omega_0 t) - \cos(-2\omega_0\tau + \omega_0 t + 2\varphi)d\tau \end{aligned}$$

Taking the Fourier transform of $[x_T(-t) * x_T(t)]$ and taking the limit inside the transform gives

$$S_{xx}(\omega) = F\left\{\lim_{T\to\infty} \frac{X^2}{2}\int_{-T}^{+T}\left[\frac{\cos(-\omega_0 t)}{2T} - \frac{\cos(-2\omega_0\tau + \omega_0 t + 2\varphi)}{2T}\right]d\tau\right\}$$

This integral is very similar to that evaluated in Example 9.3.2(b).

In the limit the second term approaches zero giving

$$\begin{aligned} S_{xx}(\omega) &= F\left\{\frac{X^2}{2}\cos\omega_0 t\right\} \\ &= \frac{X^2\pi}{2}[\delta(\omega - \omega_0) + \delta(\omega + \omega_0)] \end{aligned}$$

9.4.2 Spectrum of white noise

White noise is defined as a random signal whose power spectral density is constant at all frequencies from $\omega = -\infty$ to $\omega = +\infty$. The term comes from the analogy with white light where frequency components corresponding to all colours are present.

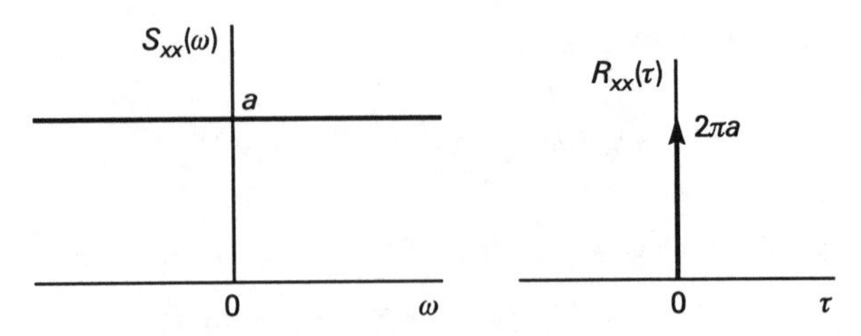

Figure 9.23 Power spectral density and autocorrelation function of white noise.

Hence for a white noise signal $x(t)$

$$S_{xx}(\omega) = a$$

where a is a constant. Taking the inverse Fourier transform gives the autocorrelation function $R_{xx}(\tau)$. Using Table 6.1 for the Fourier transform of an impulse together with the duality property, then

$$R_{xx}(\tau) = 2\pi a\delta(t)$$

Figure 9.23 shows the power spectral density and autocorrelation function of a white noise signal.

The autocorrelation function describes the signal as one where there is no correlation between its values at any two time instants (excepting $\tau = 0$). This implies there can be very rapid changes in signal amplitude given the very wide (infinite) bandwidth to the spectrum.

The white noise signal has infinite power. This can be seen as the result of the infinite area under the power spectral density function or as the result of infinite signal power represented by the amplitude $R_{xx}(0)$. Hence, the white noise signal is not a physically realisable signal; however, the concept of such a signal is very useful, as will be seen in later material.

A practical signal that approximates white noise is band limited noise ('pink' noise). In this signal the power spectral density is flat up to a certain cut-off frequency and then falls away giving a total power that is finite. If this signal is applied to a system whose pass band lies in the flat portion of the noise spectrum then the system input approximates a white noise signal.

9.4.3 Cross-spectral density

As stated in Section 9.3.4, if two signals $x(t)$ and $y(t)$ each have a periodic component of the same frequency, the cross-correlation function $R_{xy}(\tau)$ will also have a periodic component.

The Fourier transform of $R_{xy}(\tau)$ will transform this periodic component into the frequency domain. Hence, paralleling the definition of power spectral density, the cross-spectral density $S_{xy}(\omega)$ is defined as

$$S_{xy}(\omega) = \int_{-\infty}^{+\infty} R_{xy}(\tau)\,\mathrm{e}^{-\mathrm{j}\omega\tau}\,\mathrm{d}\tau \qquad (9.4.4)$$

The inverse transform gives

$$R_{xy}(\tau) = \frac{1}{2\pi}\int_{-\infty}^{+\infty} S_{xy}(\omega)\,\mathrm{e}^{+\mathrm{j}\omega\tau}\,\mathrm{d}\omega$$

Putting $\tau = 0$ gives

$$R_{xy}(0) = \frac{1}{2\pi}\int_{-\infty}^{+\infty} S_{xy}(\omega)\,\mathrm{d}\omega$$

$R_{xy}(0)$ represents the average power of the signal product $x(t)y(t)$. $S_{xy}(\omega)$, when integrated with respect to frequency, gives power; $S_{xy}(\omega)$ is a power spectral density, a cross-spectral density.

Again, paralleling the power spectral density, the cross-spectral density can be defined as

$$S_{xy}(\omega) = \lim_{T\to\infty} \frac{X_T^*(\mathrm{j}\omega)Y_T(\mathrm{j}\omega)}{2T} \qquad (9.4.5)$$

As in the case of power spectral density, some form of averaging is required if this definition is to lead to a meaningful result.

Returning to the definition given by eqn (9.4.4), a problem arises with the interpretation of this equation. Unlike the autocorrelation function, the cross-correlation function is not an even function. Hence, taking its Fourier transform gives a cross-spectral density that is complex. What interpretation can be given to complex power?

To illustrate the concept of complex power, consider again the electrical circuit of Figure 9.20 where $x(t)$ and $y(t)$ represent voltage and current signals. Suppose these signals are sinusoidal, with r.m.s. values X and Y, and it is required to calculate the power dissipated in the circuit. This power is equal to X multiplied by the component of Y that is in phase with X. Power is $XY\cos\varphi$ where φ is the phase angle between X and Y.

However, to parallel the concept of cross-spectral density, take X and Y as complex and consider the product $X * Y$ (X and Y and their components are shown in Figure 9.24):

$$\begin{aligned} X * Y &= (X_R - \mathrm{j}X_I)(Y_R + \mathrm{j}Y_I) \\ &= (X_R Y_R + X_I Y_I) + \mathrm{j}(X_R Y_I - X_I Y_R) \end{aligned}$$

The real part of this expression is the sum of the products of components that are in phase; it represents the power dissipated in the circuit. The imaginary term is the sum of products of quadrature components. Although it has the dimensions of power,

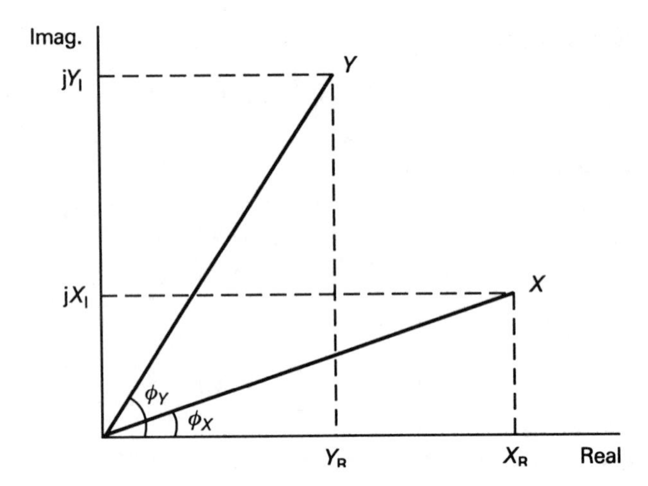

Figure 9.24 Power from complex components of a sinusoid.

there is no actual power associated with it. The quantity is referred to, by electrical power engineers, as reactive, quadrature or wattless power.

What significance does the imaginary term have in the cross-spectrum? The cross-spectrum is the Fourier transform of the cross-correlation function and the cross-correlation function contains phase information. This phase information can be obtained by considering the angle of $X * Y$. It is left as an exercise to show that this angle is $(\varphi_Y - \varphi_X)$, the phase difference φ between Y and X.

9.5 System response to random signals

A very important aspect of system theory that has been examined in earlier chapters was the calculation of a system's response to an input consisting of a deterministic signal. Two approaches to this problem were time domain analysis (via convolution) and frequency domain analysis (via the Fourier transform). This section considers whether we can use similar methods to determine system response when the input is a random signal.

Because the input and output signals to the system are random, a direct time relationship is not possible and instead we seek a relationship between correlation functions. In the frequency domain the corresponding relationships are between power and cross-spectral densities.

As an example illustrating the response of a system to a random signal, consider the problem of reducing room echoes from a loudspeaker. A microphone is used in place of the listener and the system input is a white noise signal. If there were no echoes, the autocorrelation function of the microphone signal would be an impulse. By examining the autocorrelation function, the presence of echoes can be detected and suitable steps taken to minimise them.

9.5.1 System response via correlation functions

Consider the discrete system shown in Figure 9.25. Here the input signal is $x(n)$, the output signal is $y(n)$ and $h(n)$ is the system impulse response which is also shown.

The output at any instant n is given by the convolution of the input $x(n)$ and the impulse response $h(n)$. The output $y(n)$ depends upon the present input $x(n)$ and previous inputs $x(n-1)$, $x(n-2)$ ...; these inputs are weighted by the system's impulse response. Hence, the cross-correlation function between input and output signals is going to be related to the system's impulse response.

To develop this concept further, the output $y(n)$ can be written as a convolution summation

$$y(n) = h(0)x(n) + h(1)x(n-1) + \cdots + h(r)x(n-r) \tag{9.5.1}$$

Multiplying this expression throughout by the input sample $x(n-k)$ gives

$$\begin{aligned} y(n)x(n-k) = {} & h(0)x(n)x(n-k) + h(1)x(n-1)x(n-k) \\ & + \cdots + h(r)x(n-r)x(n-k) \end{aligned}$$

Taking expectations throughout

$$\begin{aligned} E\{y(n)x(n-k)\} &= R_{xy}(k) \\ E\{x(n)x(n-k)\} &= R_{xx}(k) \\ E\{x(n-1)x(n-k)\} &= R_{xx}(k-1) \\ E\{x(n-r)x(n-k)\} &= R_{xx}(k-r) \end{aligned}$$

Hence

$$R_{xy}(k) = h(0)R_{xx}(k) + h(1)R_{xx}(k-1) + \cdots + h(r)R_{xx}(k-r)$$

Comparing this expression with eqn (9.5.1)

$$R_{xy}(k) = h(k) * R_{xx}(k) \tag{9.5.2}$$

The cross-correlation function between input and output signals is the convolution of the system's impulse response and the autocorrelation function of the input signal.

The corresponding relationship for the continuous system is

$$R_{xy}(\tau) = h(\tau) * R_{xx}(\tau) \tag{9.5.3}$$

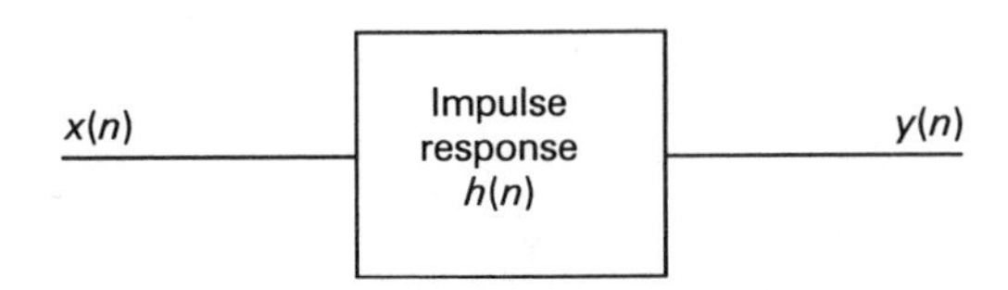

Figure 9.25 Correlation between input and output signals for a discrete system.

An important result follows if the input signal is white noise. As shown in Section 9.4.2, a white noise process has an autocorrelation function that is an impulse, $R_{xx}(\tau) = a\delta(\tau)$ Convolution of an impulse with any function leaves the function unaltered. Hence, for white noise input

$$R_{xy}(\tau) = ah(\tau) \tag{9.5.4}$$

i.e. the cross-correlation function is the system's impulse response scaled by a constant. This result is used for the experimental determination of a system's impulse response when direct measurement is impractical.

The following example illustrates the calculation of the cross-correlation function for a given system and a specific input signal.

EXAMPLE 9.5.1

The autocorrelation function of a random binary signal was determined in Example 9.3.2(a). If this signal is fed into a system having impulse response

$$h(t) = \frac{1}{T_1} e^{-t/T_1}$$

determine the cross-correlation function between the system input and output signals.

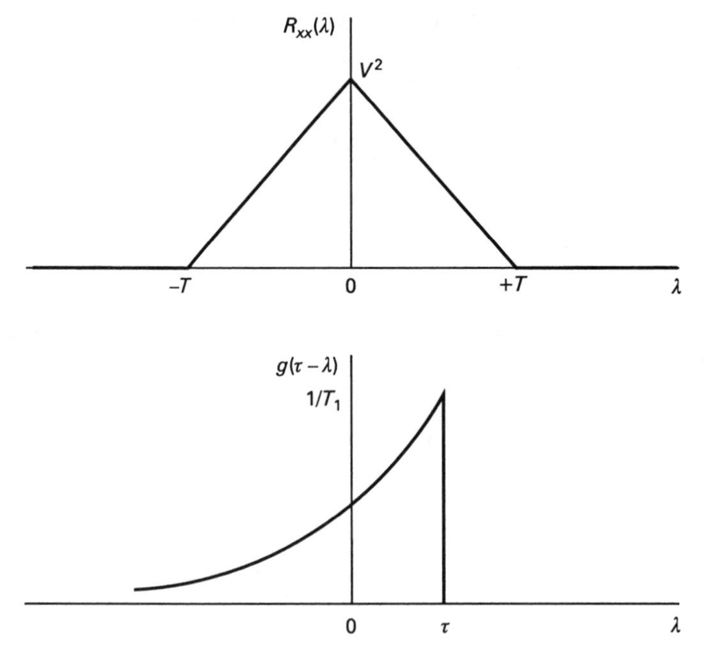

Figure 9.26 Convolution between impulse response and autocorrelation function of input signal.

SOLUTION

The cross-correlation function is given by

$$R_{xy}(\tau) = \int_{-\infty}^{+\infty} h(t-\lambda) R_{xx}(\lambda)\,\mathrm{d}\lambda$$

Figure 9.26 illustrates the signals involved in this integral for a specific value of τ.

To perform the convolution integration the associated functions must be expressed analytically over the appropriate ranges:

$$h(\tau-\lambda) = \frac{1}{T_1}\,\mathrm{e}^{-(\tau-\lambda)/T_1} \qquad \lambda \leqslant \tau$$
$$= 0 \qquad \text{otherwise}$$

$$R_{xx}(\lambda) = V^2\left(1+\frac{\lambda}{T}\right) \qquad -T \leqslant \lambda \leqslant +0$$
$$= V^2\left(1-\frac{\lambda}{T}\right) \qquad 0 \leqslant \lambda \leqslant +T$$

Because of the discontinuous nature of the functions involved, the integration must be divided into ranges. For $\lambda \leqslant T$ the integral is zero. The ranges are then

1. $-T \leqslant \lambda \leqslant 0$

$$R_{xx}(\lambda) = \int_{-T}^{\tau} V^2\left(1+\frac{\lambda}{T}\right)\frac{1}{T_1}\,\mathrm{e}^{-(\tau-\lambda)/T_1}\,\mathrm{d}\lambda$$
$$= \frac{V^2}{T_1}\,\mathrm{e}^{-\tau/T_1}\int_{-T}^{\tau}\left(1+\frac{\lambda}{T}\right)\mathrm{e}^{\lambda/T_1}\,\mathrm{d}\lambda$$
$$= V^2\,\mathrm{e}^{-\tau/T_1}\left[\mathrm{e}^{\lambda/T_1}\left(1+\frac{\lambda-T_1}{T}\right)\right]_{-T}^{\tau}$$
$$= V^2\left[\left(1+\frac{\tau-T_1}{T}\right)+\frac{T_1}{T}\,\mathrm{e}^{-(T+\tau)/T_1}\right]$$

2. $0 \leqslant \lambda \leqslant T$

$$R_{xx}(\lambda) = \int_{-T}^{0} V^2\left(1+\frac{\lambda}{T}\right)\frac{1}{T_1}\,\mathrm{e}^{-(\tau-\lambda)/T_1}\,\mathrm{d}\lambda + \int_{0}^{\tau} V^2\left(1-\frac{\lambda}{T}\right)\frac{1}{T_1}\,\mathrm{e}^{-(\tau-\lambda)/T_1}\,\mathrm{d}\lambda$$
$$= V^2\left[\left(1+\frac{\tau-T_1}{T}\right)-\frac{2T_1}{T}\,\mathrm{e}^{-\tau/T_1}+\frac{T_1}{T}\,\mathrm{e}^{-(T+\tau)/T_1}\right]$$

3. $T \leqslant \lambda$

$$R_{xx}(\lambda) = \int_{-T}^{0} V^2\left(1 + \frac{\lambda}{T}\right)\frac{1}{T_1}\,\mathrm{e}^{-(\tau-\lambda)/T_1}\,\mathrm{d}\lambda + \int_{0}^{T} V^2\left(1 - \frac{\lambda}{T}\right)\frac{1}{T_1}\,\mathrm{e}^{-(\tau-\lambda)/T_1}\,\mathrm{d}\lambda$$

$$= V^2\left(\frac{T_1}{T}\,\mathrm{e}^{(T-\tau)/T_1} + \frac{T_1}{T}\,\mathrm{e}^{-(T+\tau)/T_1} - \frac{2T_1}{T}\,\mathrm{e}^{-\tau/T_1}\right)$$

As can be seen, the calculation of the cross-correlation function, even for a simple system, is a lengthy procedure.

In this example, the expression for the cross-correlation function for $\tau > T$ can be written as

$$R_{xy}(\tau) = V^2\frac{T_1}{T}\,\mathrm{e}^{-\tau/T_1}\,(\mathrm{e}^{T/T_1} + \mathrm{e}^{-T/T_1}) - 2$$

Suppose $T/T_1 \ll 1$; then the exponentials e^{T/T_1} and e^{-T/T_1} can be approximated as

$$\mathrm{e}^{T/T_1} \approx 1 + \frac{T}{T_1} + \frac{1}{2}\left(\frac{T}{T_1}\right)^2$$

$$\mathrm{e}^{-T/T_1} \approx 1 - \frac{T}{T_1} + \frac{1}{2}\left(\frac{T}{T_1}\right)^2$$

Then

$$\mathrm{e}^{T/T_1} + \mathrm{e}^{-T/T_1} - 2 = \left(\frac{T}{T_1}\right)^2$$

and

$$R_{xy}(\tau) = V^2 T\,\frac{\mathrm{e}^{-\tau/T_1}}{T_1}$$

This illustrates the result given by eqn (9.5.4). When $T \ll T_1$, the input autocorrelation function approximates a triangular impulse having area V^2T. Hence, the process $x(t)$ under these conditions can be regarded as a white noise process. The cross-correlation function is now the system impulse response multiplied by the strength of the input impulsive autocorrelation function.

9.5.2 System response in the frequency domain

In the previous section is was shown that the relationship between the cross-correlation function and the autocorrelation function for a system is given by

$$R_{xy}(\tau) = R_{xx}(\tau) * h(\tau)$$

where $h(\tau)$ is the system's impulse response.

Taking the Fourier transform of both sides of this equation, and using the convolution property of the Fourier transform, gives

$$S_{xy}(j\omega) = S_{xx}(j\omega)H(j\omega) \tag{9.5.5}$$

where $H(j\omega)$ is the system's frequency response function which is the Fourier transform of its impulse response.

Re-arranging eqn (9.5.5)

$$H(j\omega) = \frac{S_{xy}(j\omega)}{S_{xx}(j\omega)} \tag{9.5.6}$$

This equation forms the basis of the experimental determination of a system's frequency response when the input is a random signal. This topic will be examined further in the next chapter.

The relationship between the power spectral densities at the input and output of a system is often of interest. A formal proof of this relationship requires the Fourier transforms of the input and output autocorrelation functions. The relationship between these quantities is somewhat complex and has not been derived. An intuitive relationship between input and output power spectral densities will be presented by paralleling the case for deterministic sinusoidal signals.

For such signals,

Input power at frequency ω is $|X(j\omega)|^2 = X^*(j\omega)X(j\omega)$

Output power at frequency ω is $|Y(j\omega)|^2 = Y^*(j\omega)Y(j\omega)$

However,

$$Y(j\omega) = H(j\omega)X(j\omega) \qquad Y^*(j\omega) = H^*(j\omega)X^*(j\omega)$$

$$\begin{aligned} |Y(j\omega)|^2 &= Y^*(j\omega)Y(j\omega) \\ &= H^*(j\omega)X^*(j\omega)H(j\omega)X(j\omega) \\ &= |H(j\omega)|^2|X(j\omega)|^2 \end{aligned}$$

For random signals the corresponding relationship using power spectral density functions is

$$S_{yy}(\omega) = |H(j\omega)|^2 S_{xx}(\omega) \tag{9.5.7}$$

EXAMPLE 9.5.2

(a) Show that the power spectral density corresponding to an autocorrelation function

$$R_{xx}(\tau) = \sigma^2 e^{-|\tau/T|}$$

is

$$S_{xx}(\omega) = \sigma^2 \frac{2T}{1 + (\omega T)^2}$$

(b) Verify that half the power in this signal is contained in the frequency range $\omega = 0$ to $1/T$.

(c) The signal is applied to a first order system that has a bandwidth of twice the input signal

$$H(j\omega) = \frac{1}{1 + j\omega T/2}$$

Determine the ratio output power/input power for the system.

SOLUTION

(a) From eqn (9.4.2)

$$S_{xx}(\omega) = \int_{-\infty}^{+\infty} R_{xx}(\tau)\, e^{-j\omega\tau}\, d\tau$$

The integral can be considered in two parts:

(i) For $-\infty < \tau \leqslant 0$, then

$$R_{xx}(\tau) = \sigma^2 e^{+\tau/T}$$

(ii) For $0 \leqslant \tau < +\infty$, then

$$R_{xx}(\tau) = \sigma^2 e^{-\tau/T}$$

The expression for $S_{xx}(\omega)$ becomes

$$\begin{aligned} S_{xx}(\omega) &= \sigma^2 \int_{-\infty}^{0} e^{(1/T - j\omega)\tau}\, d\tau + \sigma^2 \int_{0}^{\infty} e^{(-1/T - j\omega)\tau}\, d\tau \\ &= \sigma^2 \left[\frac{e^{(1/T - j\omega)\tau}}{1/T - j\omega} \right]_{-\infty}^{0} + \sigma^2 \left[\frac{e^{(-1/T - j\omega)\tau}}{-1/T - j\omega} \right]_{0}^{\infty} \\ &= \frac{\sigma^2 2T}{1 + (\omega T)^2} \end{aligned}$$

(b) From eqn (9.4.3), the power is given by

$$\begin{aligned} P &= \frac{1}{2\pi} \int_{-\infty}^{+\infty} S_{xx}(\omega)\, d\omega \\ &= \frac{\sigma^2}{2\pi} \int_{-\infty}^{+\infty} \frac{2T}{1 + (\omega T)^2}\, d\omega \end{aligned}$$

Writing $S_{xx}(\omega)$ as

$$S_{xx}(\omega) = \frac{2/T}{(1/T)^2 + \omega^2}$$

this can be recognised as a form of the standard integral

$$\int \frac{1}{a^2 + x^2}\, \mathrm{d}x = \frac{1}{a} \tan^{-1}\left(\frac{x}{a}\right) + \text{constant}$$

Then

$$P = \frac{\sigma^2}{2\pi} T \frac{2}{T}\left[\frac{\pi}{2} - \left(-\frac{\pi}{2}\right)\right]$$
$$= \sigma^2$$

This result can be recognised as $R_{xx}(0)$. If the frequency range is $-1/T \leqslant \omega \leqslant 1/T$ then

$$P_1 = \frac{1}{2\pi}\int_{-1/T}^{+1/T} S_{xx}(\omega)\, \mathrm{d}\omega$$
$$= \frac{\sigma^2}{2\pi} T \frac{2}{T} [\tan^{-1}(1) - \tan^{-1}(-1)]$$
$$= \frac{\sigma^2}{2}$$

(c) The output power spectral density $S_{yy}(\omega)$ is given by eqn (9.5.7)

$$S_{yy}(\omega) = |H(\mathrm{j}\omega)|^2 S_{xx}(\omega)$$

Here

$$H(\mathrm{j}\omega) = \frac{1}{1 + \mathrm{j}\omega T/2}$$

$$|H(\mathrm{j}\omega)|^2 = H(\mathrm{j}\omega)H^*(\mathrm{j}\omega)$$
$$= \frac{1}{1 + (\omega T/2)^2}$$

$$S_{yy}(\omega) = \sigma^2 \frac{2T}{1 + (\omega T)^2} \frac{1}{1 + (\omega T/2)^2}$$

$$\text{Output power} = \frac{1}{2\pi}\int_{-\infty}^{+\infty} S_{yy}(\omega)\, \mathrm{d}\omega$$

The expression for $S_{yy}(\omega)$ can be split into partial fractions:

$$S_{yy}(\omega) = \sigma^2 2\pi \left(\frac{4}{3[1 + (\omega T)^2]} - \frac{T}{3[1 + (\omega T/2)^2]}\right)$$

Each of these terms can be integrated as in part (b), giving

$$\text{Output power} = \frac{2\sigma^2}{3}$$

Hence the ratio

$$\frac{\text{Output power}}{\text{Input power}} = \frac{2}{3}$$

9.6 Summary

This chapter has extended the forms of signal description, and methods of system analysis, used in previous chapters to include the situation where the signals involved are random. These signals are important because, in the real engineering world, they arise from such complex physical processes that they must be regarded as random signals.

Because of the 'randomness' of random signals, a probabilistic description is required and some elementary probability theory has been introduced. It has been shown that the probability of a signal lying between certain amplitude levels is given by its probability density function. Convenient measures of this function are given by the moments of a random variable.

The extension of probability density to two random signals was investigated and this produced the concepts of joint probability and joint moments. The joint moment then led to the idea of correlation functions. The correlation function describes the degree of 'connectedness' between values at two time instants of a random signal (autocorrelation), or at two time instants in separate signals (cross-correlation).

Frequency domain methods have been extended to include a measure of the frequency content of random signals. This is described in terms of power spectral density and it can be derived by two methods: (i) from the Fourier transform of the time signal, (ii) from the Fourier transform of the signal's autocorrelation function. The significance of these two methods for practical spectral estimation will be considered in the next chapter.

The final section of the chapter examined the response of a system when the input is a random signal. As with deterministic signals, two approaches, corresponding to time domain and frequency domain methods, were considered. One was to relate the cross-correlation function between input and output signals to the input signal's autocorrelation function and the system's impulse response. The other method was to relate the cross-spectral density between input and output signals to the input signal's power spectral density and the system's frequency response.

Problems

9.1 A component manufacturer knows that in a batch of twelve components eight are good and four are defective.
(a) If a single component from the batch is tested, what will be the probability that it is (i) good, (ii) defective?
(b) If two components are tested, what is the probability that (i) they are both good, (ii) one of them is good, (iii) they are both good given that at least one is good?

9.2 Probability theory can be applied to the reliability of systems. For a circuit, the reliability of a component is the probability that current can flow through the component. The reliability of the overall circuit can be calculated from the reliability of the individual components. Assuming that the ability of each component to carry current forms a set of independent events, obtain the overall reliability of the circuit shown in Figure 9.27 where the reliability of the individual components is given.

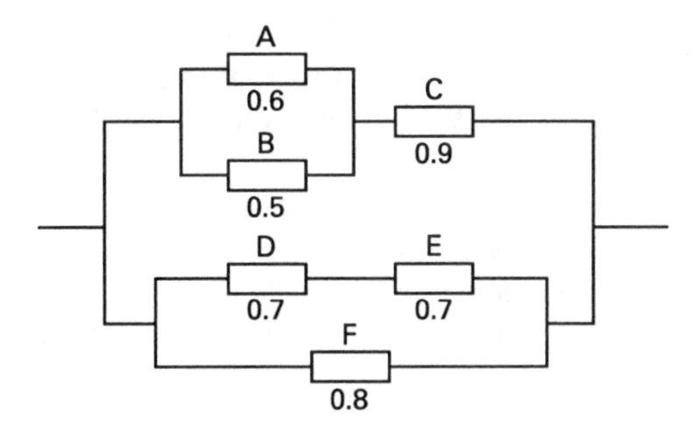

Figure 9.27

9.3 Two dice are thrown and the product of the resulting throws is taken. Taking this product as a random variable, sketch its probability density function and calculate its mean value, mean square value and variance.

9.4 Consider the random sequence generated by the coin tossing procedure of Section 9.2. Define a random variable X as the number of $+1$ values that occur in the sequence generated by 10 throws.
(a) Obtain the probability density function associated with X.
(b) Determine the mean and variance of X.

Note The number of ways of choosing k objects from a set of n objects is given by the combination formula

$$C(n, k) = \frac{n!}{k!(n-k)!}$$

In this example each of the k ways of choosing $+1$ V is associated with the probability of obtaining k values of $+1$ V and $(n-k)$ values of -1 V.

The resulting distribution is known as the binomial distribution.

9.5 (a) The Poisson distribution of a discrete random variable, X, is given by

$$P(X=k)=\mathrm{e}^{-\lambda}\,\frac{\lambda^k}{k!}$$

where λ is a parameter of the distribution and $k \geqslant 0$

(i) Show that

$$\sum_{k=0}^{\infty} P(X_k)=1$$

(ii) Determine the mean value of X.

(b) The Poisson distribution is used to model events that occur randomly in time, e.g. telephone calls at a switchboard, failure in components, traffic flow. If the average rate of occurrence of events is a, then the probability of k events occurring in time T is given by the Poisson distribution with $\lambda = aT$.

(i) Telephone calls arrive at a switchboard at an average rate of two per minute. What is the probability that there will be (1) no calls in a minute, (2) five calls in a minute?

(ii) What is the probability that there will be two or less calls in a 5 minute period?

9.6 A random signal has amplitude that is normally distributed with zero mean and unit variance. It is fed into a unit gain amplifier which saturates at levels ± 3 V.

(a) Over a long record length determine the percentage of the time that the amplifier output reaches saturation.

(b) Sketch the probability density function of the signal at the amplifier output.

Note The integral representing the area under the normal distribution cannot be evaluated analytically. Tables giving this area are available. The area is also related to the error function which is available in MATLAB.

9.7 The periodic signal of Figure 9.28 is one realisation of a stationary ergodic random signal. The equation describing the signal in the range 0 to T (where T is the periodic time) is

$$x(t)=X\,\mathrm{e}^{-t/T_1}$$

(a) Determine the probability density function, $p(x)$, for the signal.

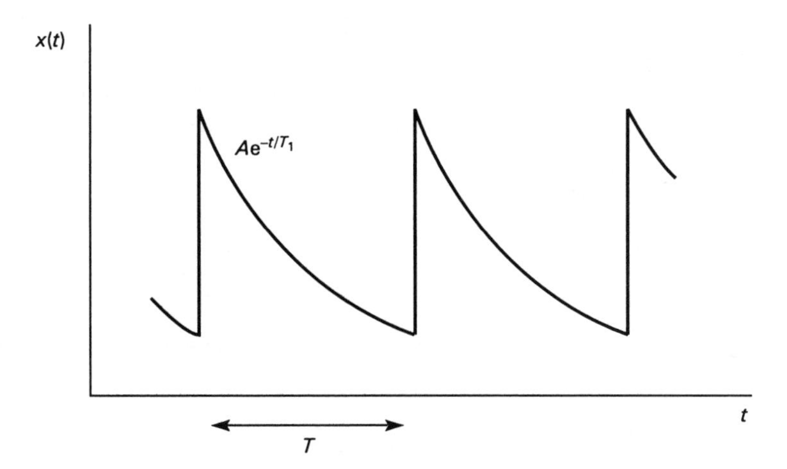

Figure 9.28

(b) Verify that

$$\int_{-\infty}^{+\infty} p(x)\,\mathrm{d}x = 1$$

(c) Calculate, from the probability density function, the mean and mean squared value of the signal and compare these to the values obtained by time averaging.

9.8 A signal can take two levels, $+a$ and $-a$, and the switching between these levels occurs randomly with time. The number of switchings, k, in a time interval, τ, is given by a Poisson distribution

$$P(k) = \frac{(\lambda\tau)^k}{k!}\,\mathrm{e}^{-\lambda\tau}$$

where λ is the average rate of switching.

Such a signal is known as a random telegraph signal. Determine its autocorrelation function. (Hint: The product, for a given τ, determining the autocorrelation function can only take values $+a^2$ and $-a^2$. The value depends on whether there are an odd number of crossings or an even number of crossings within the interval τ. Determine the probability, $P(\text{even})$, by forming a series of the probability of all the even switchings, similarly for $P(\text{odd})$.)

9.9 (a) Determine the autocorrelation function of a signal given by

$$x(t) = \sum_{k=1}^{n} a_k \cos(k\omega_0 + \varphi_k)$$

(b) Determine the autocorrelation function of a square wave, period T, that takes values $\pm a$. (i) Determine the result by direct analysis. (ii) Confirm that this agrees with the result obtained by representing the square wave as a Fourier series and using the result obtained in part (a).

9.10 A system has a transfer function

$$G(s) = \frac{1}{(1 + 0.1s)(1 + 0.5s)}$$

The input to the system is a random signal $x(t)$ whose power spectral density is given by

$$S_{xx}(\omega) = \frac{1}{1 + 0.04\omega^2}$$

Determine the total power contained in the output signal.

As problems of this nature become more complex, the mathematics involved becomes more tedious. To save on this, mathematical manipulation tables of the integral

$$I_N = \frac{1}{2\pi j}\int_{-j\infty}^{+j\infty} \left| \frac{c_{N-1}s^{N-1} + \cdots + c_1 s + c_0}{d_N s^N + \cdots + d_1 s + d_0} \right|^2 ds$$

have been tabulated. The integral relevant to this problem is for $N = 3$ and is given by

$$I_3 = \frac{c_2^2 d_0 d_1 + (c_1^2 - 2c_0 c_2) d_0 d_3 + c_0^2 d_2 d_3}{2d_0 d_3 (d_1 d_2 - d_0 d_3)}$$

Check your answer for total output power by use of this formula.

9.11 In a control system the usual object of the control is to make the output of the system $c(t)$ follow the input $r(t)$. To achieve this objective, the system is driven by an error, which, taking Laplace transforms, is $E(s) = R(s) - C(s)$. A measure of the performance of the system can be obtained by the total error power; this should be small. To ensure this, parameters of the system can be chosen to minimise the error power.

If $G(s)$ is the transfer function relating output to input

$$C(s) = G(s)R(s)$$

then

$$\begin{aligned} E(s) &= R(s) - C(s) \\ &= R(s)[1 - G(s)] \end{aligned}$$

If the system is second order

$$G(s) = \frac{\omega_n^2}{s^2 + 2\zeta\omega_n s + \omega_n^2}$$

with $\omega_n = 2$ rad/s, and the input is a random signal having power spectral density

$$S_{rr}(\omega) = \frac{1}{1 + \omega^2}$$

determine the value of the damping factor, ζ, that will give minimum error power. You are advised to use the integral I_3 given in the previous problem as an aid in the solution.

MATLAB exercises

9.12 This exercise consists of generating a long time record of a random variable and measuring its probability density function. This is then compared with the theoretical probability density function. Very large sample sets are required to produce results that closely agree with theory. The largest vector that can be produced in the student edition of MATLAB is of length 8192.

(a) Generate a long vector of uniformly distributed random variables. Make use of the command `hist` to produce and plot a probability density function. On the same axes, add the plot of the theoretical probability density function.
(b) Repeat (a) using normally distributed random variables having zero mean and unit variance.
(c) Repeat (a) using 1 cycle of a sinusoidal signal. This can be regarded as one realisation of a sinusoidal signal with random phase.
(d) Repeat (a) using the exponential signal of Problem 9.7. This can be regarded as one realisation of the exponential signal with a random time origin.

Note Some care is needed in processing the results of the function `hist` to produce the correct scales for a normalised (unit area) probability density function. The command `[n,x]=hist(y,nb)` returns the vectors `n` and `x` containing the frequency counts and bin locations. Use can be made of the definition of probability density function $p(x)$,

$$p(x)\Delta x = n/N$$

where N is the total number of points in the vector representing the random signal. The width Δx is given by the range of the signal divided by the number of bins `nb`. For the signals considered in parts (a), (b) and (c), the range is defined. However, for the signal in part (b) the theoretical range is infinite. In practice the probability of the signal lying outside the range $\pm 3\sigma$ is small and the range can be restricted to this. However, the command `[n,x]=hist(y,nb)` fixes the width of the bins according to the maximum and minimum values in `y`. The command `[n,x]=hist(y,x)` can be used where the bins are specified in the vector `x`.

9.13 In this exercise the same processes are generated as in the previous exercise. Again, the probability density functions are measured and compared with theory. However, the probability density functions are obtained from ensemble measurements rather than time measurements.

(a) Produce a single random variable with uniform distribution to represent the value of a random time signal at a given time instant. By incorporating this procedure into a 'for loop', a long sequence of ensemble values can be produced. To make use of the `hist` function, these have to form the elements of a vector. However, as the length of the vector is restricted to 8192 elements, and ensembles of greater length are required, the whole procedure is repeated by incorporating it into another 'for loop'. Use the command `[n,x]=hist(y,x)` to determine the frequency count in bins located by the vector `x` and keep a running total of `n` over a fixed number of runs. Averaging and scaling the results will produce a normalised probability density function. Plot and compare with the theoretical function.

(b) Repeat (a) using normally distributed random variables having zero mean and unit variance.

(c) Repeat (a) using a sinusoidal signal. To make the variables elements of an ensemble, a fixed time instant is used, but a random phase angle is introduced that is uniformly distributed in the range 0 to 2π.

(d) Repeat (c) using the exponential signal. This time the variable for each ensemble is produced by time shifting the signal by a random time interval uniformly distributed in the range 0 to T where T is the length of the signal.

9.14 This exercise examines random processes from the viewpoint of stationarity and ergodicity and looks at expected values derived from time and ensemble averages. Each of the four functions given has output vectors that represent one realisation of a random process. The length of the vector is set by `N`, but this cannot exceed 8192 elements.

```
% Random process 1
  function y=rp1(N)
  u=randn(1,N);
```

```
  p=-0.99;
  num=[1 0];
  den=[1 p];
  y=dlsim(num,den,u);
  y=y';

% Random process 2
  function y=rp2(N)
  y=zeros(1,N);
  x=sign(rand(1,N)-0.5);
  for k=[2:N];
      y(k)=x(k)+y(k-1);
  end

% Random process 3
  function y=rp3(N)
  t=[1:N];
  y=randn(1,N)+0.5*sin(2*pi*t/N);

% Random process 4
  function y=rp4(N)
  y0=round(4*(rand-0.5));
  y=y0+rand(1,N);
```

(a) For each process, generate four realisations and display all these by use of `subplot`. Decide by inspection of these four realisations whether each process is stationary/ergodic.

The plots are most efficiently produced by forming a 4 × `N` matrix of the points in the realisations. However, because of the restricted matrix size, this would restrict the number of points, `N`, to about 2000, which may not be large enough to determine the stationarity/ergodicity of the process.

In order to compare the realisations, it is better if the scalings for the subplots are identical. This can be achieved by use of the `axis` command.

(b) For each of the processes, measure the mean and standard deviation by both time and ensemble averaging (the commands `mean` and `std` are useful here). Measure the ensemble averages at several time points and, from these results, decide on the stationarity/ergodicity of the processes.

(c) Examine the MATLAB code for the generation of each process. Decide on this basis whether the processes are stationary/ergodic and, if possible, calculate theoretical time/ensemble averages.

9.15 This exercise simulates the amplitude limiting of a random signal as examined theoretically in Problem 9.6. It measures the probability density function and variance of the signal and compares these results with theory.

(a) Produce a random signal consisting of a long vector of random variables having zero mean and uniformly distributed between levels $+1$ and -1. Simulate the action of an amplitude limiter and limit this signal to levels ± 0.5 V.
(b) Measure the percentage of time that the signal is at the limiting levels and compare this with theory.
(c) Display the probability density functions of the signal both before and after limiting and plot on the same display as the theoretical probability density functions.
(d) Measure the variance of the signal before and after limiting and compare with theoretical values.
(e) Repeat parts (a) to (d) for a signal consisting of a vector of normally distributed random variables. Keep the limits at ± 0.5 V and make the random variables zero mean and 0.5 standard deviation.

Note The command `find` can be used to determine elements of a matrix that are outside given limits – these can then be replaced by limit values.

The area under the normal distribution cannot be evaluated analytically. However, the error function, `erf`, is available to aid in the evaluation. To evaluate the variance of the limited signal, use can be made of numerical integration using the `quad` function.

9.16 This exercise illustrates the central limit theorem.

Random variables having a uniform distribution are added and the distribution of the resulting sum compared with a normal distribution.

Generate a long vector of uniformly distributed random variables. Repeat this procedure 12 times and keep a running total of the corresponding elements to produce a random process that is the sum of 12 uniformly distributed processes.

Measure the mean and variance of the resulting random process and compare with theoretical values (mean and variance of the sum of independent random variables). Plot the probability density function of the resulting random variable and compare with a normal distribution using theoretical values of mean and variance.

9.17 This exercise simulates the random processes of Section 9.3 and obtains their joint probability density functions and correlation functions.

(a) Generate a long vector $x(n)$ with elements that are random variables representing the outcome of the 'coin-tossing process'. Each random variable should have equal probabilities of taking values $+1$ or -1.
(b) Taking $x(n)$ as the input to a discrete system, generate the system output $y(n)$ according to

$$y(n) = \frac{x(n) + x(n-1)}{2}$$

(c) Display $y(n)$ and $x(n)$ as subplots.

(d) Obtain the joint probability density functions between adjacent samples in the processes $x(n)$ and $y(n)$. Also obtain the joint probability density function between the variables $x(n)$ and $y(n)$, $x(n-1)$ and $y(n-1)$, $x(n-1)$ and $y(n)$, $x(n)$ and $y(n-1)$.

(e) From the joint probability density functions obtained in part (d), determine the autocorrelation functions for $x(n)$ and $y(n)$ and the cross-correlation function between $x(n)$ and $y(n)$.

(f) Determine the correlation functions of part (e) by time averaging.

(g) Verify that the measured cross-correlation function is the convolution of the system impulse response and the input autocorrelation function.

Note The binary process can be generated by use of the command `find` on a vector whose elements are random variables. Elements are identified whose values are above, or below, the mean value and these are replaced by values of $+1$ or -1.

The command `stem` can be used to plot the discrete variables. The system response can be obtained by means of the command `dlsim` or directly from the difference equation describing the system.

The joint probability can be obtained with the aid of the logical operator `&`. For example, it can be used to determine when corresponding elements in two vectors both have value $+1$.

9.18 This exercise investigates the cross-correlation function between signals that form the input and output signals of a system.

(a) Produce a long vector whose elements consist of normally distributed random variables with zero mean and unit variance. Assume this represents a signal with a sampling interval of 0.05 second.

(b) Obtain the autocorrelation function (for both positive and negative shifts) of the signal generated in part (a) and compare with the theoretical autocorrelation function.

(c) Use the random process generated in part (a) as the input to a system having transfer function

$$G_1(s) = \frac{1}{1+s}$$

(d) Obtain the cross-correlation function between input and output signals for the system given in part (c). Plot this and compare with the system's impulse response.

(e) Modify the input to the system G_1 by making it band limited noise. This can be done by feeding the original signal into a system having transfer function

$$G_2(s) = \frac{1}{1+0.3s}$$

Obtain the autocorrelation functions of the signals at the outputs of systems G_1 and G_2 and the cross-correlation function between the input and output of the system G_1.

(f) Plot the cross-correlation function and compare it with the signal obtained by the convolution of the input autocorrelation function and the impulse response of G_1.

(g) It can be shown that the autocorrelation function of the output of system G_1 is the convolution between the cross-correlation function between input and output of G_1 and the impulse response of G_1 with a reversed time scale

$$R_{yy}(\tau) = R_{xy}(\tau) * h(-\tau)$$

Obtain the output autocorrelation function by this method and compare with that obtained in part (e).

10

Spectral Analysis

10.1 Introduction

In the previous chapter the concepts of power spectral density and cross-spectral density were introduced. Usually in physical systems any random signals are produced by processes that are so complex that it is not possible to predict the spectra of these signals theoretically. The spectra have to be estimated by measurements on the random signals. Some examples of physical systems where spectral estimates are required are as follows:

1. The power spectrum of the displacement of a road wheel due to irregularities in road surfaces. This spectrum is required in the design of vehicle suspension systems in order to produce satisfactory passenger comfort and vehicle handling characteristics.
2. The power spectrum of vibration produced by rotating machinery. Excessive vibration causes undue wear and may be indicative of imbalance or other faults in the machinery. By identifying the frequency components present in the measured spectrum, the causes of the vibration can be identified.

3. Identification of noise sources in the driving compartments of motor vehicles. By measuring the cross-spectra between the cabin noise and engine noise/road noise/wind noise, it is possible to identify the components contributing to the noise.
4. Identification of system models. As shown in the previous chapter, the frequency response of a system can be estimated by the ratio of cross-spectral density/power spectral density measured from signals at the input and output of the system. This enables a frequency response model to be obtained from normal operating signals, or from small amplitude injected perturbation signals. This is useful for obtaining models of systems where it is difficult to separate the system from its environment and to do laboratory measurements, e.g. estimation of models of aircraft in flight.

This chapter will consider the factors involved in the measurement of power spectral density, cross-spectral density and the estimation of frequency response functions.

10.2 Power spectrum measurement – basic principles

The methods of power spectrum estimation considered in most of this chapter will be discrete methods. However, the basic principles involved in the measurement are the same regardless of the method used. Hence, in this section these principles will be explained via an analogue method which gives a clearer understanding of the problems involved than do the corresponding principles for discrete methods.

From the definition of power spectral density the power contained in an infinitesimal range $d\omega$ is $S_{xx}(\omega)\,d\omega$. If the power of the signal in a small frequency range $\Delta\omega$ is measured, then this power is approximately $S_{xx}(\omega)\Delta\omega$. This is the basis of the measurement scheme shown in Figure 10.1.

The signal to be analysed, $x(t)$, is fed to an ideal bandpass filter. This filter has bandwidth $\Delta\omega$ and is centred on frequency ω_0. The resultant output signal $x_o(t)$ is squared to obtain the instantaneous power in the band $\Delta\omega$. $x_o^2(t)$ is then averaged over time T to obtain the mean power P_m. Then $\hat{S}_{xx}(\omega_o) = P_m/\Delta\omega$. (Note that the 'hat', ^, over $S_{xx}(\omega)$ denotes an estimated quantity.)

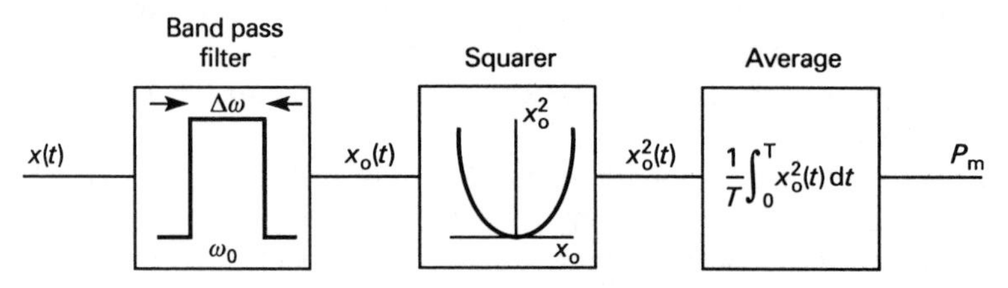

Figure 10.1 Basic power spectral density measurement.

This process would have to be repeated with the filter centred at frequencies $\omega_0 + \Delta\omega$, $\omega_0 + 2\Delta\omega$, etc., in order to cover the complete frequency range of the signal. Alternatively, the whole of the system of Figure 10.1 could be repeated in a parallel arrangement, again with the filters set to the appropriate centre frequencies.

The problems associated with this measurement scheme centre on two areas:

1. The ideal bandpass filter characteristic is unrealisable. Given that a realisable filter must be used, what should be the shape of its frequency response and what should be its effective bandwidth?
2. How should the averaging time T be chosen?

Considering the first problem, and for the moment assuming that an ideal bandpass characteristic could be obtained, what determines the bandwidth $\Delta\omega$?

Consider the power spectral density of a signal consisting of two sinusoids, frequencies ω_1 and ω_2. If the filter bandwidth $\Delta\omega$ is less than the difference $|\omega_2 - \omega_1|$ then both sinusoids will produce distinct components in the measured spectrum. If, however, $\Delta\omega > |\omega_2 - \omega_1|$ then the two sinusoids will be averaged to produce one component at the filter mid frequency.

Hence, the bandwidth of the filter determines the frequency resolution available. On this basis, $\Delta\omega$ should be as small as possible, but this requirement poses other problems. The action of the filter is multiplication in the frequency domain and this is equivalent to convolution in the time domain. Also, using the scaling property of the Fourier transform (Table 6.2), compressing the filter characteristic in the frequency domain (reducing $\Delta\omega$), is equivalent to increasing the effective length of its impulse response in the time domain. In order to obtain this time convolution a long signal record is required and in practice this may not be available.

Consider now the shape of the filter characteristic. In Chapter 8 it was shown that high cut-off rates in filters are accompanied by ripples in their frequency responses. Hence, a more practical bandpass filter could either (a) have a high cut-off rate and accept these ripples, or (b) reduce the cut-off rate and reduce the accompanying ripples. Figure 10.2 shows these alternatives.

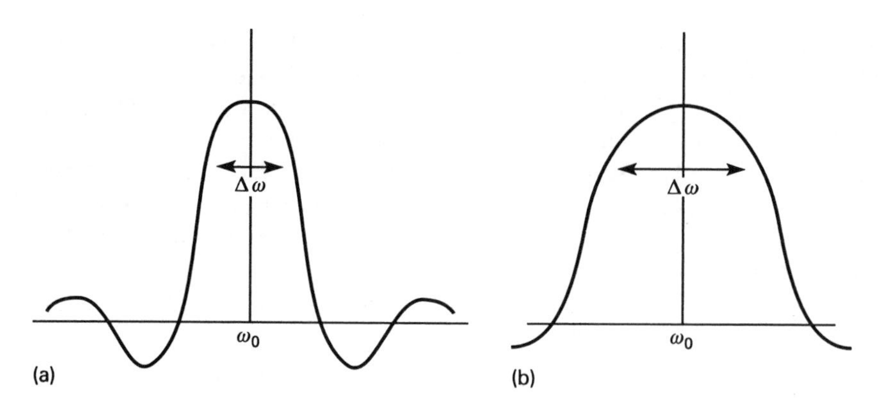

Figure 10.2 Alternative bandpass filter characteristics.

Because practical bandpass filters do not have flat-top characteristics, $\Delta\omega$ must be taken as the bandwidth between the -3 dB points. Figure 10.2(a) shows a bandpass filter with a higher cut-off rate and, hence, a narrow bandwidth; however, it produces ripples in the overall response. The characteristic of the filter of Figure 10.2(b) has negligible ripple in its response, but this has been achieved at the expense of a lower cut-off rate and wider bandwidth.

Hence, good resolution (small bandwidth $\Delta\omega$) can only be obtained at the expense of ripples in the filter's response. These ripples, or sidelobes, can have a detrimental effect on the measured spectrum. A strong frequency component at the peak of a sidelobe would appear at centre frequency ω_0 and completely mask a weaker component at this frequency. This effect is known as leakage.

The second problem listed earlier was the choice of averaging time T. Referring to Figure 10.1, if the input signal is random, then the signal $x_o(t)$ produced after the filtering and squaring is also random. It represents the *instantaneous* power in the chosen frequency band and this is varying in a random manner. The averaging operation produces the mean power; however, it is only the mean power if T is infinite. This is clearly not possible and a finite T will produce some variability in the measured spectrum. Hence, some compromise is required between undue record lengths and variability in the final result.

This section has presented the (often conflicting) requirements for the measurement of power spectra. In practice, the optimum filter characteristics and averaging time depend on the particular spectrum being estimated. If this is completely unknown, then the parameters of the measurement will have to be modified over a series of tests in order to produce optimum results.

Many of the considerations of this section will be re-examined in a different guise in Section 10.4, when digital measuring techniques will be examined.

10.3 Parameter estimation

The parameters describing a random signal are defined by integration (or summation) over an infinite time period. In practice one can only measure over a finite time interval and only an estimate of a parameter value can be obtained. Often there is more than one method of obtaining a value that is used as an *estimator* of the true value. Because different estimators can be used for the same parameter, it does not mean that there is one correct estimator. However, one estimator may have more desirable properties than another estimator and in this section some of these properties are discussed.

Consider the problem of estimating the mean value of a stationary random signal $x(t)$. The true mean is defined as

$$\mu = \lim_{T\to\infty} \frac{1}{T}\int_0^T x(t)\,\mathrm{d}t$$

An estimated quantity is denoted by putting a 'hat', ^, symbol over the quantity, e.g. $\hat{\mu}$ is an estimate of μ. If only a finite length, T, of record is available, then an estimator of μ could be

$$\hat{\mu} = \frac{1}{T}\int_0^T x(t)\,\mathrm{d}t$$

Suppose the mean value of $x(t)$ is zero. Figure 10.3 shows estimates based on three different portions of the signal.

Each of these estimates will differ (μ_1 positive, μ_2 approximately zero, μ_3 negative) so what can be said about the properties of this estimator? Suppose a large number of estimates were produced from different finite time records. If, from these estimates, a probability density function $p(\mu)$ were produced, what would be its desirable features to indicate a good estimator? One would expect the probability density function to centre around the true value of the parameter and it would be desirable if its spread around this value were small.

Figure 10.4 shows a probability density function where the expected value of the estimator, $E\{\hat{\mu}\}$, does not equal the true value μ – the estimate is said to be *biased*. In general, for any estimator $\hat{\theta}$

$$\text{Bias} = E\{\hat{\theta}\} - \theta$$

If the bias is zero, the estimator is said to be *unbiased*.

Bias is not the only consideration when comparing estimators. Figure 10.5 shows two different estimators of the same parameter.

Figure 10.5(a) shows an unbiased estimator which has a large variance, while the estimator shown in Figure 10.5(b) is biased but has a small variance. If an estimate based on a single measurement were made, estimator $\hat{\theta}_2$ (although biased) would have a greater probability of producing less error than estimator $\hat{\theta}_1$. Also, if the bias in an estimator is known, a correction can be applied to account for this.

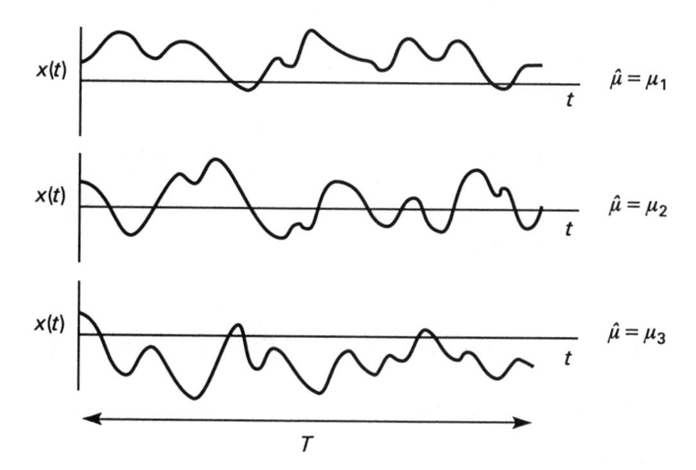

Figure 10.3 Three different estimates of the mean value of a signal.

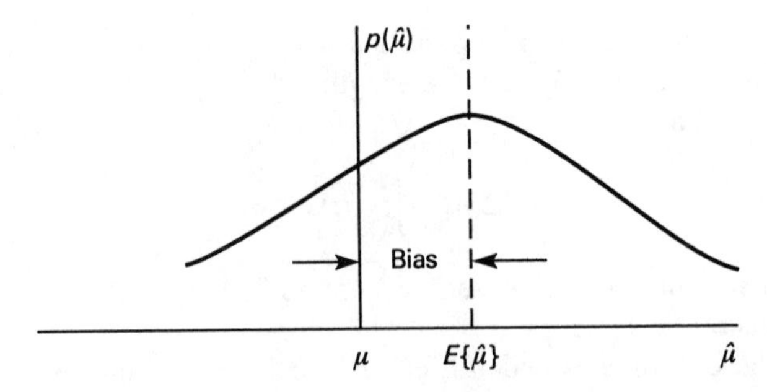

Figure 10.4 Probability density function for a biased estimator.

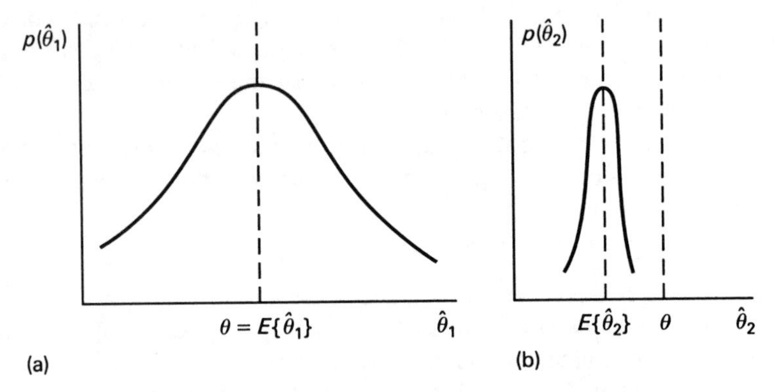

Figure 10.5 Two different estimators for the same parameter.

As both bias and variance contribute to the total error of an estimator, a measure of goodness that can be used is the mean square error:

$$\text{Mean square error} = E\{(\hat{\theta} - \theta)^2\}$$

An estimator having a smaller mean square error than another estimator is said to be a more *efficient* estimator.

Returning to the example involving the estimation of mean value, if probability densities were produced for different record lengths, one would hope the variance of the estimator would decrease as the record length increases. An estimator with this property is termed a *consistent* estimator.

10.4 Digital techniques for spectral estimation

In the previous chapter two methods of obtaining spectral density were considered:

1. Via the autocorrelation function – this method is known as the correlogram method.

2. Directly from the Fourier transform of the time signal – this method is known as the periodogram method.

Practical spectral measurement techniques have been based on both these methods. Historically the periodogram method was the earliest used. However, as stated in the previous chapter, some form of averaging is required to obtain meaningful results. Early workers did not realise this and the results were not very encouraging. The correlogram method does not suffer from the statistical variability of the periodogram and became the standard technique. Even when theoretical methods of smoothing the periodogram were developed, the number of arithmetic operations involved was too great to make this a feasible method for the computers then available. However, with the advent of the fast Fourier transform algorithm, the periodogram method became a viable technique and is now recognised as the standard method. This is the method that will be considered in the next section.

10.4.1 Estimation via the periodogram

In the previous chapter the power spectral density of a signal $x(t)$ was defined as

$$S_{xx}(\omega) = \lim_{T\to\infty} \frac{|X(\mathrm{j}\omega)|^2}{2T}$$

where $X(\mathrm{j}\omega)$ is the Fourier transform of $x(t)$ over a range $-T$ to T. Some form of averaging is required to reduce the statistical variability of the result.

In a practical measurement scheme T is finite and leaving aside the question of averaging an estimate of $S_{xx}(\omega)$ would be

$$\hat{S}_{xx}(\omega) = \frac{|X(\mathrm{j}\omega)|^2}{2T} \tag{10.4.1}$$

For a digital measuring scheme, the Fourier transform of the signal must be replaced by the discrete Fourier transform of the sampled signal. In Section 6.5.1 it was shown that, if there are no aliasing effects, the discrete Fourier transform $X(k\omega_0)$ is related to the continuous transform by

$$X(\mathrm{j}\omega) = \Delta t X(k\omega_0)$$

where Δt is the sampling time and $\omega_0 = 2\pi/N\Delta t$, N being the number of sample values used in the transform (note that $N\Delta t$ is the length of the record). The discrete equivalent of eqn (10.4.1) becomes

$$\hat{S}_{xx}(k\omega_0) = \frac{2\Delta t}{N}|X(k)|^2 \tag{10.4.2}$$

The factor of 2 has been included to represent all the power at positive frequencies.

A block diagram outlining a discrete measuring scheme is shown in Figure 10.6.

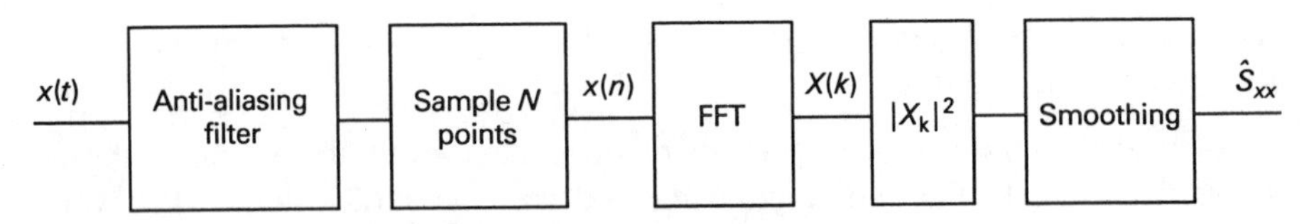

Figure 10.6 Discrete power spectral density measurement.

Aliasing has been described in Section 6.5.1 and it is assumed that the anti-aliasing filter removes aliasing components completely. The digital scheme can be compared to the analogue method of Figure 10.1. The conversion to the frequency domain by use of the fast Fourier transform is equivalent to the use of a bank of bandpass filters in the analogue measurement. The squaring operations to obtain power are equivalent in both the time and frequency domains. Both schemes use some form of smoothing to reduce the statistical variability of the estimates.

10.4.2 Effect of finite time record

To illustrate the digital method described in the previous section, consider $x(t)$ to be a deterministic signal. This removes the variability associated with a random input but still leaves many important issues to be considered. The simplest signal is one where all the power is concentrated at a single frequency. Let

$$x(t) = X \cos \omega_0 t$$

The power spectral density of this signal has been derived in the previous chapter:

$$S_{xx}(\omega) = \frac{\pi X^2}{2} [\delta(\omega - \omega_0) + \delta(\omega + \omega_0)]$$

This represents two impulses, each of area $\pi X^2/2$ at frequencies $\omega = \pm\omega_0$. In order to compare this with the result obtained by digital methods, two modifications are required. The power is taken to be at positive frequencies only and the units are changed from (amplitude)^2/rad/s to (amplitude)^2/Hz. This gives

$$S_{xx}(f) = \frac{X^2}{2} \delta(f - f_0)$$

The spectrum from the digital approach can be obtained from eqn (10.4.2). It is assumed that there are an integral number of cycles of the waveform in the record length T, $T = m/f_0$, where m is an integer. Because the waveform is relatively simple in form, the discrete Fourier transform can be calculated directly giving

$$\begin{aligned} X(k) &= \frac{XN}{2} \qquad && k = m/T \\ &= 0 && \text{otherwise} \end{aligned}$$

Here T is the record length, $T = N\Delta t$.

Substituting into eqn (10.4.2)

$$\hat{S}_{xx}(k) = \frac{2\Delta t}{N}\frac{X^2N^2}{4}$$

$$= \frac{X^2}{2}N\Delta t \quad \text{for } k = m/T$$

The result obtained by digital analysis appears to differ by a factor $N\Delta t$ from the theoretical result. However, this is not the case; the theoretical result gives the power spectral density as an impulse and $X^2/2$ is the strength or *area* of this impulse (its amplitude is infinite). For the digital analysis $X^2N\Delta t/2$ is the *amplitude* of the result. If this amplitude were taken as constant over a frequency band $\Delta f = 1/N\Delta t$ then the area of this band would be $X^2/2$ as in the theoretical result. Comparing with the analogue measurement technique, each discrete frequency can be regarded as a bandpass filter of bandwidth equal to 1/(record length). This idea will be investigated further in Section 10.4.3.

In the preceding digital analysis it was assumed that the record length T contained an integral number of cycles of the signal. Consider now the same signal

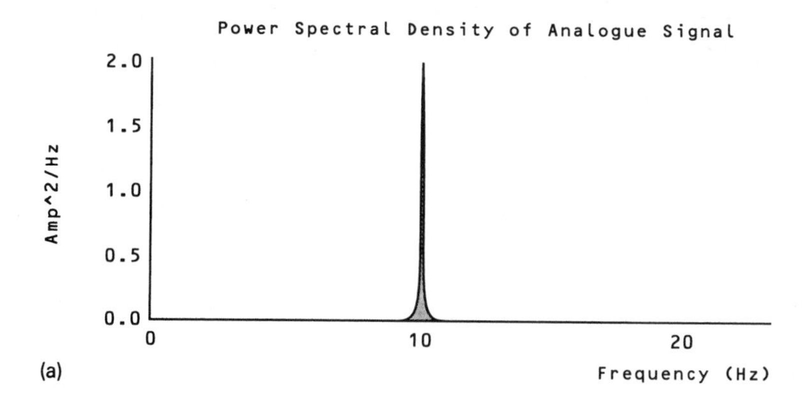

(a)

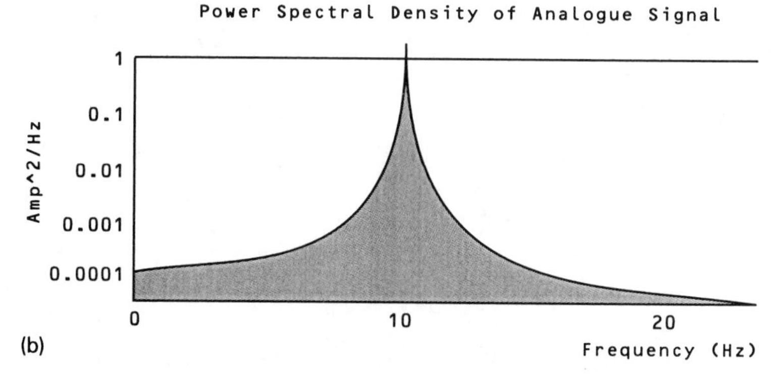

(b)

Figure 10.7 Spectrum of a single sinusoid showing leakage.

with the record length chosen not to be an integral number of cycles. To make this a specific example, suppose $X = 1$, $f_0 = 10.05$ Hz and the record length is 10 seconds; this gives 1.005 cycles in the record. To calculate the resulting power spectral density is now much more difficult. Figure 10.7 shows the resulting spectrum where the results have been obtained by computer simulation. Figure 10.7(a) shows the results using a linear power spectral density scale and 10.7(b) shows the same results on a log scale.

The estimated spectrum has completely changed from that obtained when the record length was a whole number of cycles. Instead of the power in the signal being concentrated at the signal frequency, it is now spread over a range of frequencies. This is more apparent on the log plot of Figure 10.7(b).

The total power of the signal has not changed by altering the record length; hence all the power that was concentrated at one frequency can be considered to have 'leaked' to other frequencies. The effect is known as *leakage*.

There are two viewpoints that can be taken to explain leakage. The first will be given here; the second, wider viewpoint, introduces the concept of windows, and will be explained in the next section.

In Section 6.4.3 it was seen that the discrete Fourier transform is equivalent to a discrete Fourier series. If N samples of an aperiodic signal are used in the discrete transform this is treated as a periodic waveform with period N. Figure 10.8 illustrates the effect of this on a sinusoidal signal. (A continuous waveform has been used for illustration on the assumption that aliasing effects are negligible.)

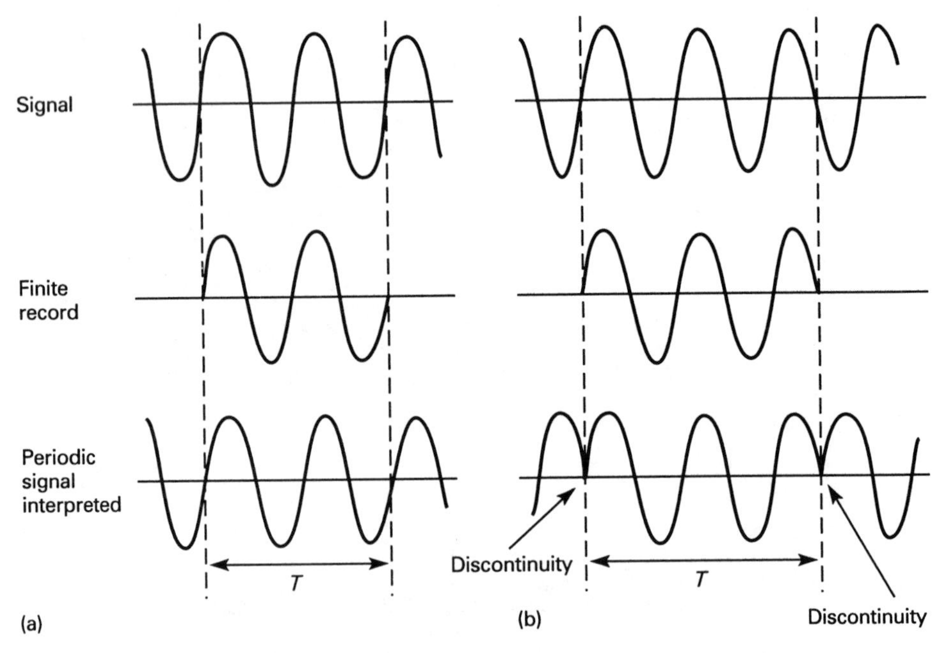

Figure 10.8 Discontinuities produced by finite record length.

In Figure 10.8(a) the record length T chosen contains a whole number of cycles of the signal. The periodic signal interpreted is unchanged from the original. In Figure 10.8(b) the record length contains $2\frac{1}{2}$ cycles of the signal. The periodic signal interpreted has discontinuities as shown. It is these discontinuities that produce the additional components in the Fourier transform and account for the leakage effects.

10.4.3 Time windows

An alternative method of explaining leakage effects can be understood by reference to Figure 10.9. Here $x(t)$ represents the complete time record and the finite record used for the measurement is obtained by multiplying $x(t)$ by the function $w(t)$. $w(t)$ is defined as

$$
\begin{aligned}
w(t) &= 1 \qquad 0 \leqslant t \leqslant T \\
&= 0 \qquad \text{otherwise}
\end{aligned}
$$

The record to be analysed is $w(t)x(t)$ and the Fourier transform of this record is required. However, the convolution property of the Fourier transform gives

$$
\begin{aligned}
F\{w(t)x(t)\} &= F\{w(t)\} * F\{x(t)\} \\
&= W(f) * X(f)
\end{aligned}
$$

where

$X(f)$ is the Fourier transform of the signal
$W(f)$ is the Fourier transform of $w(t)$

$w(t)$ is known as a *time window* and, because of its shape, this particular window is known as a *rectangular* or *box car* window. The Fourier transform of such a time

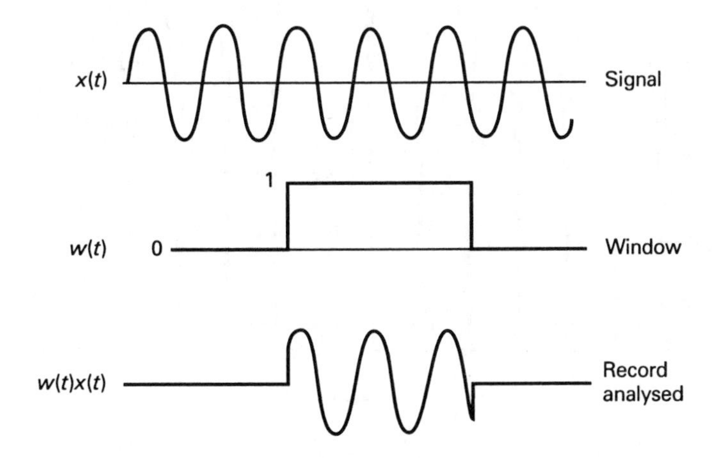

Figure 10.9 Finite record length as time window.

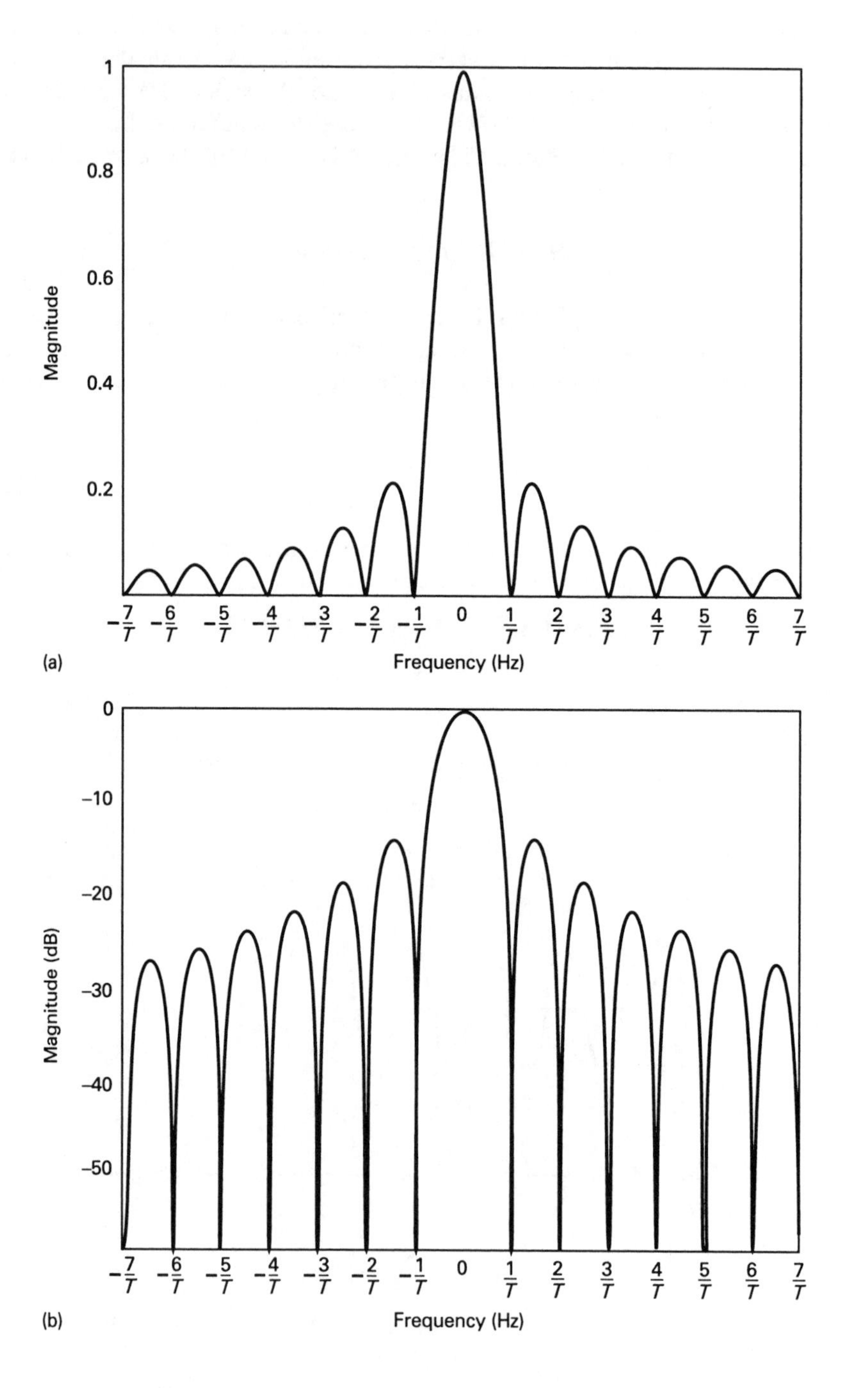

Figure 10.10 Fourier transform of rectangular window.

function has already been derived in Section 6.2.1. The pulse considered there was symmetric about the time origin, and the window $W(t)$ can be considered to be such a pulse delayed by time $T/2$. Using the time shift property, this gives a Fourier transform

$$W(f) = \frac{T\sin(2\pi fT/2)}{2\pi fT/2}\,\mathrm{e}^{-\mathrm{j}2\pi fT/2} \tag{10.4.3}$$

The magnitude of $W(f)$ is plotted in Figure 10.10. Figure 10.10(a) shows the response using a linear amplitude scale, and Fig 10.10(b) that using a logarithmic scale.

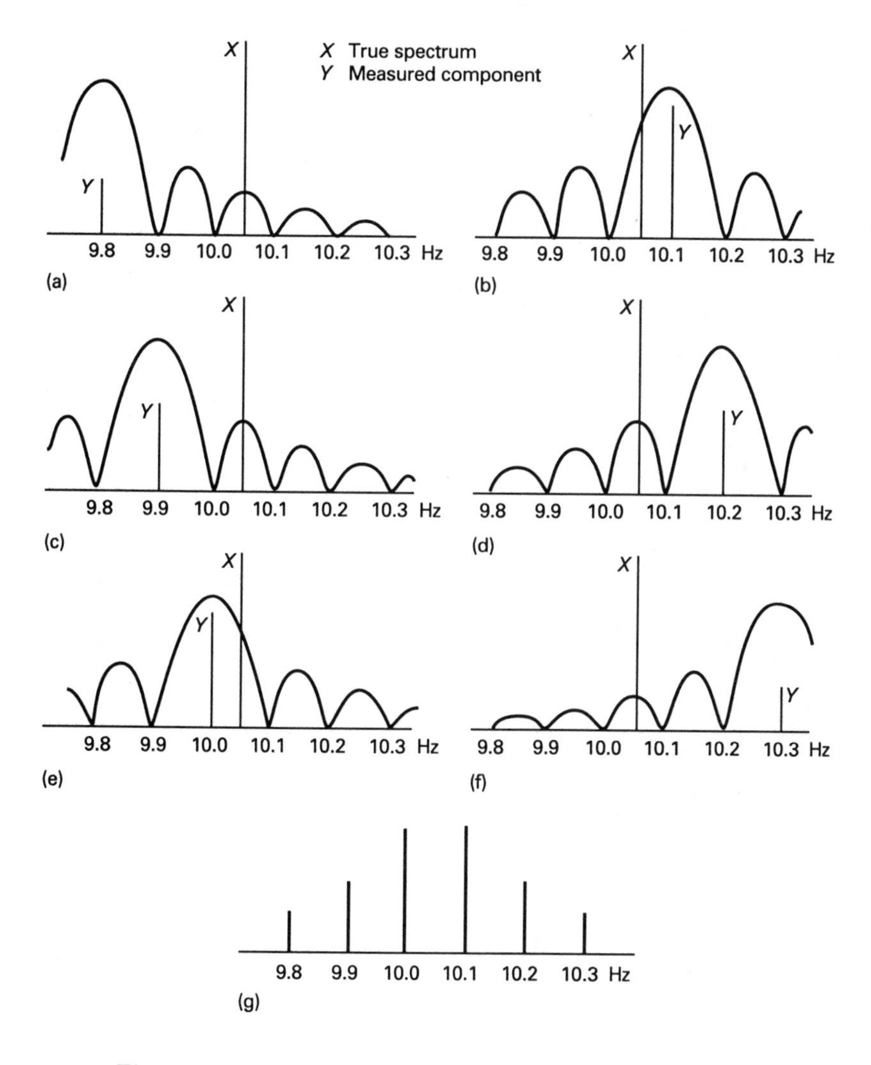

Figure 10.11 Generation of measured components in spectrum.

The Fourier transform of the finite record is obtained by convolution of $W(f)$ with the Fourier transform of the signal. However, because of the sampling, estimates can only be obtained at discrete frequencies k/T, $k=0, 1, \ldots, N/2$. Hence, at a specific frequency k/T the measured transform is obtained by multiplying the true transform by $W(f-k/T)$. This explains the leakage effect as shown in Figure 10.7. The values used for Figure 10.7 have been used again here. The input signal is a single sinusoid, amplitude $X=1$, frequency 10.05 Hz. The record length is 10 seconds, giving a spacing between components of 0.1 Hz.

Figure 10.11 shows how the measured components at frequencies 9.8, 9.9. 10.0, 10.1, 10.2 and 10.3 Hz are obtained. At each frequency convolution of the window characteristic with the input signal at 10.05 Hz gives the contribution to the measured spectrum. Because the input is at a single frequency, this is equivalent to multiplication of the amplitude of the input signal by the amplitude of the window $|W(f-10.5)|$. Figure 10.11(a)–(f) shows how each component in the measured

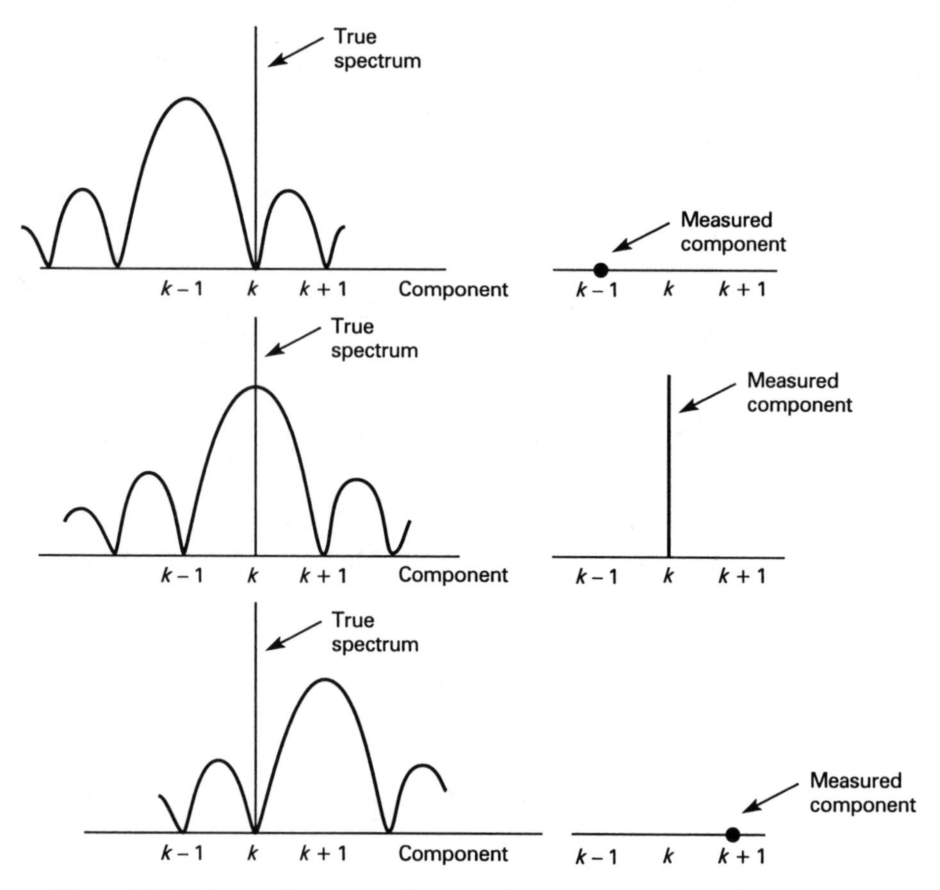

Figure 10.12 Measured spectrum when record contains an integer number of cycles.

spectrum is obtained; Figure 10.11 (g) shows the complete measured spectrum from 9.8 Hz to 10.3 Hz. To obtain the associated power density spectrum, the magnitude of these components must be substituted into eqn (10.4.2).

It should be noted that, if the record length T contains an integer number of cycles, then the input frequency must be a multiple of $1/T$. Any measured frequency will also be a multiple of $1/T$ and there will be no leakage into these measured frequencies as the input frequency will always occur at zero value of the window function. Figure 10.12 illustrates this situation for three measured frequencies.

The effects of leakage become more complex when more than one frequency component is present in the input signal. Leakage from one component can distort the measured power of another component. This situation is illustrated in the following example.

EXAMPLE 10.4.1

Compare the measured spectra for a signal consisting of two sinusoidal components as follows:

(a) Sinusoid 1 has amplitude 1 unit and frequency 9.95 Hz.
Sinusoid 2 has amplitude 0.05 unit and frequency 20 Hz.

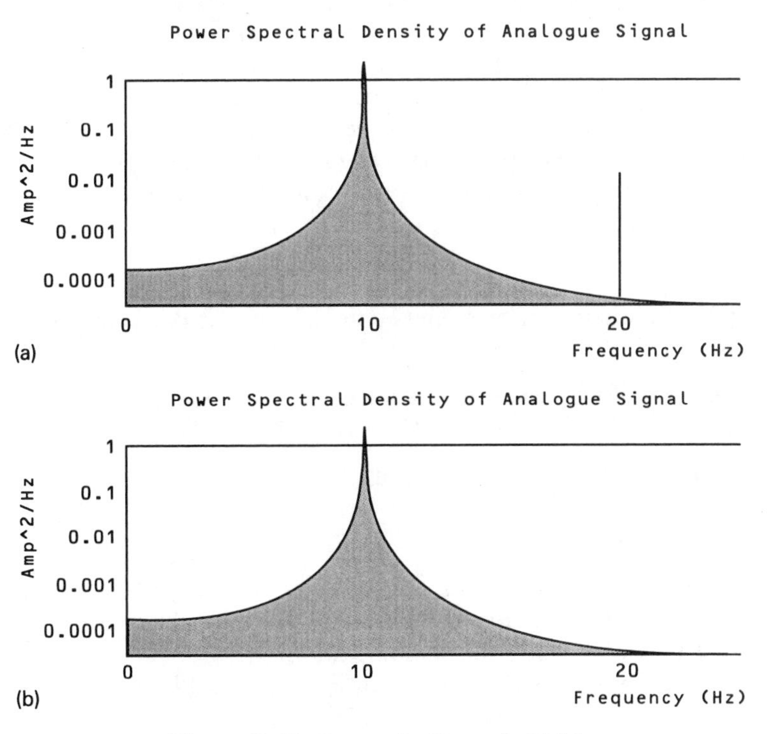

Figure 10.13 Spectra for Example 10.4.1.

(b) Sinusoid 1 has amplitude 1 unit and frequency 9.95 Hz.
Sinusoid 2 has amplitude 0.05 unit and frequency 10.5 Hz.

In both cases the record length is 10 seconds.

SOLUTION

The measured spectrum for case (a) is shown in Figure 10.13(a). As 20 Hz is an integer multiple of $1/T$, where $T = 10$ seconds, there is no leakage from this component and it appears as a single line in the spectrum. As 9.95 Hz is not an integer multiple of $1/T$, it produces leakage as shown. This leakage interferes with the component at 20 Hz, but it is not sufficient to mask it and the 20 Hz component is clearly shown.

In case (b) 10.5 Hz is also an integer multiple of 1/T and no leakage occurs from this component. However, leakage still occurs from the 9.95 Hz component and this masks the 10.5 Hz component, making it barely discernible. This case is plotted in Figure 10.13(b); it is also re-plotted in Figure 10.19(a) on a more compact frequency scale.

Another important consideration when more than one frequency component is present is that of *resolution*. This can be defined as the minimum frequency separation between two equal amplitude components such that these components appear as distinct in the measured spectrum.

If the effective bandpass filters produced by the discrete Fourier transform were flat topped and of width $\Delta f = 1/T$, then the resolution would be $1/T$ (see Section 10.2). Because the filter is not flat topped some other measure of bandwidth is required. One commonly used measure is the bandwidth between the −3 dB points. In fact this works out to be approximately $1/T$. Hence, the longer the time record, the better the resolution. The resolution of two sinusoidal components is investigated in the following example.

EXAMPLE 10.4.2

Investigate the resolution of two unit amplitude sinusoidal components for the following two cases:

(a) Frequencies of the components 9.94 Hz and 10.15 Hz.
(b) Frequencies of the components 10.01 Hz and 10.09 Hz.

In both cases the record length is 10 seconds.

SOLUTION

(a) The spectrum for this case is shown in Figure 10.14(a). Two peaks are visible at 9.9 Hz and 10.2 Hz. (Note that components can only appear at frequencies that are multiples of $1/T$, i.e. multiples of 0.1 Hz.)
(b) The spectrum for this cases is shown in Figure 10.14(b). Equal components are visible at 10.0 Hz and 10.1 Hz. However, it is not possible to determine whether these are due to separate components or one component at 10.05 Hz.

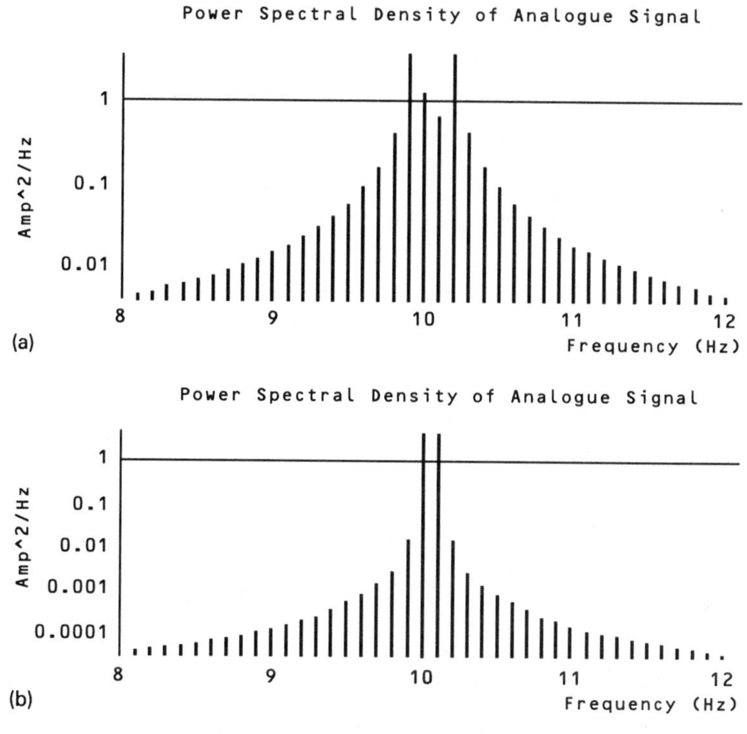

Figure 10.14 Spectra for Example 10.4.2.

In part (a) the separation of the components was $10.16 - 9.94 = 0.22$ Hz, greater than the theoretical resolution of 0.1 Hz. In part (b) the difference was $10.09 - 10.01 = 0.8$ Hz, less than the theoretical resolution.

10.4.4 The Hanning window

In the previous section two important consequences of leakage were discussed. Leakage from close frequency components can cause a loss of resolution in the estimated spectrum and leakage from more distant strong components can mask or distort weaker components. An explanation given for leakage was the interpretation by the discrete Fourier transform (DFT) of the measured data as one period of a periodic waveform. Discontinuities at the ends of the data now produce additional frequency components which account for the leakage.

One method of removing the discontinuity would be to 'taper' the measured data to zero at each end of the record. From a windowing viewpoint this can be interpreted as not using a rectangular window, but one that tapers to zero at each end. Figure 10.15 illustrates this idea.

The reader may object to this procedure on the grounds that the tapering amounts to a distortion of the true data and this is bound to lead to errors in the

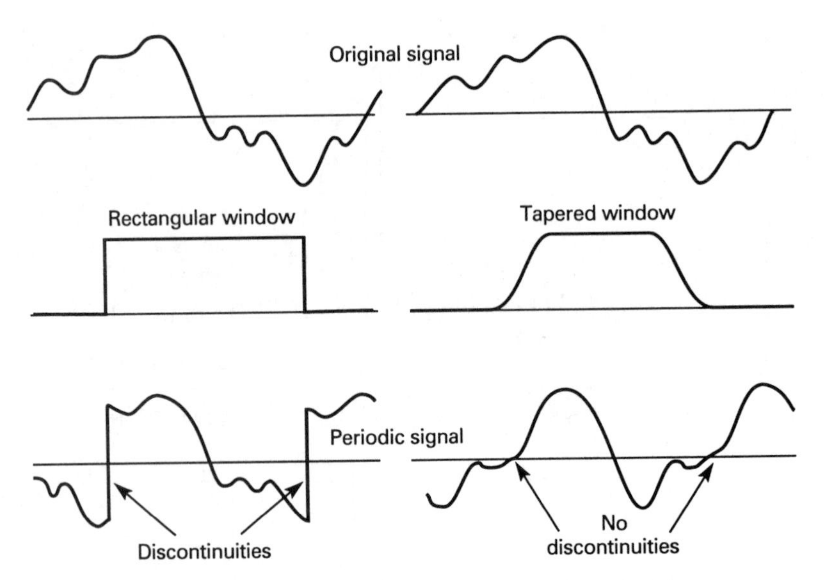

Figure 10.15 Tapering of record to avoid discontinuities.

estimated spectrum. This is true, but there are already errors due to leakage; if one can obtain a suitable 'trade-off' between the errors from these two sources perhaps the overall effect will be an improvement in the estimated spectrum. To investigate the overall effect, the tapering has to be considered in more detail.

The tapering of the data can take various forms: one can taper quite sharply towards the ends of the data or one can taper more gradually over the whole data record. These differences can lead to a number of different windows, but only one of these will be considered here – the *Hanning window*. This is described by the equation

$$w_{\mathrm{H}}(t) = \frac{1}{2}\left(1 - \cos\frac{2\pi t}{T}\right) \qquad 0 \leqslant t \leqslant T \qquad (10.4.4)$$
$$= 0 \qquad \text{otherwise}$$

The window is shown in Figure 10.16 and, as can be seen, the tapering extends over the whole range of the data and takes the form of a cosine function. When considering the rectangular window, its effect in the frequency domain is obtained by considering the convolution of the Fourier transform of the window with that of the signal. The same approach can be adopted with the Hanning window and one requires the Fourier transform $W_{\mathrm{H}}(f)$ of the time window $w_{\mathrm{H}}(t)$. This can be obtained directly, but it is instructive to consider the window as shown in Figure 10.17.

Here the Hanning window has been considered as the sum of a rectangular window plus a truncated cosine wave (a factor of $\frac{1}{2}$ has to be included). The corresponding Fourier transforms have to be added and that of the rectangular window has already been derived. The truncated cosine can be considered as the

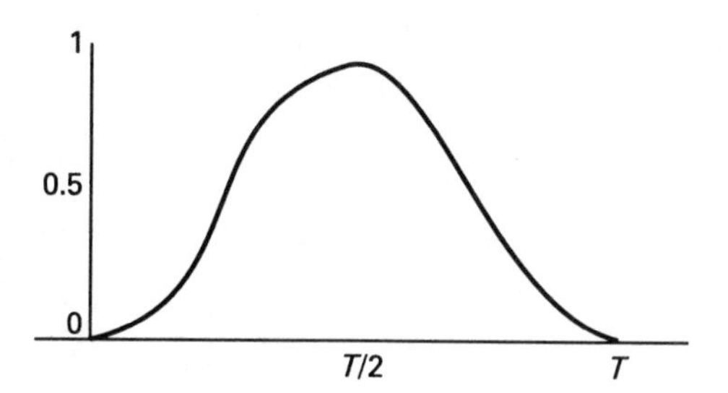

Figure 10.16 Hanning time window.

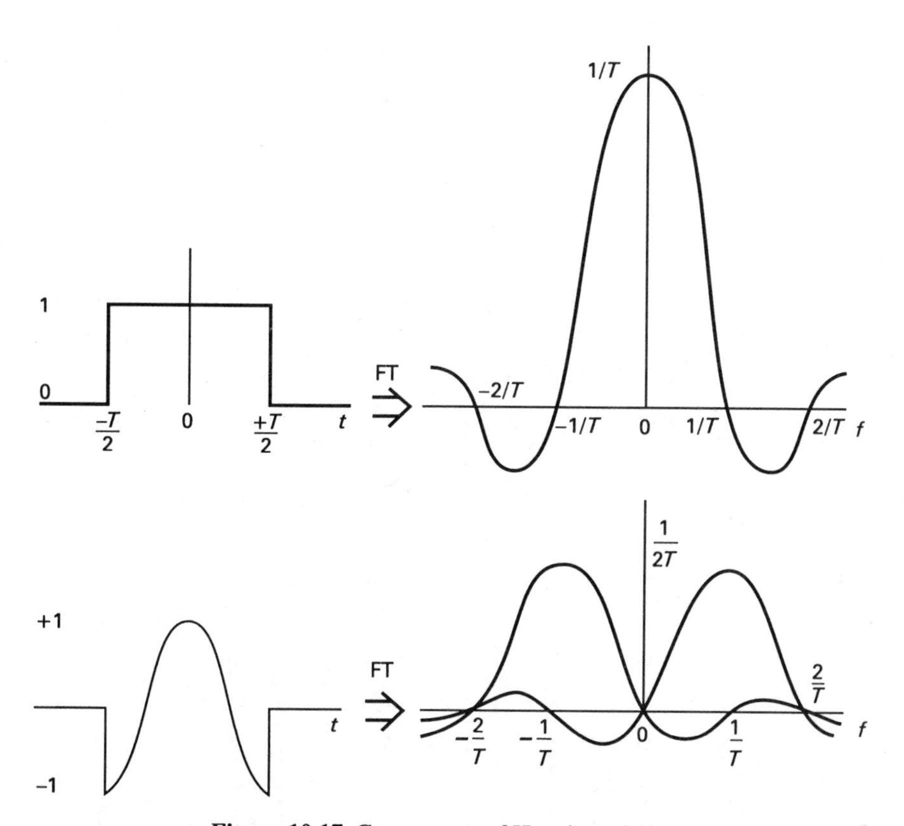

Figure 10.17 Components of Hanning window.

product of a cosine wave and a rectangular window. Its transform is the convolution of the transform of the rectangular window and the transform of a cosine waveform. As the latter consists of two impulses at frequencies $(f+1/T)$ and $(f-1/T)$, the transform of the truncated cosine becomes

$$\tfrac{1}{2}W_R(f-1/T)+\tfrac{1}{2}W_R(f+1/T)$$

where $W_R(f)$ is the transform of the rectangular window. Hence the Fourier transform of the Hanning window $W_H(f)$ is given by

$$W_H(f) = \tfrac{1}{2}W_R(f) + \tfrac{1}{4}W_R(f - 1/T) + \tfrac{1}{4}W_R(f + 1/T) \tag{10.4.5}$$

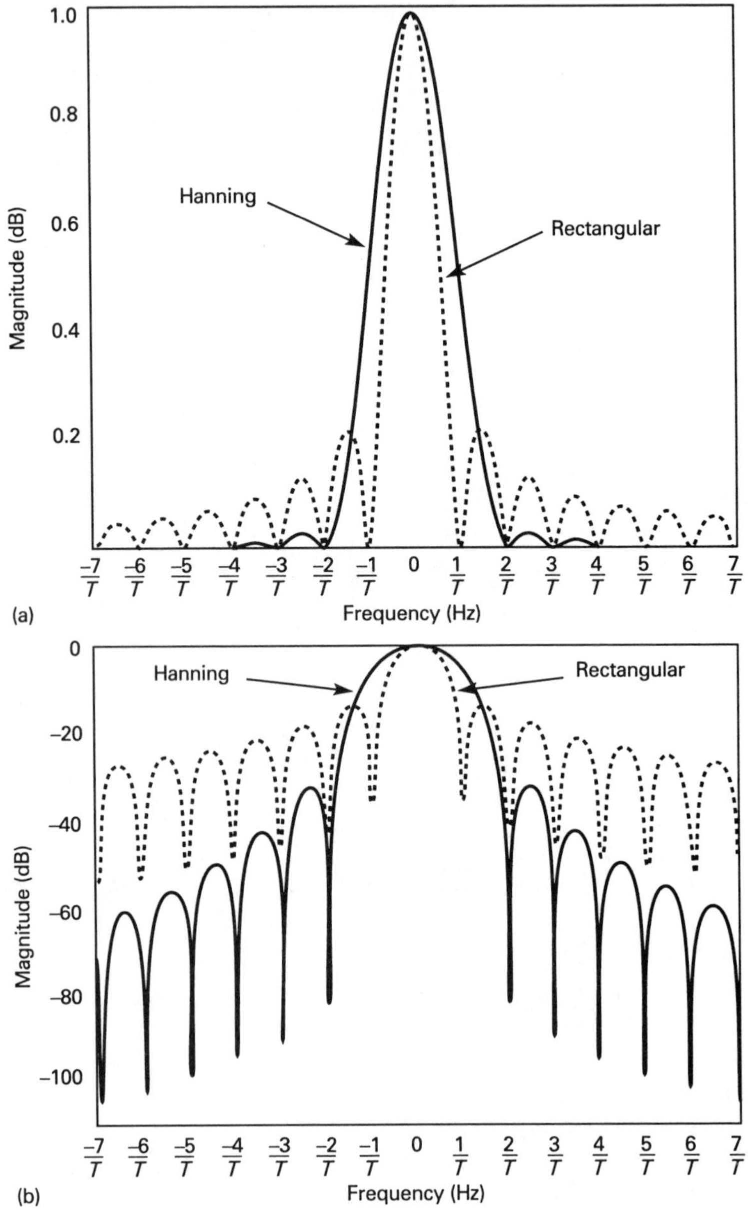

Figure 10.18 Fourier transform of Hanning window.

It should be noted that this derivation has assumed the window to be symmetrical about $t = 0$. Hence no phase term is present. As with the rectangular window, the time delay of $T/2$ produces a phase term $e^{-j\omega T/2}$.

The magnitude of $W_H(f)$ is plotted in Figure 10.18 together with $|W_R(f)|$ for comparison purposes. Figure 10.18(a) shows the plots on a linear magnitude scale and Figure 10.18(b) those on a logarithmic scale. As can be seen, the Hanning window produces a much higher decay rate for the sidelobes than the rectangular window. This is a desirable result as it means the leakage effect from distant frequency components is very much reduced. However, a less desirable result is the broadening of the centre lobe, giving a bandwidth of approximately twice the rectangular window. This gives a loss of frequency resolution.

Results obtained using a Hanning window are compared with those obtained using a rectangular window by consideration of the same spectra as used in Examples 10.4.1 and 10.4.2.

EXAMPLE 10.4.3

(a) In Example 10.4.1 the effect of leakage was considered by investigating the masking of a weak component by a nearby stronger component. Repeat part (b) of this example using a Hanning window.

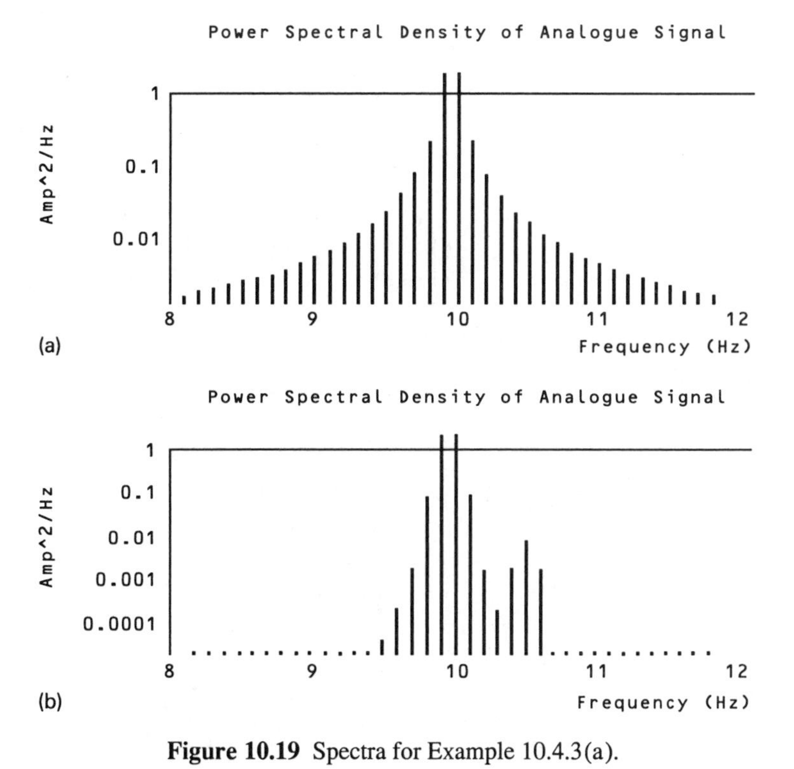

Figure 10.19 Spectra for Example 10.4.3(a).

(b) In Example 10.4.2 the effect of frequency resolution was considered. Repeat part (b) of this example using a Hanning window.

SOLUTION

(a) The components present were two sinusoids, one of amplitude 1 unit and frequency 9.95 Hz, the other of amplitude 0.05 unit and frequency 10.50 Hz. The resulting spectrum, obtained using a rectangular window, is shown in Figure 10.19(a). The component at 10.5 Hz is barely visible.

The results using a Hanning window are shown in Figure 10.19(b). Now a clear peak is visible at 10.5 Hz indicating a component at this frequency. From the plot it is also seen that the spread of the spectrum due to leakage has been very much reduced.

(b) The components present were two sinusoids of equal amplitudes, having frequencies of 9.94 Hz and 10.16 Hz. Figure 10.20(a) is a repeat of Figure 10.14(a) and shows that, using a rectangular window, two components can be clearly identified in the spectrum.

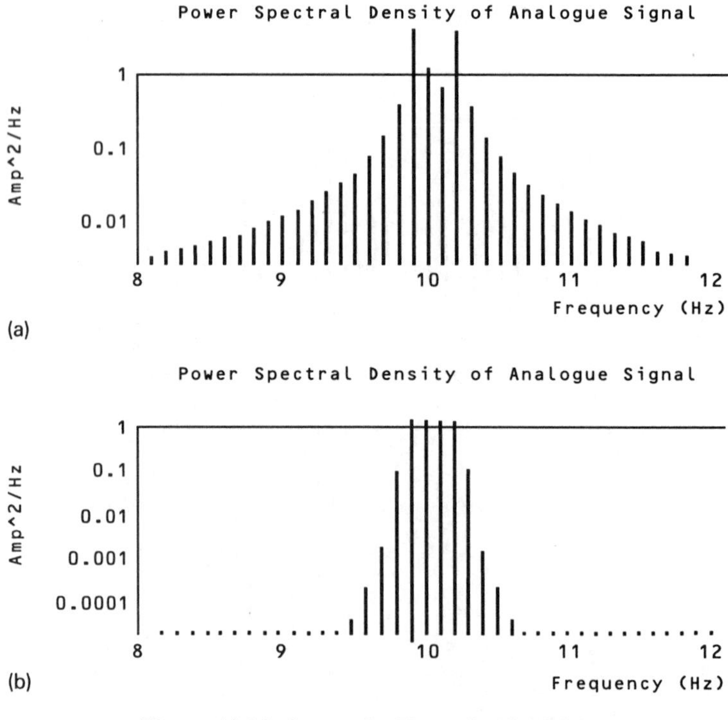

Figure 10.20 Spectra for Example 10.4.3(b).

The results using a Hanning window are shown in Figure 10.20(b). There is now no evidence of two distinct peaks. Use of the Hanning window has resulted in a loss of resolution.

As stated earlier, there are a variety of ways in which the tapering of the time window can be made – each of these leads to a different window type. It is beyond the scope of this book to investigate or list the properties of these windows. The reason why so many windows are defined concerns the many different requirements that can arise in spectral analysis. Some of these are as follows:

1. Good resolution.
2. Low interference between spectral components. Depending on the relative spacing of these components, a range of windows is required.
3. Good signal to noise performance for the detection of weak signals in background noise.
4. Accurate power measurement of a single sinusoidal frequency of varying frequency.
5. Accurate relative power level measurement of a number of sinusoidal signals.

10.4.5 Power spectra of random signals

In the previous section the signals considered were deterministic. In this section the discussion is broadened to include the problems involved with the power spectral estimation of random signals. The leakage problem associated with finite record length still exists. However, as the signal is random, it now becomes an average effect. It can be shown that

$$E\{\hat{S}_{xx}(f)\} = |W(f)|^2 * S_{xx}(f) \tag{10.4.6}$$

This shows that, unless $W(f)$ takes the form of an impulse (infinite record length), the power spectral estimate is biased. The nature of the bias depends upon the window used and the choice depends upon the same factors that were discussed in the previous sections. The following example compares the use of rectangular and Hanning windows in the estimation of the power spectral density of a random signal.

EXAMPLE 10.4.4

A random process is formed by the output of a second order system when the input is white noise. The system is lightly damped and has a frequency response function given by

$$H(\mathrm{j}\omega) = \frac{\omega_\mathrm{n}^2}{(\mathrm{j}\omega)^2 + 2\zeta\omega_\mathrm{n}(\mathrm{j}\omega) + \omega_\mathrm{n}^2}$$

where $\omega_\mathrm{n} = 2 \times \pi \times 10$ and $\zeta = 0.03$.

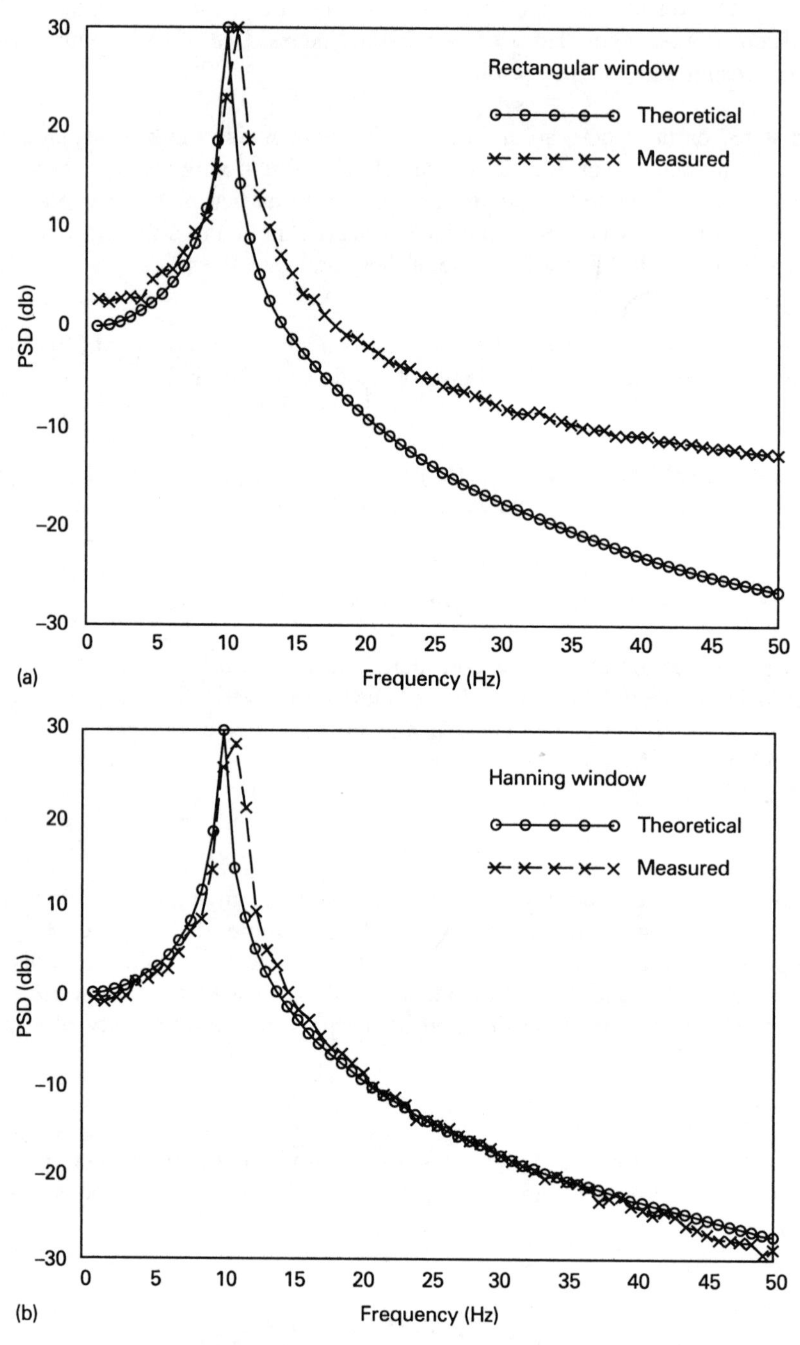

Figure 10.21 Spectra for Example 10.4.4.

Using a measurement time of 51.2 seconds, determine the power spectral density of the system output using both rectangular and Hanning windows.

SOLUTION

The power spectral density at the output of this system can be determined theoretically (see Section 9.5.2). In Figure 10.21(a) the theoretical results are shown together with the measured results using a rectangular window. Figure 10.21(b) shows the theoretical and measured results using a Hanning window.

The results using the rectangular window are in considerable error over a large part of the frequency range. This is due to the leakage from the strong components that occur around the resonant frequency of the system. The results using the Hanning window show better agreement with the theoretical results over most of the frequency range. However, the wider bandwidth of the Hanning window has resulted in a broadening of the peak of the spectral estimate.

In the previous example, averaging was required in order to produce a statistically stable result. The simulation was repeated a number of times and the results obtained at each frequency were averaged. This is the smoothing process that was represented in Figure 10.6 and the averaging has reduced the *variance* of the result.

Intuitively, it may be thought that the length of the data record would influence the variance and that, as the record length was increased, the variance would decrease. However, this is not the case, but a formal proof of this statement requires more advanced knowledge of random variables than has been presented here. A heuristic argument proceeds as follows.

Assuming a rectangular window, the estimate of the spectral density is given by eqn (10.4.2):

$$\hat{S}_{xx}(kf_0) = 2\Delta t \, \frac{|X(k)|^2}{N}$$

As N increases, there are two effects on the variance of S_{xx}

1. The variance decreases owing to the factor N in the denominator.
2. As N increases, the length of the record $N\Delta t$ increases and the bandwidth $1/N\Delta t$ of the equivalent bandpass filter decreases. The spectrum is being examined with increasing resolution, i.e. smaller effective bandwidth of the equivalent bandpass filter. As this filter averages the spectrum across its bandwidth, the smaller the bandwidth the less the averaging and the greater the resulting variance in the estimate. Hence, the variance of the measured spectrum increases as the bandwidth decreases (N increases).

More formal methods show that as $N \rightarrow \infty$ these two effects cancel and the variance remains constant. Unfortunately this constant variance is not small, but is equal to $E\{S_{xx}(f)\}$, the mean value of the measured spectrum. The periodogram is not a consistent estimator and some form of smoothing is required to reduce its variability.

10.4.6 Smoothing of spectral estimates

As discussed in the previous section, the periodogram is not a consistent estimator of power spectral density and some form of smoothing is required to reduce the variance of the estimate. There are various types of smoothing that can be used and only the two most common methods will be described here.

The Bartlett method is illustrated in Figure 10.22(a). The complete time record is subdivided into P non-overlapping segments each of equal length. For each of these segments a periodogram is obtained. The estimate, at a given frequency, is obtained by averaging the results, at that frequency, from the P periodograms:

$$\hat{S}_B(kf_0) = \frac{1}{P}\sum_{i=1}^{P}\hat{S}_i(kf_0)$$

where $\hat{S}_i(kf_0)$ is the power spectral density estimate of the kth frequency component from the ith periodogram.

It can be shown that the variance of the estimate is inversely proportional to P, the number of segments. However, the greater the number of segments, in a given

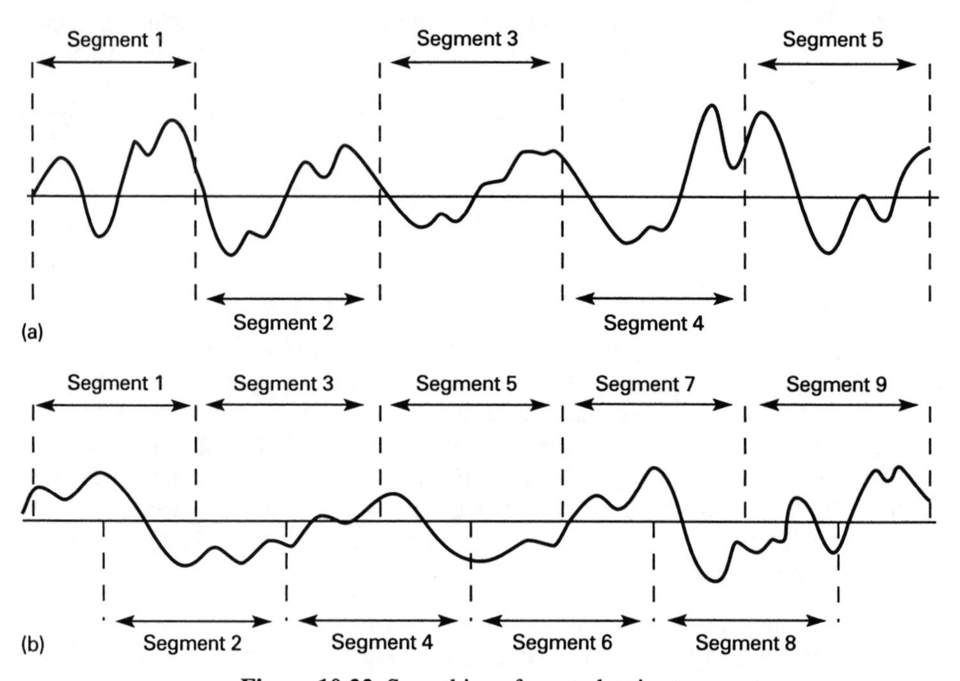

Figure 10.22 Smoothing of spectral estimates.

length record, the shorter the length of each segment and this leads to a lack of resolution.

More segments can be obtained from the same length record if the segments are overlapped. The Welch method uses a 50% overlap, as shown in Figure 10.22(b). As more segments are used than in the Bartlett method, a reduced variance is obtained. However, because of the overlap, the estimates from each segment are not independent and the reduction in variance is not as great as $1/P$.

The following example illustrates the effect of smoothing.

EXAMPLE 10.4.5

For the random process defined in Example 10.4.4, investigate the effect of the number of data blocks on the smoothed spectral estimate.

SOLUTION

Figure 10.23 shows the estimated power spectral densities together with the theoretical values. The smoothing is for 1, 5 and 20 blocks in Figure 10.23(a), (b) and (c) respectively. The decrease in variance is evident as the number of blocks is increased. In order not to confuse two effects, the length of the record has been increased as the number of blocks increases; hence there is no loss of resolution accompanying the decrease in variance.

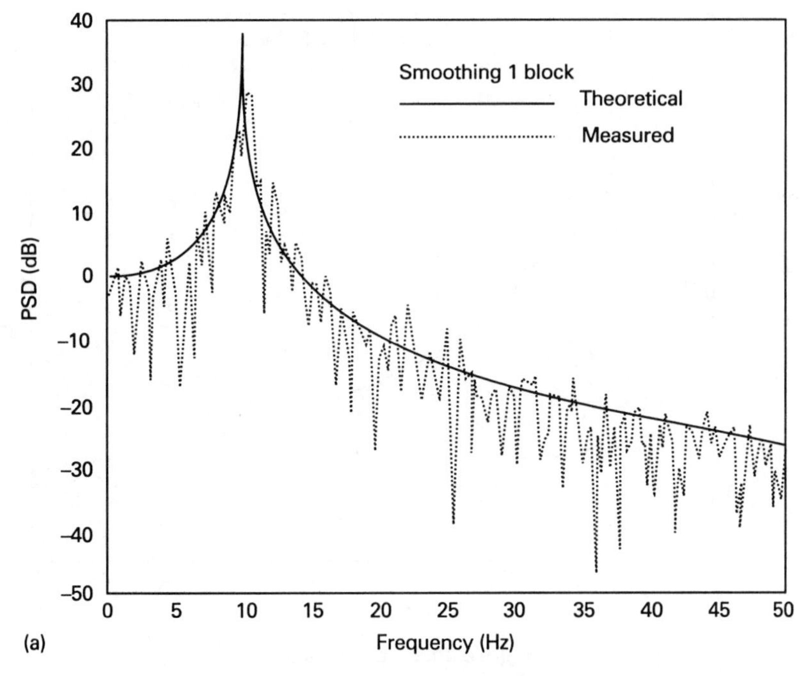

Figure 10.23 Spectra for Example 10.4.5.

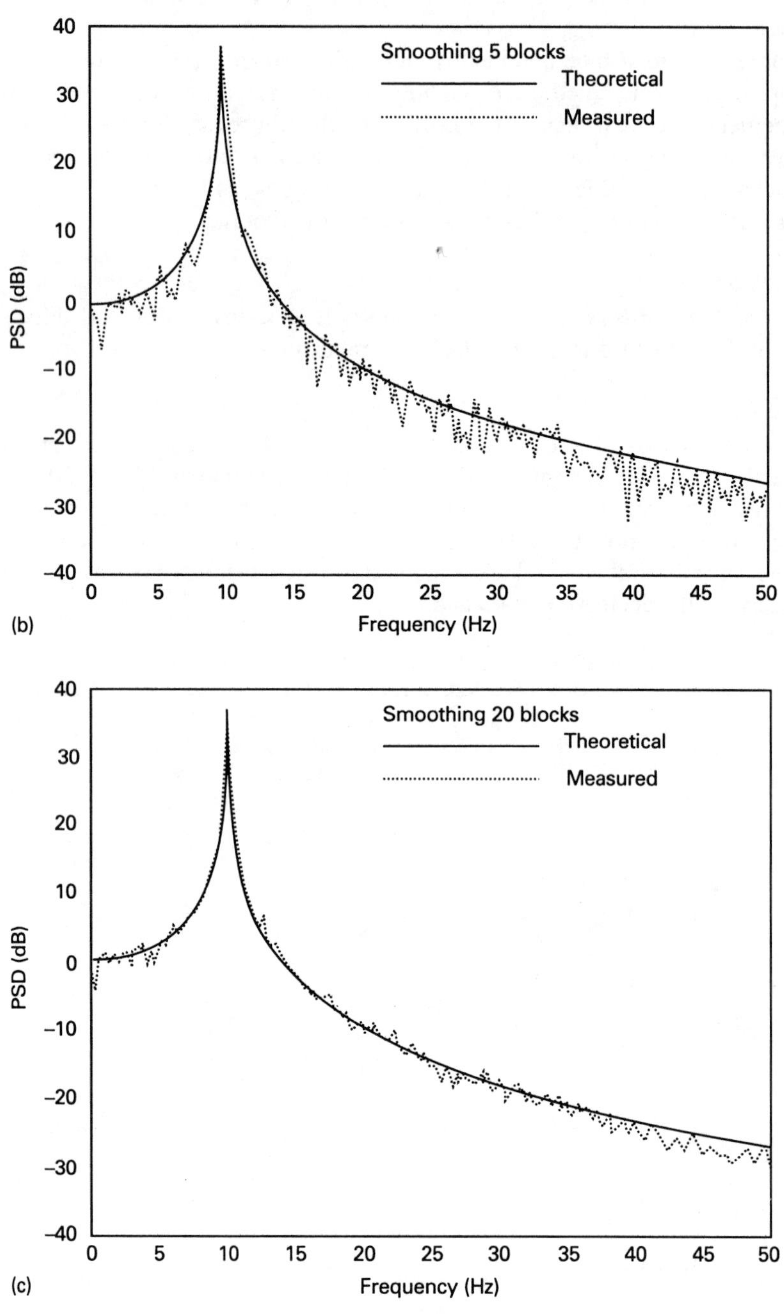

Figure 10.23 *Continued*

10.5 Cross-spectrum estimation

Paralleling the reasoning in section 9.4.1, an estimator for cross-spectral density is

$$\hat{S}_{xy}(k) = 2\Delta t \frac{X^*(k)Y(k)}{N} \tag{10.5.1}$$

A scheme based on this formula is shown in Figure 10.24.

The same problems occur, as in power spectral density estimation, owing to the finite length of the record available. This causes a 'windowing' effect giving bias to the cross-spectral estimates. As with power spectral estimation by choice of suitable windows, bias can be traded for resolution.

With cross-spectral estimates, it is important that $X(k)$ and $Y(k)$ should give the correct phase information. However, the equivalent bandpass filter, owing to the window, not only averages magnitude over its effective bandwidth, but also averages phase. If the relative phase of the two signals is changing rapidly, this can lead to considerable error in the estimated cross-spectrum.

Again, as with the power spectral density estimates, some form of averaging is required to reduce the statistical variability of the estimate. However, reduction of the variance of the estimate not only depends upon the number of blocks used in the averaging, but also depends upon a quantity known as the *coherency.*

10.5.1 Coherency

The coherency between two signals $x(t)$ and $y(t)$ is a frequency domain measure of the degree of correlation between the signals. If the two signals are independent, at a given frequency, the coherency between them is zero at this frequency. If one signal

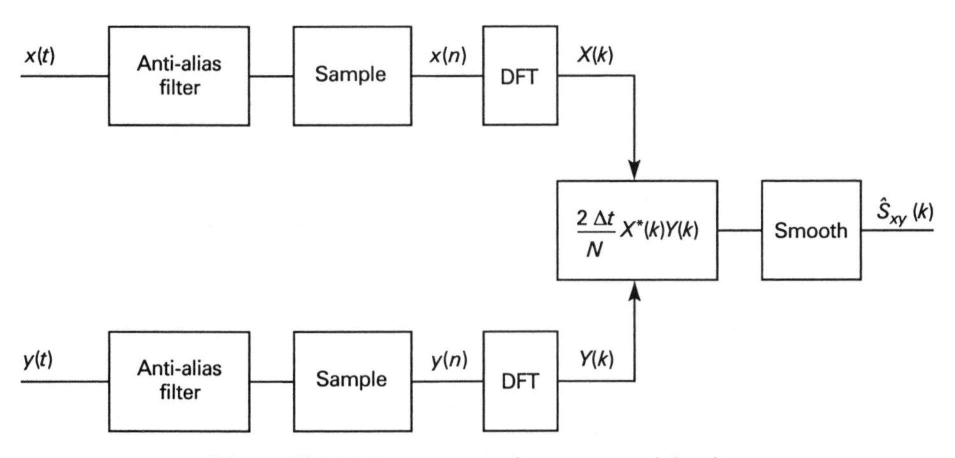

Figure 10.24 Measurement of cross-spectral density.

can be related to the other by considering them as input and output of a linear time invariant system, then the coherency is one.

More generally, $x(t)$ and $y(t)$ can be related as shown in Figure 10.25. Here $z(t)$ is the output of a linear time invariant system with $x(t)$ as the input. However, $z(t)$ is not a measurable output; a noise signal, $n(t)$, uncorrelated with $x(t)$, is added to $z(t)$ to form the measurable output $y(t)$. A measure of coherency is the ratio of noise power to signal power, $S_{nn}(f)/S_{zz}(f)$ at any frequency. However, this is not a very convenient measure, as it can vary between zero and infinity. A more convenient measure can be obtained by defining the squared coherency function, $\gamma_{xy}^2(f)$, as

$$\gamma_{xy}^2(f) = \frac{1}{1 + S_{nn}(f)/S_{zz}(f)} \tag{10.5.2}$$

This definition has the advantage that the coherency defined can only vary between 0 and 1. When there is no noise, $S_{nn}(f) = 0$, $\gamma_{xy}^2(f) = 1$. When $S_{nn}(f) \gg S_{zz}(f)$, $\gamma_{xy}^2(f) \rightarrow 0$.

However, the definition given by eqn (10.5.2) is not very practical, as both $S_{nn}(f)$ and $S_{zz}(f)$ are not directly measurable. The definition can be framed in terms of more measurable quantities if eqn (10.5.2) is re-arranged as follows (note that, for convenience, the dependence of the quantities on frequency has been dropped).

Noting that $x(t)$ and $n(t)$ are uncorrelated, then

$$S_{xy} = HS_{xx} \tag{10.5.3}$$

$$S_{yy} = S_{zz} + S_{nn} \tag{10.5.4}$$

$$S_{zz} = |H|^2 S_{xx} \tag{10.5.5}$$

From eqn (10.5.4)

$$\frac{S_{nn}}{S_{zz}} = \frac{S_{yy}}{S_{zz}} - 1$$

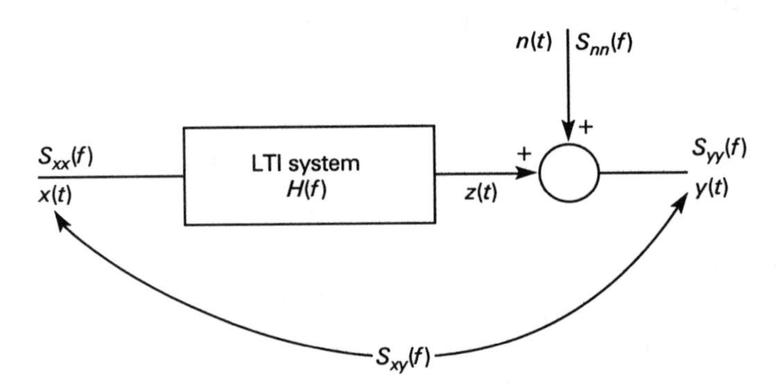

Figure 10.25 Block diagram representation for coherency.

Substituting into the expression for coherency, eqn (10.5.2),

$$\gamma^2 = \frac{S_{zz}}{S_{yy}} \tag{10.5.6}$$

From eqn (10.5.3)

$$|H|^2 = \frac{|S_{xy}|^2}{S_{xx}^2}$$

and substituting into eqn (10.5.5)

$$S_{zz} = \frac{|S_{xy}|^2}{S_{xx}}$$

Substituting into eqn (10.5.6) gives the required expression for coherency:

$$\gamma^2 = \frac{|S_{xy}|^2}{S_{xx}S_{yy}} \tag{10.5.7}$$

Care is needed when interpreting this expression from measured power and cross-spectra. Using eqns (10.4.2) and (10.5.1) and recognising that the constants will cancel,

$$\hat{S}_{xy} = \frac{X^*Y}{T}$$

$$|\hat{S}_{xy}|^2 = \frac{X^*YXY^*}{T^2}$$

$$\hat{S}_{xx} = \frac{X^*X}{T} \quad \hat{S}_{yy} = \frac{Y^*Y}{T}$$

Substituting into eqn (10.5.7) gives the estimated coherency as

$$\gamma^2 = \frac{X^*YXY^*/T^2}{(X^*X/T)(Y^*Y/T)} = 1$$

The coherency is always unity regardless of the noise present! This result arises from using unsmoothed estimates for power and cross-spectral densities. Smoothing must be performed over two blocks at least in order to obtain a meaningful result from eqn (10.5.7).

The cross-spectral density shows that two processes are related; the coherency function gives a measure of the correlation on a scale zero to one.

The following example illustrates the effect of quantisation on the coherency between two signals.

EXAMPLE 10.5.1

Analogue to digital converters (ADCs) are used to sample continuous signals and to convert the samples into coded form suitable for interpretation by digital computers. Because of the finite length of these codes, the samples must be quantised into a finite number of amplitude levels. This example investigates the effect of this quantisation on the coherency between the input and output signals.

Figure 10.26 shows a second order, lightly damped system. At the output is an 8 bit ADC covering the voltage range -10 V to $+10$ V. The input signal $x(t)$ is uniformly distributed white noise with an r.m.s. value of 3 V. Determine the measured coherency between the system input and output signals.

SOLUTION

The total record length used was 100 seconds and this was subdivided into eight segments. Using a Hanning window and averaging over these segments, the measured coherency was as shown in Figure 10.27(a).

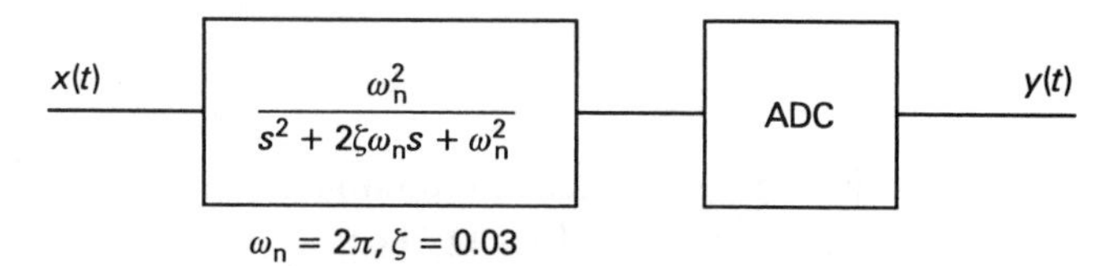

Figure 10.26 System for Example 10.5.1.

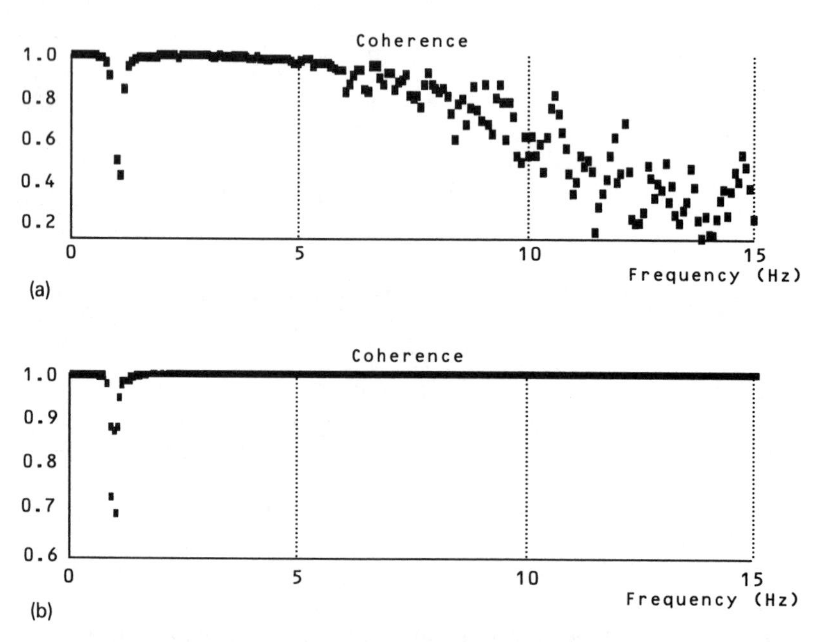

Figure 10.27 Coherency measurements for Example 10.5.1.

As can be seen, there is a dip in the coherency at 1 Hz and a falling off and breaking up of the coherency above 6 Hz. The latter effect is due to quantisation. As the frequency increases, the output of the system falls and the effect of quantisation becomes more significant. The output signal eventually falls to a level comparable to the width of one quantisation level ($20/2^8 = 0.078$ V).

In Figure 10.27(b) the ADC has been removed from the system and, as can be seen, the coherency is now unity above 6 Hz. There is still, however, the dip in the coherency around 1 Hz, the resonant frequency of the system. This is due to the rapidly changing phase, at this frequency, introducing errors into the estimates of the cross-spectral density. To reduce this error, the record length would have to be greatly increased in order to reduce the bandwidth of the effective filter.

10.6 Frequency response estimation

The frequency response function of a system was introduced in Section 5.2.1. This function is a very important form of system description and one reason why it is widely used is the relative ease in which it can be measured. In many of the problems of Chapter 5 the frequency response function was derived analytically, either using reactance methods or from the differential equations of the physical system. In practice this approach is only possible in relatively simple systems. Often, in complex systems, the dynamics of the system are not well enough understood to enable an analytical model to be derived. In these situations it is often possible to estimate the system's frequency response function by measurement of its input and output signals.

Theoretically, the measurement of the frequency response function is very simple. An oscillator produces a sinusoidal signal of known frequency and amplitude and this signal forms the system's input. Measurement of the output signal's magnitude and phase, relative to the input, gives the magnitude and angle of the system's frequency response function at one frequency. Repeating this procedure over a given frequency range produces the system's frequency response.

Often the procedure outlined in the previous paragraph is not applicable in practical situations. Even if applied, it may not give very satisfactory results. Some of the reasons for this are as follows:

1. Noise present in the system.
2. The system contains non-linearities.
3. Difficulties in injecting a sinusoidal signal into the system, especially in non-electrical systems.
4. Drift in the system's operating point during the test.
5. Limitations of test time due to wear, cost or safety requirements.

6. The nature of the system precludes off-line laboratory testing. Testing must be on-line using normal operating signals or injected perturbation signals.
7. Damaging resonances may be excited within the system.

Some of these difficulties will be considered in more detail in the following sections where specific testing methods will be examined.

10.6.1 Single frequency testing

A possible measurement scheme is shown in Figure 10.28. $n(t)$ represents the effect of noise in the system; it is not directly measurable, but is assumed to be uncorrelated with the input $x(t)$. The device measuring the output responds to r.m.s. values and it responds to the signal $y(t)$ where $y(t) = z(t) + n(t)$. The mean squared value of $y(t)$ is given by

$$\begin{aligned} \overline{y^2(t)} &= \overline{[z(t)+n(t)]^2} \\ &= \overline{z^2(t)} + \overline{n^2(t)} + \overline{2z(t)n(t)} \end{aligned}$$

As the noise signal $n(t)$ has been assumed to be uncorrelated with the system output $z(t)$, the last term in this expression will be zero. However, the result would still be in error owing to the contribution $n^2(t)$. The noise would also cause errors in the measurement of the phase.

As $n(t)$ is uncorrelated with $x(t)$, then, if the cross-correlation function between $x(t)$ and $y(t)$ were measured, the term involving $n(t)$ would become negligible providing the measurement time was sufficiently long. Because the input is sinusoidal, there is no need to obtain the complete cross-correlation function for all delays. If it were measured for two delay values, the simultaneous equations so

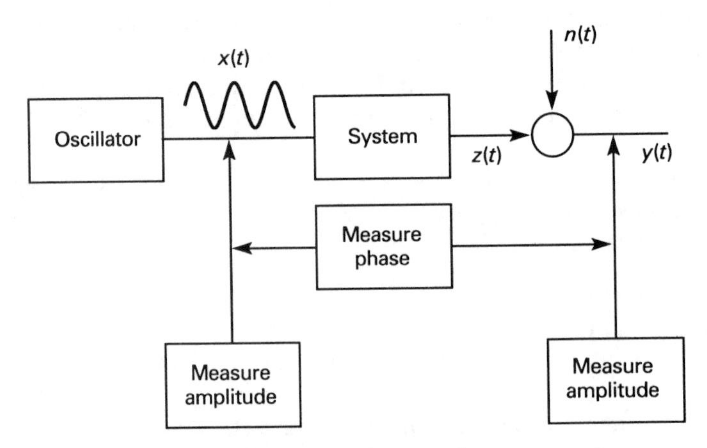

Figure 10.28 Measurement of system frequency response.

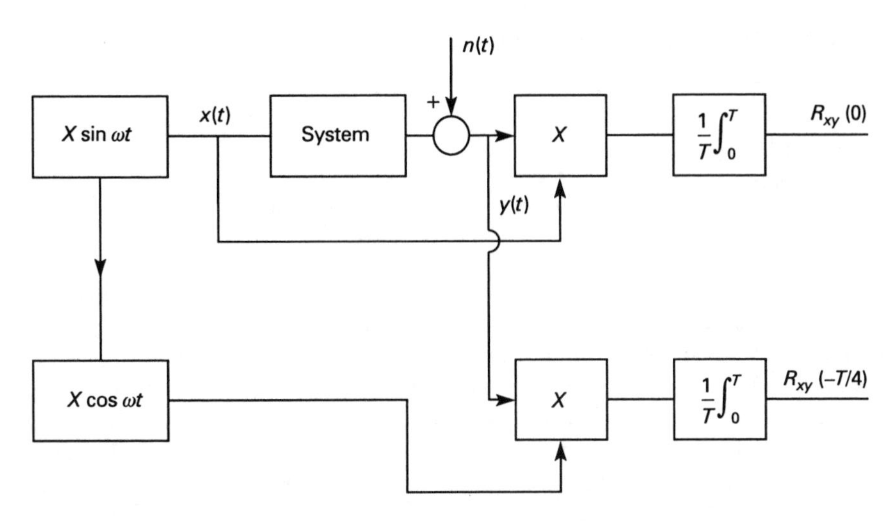

Figure 10.29 Correlation method for frequency response measurement.

formed could be solved to give the magnitude and phase of the frequency response function. The two most convenient delays to choose are those corresponding to phase shifts of 0° and 90°. Figure 10.29 shows the implementation of such a scheme.

Because the input is periodic, the cross-correlation function need only be obtained over a complete period, T. Then if

$$y(t) = Y\sin(\omega t + \varphi) + n(t)$$

the cross-correlation functions corresponding to zero delay and a delay $-T/4$ (90° phase shift) are

$$R_{xy}(0) = \frac{1}{T}\int_0^T X\sin(\omega t)Y\sin(\omega t + \varphi)\,\mathrm{d}t + R'_{xn}(0)$$

$$R_{xy}(-T/4) = \frac{1}{T}\int_0^T X\cos(\omega t)Y\sin(\omega t + \varphi)\,\mathrm{d}t + R'_{xn}(-T/4)$$

Here R'_{xn} represents the cross-correlation function between $x(t)$ and $n(t)$ evaluated over time T.

Performing the integrations gives

$$R_{xy}(0) = \frac{XY}{2}\cos\varphi + R'_{xn}(0)$$

$$R_{xy}(-T/2) = \frac{XY}{2}\sin\varphi + R'_{xn}(-T/4)$$

Noting that $R_{xx}(0) = X^2/2$ and assuming that $R_{xn} = 0$, then

$$\frac{R_{xy}(0)}{R_{xx}(0)} = \frac{Y}{X}\cos\varphi \qquad \text{real part of frequency response function}$$

$$\frac{R_{xy}(-T/4)}{R_{xx}(0)} = \frac{Y}{X}\sin\varphi \qquad \text{imaginary part of frequency response function}$$

To obtain these results, it has been assumed that $R'_{xn} = 0$. However, this is only true if the correlation is performed over infinite time. It is beyond the scope of this book to consider the effect of finite correlation time. However, it can be shown that, as this time is increased, the noise term decreases. In the calculation of the cross-correlation function, the integration has been taken over 1 cycle of the input signal. However, it could be performed over any integral number of cycles; this gives a convenient method of reducing the error due to noise at the expense of measurement time.

It should be noted that, when the frequency response function was introduced in Section 5.2.1, it was emphasised that the system output was only sinusoidal after the transient has decayed to zero. It is important that this condition is fulfilled when any frequency response measurement is made. In practice, this means the input signal must be applied for a time equal to approximately five times the largest system time constant before the measurement over an integral number of cycles is made.

The method of frequency response measurement just outlined is incorporated into commercial instruments known as transfer function analysers (TFAs). Modern instruments use extensive digital techniques and then become digital transfer function analysers (DTFAs). These instruments can include facilities such as automatic frequency incrementing, presentation of results as real and imaginary parts, magnitude and angle form, log magnitude form, display and printing of results in graphical form.

Advantages of single frequency testing are as follows:

1. All the power in the input signal is concentrated at one frequency; this gives good noise immunity.
2. Noise performance can easily be improved at the expense of measurement time.
3. By correlating the system output with harmonics of the input frequency, the harmonic content of the output signal can be measured. The harmonic content is a useful indication of system non-linearity.

Disadvantages of single frequency testing are as follows:

1. The length of time required to obtain a complete frequency response. This is not only the time for the requisite number of integrating cycles at each measurement frequency, but also the settling time required each time the frequency is changed. This excessive measuring time may cause undue wear on the system and it may be difficult to keep the operating point fixed during this time.

2. Because of the concentration of power at a single frequency, resonances may be excited in the system, which may drive it into non-linear regions or cause physical damage.

The excessive measurement time associated with single frequency testing can be reduced by use of multi-frequency periodic test signals. These will be considered in the following section.

10.6.2 Periodic multi-frequency test signals

The long measurement time associated with a single frequency test signal would be reduced if all the required frequency components could be fed simultaneously into the system. One could envisage the scheme shown in Figure 10.29 extended to include a bank of oscillators (sine and cosine) at frequencies ω_o, $2\omega_o$, $3\omega_o$ ... plus a bank of correlators to identify the output at each component frequency. Because there is no correlation between sinusoids of different harmonic frequencies, when measured over an integral number of cycles, only the component of interest would appear at the output of each correlator. To implement such a scheme by analogue means would be impractical. However, a digital implementation is quite feasible.

The input test signal could be synthesised digitally from the required frequency components and then fed via a digital to analogue converter (DAC) to form the input of the system under test. The correlator would perform the digital equivalent of the equations

$$a_n = \frac{1}{T}\int_0^T y(t)\cos n\omega t\,\mathrm{d}t$$

$$b_n = \frac{1}{T}\int_0^T y(t)\sin n\omega t\,\mathrm{d}t$$

Table 10.1 Real and imaginary parts of frequency components

	$X(k)$	
k	*Real*	*Imaginary*
0	$X(0)$	0
1	$X_R(1)$	$X_I(1)$
2	$X_R(2)$	$X_I(2)$
⋮	⋮	⋮
$N/2$	$X_R(N/2)$	0
⋮	⋮	⋮
$N-2$	$X_R(2)$	$-X_I(2)$
$N-1$	$X_R(1)$	$-X_I(1)$

These equations define the coefficients in the Fourier series representation of $y(t)$ (see Section 5.4.1). If the sampled system output is used, then, assuming no aliasing, they can be replaced by the DFT equivalents. The method then becomes very similar to the scheme for power spectral density measurement shown in Figure 10.6. There is now no need to take the modulus squared of the result and real and imaginary parts (gain and phase) are obtained. However, the big difference is that the measurement time is always over *an integral number of cycles* and leakage effects are completely removed.

Before considering the overall scheme for this method of frequency response measurement, it is worth considering in more detail the construction of the multi-frequency test signal. The discrete form of this signal will be

$$x(n) = X_1 \sin\left(n \frac{2\pi}{N} + \varphi_1\right) + X_2 \sin\left(2n \frac{2\pi}{N} + \varphi_2\right) + \cdots + X_k \sin\left(kn \frac{2\pi}{N} + \varphi_k\right)$$

where X_k and φ_k are specified by the user. Although the signal could be produced directly from this form, it is much easier, and faster, to construct it from the inverse DFT (or FFT). This algorithm will already be present in the analyser in order to evaluate the components of the output waveform. The real and imaginary parts of each component are specified as in Table 10.1.

The d.c. component, $X(0)$, is usually set to zero. Using the definition of the DFT

$$X(k) = \sum_{n=0}^{N-1} x(n)\, e^{-jnk2\pi/N}$$

then

$$X(N-k) = \sum_{n=0}^{N-1} x(n)\, e^{-jn2\pi}\, e^{+jnk2\pi/N} = X^*(k)$$

This property must be taken into account when constructing the table. Even though frequency components greater than $N/2$ are not going to be present in the test waveform, their real and imaginary parts must be entered according to the relationship $X(N-k) = X^*(k)$. If this procedure is not followed, when the time signal is obtained from the inverse transform it will contain an imaginary component.

The real and imaginary parts of a component relate to its magnitude, M, and phase, φ, as

$$X_R = M \cos \varphi$$

$$X_I = M \sin \varphi$$

Rather than specify the magnitude of a component, it is often required to set the power of a component P_k. Then for the kth component

$$M_k = \sqrt{2P_k}$$

After taking the inverse transform, the values of $x(n)$ must be 'clocked' to the DAC to form the test signal. The clock time becomes the sampling time Δt, determining the frequency of the fundamental, $f_0 = 1/N\Delta t$. A block diagram of the complete scheme is shown in Figure 10.30.

At any frequency $\omega = k2\pi/N\Delta t$

$$|H(j\omega)| = \left|\frac{Y(k)}{X(k)}\right|$$

$$\text{angle } H(j\omega) = \text{angle } Y(k) - \text{angle } X(k)$$

The choice of the frequencies, amplitudes and phases of the test signal components depends upon the known characteristics of the system under test.

Amplitudes and frequencies of components

The total power in the input signal is spread among the individual frequency components. By choosing what components are to be present and their amplitudes, the power spectral density of the input signal is determined. If some prior knowledge of the system is available, the power spectral density of the input signal can be chosen to best advantage. The resolution of the components can be specified: wide spacing for flat parts of the spectrum, narrow spacing around any peaks.

If an estimate of the spectrum of any noise disturbance is available then this can affect the choice of the amplitudes of the components. More power can be put into the components in the region of the spectrum where the noise power is greatest.

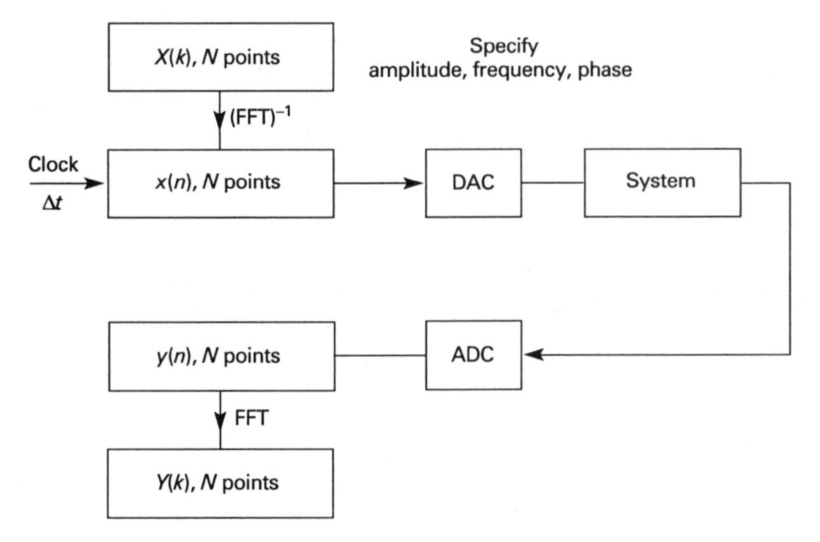

Figure 10.30 Periodic multi-frequency test scheme.

Phases of components

The system phase estimated is that of the output component with respect to the corresponding input component. On this basis, the absolute phase of the input components does not affect the estimate and it can be chosen arbitrarily. However, alteration of the relative phases of the components will alter the shape of the time signal produced and in particular it will alter its peak–peak value.

Consider a simple example of a signal consisting of two components only, a fundamental plus a third harmonic, both of equal unit amplitude:

$$x(t) = \sin \omega t + \sin (3\omega t + \varphi)$$

Figure 10.31 shows this signal for two values of φ, the phase angle of the harmonic component. In Figure 10.31(a) $\varphi = 0$ and the peak–peak amplitude of $x(t)$ is 3 units. In Figure 10.31(b) $\varphi = \pi$ and the peak–peak amplitude is 4 units. It should be noted that the power in the signal depends only upon the amplitudes of the components and not upon their phases; the power in the signal $x(t)$ for both phase values is identical.

The importance of the peak–peak value of the test signals lies in the fact that all practical systems are non-linear. To remain in a linear region, it is required to keep the peak–peak excursion of the input signal to a minimum. However, in order to obtain a high signal power to noise power ratio, it is important to make the signal power as large as possible. The ratio of signal power to peak–peak value for the signal is given by the *peak factor* of the signal. This is defined as

$$\text{Peak factor} = \frac{x_{\max} - x_{\min}}{2\sqrt{(2)}x_{\text{rms}}}$$

The factor $2\sqrt{2}$ is used to normalise the factor to unity for a sine wave signal.

How should the phases of the components of a multi-frequency signal be chosen to give minimum peak factor? There is no analytical method of obtaining an answer to this question. However, a method of choosing the phase that usually results in low peak factor is given by Schroeder phases. The phase angles are given by the following formula:

$$\varphi_n = 2\pi \sum_{i=1}^{n} iP_i$$

where P_i is the relative power associated with the ith component. For the case where the power is distributed equally among N components, $P_i = 1/N$ and

$$\begin{aligned}\varphi_n &= 2\pi \sum_{i=1}^{n} \frac{i}{N} \\ &= \frac{2\pi}{N} \sum_{i=1}^{n} i\end{aligned}$$

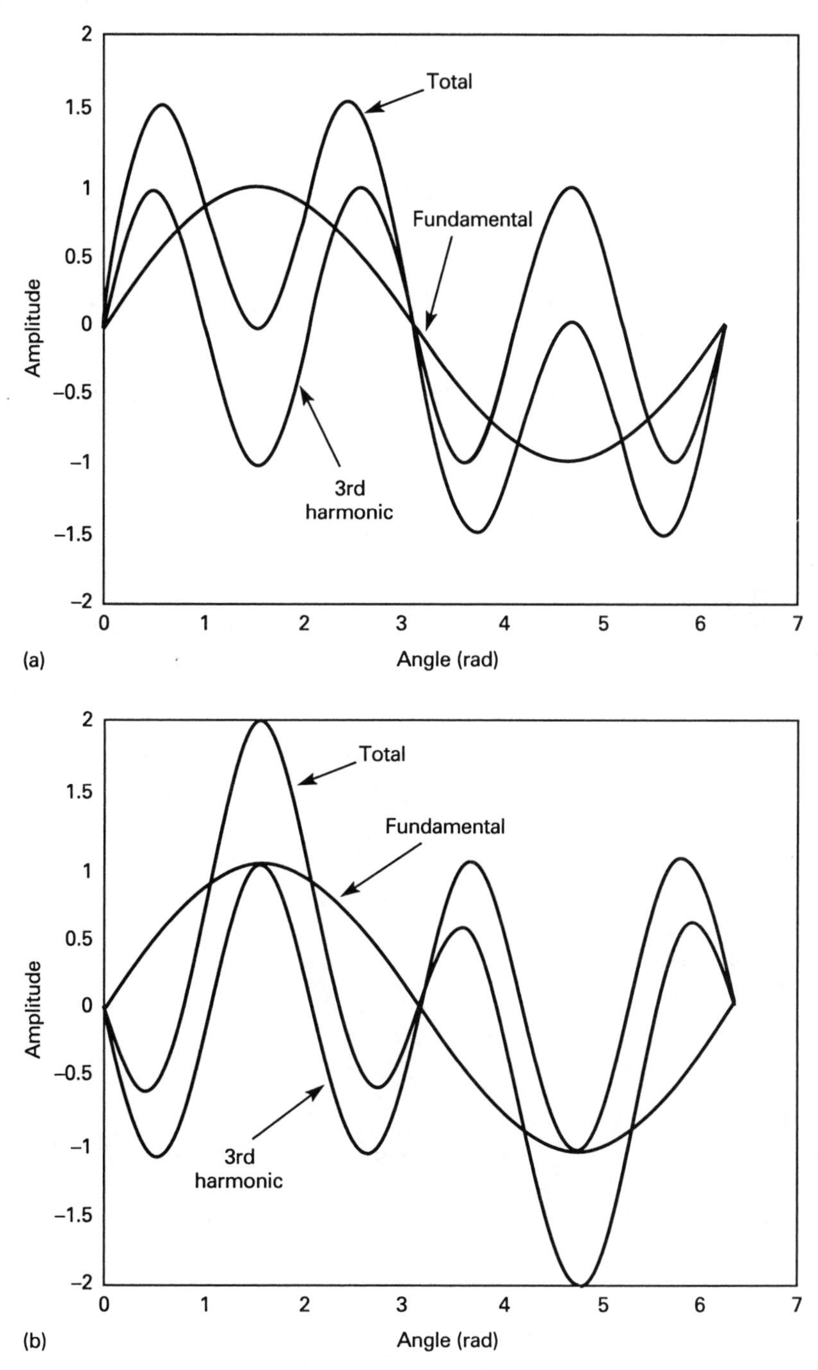

Figure 10.31 Effect of phase on peak–peak values.

The summation is the sum of an arithmetic progression giving

$$\varphi_n = \frac{\pi}{N} n(n+1)$$

$$= \frac{n^2\pi}{N} + \frac{n\pi}{N}$$

The second term gives a phase shift proportional to frequency, a linear phase term. This would give a time shift to the waveform without changing its shape; it can be ignored.

The following example illustrates the construction of a multi-frequency periodic signal.

EXAMPLE 10.6.1

A multi-frequency test signal is to consist of a 1 Hz fundamental plus 29 successive harmonics. The signal is to have a total power of 1 volt^2 and this power is to be equally distributed among the components. Obtain the time signal for the case when all the components have zero phase and for the case when the components have Schroeder phases. Calculate the peak factor of the signal in both cases.

SOLUTION

As the power is equally distributed among the components, the power in each component is 1/30 volt^2 and the peak value of each component is

$$M_n = \sqrt{(2/30)}$$

$$= 0.258 \text{ V}$$

The time signal is given by

$$x(t) = \sum_{n=1}^{30} 0.258 \sin(n\omega t + \varphi_n)$$

where $\omega = 2\pi$ and φ_n is to be chosen.

The time signal for all $\varphi_n = 0$ is shown in Figure 10.32(a) and it has a peak–peak value of 14 V.

The Schroeder phases are determined by the expression

$$\varphi_n = \frac{n^2\pi}{N}$$

where $N = 30$. The time signal for this condition is shown in Figure 10.32(b) and it has a peak–peak value of 5.83 V.

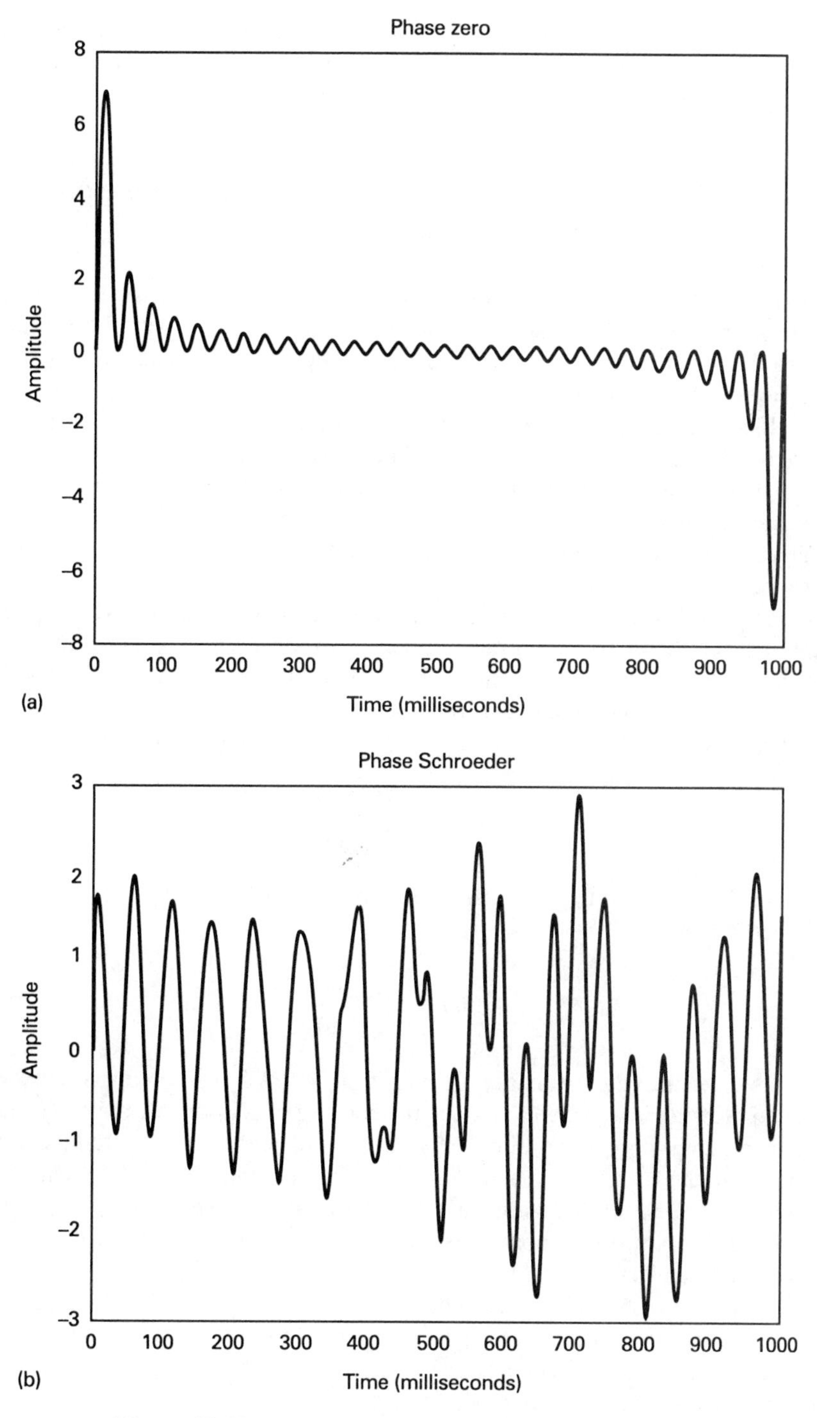

Figure 10.32 Multi-frequency signals for Example 10.6.1.

Both signals have a mean square value of unity, giving r.m.s. values also of unity. The peak factors are then 4.95 and 1.96 for zero phase and for Schroeder phased signals respectively.

10.6.3 Frequency response function from random input signals

The previous two sections have investigated the measurement of frequency response by single sine wave testing and by the use of multi-frequency periodic test signals. There are, however, situations where it is impractical to inject such signals into the system. For example, suppose it was required to obtain the frequency response of a motor vehicle suspension system. The wheels of the vehicle could be placed on mechanical vibrators and the appropriate sinusoidal test signal fed to these vibrators. However, in this 'static' situation, the characteristics of the tyres are very different to their characteristics on the road and the frequency response obtained may not be applicable to road conditions. The response could be measured by 'on road' testing, but now the input to the system is random road noise.

In many industrial processes it is unattractive commercially to take the plant off-line and measurements to obtain the frequency response must be performed on-line using normal operating signals. In such a situation the operating signals are so complex that they can be regarded as random signals.

The relationship between the required spectral densities is given by eqn (9.5.6)

$$H(\mathrm{j}\omega) = \frac{S_{xy}(\mathrm{j}\omega)}{S_{xx}(\mathrm{j}\omega)}$$

As only finite lengths of data are available, only estimates of $S_{xx}(\mathrm{j}\omega)$ and $S_{xy}(\mathrm{j}\omega)$ can be obtained. Then

$$\hat{H}(\mathrm{j}\omega) = \frac{\hat{S}_{xy}(\mathrm{j}\omega)}{\hat{S}_{xx}(\mathrm{j}\omega)}$$

Errors enter into the estimation of both $S_{xx}(f)$ and $S_{xy}(f)$ owing to leakage and suitable windowing is required to reduce the ensuing bias in the estimate $\hat{H}(f)$.

With random signals, an effect of finite record length is to introduce variance into the estimate of $S_{xx}(f)$ and $S_{xy}(f)$. The variance in $S_{xx}(f)$ can have a very significant effect on the estimate $\hat{H}(f)$. If, at a given frequency, $S_{xx}(f)$ is very small, because it is in the denominator of the expression for $H(f)$, it can give very large errors to the estimate $\hat{H}(f)$.

The variance of $S_{xx}(f)$ can be reduced by smoothing over blocks as in Section 10.4.5. However, it is important that the input signal $x(t)$ is such that it contains significant power at all frequencies over which the estimate $\hat{H}(f)$ is made. Such an input signal is known as *persistently exciting*.

External noise to the system will also increase the variance of the estimate $\hat{H}(f)$. As in cross-spectral estimation, the coherency gives an indication of the confidence that can be placed in the result and whether smoothing over more data blocks is required.

An expression for the estimate $\hat{H}(f)$ using the DFT of input and output signals can be obtained via eqn (10.7.1). It should be noted that all the constants cancel when taking the ratio giving

$$\hat{H}(f) = \frac{\sum_{i=1}^{L} X_i^* Y_i}{\sum_{i=1}^{L} X_i^* X_i}$$

where X_i, Y_i are the DFTs of the ith blocks of input and output data and L is the number of blocks.

The following example illustrates the estimation of a system's frequency response function.

EXAMPLE 10.6.2

Figure 10.33 shows a first order system where the input $x(t)$ is a white noise process having unit variance. The output is contaminated by a random process $n(t)$ consisting of two components. One is a white noise process having variance 0.0016; the other is formed by passing a white noise process of variance 0.16 through a bandpass filter having a pass band extending from 0.5 Hz to 2 Hz.

Estimate the system's frequency response function, making measurements of the coherency.

SOLUTION

With a sampling time of 0.05 seconds, a block length of 1024 samples was used. This gives a block length of 51.2 seconds and a frequency resolution of 0.0195 Hz. Hanning windows were used on both input and output records. The 'banded' noise was produced by passing white noise through a fourth order, Chebyshev type 1 filter, with 1 dB of ripple in its pass band and having cut-off frequencies of 0.5 Hz and 2 Hz.

Initially, the results were smoothed over three records, each of length 51.2 seconds. Figure 10.34 shows the results obtained. The magnitude results show some variability, but are recognisable as those of a first order system. The phase results are

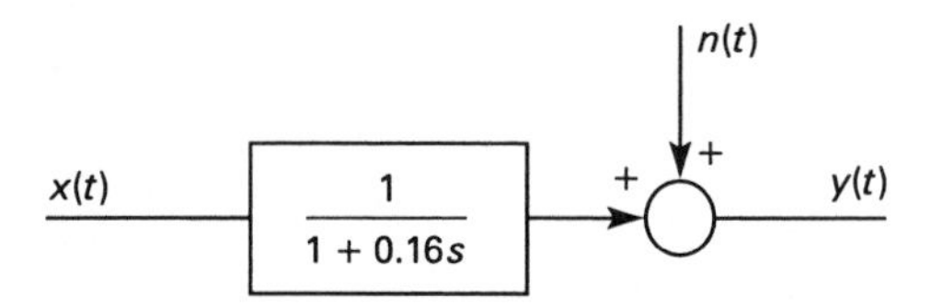

Figure 10.33 Block diagram for Example 10.6.2.

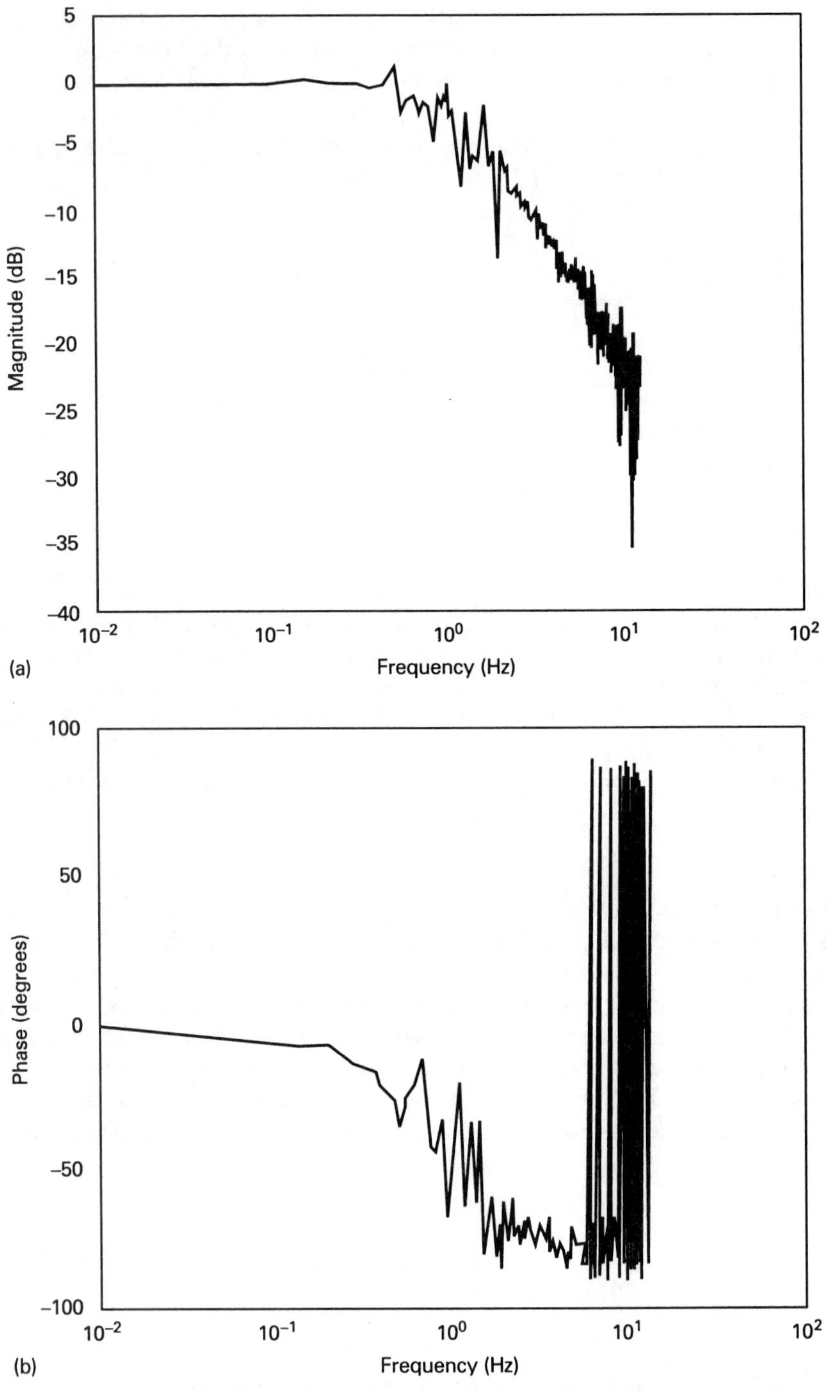

Figure 10.34 Frequency response for Example 10.6.2, smoothing over three blocks.

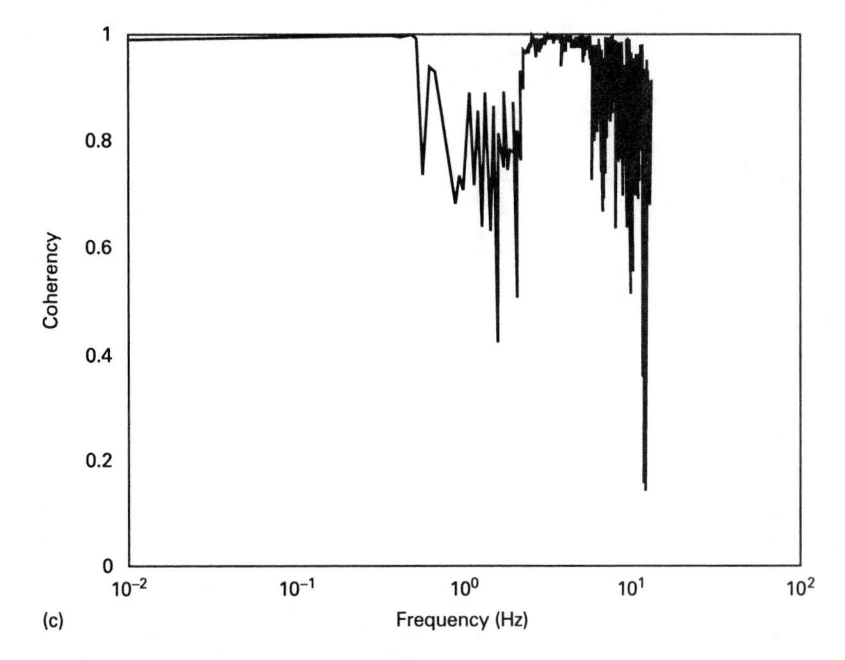

Figure 10.34 *Continued*

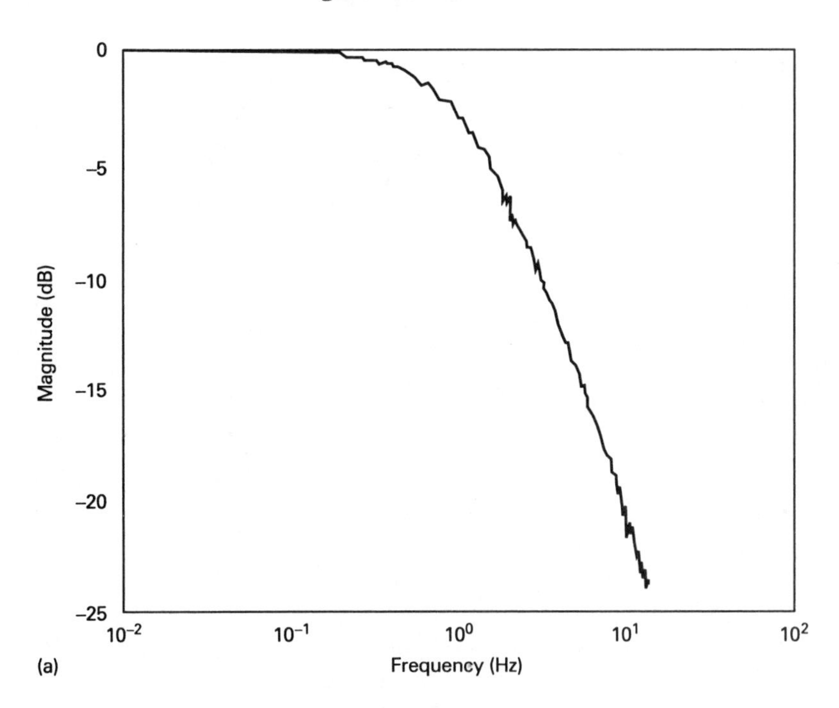

Figure 10.35 Frequency response for Example 10.6.2, smoothing over 100 blocks.

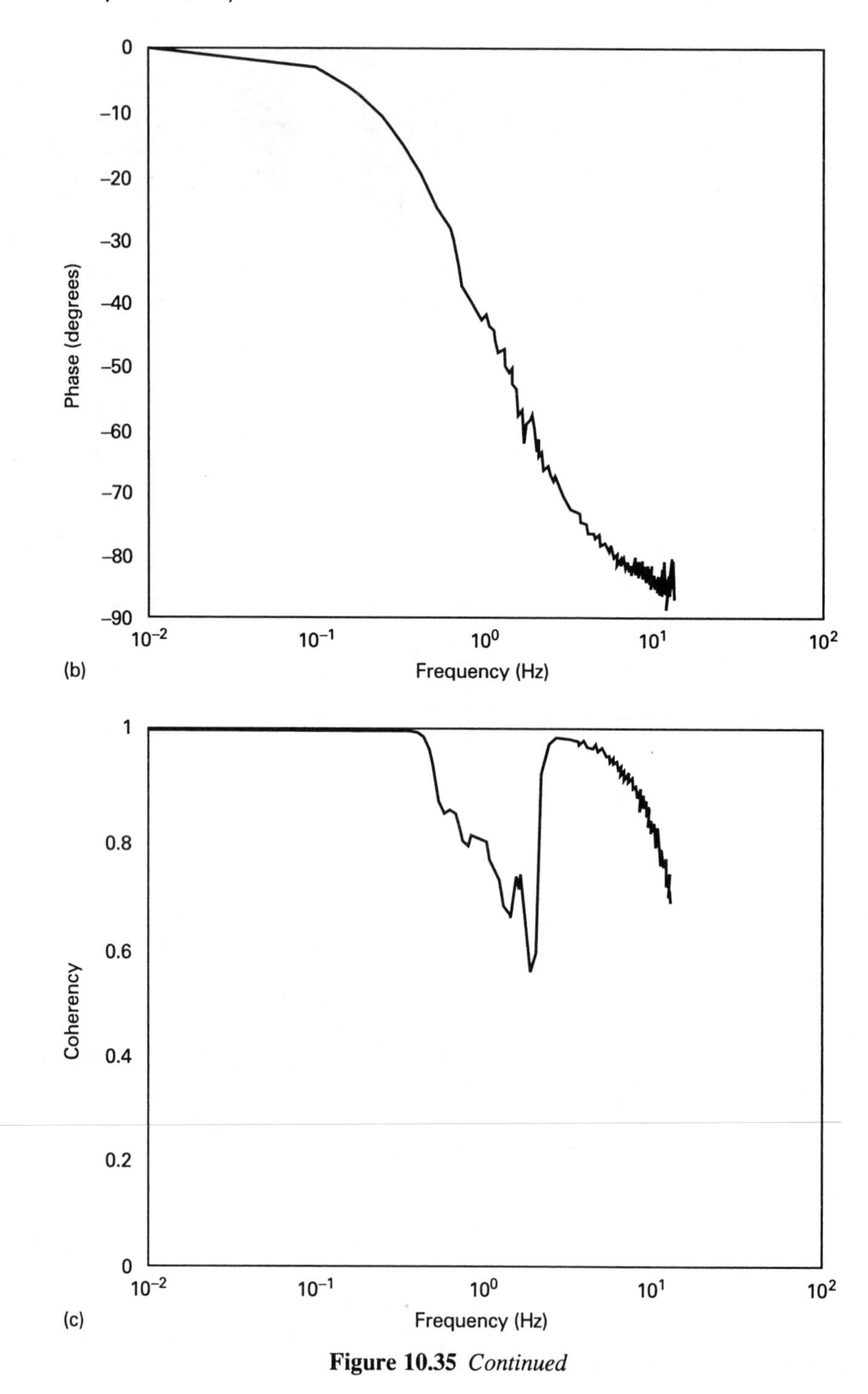

Figure 10.35 *Continued*

very bad, the plot giving no information at higher frequencies. The measured coherency reflects these results. Up to approximately 0.5 Hz the coherency is unity, showing that the output signal is large enough to give a good signal to noise ratio. The coherency then falls owing to the 'banded' noise at the output. It then recovers to around unity, but, as the system gain falls, the signal to noise ratio worsens, giving very poor coherency figures.

Figure 10.35 shows the corresponding plots when the records were smoothed over 100 blocks (again the block length remains constant at 51.2 seconds). The magnitude and phase results are now much more acceptable. This is reflected in the improved coherency figures; the coherency does dip to 0.6, but remains above 0.8 throughout most of the frequency range.

10.7 Summary

To obtain the spectral content of any signal that contains a random component, an infinite time record is required. In practice only a finite record length is available and it is only possible to obtain an estimate of the spectral content. This chapter has investigated the problems involved in obtaining this estimate.

The chapter introduced spectral estimation by considering an analogue method that used a bandpass filter to pick out the required spectral component. Although analogue, this method highlighted the problems present in any spectral analysis method:

1. Obtaining the narrow bandwidth required to obtain satisfactory frequency resolution.
2. Preventing ripples in the stop band of the filter.
3. Reducing the statistical variation of the estimate.

These problems are interrelated and any practical estimation scheme must compromise with some 'trade-off' in their solutions.

A digital estimation scheme using the periodogram was introduced. It was shown that a finite record length gave rise to errors in the estimated spectrum due to leakage effects and these effects could be reduced by use of suitable windows. When the record contains a random component there are errors due to the statistical variation of the estimate. These can be reduced by some form of smoothing. One smoothing method is to divide the record into a number of smaller records and to average at each frequency over the spectra of these records.

The estimation of the cross-spectra between two time records was considered. It was shown that the problems involved were similar to those arising in the estimation of autospectra. However, the variance of the estimated spectra also depended upon the coherency between the signals and this could be interpreted as a measure of the noise in the system relating the two signals.

The last section of the chapter considered the estimation of frequency response. Three methods of estimation were investigated: use of a single sine wave, use of periodic non-sinusoidal signals, use of random signals. The relative merits and disadvantages of these methods were discussed.

MATLAB exercises

10.1 This exercise demonstrates the use of spectral analysis for a single sinusoid.

(a) Produce a long vector representing 10 complete cycles of a single sine wave, amplitude 2 V, frequency 10 Hz. Choose a suitable sampling interval for the representation.

Obtain the power spectral density of this waveform, plot and compare with the theoretical spectrum. Plot the spectrum for positive frequency only and for positive and negative frequencies, showing in each case the correct frequency scales.

(b) Repeat part (a), but modify the sampling time such that 9.5 cycles are contained in the record. As well as plotting the spectrum on a linear amplitude scale, plot also on a dB scale. Verify that the total power in the spectrum equals the power in the time waveform.

Note The formula for the calculation of power spectral density is

$$S_{xx}(f) = \frac{\Delta t|X(f)|^2}{N}$$

where Δt is the sampling time and N is the number of points in the transform. Remember that power is contained at both positive and negative frequencies and the power contained at a single frequency is $S_{xx}(f)\Delta f$ where Δf is the frequency spacing.

The command `fftshift` is useful when a plot showing both positive and negative frequencies is required.

To show a line spectrum, the command `stem` can be used instead of `plot`.

10.2 This exercise illustrates how the action of a rectangular window in the time domain can be interpreted as a convolution in the frequency domain.

Construct a vector representing a single sine wave of unit amplitude, frequency 5.5 Hz, using a sample time of 1/1000 s. The vector should be of length 1000 (representing $5\frac{1}{2}$ cycles). Pad this record with 1000 zeros and obtain the power spectral density of the resulting time function. Plot this power spectral density, as one of two subplots, over the frequency range -10 Hz to $+10$ Hz.

Construct a vector representing the same sine wave, but make the vector 2000 points in length to represent 11 complete cycles. Construct a vector consisting of a 1000 point box car window padded with another 1000 zero points. Take the DFTs of both these time signals and perform a frequency domain convolution between them. Express the result as a power spectral density and display as a second subplot. Compare the power spectral densities obtained by the two methods.

Note As speed of operation is not a consideration here, the DFTs have been taken over 2000 points for convenience rather than having the number of points as a power of 2. The zero padding of the records is necessary to give the same frequency intervals, both for the DFT convolution and for comparison of the results. (See Section 6.4.3 for the effect of zero padding.)

To obtain responses centred around zero frequency, the command `fftshift` can be used. The command `conv` can be used to perform the required convolution. Theoretically, a circular convolution is required (see Section 6.4.3). However, as the signals are band limited, their transforms fall to zero at the ends of the records and the resulting convolution is correct over the frequency range used. Note that a factor of $1/N$ is required to scale the convolution when it is performed in the frequency domain.

10.3 This exercise investigates the properties of time windows. Any window can be investigated, but it is suggested that you include the box car window as a reference. Initially the box car, triangular and Hanning windows could be investigated, as the points representing these windows are available using MATLAB functions. A window length of 100 points is suitable.

(a) Plot the window functions considered all on the same graph and compare the window characteristics in the time domain.
(b) By padding the window function with zeros, determine by use of the DFT its frequency domain magnitude characteristics. Plot all the window characteristics, using a dB scale, on the same graph. Choose a frequency scale to show about five sidelobes each side of the main lobe.
(c) Determine the bandwidth to the -3 dB points for all the windows considered. Quote this as a normalised value by dividing by the bin width.
(d) Determine the highest sidelobe level for each of the windows. This is usually quoted in dB with reference to the main lobe.
(e) By plotting the window characteristic over its full frequency range using a logarithmic frequency scale, determine the sidelobe fall-off in dB/octave.
(f) Determine the scallop loss for each of the windows. This is the additional loss that occurs when a single frequency signal lies mid way between two

bin frequencies, compared to that when it lies at an adjacent bin frequency. The scallop loss can be obtained by two methods:

(i) Direct measurement from the window frequency domain characteristics.
(ii) By measurement of the DFT of a single frequency that lies mid way between two bin frequencies. The appropriate window is used on the data and the magnitude of the result compared with that obtained when the single frequency is a bin frequency.

(g) Determine the window's processing gain. Regarding the window as a filter in the frequency domain, the processing gain is a measure of how well the filter can detect a single frequency signal in a white noise background. It can be defined as

$$\mathrm{PG} = \frac{\text{signal to noise ratio at output}}{\text{signal to noise ratio at input}}$$

For each window considered generate a single sine wave, window the data over a complete period and determine its power spectral density. Hence, obtain the measured signal output power at the equivalent filter centre frequency. Repeat this procedure for a unit variance white noise signal. As the theoretical input signal power and noise power are known, the processing gain can be obtained. The processing gain can be shown to be given by

$$\mathrm{PG} = \frac{1}{N} \frac{\left(\sum_{n=0}^{N-1} w(n) \right)^2}{\sum_{n=0}^{N-1} w^2(n)}$$

where $w(n)$ is the window function and N is the window length.

Calculate this for each window and compare with the measured values.

10.4 This exercise examines the frequency resolution that can be obtained in the power spectral density estimate of a signal consisting of the sum of two sinusoids and compares the available resolution using box car and Hanning windows.

Using a vector of length 1000 and a sampling time of 1 ms, generate a signal consisting of the sum of two sinusoids. The sinusoids should be of equal amplitude, but arrange your program so that their frequencies and phases can be easily varied. The frequencies chosen should be in the region of 10 Hz.

Using a box car window, determine the minimum frequency separation that enables two peaks to be detected in the power spectrum of the signal. Investigate how this minimum separation varies with the frequencies and phases of the component sinusoids. Repeat using a Hanning window and compare the minimum frequency separations obtained.

Compare the measured resolutions obtained with the bandwidth of the windows as obtained in Exercise 10.3.

Note The two components can be resolved if two peaks can be detected in the spectrum. The available resolution depends on the relative phases of the components and their frequencies in relation to a bin frequency.

10.5 This exercise illustrates how the spectral leakage from one sinusoidal component can mask the presence of a second component.

Construct a vector of length 1000 points representing a signal formed by the sum of two component sinusoids. The sampling interval can be taken as 1/1000 s.

The first component is a sine wave of unit amplitude and of frequency 10.25 Hz. The second component is sinusoidal with frequency 15 Hz and has amplitude and phase that can be set within the program.

(a) Display the power spectral density of this signal and note the effect of varying the amplitude and phase of the 15 Hz component upon the 'detectability' of this component in the spectrum.
(b) Remove the 15 Hz component completely and measure the amplitude and phase of the leakage component at 15 Hz due to the 10.25 Hz sine wave. Re-introduce the 15 Hz component in anti-phase to the measured component and with amplitude equal to that of the measured component.
(c) Repeat part (a), but use a Hanning window on the signal prior to obtaining its power spectral density. Compare results with and without the Hanning window.

10.6 This exercise investigates the relationship between variance and record length of the power spectral density (PSD) estimate of a random signal.

(a) Produce a vector of length $N = 64$, consisting of normally distributed random variables having zero mean and unit variance. Obtain the PSD of this signal. Repeat for $N = 128$, $N = 256$, $N = 512$. Display all these PSDs by means of `subplot` and make a visual comparison of their mean values and their variances.
(b) Obtain the mean and variance of each of the PSDs derived in part (a) by averaging over an ensemble (1000 say) of signals. Compare the mean value obtained with the theoretical PSD of the signal. Comment on the magnitude of the variance of the PSD estimates in comparison to their mean values and note the fact that the variance does not decrease as record length increases.
(c) For $N = 64$ smooth the PSD estimates by averaging over L blocks. Obtain the variance of the smoothed estimates by measuring over an ensemble of signals and plot against L.

10.7 This exercise illustrates the use of the cross-spectral density function to determine the time delay of a system.

Produce a vector of length 1074 consisting of unit variance, zero mean random variables. From this vector, obtain two more vectors, x_2 formed from elements 1 to 1024, x_1 formed from elements 51 to 1074. x_1 and x_2 now represent two time signals where x_2 is a delayed version of x_1.

(a) Form the cross-spectral density function between the two signals, x_1 and x_2, and, from its angle, determine the time delay between the signals.
(b) Form the signal $x = x_1 + 0.1x_2$. Obtain the cross-spectral density function between x_1 and x. Plot the angle of this function and account for the shape of the plot. Again determine the time delay between the signals x_1 and x_2.

Note Smoothing will be required to obtain the estimates of cross-spectral density. Repeat the experiment about 500 times and take ensemble averages. The delay can be obtained as a number of samples or as a time delay by assigning a sampling interval.

10.8 This exercise demonstrates coherency as a measure of the cross-talk between two communication channels. Figure 10.36 is a simplified block diagram representing two such channels.

The transmittances $u_1 \longrightarrow y_1$ and $u_2 \longrightarrow y_2$ are both unity. However, there is a contribution to the output y_2 from u_1 via a low-pass filtering action. The object of the exercise is to simulate this system and to measure the coherency functions $\gamma^2_{u1\,y1}$, $\gamma^2_{u1\,y2}$, $\gamma^2_{u2\,y1}$, $\gamma^2_{u2\,y2}$.

The input signals u_1 and u_2 are random signals, each consisting of a sequence of 1024 zero mean, unit variance, normally distributed random variables. Taking a sampling time of 1 ms, this gives a half sampling frequency of 512 Hz. The low-pass filter can be chosen to have a cut-off frequency of 100 Hz. The function `cheby1` will give a suitable filter, but the order is restricted by the limited matrix size of the Student Edition of MATLAB.

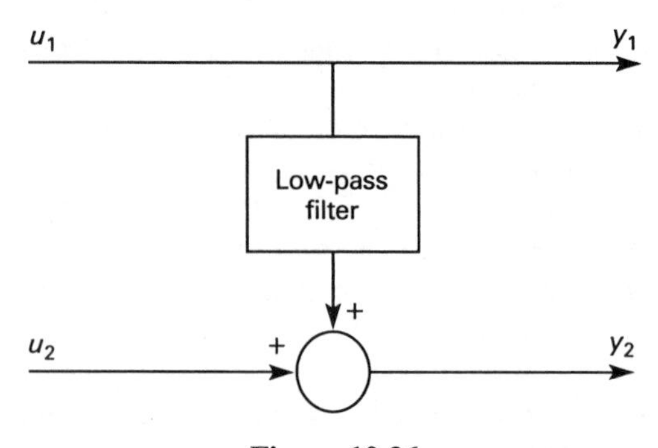

Figure 10.36

From the FFTs of input and output signals, calculate the power and cross-spectral densities and, hence, the required coherency functions. Plot all these against frequency and display using `subplot`.

Note Although 1024 points are used for the FFT analysis, extra points will have to be included in the signals to allow the transients in the filter to decay. To obtain consistent coherency estimates, some ensemble averaging will be required on the power and cross-spectral density estimates. There is no need to include any constants in the expressions for spectral densities, as these will cancel when obtaining the coherency functions. In order to aid comparison of the coherency functions, it is useful to fix the axis scaling to the range 0 to 1.

10.9 This exercise investigates the production of multi-frequency test signals.

Write a program to produce a multi-frequency test signal, as described in Section 10.6.2, making use of the inverse fast Fourier transform as shown in Figure 10.30. Make the sequence length $N = 1024$ points and assume that this represents a fundamental period of 1 second giving a fundamental frequency of 1 Hz. Remember that when specifying the frequency content, for the kth harmonic, $X(N-k) = X^*(k)$, otherwise the inverse transform will not produce a real time signal.

(a) Produce signals with the harmonic content as specified. For each signal obtain its peak factor when the phases are (i) zero for all components, (ii) uniformly random between $-\pi$ and $+\pi$, (iii) Schroeder. The signals are

(1) as in Example 10.6.1;
(2) a bandpass signal where the total power is equally distributed between harmonics 10 to 19;
(3) a signal consisting of 'prime harmonics' only. This is a signal where the total power is equally distributed between harmonics 2, 3, 5, 7, 11, 13, 17, 19, 23. Note that there is no power in the fundamental.

Signals of this form are useful when harmonic distortion due to non-linearities occurs in a system. Because no two components of the signal are harmonically related, harmonic distortion of one component will not cause errors in the measurement of other components.

(b) Consider the following algorithm for a reduction in the peak value of a signal:

(1) Produce a signal with the required harmonic content, the phases of the components being Schroeder phases.
(2) Clip this waveform to 90% of its peak positive and negative values.
(3) Using the FFT, obtain the phases of the components in the clipped signal.

(4) Construct a new test signal with the required harmonic amplitudes. However, the phases of the components are now set to the values obtained in (3).

Write a program to implement this algorithm and test it on the signals produced in part (a).

10.10 This exercise investigates the estimation of a system's frequency response function when the system input is a random signal. In particular, the 'trade-off' between resolution and variance for a given record length is investigated.

Simulate an undamped second order system with $\omega_n = 2\pi$ and $\zeta = 0.2$. For the input signal use a long vector or normally distributed random variables having zero mean and unit variance. Taking the sampling time to be 0.02 seconds, determine, using `lsim`, the corresponding system output vector.

Subdivide both input and output vectors into a set of L shorter length vectors. From the DFTs of these segments, determine an estimate of the system's frequency response as

$$\hat{G}(f) = \frac{\sum_{i=1}^{L} X_i^*(f) Y_i(f)}{\sum_{i=1}^{L} X_i^*(f) X_i(f)}$$

where $X_i(f)$ and $Y_i(f)$ are the DFTs of the ith input and output segment respectively.

Plot the magnitude of $\hat{G}(f)$ (dB) against frequency on a log scale and also display the theoretical response $G(f)$. Repeat the above procedure over an ensemble of records. By using the `hold` command in the graph plotting, superimpose the estimated frequency response, $\hat{G}(f)$, for each realisation of the ensemble. This enables a visual representation of the variance of the estimate to be obtained.

Investigate the effect of record length on the resolution of the peak, the bias of the estimate and the variance of the estimate. Use both rectangular and Hanning windows.

APPENDIX

An Introduction to Complex Numbers

Cartesian representation
Complex conjugate
Polar representation
Exponential representation
Roots of a complex number

This appendix is not intended for the student who has no knowledge whatsoever of complex numbers. It is meant to be used as a 'refresher' and to help with specific points that may arise in the text.

Complex numbers involve $\sqrt{-1}$ and throughout this book this has been represented by the symbol j. Hence

$$j = \sqrt{-1}, \qquad j^2 = -1$$

j does not have a numerical value, it is not a number as there is no number that when squared, would result in a negative quantity.

An interpretation of j can be obtained graphically. With reference to Figure A.1, real numbers are represented on the real axis and the number +1 is as shown. Multiplication by −1 (which equals j^2) produces the point −1 on the negative real axis. This can be interpreted as a rotation through 180°. It therefore seems reasonable to interpret $j = \sqrt{-1}$ as representing a rotation through 90°. This gives the point j1 and further multiplications by j produce the point −j1. Hence the vertical axis in this context is known as the imaginary axis.

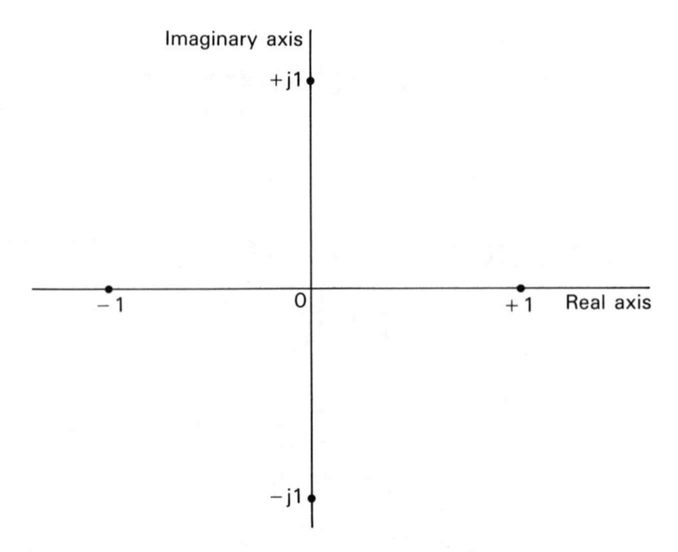

Figure A.1 Interpretation of $\sqrt{-1}$.

Cartesian representation

Numbers of the form jc, jd (where c and d are real) are referred to as imaginary numbers. When combined with real numbers giving the form $(a+jb)$ then they produce a complex number. These can be interpreted graphically as shown in Figure A.2. The complex number $z=x+jy$ can be considered as formed from the two distances x and y, along the real axis as shown. However because of the j multiplier the distance y is rotated through 90° to give the complex number as shown.

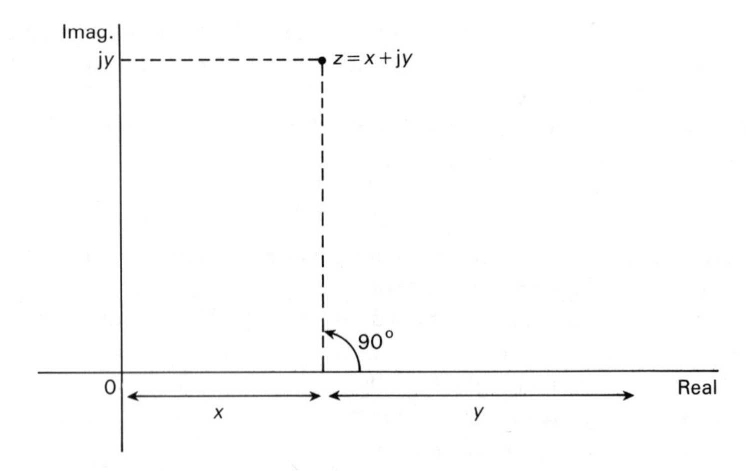

Figure A.2 Cartesian representation of complex numbers.

This type of representation, with the axes referencing the real and imaginary parts of the complex number, is known as an Argand diagram. Alternatively the real and imaginary axes define a plane known as the complex plane. The notation is used

$$x = \mathrm{Re}\{z\} \qquad x \text{ is the real part of } z$$
$$y = \mathrm{Im}\{z\} \qquad y \text{ is the imaginary part of } z$$

where $z = x + jy$.

Addition and subtraction of complex numbers is carried out by performing the operations separately on the real and imaginary parts.

For example

$$(2 - j3) + (5 + j2) = (2 + 5) + j(-3 + 2) = 7 - j1$$
$$(2 - j3) - (5 + j2) = (2 - 5) + j(-3 - 2) = -3 - j5$$

Multiplication of complex numbers is accomplished by use of the distributive rule.

$$(2 + j3) \times (4 + j5)$$
$$= 2 \times 4 + j2 \times 5 + j3 \times 4 + j^2 3 \times 5$$
$$= (8 - 15) + j(10 + 12) \text{ (Remembering that } j^2 = -1)$$
$$= -7 + j22$$

Division of complex numbers is not quite so straightforward and it will be dealt with later.

Complex conjugate

The complex conjugate is a useful concept that simplifies the manipulation of complex numbers.

The conjugate of a complex number is obtained by changing the sign of the imaginary part of the number. If

$$z = x + jy$$

then

$$z^* = x - jy$$

where z^* denotes the complex conjugate. On the complex plane z^* represents a point that is the mirror image of the point z in the real axis.

The following relationships involving the complex conjugate are easily proved.

$$z + z^* = 2\,\mathrm{Re}\{z\}$$
$$z - z^* = j2\,\mathrm{Im}\{z\}$$
$$zz^* = [\mathrm{Re}\{z\}]^2 + [\mathrm{Im}\{z\}]^2$$

Polar representation

In the cartesian representation the position of a number on the complex plane has been referenced by use of its real and imaginary parts. Its position can also be referenced by use of polar form. Referring to Figure A.3 the complex number z can be represented in polar form as $R\angle\theta$.

R is the distance of the number from the origin.

$$R = \sqrt{(x^2 + y^2)}$$

and is known as the magnitude or modulus (written $|z|$) of the number.

θ is the angle the line joining the number to the origin makes with the positive real axis (positive angles are taken as counter-clockwise rotation).

$$\theta = \tan^{-1}\frac{y}{x}$$

θ is known as the angle or argument of the number.

The conjugate of a number $R\angle\theta$ is easily seen to be $R\angle -\theta$.

From Figure A.3 it can be seen that a complex number can easily be converted from polar to cartesian form. Referring to this figure, $x = R\cos\theta$ and $y = R\sin\theta$. Hence z can be written.

$$z = R(\cos\theta + \mathrm{j}\sin\theta) \tag{A.1}$$

One of the advantages of the polar form is that it is much more convenient for performing multiplication than the cartesian form.

if

$$z_1 = R_1(\cos\theta_1 + \mathrm{j}\sin\theta_1)$$
$$z_2 = R_2(\cos\theta_2 + \mathrm{j}\sin\theta_2)$$

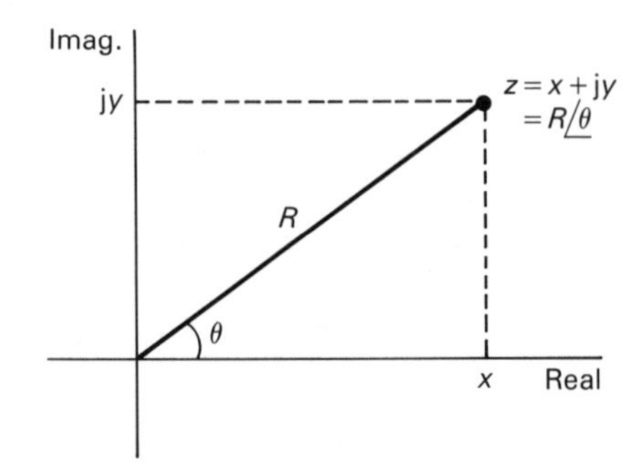

Figure A.3 Polar representation of complex numbers.

then

$$z_1z_2 = R_1R_2(\cos\theta_1 + j\sin\theta_1)(\cos\theta_2 + j\sin\theta_2)$$
$$= R_1R_2[\cos\theta_1\cos\theta_2 - j\sin\theta_1\sin\theta_2 + j(\sin\theta_1\cos\theta_2 + \cos\theta_1\sin\theta_2)]$$
$$= R_1R_2[\cos(\theta_1 + \theta_2) + j\sin(\theta_1 + \theta_2)]$$

Multiplication of complex numbers in polar form is performed by a multiplication of moduli and an addition of angles. This rule can be extended to any number of variables.

It follows that a complex number can be raised to a positive power

$$z^n = \{R(\cos\theta + j\sin\theta)\}^n$$
$$= R^n(\cos n\theta + j\sin n\theta) \tag{A.2}$$

This result is known as De Moivre's theorem.

Also from the rule for multiplication it follows that

$$zz^* = R\angle\theta\, R\angle{-\theta}$$
$$= |z|^2 \tag{A.3}$$

The polar form of representation is useful for performing the operation of division. Division can be regarded as a multiplication by an inverse to give unity ($1\angle 0°$).

$$z\frac{1}{z} = 1\angle 0° \tag{A.4}$$

Equating moduli on left- and right-hand sides of equation (A.4)

$$\left|z\frac{1}{z}\right| = 1$$

but

$$\left|z\frac{1}{z}\right| = |z|\left|\frac{1}{z}\right|$$

hence

$$\left|\frac{1}{z}\right| = \frac{1}{|z|}$$

Equating angles

$$\angle z + \angle 1/z = 0°$$
$$\angle 1/z = -\angle z°$$

Hence division of complex numbers

$$z_1 = |z_1|\angle\theta_1$$
$$z_2 = |z_2|\angle\theta_2$$

gives

$$\frac{z_1}{z_2} = \left|\frac{z_1}{z_2}\right| \angle \theta_1 - \theta_2 \tag{A.5}$$

Using the rules for multiplication and division it follows that if

$$z = \frac{z_1 z_2 \ldots\ldots}{z_a z_b \ldots\ldots}$$

then

$$|z| = \frac{|z_1||z_2|\ldots\ldots}{|z_a||z_b|\ldots\ldots}$$

$$\angle z = \angle z_1 + \angle z_2 + \ldots\ldots$$
$$- \angle z_a - \angle z_b \ldots\ldots$$

Exponential representation

The exponential function e^x can be expressed as a power series

$$e^x = 1 + x + \frac{x^2}{2!} + \ldots\ldots$$

Assuming that the series holds for an imaginary exponent, $j\theta$, then it can be expressed in terms of the series expansions for sine and cosine giving

$$e^{j\theta} = \cos\theta + j\sin\theta$$

This equation is known as Euler's equation and multiplying both sides by the magnitude R gives

$$Re^{j\theta} = R(\cos\theta + j\sin\theta)$$

This is the representation of the complex number given by eqn (A.1). Hence

$$z = Re^{j\theta} \tag{A.6}$$

and this is the exponential form of representation.

The exponential form of representation, like the polar form, is very useful for performing multiplication and division. If

$$z_1 = R_1 e^{j\theta_1}, \qquad z_2 = R_2 e^{j\theta_2}$$

then

$$z_1 z_2 = R_1 R_2\, e^{j(\theta_1 + \theta_2)}$$

$$\frac{z_1}{z_2} = \frac{R_1}{R_2}\, e^{j(\theta_1 - \theta_2)}$$

Roots of a complex number

It can be shown that De Moivre's theorem (eqn A.2) holds for non-integer powers

$$z^{1/N} = \{R(\cos\theta + \mathrm{j}\sin\theta)\}^{1/N}$$

However, remembering that sine and cosine are periodic functions, $\sin\theta = \sin(\theta + 2k\pi)$, k integer, then

$$z^{1/N} = R^{1/N}\left[\cos\frac{(\theta + 2k\pi)}{N} + \mathrm{j}\sin\frac{(\theta + 2k\pi)}{N}\right]$$

$$k = 0, 1, \ldots\ldots (N-1)$$

This can also be expressed in exponential form

$$z^{1/N} = R^{1/N}\mathrm{e}^{\mathrm{j}[(\theta+2k\pi)/N]} \qquad k = 1, 2, \ldots\ldots (N-1)$$

These two equations express the result that there are N values for the Nth root of a complex number.

Bibliography

The following books provide a general introduction to the subject of signals and systems.

Carlson, G. E. (1995) *Signal and Linear System Analysis*, New York: Wiley.
Gabel, R. A. and Roberts, R. (1987) *Signals and Linear Systems*, New York: Wiley.
Kamen, E. (1987) *Introduction to Signals and Systems*, New York: Macmillan.
Lynn, P. A. (1986) *Electronic Signals and Systems*, Basingstoke: Macmillan.
Lynn, P. A. and Fuerst, W. (1994) *Introductory Digital Signal Processing with computer applications*, Chichester: Wiley.
McGillem, C. D. and Cooper, G. R. (1991) *Continuous and Discrete Signal and System Analysis*, Philadelphia: Holt, Rinehart and Winston.
O'Flynn, M. F. and Moriarity, G. M. (1987) *Linear Systems*, New York: Wiley.
Poularikas, A. D. and Seely, S. (1988) *Elements of Signals and Systems*, Boston: PWS-Kent.
Shearer, J. L. and Kulakowski, B. T. (1990) *Dynamic Modelling and Control of Engineering Systems*, New York: Macmillan.

The following books provide further reading on particular topic areas.

Bendat, J. S. and Piersol, A. G. (1986) *Random Data Analysis and Measurement Procedures*, New York: Wiley.
Chen, W. K. (1986) *Passive and Active Filters*, New York: Wiley.
Connor, F. R. (1982) *Modulation*, London: Edward Arnold.
Papoulis, A. (1991) *Probability, Random Variables and Stochastic Processes*, New York: McGraw-Hill.
Williams, C. W. (1986) *Designing Digital Filters*, Englewood Cliffs, N.J.: Prentice Hall.

Index